Geophysical Monograph Series

Geophysical Monograph Series

225 **Active Global Seismology** *Ibrahim Cemen and Yucel Yilmaz (Eds.)*
226 **Climate Extremes** *Simon Wang (Ed.)*
227 **Fault Zone Dynamic Processes** *Marion Thomas (Ed.)*
228 **Flood Damage Survey and Assessment: New Insights from Research and Practice** *Daniela Molinari, Scira Menoni, and Francesco Ballio (Eds.)*
229 **Water-Energy-Food Nexus – Principles and Practices** *P. Abdul Salam, Sangam Shrestha, Vishnu Prasad Pandey, and Anil K Anal (Eds.)*
230 **Dawn–Dusk Asymmetries in Planetary Plasma Environments** *Stein Haaland, Andrei Rounov, and Colin Forsyth (Eds.)*
231 **Bioenergy and Land Use Change** *Zhangcai Qin, Umakant Mishra, and Astley Hastings (Eds.)*
232 **Microstructural Geochronology: Planetary Records Down to Atom Scale** *Desmond Moser, Fernando Corfu, James Darling, Steven Reddy, and Kimberly Tait (Eds.)*
233 **Global Flood Hazard: Applications in Modeling, Mapping, and Forecasting** *Guy Schumann, Paul D. Bates, Giuseppe T. Aronica, and Heiko Apel (Eds.)*
234 **Pre-Earthquake Processes: A Multidisciplinary Approach to Earthquake Prediction Studies** *Dimitar Ouzounov, Sergey Pulinets, Katsumi Hattori, and Patrick Taylor (Eds.)*
235 **Electric Currents in Geospace and Beyond** *Andreas Keiling, Octav Marghitu, and Michael Wheatland (Eds.)*
236 **Quantifying Uncertainty in Subsurface Systems** *Celine Scheidt, Lewis Li, and Jef Caers (Eds.)*
237 **Petroleum Engineering** *Moshood Sanni (Ed.)*
238 **Geological Carbon Storage: Subsurface Seals and Caprock Integrity** *Stephanie Vialle, Jonathan Ajo-Franklin, and J. William Carey (Eds.)*
239 **Lithospheric Discontinuities** *Huaiyu Yuan and Barbara Romanowicz (Eds.)*
240 **Chemostratigraphy Across Major Chronological Eras** *Alcides N.Sial, Claudio Gaucher, Muthuvairavasamy Ramkumar, and Valderez Pinto Ferreira (Eds.)*
241 **Mathematical Geoenergy: Discovery, Depletion, and Renewal** *Paul Pukite, Dennis Coyne, and Daniel Challou (Eds.)*
242 **Ore Deposits: Origin, Exploration, and Exploitation** *Sophie Decree and Laurence Robb (Eds.)*
243 **Kuroshio Current: Physical, Biogeochemical, and Ecosystem Dynamics** *Takeyoshi Nagai, Hiroaki Saito, Koji Suzuki, and Motomitsu Takahashi (Eds.)*
244 **Geomagnetically Induced Currents from the Sun to the Power Grid** *Jennifer L. Gannon, Andrei Swidinsky, and Zhonghua Xu (Eds.)*
245 **Shale: Subsurface Science and Engineering** *Thomas Dewers, Jason Heath, and Marcelo Sánchez (Eds.)*
246 **Submarine Landslides: Subaqueous Mass Transport Deposits From Outcrops to Seismic Profiles** *Kei Ogata, Andrea Festa, and Gian Andrea Pini (Eds.)*
247 **Iceland: Tectonics, Volcanics, and Glacial Features** *Tamie J. Jovanelly*
248 **Dayside Magnetosphere Interactions** *Qiugang Zong, Philippe Escoubet, David Sibeck, Guan Le, and Hui Zhang (Eds.)*
249 **Carbon in Earth's Interior** *Craig E. Manning, Jung-Fu Lin, and Wendy L. Mao (Eds.)*
250 **Nitrogen Overload: Environmental Degradation, Ramifications, and Economic Costs** *Brian G. Katz*
251 **Biogeochemical Cycles: Ecological Drivers and Environmental Impact** *Katerina Dontsova, Zsuzsanna Balogh-Brunstad, and Gaël Le Roux (Eds.)*

252 **Seismoelectric Exploration: Theory, Experiments, and Applications** *Niels Grobbe, André Revil, Zhenya Zhu, and Evert Slob (Eds.)*
253 **El Niño Southern Oscillation in a Changing Climate** *Michael J. McPhaden, Agus Santoso, and Wenju Cai (Eds.)*
254 **Dynamic Magma Evolution** *Francesco Vetere (Ed.)*
255 **Large Igneous Provinces: A Driver of Global Environmental and Biotic Changes** *Richard. E. Ernst, Alexander J. Dickson, and Andrey Bekker (Eds.)*
256 **Coastal Ecosystems in Transition: A Comparative Analysis of the Northern Adriatic and Chesapeake Bay** *Thomas C. Malone, Alenka Malej, and Jadran Faganeli (Eds.)*
257 **Hydrogeology, Chemical Weathering, and Soil Formation** *Allen Hunt, Markus Egli, and Boris Faybishenko (Eds.)*
258 **Solar Physics and Solar Wind** *Nour E. Raouafi and Angelos Vourlidas (Eds.)*
259 **Magnetospheres in the Solar System** *Romain Maggiolo, Nicolas André, Hiroshi Hasegawa, and Daniel T. Welling (Eds.)*
260 **Ionosphere Dynamics and Applications** *Chaosong Huang and Gang Lu (Eds.)*
261 **Upper Atmosphere Dynamics and Energetics** *Wenbin Wang and Yongliang Zhang (Eds.)*
262 **Space Weather Effects and Applications** *Anthea J. Coster, Philip J. Erickson, and Louis J. Lanzerotti (Eds.)*
263 **Mantle Convection and Surface Expressions** *Hauke Marquardt, Maxim Ballmer, Sanne Cottaar, and Jasper Konter (Eds.)*
264 **Crustal Magmatic System Evolution: Anatomy, Architecture, and Physico-Chemical Processes** *Matteo Masotta, Christoph Beier, and Silvio Mollo (Eds.)*
265 **Global Drought and Flood: Observation, Modeling, and Prediction** *Huan Wu, Dennis P. Lettenmaier, Qiuhong Tang, and Philip J. Ward (Eds.)*
266 **Magma Redox Geochemistry** *Roberto Moretti and Daniel R. Neuville (Eds.)*
267 **Wetland Carbon and Environmental Management** *Ken W. Krauss, Zhiliang Zhu, and Camille L. Stagg (Eds.)*
268 **Distributed Acoustic Sensing in Geophysics: Methods and Applications** *Yingping Li, Martin Karrenbach, and Jonathan B. Ajo-Franklin (Eds.)*
269 **Congo Basin Hydrology, Climate, and Biogeochemistry: A Foundation for the Future (English version)** *Raphael M. Tshimanga, Guy D. Moukandi N'kaya, and Douglas Alsdorf (Eds.)*
269 **Hydrologie, climat et biogéochimie du bassin du Congo: une base pour l'avenir (version française)** *Raphael M. Tshimanga, Guy D. Moukandi N'kaya, et Douglas Alsdorf (Éditeurs)*
270 **Muography: Exploring Earth's Subsurface with Elementary Particles** *László Oláh, Hiroyuki K. M. Tanaka, and Dezsö Varga (Eds.)*
271 **Remote Sensing of Water-Related Hazards** *Ke Zhang, Yang Hong, and Amir AghaKouchak (Eds.)*
272 **Geophysical Monitoring for Geologic Carbon Storage** *Lianjie Huang (Ed.)*
273 **Isotopic Constraints on Earth System Processes** *Kenneth W. W. Sims, Kate Maher, and Daniel P. Schrag (Eds.)*
274 **Earth Observation Applications and Global Policy Frameworks** *Argyro Kavvada, Douglas Cripe, and Lawrence Friedl (Eds.)*
275 **Threats to Springs in a Changing World: Science and Policies for Protection** *Matthew J. Currell and Brian G. Katz (Eds.)*
276 **Core-Mantle Co-Evolution: An Interdisciplinary Approach** *Takashi Nakagawa, Taku Tsuchiya, Madhusoodhan Satish-Kumar, and George Helffrich (Eds.)*

Geophysical Monograph 276

Core-Mantle Co-Evolution
An Interdisciplinary Approach

Takashi Nakagawa
Taku Tsuchiya
Madhusoodhan Satish-Kumar
George Helffrich

Editors

This Work is a co-publication of the American Geophysical Union and John Wiley and Sons, Inc.

WILEY

This edition first published 2023
© 2023 American Geophysical Union

All rights reserved. No part of this publication may be reproduced, stored in a retrieval system, or transmitted, in any form or by any means, electronic, mechanical, photocopying, recording or otherwise, except as permitted by law. Advice on how to obtain permission to reuse material from this title is available at http://www.wiley.com/go/permissions.

Published under the aegis of the AGU Publications Committee

Matthew Giampoala, Vice President, Publications
Carol Frost, Chair, Publications Committee
For details about the American Geophysical Union, visit us at www.agu.org.

The right of Takashi Nakagawa, Taku Tsuchiya, Madhusoodhan Satish-Kumar, and George Helffrich to be identified as the editors of this work has been asserted in accordance with law.

Registered Office
John Wiley & Sons, Inc., 111 River Street, Hoboken, NJ 07030, USA

Editorial Office
111 River Street, Hoboken, NJ 07030, USA

For details of our global editorial offices, customer services, and more information about Wiley products visit us at www.wiley.com.

Wiley also publishes its books in a variety of electronic formats and by print-on-demand. Some content that appears in standard print versions of this book may not be available in other formats.

Limit of Liability/Disclaimer of Warranty
While the publisher and authors have used their best efforts in preparing this work, they make no representations or warranties with respect to the accuracy or completeness of the contents of this work and specifically disclaim all warranties, including without limitation any implied warranties of merchantability or fitness for a particular purpose. No warranty may be created or extended by sales representatives, written sales materials or promotional statements for this work. The fact that an organization, website, or product is referred to in this work as a citation and/or potential source of further information does not mean that the publisher and authors endorse the information or services the organization, website, or product may provide or recommendations it may make. This work is sold with the understanding that the publisher is not engaged in rendering professional services. The advice and strategies contained herein may not be suitable for your situation. You should consult with a specialist where appropriate. Further, readers should be aware that websites listed in this work may have changed or disappeared between when this work was written and when it is read. Neither the publisher nor authors shall be liable for any loss of profit or any other commercial damages, including but not limited to special, incidental, consequential, or other damages.

Library of Congress Cataloging-in-Publication Data applied for:

9781119526902 (Hardback); 9781119526957 (Adobe PDF); 9781119526940 (ePub); 9781119526919 (oBook)

Cover design by Wiley
Cover image: Courtesy of Chihiro Shiraishi

Set in 10/12pt TimesNewRomanMTStd by Straive, Chennai, India

SKY10047868_051623

CONTENTS

List of Contributors vii

Preface ix

Part I Structure and Dynamics of the Deep Mantle: Toward Core-Mantle Co-Evolution

1 Neutrino Geoscience: Review, Survey, Future Prospects 3
   William F. McDonough and Hiroko Watanabe

2 Trace Element Abundance Modeling with Gamma Distribution for Quantitative Balance
   Calculations 17
   *Sanshiro Enomoto, Kenta Ueki, Tsuyoshi Iizuka, Nozomu Takeuchi, Akiko Tanaka, Hiroko Watanabe,
   and Satoru Haraguchi*

3 Seismological Studies of Deep Earth Structure Using Seismic Arrays in East, South, and Southeast Asia,
   and Oceania 31
   Satoru Tanaka and Toshiki Ohtaki

4 Preliminary Results from the New Deformation Multi-Anvil Press at the Photon Factory: Insight
   into the Creep Strength of Calcium Silicate Perovskite 59
   *Andrew R. Thomson, Yu Nishihara, Daisuke Yamazaki, Noriyoshi Tsujino, Simon A. Hunt,
   Yumiko Tsubokawa, Kyoko Matsukage, Takashi Yoshino, Tomoaki Kubo, and David P. Dobson*

5 Deciphering Deep Mantle Processes from Isotopic and Highly Siderophile Element Compositions
   of Mantle-Derived Rocks: Prospects and Limitations 75
   Katsuhiko Suzuki, Gen Shimoda, Akira Ishikawa, Tetsu Kogiso, and Norikatsu Akizawa

6 Numerical Examination of the Dynamics of Subducted Crustal Materials with Different Densities 103
   Taku Tsuchiya, Takashi Nakagawa, and Kenji Kawai

Part II Core-Mantle Interaction: An Interdisciplinary Approach

7 Some Issues on Core-Mantle Chemical Interactions: The Role of Core Formation Processes 117
   Shun-ichiro Karato

8 Heat Flow from the Earth's Core Inferred from Experimentally Determined Thermal Conductivity
   of the Deep Lower Mantle 133
   Yoshiyuki Okuda and Kenji Ohta

9 Assessment of a Stable Region of Earth's Core Requiring Magnetic Field Generation
   over Four Billion Years 145
   Takashi Nakagawa, Shin-ichi Takehiro, and Youhei Sasaki

10   Inner Core Anisotropy from Antipodal PKIKP Traveltimes 165
Hrvoje Tkalčić, Thuany P. Costa de Lima, Thanh-Son Phạm, and Satoru Tanaka

11   Recent Progress in High-Pressure Experiments on the Composition of the Core 191
Ryosuke Sinmyo, Yoichi Nakajima, and Yasuhiro Kuwayama

12   Dynamics in Earth's Core Arising from Thermo-Chemical Interactions with the Mantle 219
Christopher J. Davies and Sam Greenwood

Index 259

LIST OF CONTRIBUTORS

Norikatsu Akizawa
Atmosphere and Ocean Research Institute
The University of Tokyo
Kashiwa, Japan

Thuany P. Costa de Lima
Research School of Earth Sciences
The Australian National University
Canberra, Australia

Christopher J. Davies
School of Earth and Environment
University of Leeds
Leeds, UK

David P. Dobson
Department of Earth Sciences
University College London
London, UK

Sanshiro Enomoto
Kavli Institute for the Physics and Mathematics of the Universe
The University of Tokyo
Kashiwa, Japan
and
Department of Physics
University of Washington
Seattle, Washington, USA

Sam Greenwood
School of Earth and Environment
University of Leeds
Leeds, UK

Satoru Haraguchi
Research Institute for Marine Geodynamics
Japan Agency for Marine-Earth Science and Technology
Yokosuka, Japan
and
Earthquake Research Institute
The University of Tokyo
Tokyo, Japan

Simon A. Hunt
Department of Materials
University of Manchester
Manchester, UK

Tsuyoshi Iizuka
Department of Earth and Planetary Sciences
The University of Tokyo
Tokyo, Japan

Akira Ishikawa
Department of Earth and Planetary Sciences
Tokyo Institute of Technology
Tokyo, Japan

Shun-ichiro Karato
Department of Earth and Planetary Sciences
Yale University
New Haven, Connecticut, USA

Kenji Kawai
Department of Earth and Planetary Sciences
The University of Tokyo
Tokyo, Japan

Tetsu Kogiso
Graduate School of Human and Environmental Studies
Kyoto University
Kyoto, Japan

Tomoaki Kubo
Department of Earth and Planetary Sciences
Kyushu University
Fukuoka, Japan

Yasuhiro Kuwayama
Department of Earth and Planetary Sciences
The University of Tokyo
Tokyo, Japan

Kyoko Matsukage
Department of Natural and Environmental Science
Teikyo University of Science
Uenohara, Japan

William F. McDonough
Research Center for Neutrino Science
Tohoku University
Sendai, Japan
and
Department of Earth Sciences
Tohoku University
Sendai, Japan
and
Department of Geology
University of Maryland
College Park, Maryland, USA

Takashi Nakagawa
Department of Planetology
Kobe University
Kobe, Japan
and
Department of Earth and Planetary System Science
Hiroshima University
Higashi-Hiroshima, Japan

Yoichi Nakajima
Department of Physics
Kumamoto University
Kumamoto, Japan

viii LIST OF CONTRIBUTORS

Yu Nishihara
Geodynamics Research Center
Ehime University
Matsuyama, Japan

Kenji Ohta
Department of Earth and Planetary Sciences
Tokyo Institute of Technology
Tokyo, Japan

Toshiki Ohtaki
Geological Survey of Japan
National Institute of Advanced Industrial Science
and Technology
Tsukuba, Japan

Yoshiyuki Okuda
Department of Earth and Planetary Sciences
Tokyo Institute of Technology
Tokyo, Japan

Thanh-Son Phạm
Research School of Earth Sciences
The Australian National University
Canberra, Australia

Youhei Sasaki
Department of Information Media
Hokkaido Information University
Ebetsu, Japan

Gen Shimoda
Geological Survey of Japan
National Institute of Advanced Industrial Science
and Technology
Tsukuba, Japan

Ryosuke Sinmyo
Department of Physics
Meiji University
Kawasaki, Japan

Katsuhiko Suzuki
Submarine Resources Research Center
Japan Agency for Marine-Earth Science and Technology
Yokosuka, Japan

Shin-ichi Takehiro
Research Institute for Mathematical Sciences
Kyoto University
Kyoto, Japan

Nozomu Takeuchi
Earthquake Research Institute
The University of Tokyo
Tokyo, Japan

Akiko Tanaka
Geological Survey of Japan
National Institute of Advanced Industrial Science
and Technology
Tsukuba, Japan

Satoru Tanaka
Research Institute for Marine Geodynamics
Japan Agency for Marine-Earth Science and Technology
Yokosuka, Japan

Andrew R. Thomson
Department of Earth Sciences
University College London
London, UK

Hrvoje Tkalčić
Research School of Earth Sciences
The Australian National University
Canberra, Australia

Yumiko Tsubokawa
Department of Earth and Planetary Sciences
Kyushu University
Fukuoka, Japan

Taku Tsuchiya
Geodynamics Research Center
Ehime University
Matsuyama, Japan

Noriyoshi Tsujino
Institute for Planetary Materials
Okayama University
Misasa, Japan

Kenta Ueki
Research Institute for Marine Geodynamics
Japan Agency for Marine-Earth Science and Technology
Yokosuka, Japan

Hiroko Watanabe
Research Center for Neutrino Science
Tohoku University
Sendai, Japan

Daisuke Yamazaki
Institute for Planetary Materials
Okayama University
Misasa, Japan

Takashi Yoshino
Institute for Planetary Materials
Okayama University
Misasa, Japan

PREFACE

The Earth's deep interior, being physically inaccessible, is difficult to study directly. The necessarily indirect methods used in its study are best pursued collaboratively in order to bring all possible sources of knowledge to bear on the topic, hence the need for interdisciplinary research. Over recent decades, there have been advances in investigating the dynamics of the Earth's deep interior. In terms of experimental and observation work, there have been innovations in high-temperature and high-pressure experiments (employing diamond anvil cells and synchrotron radiation facilities), dramatically improved geochemical analyses aided by particle physics detectors, and seismic wave observations and theory. In terms of computational work, methodological innovations and increased computational power have facilitated theoretical calculations of mineral properties and fluid dynamical simulations at the micro and macro scale.

This monograph describes results of the research project "Core-Mantle Co-Evolution" that was selected by the Ministry of Education, Culture and Sports (MEXT) in Japan. It was a component of a national program of innovative research projects intended to apply technological innovations in an interdisciplinary framework to contemporary research questions, in this case the composition, dynamics and evolution of the Earth's deep interior.

Recent observational and experimental investigations have significantly advanced our understanding of the structure and constituent materials of the deep Earth. However, details of the chemical composition of the mantle, accounting for 85% of the volume of the entire Earth, and light elements expected to exist in the core, corresponding to the remaining 15%, have remained unclear even after 60 years of research in various fields of science related to deep Earth.

Seismological evidence suggests vigorous convection at the core-mantle boundary region, whereas geochemical signatures suggest the presence of stable regions that hold the chemical fingerprints of early Earth's formation 4.6 billion years ago. In addition, the concentrations and disposition of various radioactive isotopes that act as the heat sources driving various dynamic behaviors of the deep Earth are also still largely unknown. With this backdrop the "Core-Mantle Co-Evolution" project attempted to tackle the unresolved mysteries of deep Earth science through comprehensive investigations of the interactions between the core and mantle by combining high-pressure and high-temperature experiments, geochemical analyses at different scales, high-resolution geophysical observations, and large-scale numerical simulations.

The 12 chapters comprise interdisciplinary contributions by internationally recognized researchers in Earth's deep interior. Part I summarizes recent research on the structure, composition, and dynamics of the Earth's deep mantle, which is the primary source of evidence for core-mantle interaction. The first two chapters introduce the geoneutrino observations that are expected to reveal the distribution of heat sources in the Earth's deep mantle. Chapter 1 (McDonough and Watanabe) is a comprehensive review on the present status of geoneutrino observations. This is followed by Chapter 2 (Enomoto et al.) on delineating the heat source distribution in the deep mantle. The structure of seismic wave velocities at the deepest part of the mantle is described in Chapter 3 (Tanaka and Ohtaki), which evaluates the structure of the edge of a large-scale low-velocity region in the western Pacific region based on the deployment and observation of a seismic array in Thailand. Recent developments on experimental facilities for deformation at lower mantle conditions are elegantly presented in Chapter 4 on bridgmanite (Thomson et al.), which is the major mineral phase in the lower mantle. Next, in Chapter 5 (Suzuki et al.), a comprehensive geochemical and isotopic analysis of mantle-derived rocks forms the basis of a discussion on core-mantle co-evolution where the authors suggest ways to decipher the isotope signature to fingerprint any core-mantle interaction. Finally, in Chapter 6 (Tsuchiya et al.), the density structure of the deep mantle and its modeling of mantle dynamics are presented to provide a feasible interpretation of observational and experimental studies.

Part II summarizes the results of studies that invoke core-mantle interaction from direct inferences. Chapter 7 (Karato) introduces the results of a high-pressure mass diffusion experiment that found how iron originated from the core seeps into the mantle side, suggesting the possibility of an actual chemical interaction between the core and mantle. Chapter 8 (Okuda and Ohta) looks at heat flow across the core-mantle boundary, the most important constraint for core-mantle co-evolution, which is measured under high-pressure and high-temperature conditions. Chapter 9 (Nakagawa et al.) shows that heat flow is also a crucial parameter in the fluid dynamics theory on the formation of stably stratified region below the core-mantle boundary. A review of seismic velocity structure in the inner core is presented in Chapter 10 (Tkalčić et al.), while Chapter 11 (Sinmyo et al.) describes the continuing search for light-element candidates for the Earth's core based on high-temperature and high-pressure experiments.

Finally, Chapter 12 (Davies and Greenwood) presents a comprehensive discussion on the generation and maintenance mechanisms of the Earth's magnetic field based on chemical interactions between the core and mantle.

In assembling this monograph, our goal has been to show how an interdisciplinary approach can reveal what happens in Earth's deep interior. We hope that the next generation of brilliant researchers will build on the research results presented here and continue to produce innovative results that will further elucidate the dynamics and evolution of the Earth's deep interior.

We express sincere thanks to everyone who has contributed to this monograph as chapter authors and reviewers. We also thank Kate Lajtha, Editor in Chief AGU Books, and her team for their encouragement with this project, which had to pass through the tough period of pandemic-related delays. We take this opportunity to thank the authors and publishers for their patience in completing this book.

Takashi Nakagawa
Kobe University and Hiroshima University, Japan
Taku Tsuchiya
Ehime University, Japan
Madhusoodhan Satish-Kumar
Niigata University, Japan
George Helffrich
Tokyo Institute of Technology, Japan

Part I
Structure and Dynamics of the Deep Mantle: Toward Core-Mantle Co-Evolution

1

Neutrino Geoscience: Review, Survey, Future Prospects

William F. McDonough[1,2,3] and Hiroko Watanabe[1]

ABSTRACT

The Earth's surface heat flux is 46 ± 3 TW (terawatts, 10^{12} watts). Although many assume we know the Earth's abundance and distribution of radioactive HPEs (i.e., U, Th, and K), estimates for the mantle's heat production varying by an order of magnitude and recent particle physics findings challenge our dominant paradigm. Geologists predict the Earth's budget of radiogenic power at 20 ± 10 TW, whereas particle physics experiments predict $15.3^{+4.9}_{-4.9}$ TW (KamLAND, Japan) and $38.2^{+13.6}_{-12.7}$ TW (Borexino, Italy). We welcome this opportunity to highlight the fundamentally important resource offered by the physics community and call attention to the shortcomings associated with the characterization of the geology of the Earth. We review the findings from continent-based, physics experiments, the predictions from geology, and assess the degree of misfit between the physics measurements and predicted models of the continental lithosphere and underlying mantle. Because our knowledge of the continents is somewhat uncertain ($7.1^{+2.1}_{-1.6}$ TW), models for the radiogenic power in the mantle (3.5–32 TW) and the bulk silicate Earth (BSE; crust plus mantle) continue to be uncertain by a factor of ~10 and ~4, respectively. Detection of a geoneutrino signal in the ocean, far from the influence of continents, offers the potential to resolve this tension. Neutrino geoscience is a powerful new tool to interrogate the composition of the continental crust and mantle and its structures.

1.1. INTRODUCTION

Core-mantle evolution involves understanding Earth's differentiation processes, which established the present-day distribution of the HPEs, and its dynamic consequences (i.e., the radiogenic heat left in the mantle powering mantle convection, plate tectonics, and the geodynamo). The energy to drive the Earth's engine comes from two different sources: primordial and radiogenic.

Primordial energy represents the kinetic energy inherited during accretion and core formation. Radiogenic energy is the heat of reaction from nuclear decay. We do not have a constraint on the proportion of these different energy sources driving the present-day Earth's dynamics. In turn, this means that we do not have sufficient constraint on the thermal evolution of the planet, aside from first-order generalities. You might ask, is this important? We ask the question – how much energy (and time) is left to keep the Earth habitable?

We understand that the Earth started out hot due to abundant accretion energy, the gravitational energy of sinking metal into the center, a giant impact event for the formation of the Earth's Moon, and energy from short-lived (e.g., ^{26}Al and ^{60}Fe) and long-lived (K, Th, and U) radionuclides. From this hot start the planet should quickly lose some of its initial energy, although

---

[1]*Research Center for Neutrino Science, Tohoku University, Sendai, Japan*
[2]*Department of Earth Sciences, Tohoku University, Sendai, Japan*
[3]*Department of Geology, University of Maryland, College Park, Maryland, USA*

Core-Mantle Co-Evolution: An Interdisciplinary Approach, Geophysical Monograph 276, First Edition.
Edited by Takashi Nakagawa, Taku Tsuchiya, Madhusoodhan Satish-Kumar, and George Helffrich.
© 2023 American Geophysical Union. Published 2023 by John Wiley & Sons, Inc.
DOI: 10.1002/9781119526919.ch01

the amount and rate are unknowns. There are many significant unknowns regarding the thermal evolution of the Earth: (1) the nature and presence (or absence) and lifetime of an early atmosphere, which has a thermal blanketing effect; (2) the compositional model for the Earth, particularly the absolute abundances of the HPEs (K, Th, and U); (3) the cooling rate of the mantle (present-day estimates: 100 ± 50 K/Ga); and (4) the rate of crust formation and thus extraction of HPEs from the mantle.

The recent recognition (Krauss et al., 1984) and subsequent detection (Araki et al., 2005) of the planet's geoneutrino emission have opened up a new window into global scale geochemistry of the present-day Earth. The measurement of this flux presents Earth scientists with a transformative opportunity for new insights into the composition of the Earth and its energy budget. For the most part, solid Earth geophysics measures and parameterizes the present-day state of the planet. In contrast, solid Earth geochemistry measures and parameterizes its time-integrated state, mostly on a hand sample scale and then extrapolates these insights to larger scales. The advent of measuring the Earth's geoneutrino flux allows us, for the first time, to get a global measure of its present-day amount of Th and U.

This chapter is organized as follows: we provide the rationale for the field of neutrino geoscience and define some terms (section 1.2). We review the existing and developing detectors, the present-day detection methods, and future technologies (section 1.3). We discuss the latest results from the physics experiments (section 1.4). We present the range of compositional models proposed for the Earth (section 1.5) followed by a discussion of the geological prediction of the geoneutrino fluxes at various detectors (section 1.6). We finish with a discussion on determining the radioactive power in the mantle (section 1.7) and future prospects (section 1.8).

1.2. NEUTRINO GEOSCIENCE

The field of neutrino geoscience focuses on constraining the Earth's abundances of Th and U and with these data we can determine: (1) the absolute concentration of refractory elements in the Earth and from that determine the BSE's composition (crust plus mantle), and (2) the amount of radiogenic power in the Earth driving the planet's major dynamic processes (e.g., mantle convection, plate tectonics, magmatism, and the geodynamo). These two constraints set limits on the permissible models for the composition of the Earth and its thermal evolutionary history.

First, the refractory elements are in constant relative abundances in all chondrites. There are 36 of these elements (e.g., Al, Ca, Sr, Zr, REE, Th, and U) and by establishing the absolute abundance of one defines all abundances, since refractory elements exist in constant ratios to each other (McDonough & Sun, 1995). Most of these elements are concentrated in the bulk silicate Earth, but not all (e.g., Mo, W, Ir, Os, Re, etc.) and these latter ones are mostly concentrated in the metallic core. Knowing the Earth's abundance of Ca and Al, two of the eight most abundant elements (i.e., O, Fe, Mg, Si, Ca, Al, Ni, and S) that make up terrestrial planets (i.e., 99%, mass and atomic proportions) define and restrict the range of accepted compositional models of the bulk Earth and BSE.

Second, the decay of ^{40}K, ^{232}Th, ^{238}U, and ^{235}U (i.e., HPE) provides the Earth's radiogenic power, accounts for 99.5% of its total radiogenic power, and is estimated to be 19.9 ± 3.0 TW (1 TW = 10^{12} watts). This estimate, however, assumes a specific BSE model composition (McDonough & Sun, 1995; Palme & O'Neill, 2014). It must be noted that there is no consensus on the composition of the BSE, and so predictions from competing compositional models span from about 10–38 TW (Agostini et al., 2020; Javoy et al., 2010). This uncertainty in our present state of knowledge means that the field of neutrino geoscience plays a crucial role in resolving fundamental questions in Earth sciences.

1.2.1. Background Terms

The field of neutrino geoscience spans the disciplines of particle physics and geoscience, including geochemistry and geophysics. The following list of terms are offered to support this interdisciplinary research field.

Alpha (α) *decay*: a radioactive decay process that reduces the original nuclide (X) by four atomic mass units by the emission of a $^4_2$He nucleus and reaction energy (Q). Commonly, the α particle is emitted with between 4 and 9 MeV (1 MeV = 10^6 eV) of discrete kinetic energy. The basic form of α decay is as follows:

$$\text{Alpha} \quad (\alpha) \quad {}^A_Z X \to {}^{A-4}_{Z-2} X' + \alpha + Q \quad (1.1)$$

Beta decay: a radioactive decay process that transforms the original nuclide (X) into an isobar (same mass A) with the next lower proton number (Z) during either electron capture (*EC*) or β^+ decays or, alternatively, the next higher proton number (Z) during β^- decay. During each decay, there is an exchange of two energetic leptons (i.e., beta particles) and reaction energy (Q). Basic forms are as follows:

$$\begin{aligned}
\text{Beta Minus} \quad (\beta^-) & \quad {}^A_Z X \to {}^A_{Z+1} X' + e^- + \bar{\nu}_e + Q, \\
\text{Electron Capture} \quad (EC) & \quad {}^A_Z X + e^- \to {}^A_{Z-1} X' + \nu_e + Q, \\
\text{Beta Plus} \quad (\beta^+) & \quad {}^A_Z X \to {}^A_{Z-1} X' + e^+ + \nu_e + Q
\end{aligned}$$
$$(1.2)$$

Beta particles: first-generation energetic leptons, either matter leptons (electrons and neutrinos: e^- and ν_e) or antimatter leptons (positrons and antineutrinos: e^+ and $\bar{\nu}_e$).

Chondrite: an undifferentiated stony meteorite containing chondrules (flash-melted spheres, sub-mm to several mm across), matrix [fine grained (micron scale) aggregate of dust and crystals], and sometimes Ca-Al-inclusions and other refractory phases. They are typically mixtures of silicates and varying amounts of Fe-Ni alloys and classified into groups based on their mineralogy, texture, and redox state. Three dominant groups are the carbonaceous, ordinary, and enstatite type chondrites, from most oxidized to reduced, respectively. Isotopic observations are also used to create a twofold classification of chondrites and related meteorites (i.e., the NC and CC groups). The NC (non carbonaceous) group includes enstatite and ordinary chondrites and is believed to have formed in the inner solar system inside of Jupiter. The CC (carbonaceous) group includes carbonaceous chondrites and is believed to have formed in the outer solar system from Jupiter and beyond. The CI carbonaceous chondrite type (the sole chondrite type lacking chondrules) is considered most primitive because its element abundances matches that of the solar photosphere 1:1 over 6 orders of magnitude, except for the noble and H-C-N-O gases.

Geoneutrinos: naturally occurring electron antineutrinos ($\bar{\nu}_e$, with the over-bar indicating it is an antimatter particle), mostly, and electron neutrinos (ν_e), much less so, produced during β^-, and [EC and β^+] decays, respectively. The interaction cross-sections, which scale with their energy, for the detectable geoneutrinos (i.e., Th and U) is on the order of 10^{-47} m^2. Consequently, these particles rarely interact with matter in the Earth. The Earth's geoneutrino flux is 10^{25} $\bar{\nu}_e$ s^{-1} (McDonough et al., 2020). Each neutrino leaving the Earth removes a portion of the Earth's radiogenic heat (Q).

Heat-producing elements (HPEs): potassium, thorium, and uranium (i.e., K, Th, and U, or more specifically ^{40}K, ^{232}Th, ^{235}U, and ^{238}U) account for $\sim$99.5% of the Earth's radiogenic heating power.

Inverse Beta Decay (IBD): a nuclear reaction used to detect electron antineutrinos in large underground liquid scintillation detectors that are surrounded by thousands of photomultiplier tubes. The reaction [$\bar{\nu}_e + p \rightarrow e^+ + n$] involves a free proton (i.e., H atom) capturing a through going $\bar{\nu}_e$ and results in two flashes of light close in space and time. The first flash of light involves $e^+ - e^-$ annihilation (order a picosecond following $\bar{\nu}_e + p$ interaction) and the second flash ($\sim$0.2 ms later) comes from the capture by a free proton of the thermalizing neutron. This coincidence of a double light flash in space and time, with the second flash having 2.2 MeV light, reduces the background by a million-fold. This reaction requires the $\bar{\nu}_e$ to carry sufficient energy to overcome the reaction threshold energy of $E^{thr}_{\bar{\nu}_e} = 1.8$ MeV. Thus, the IBD reaction restricts us to detecting only antineutrinos from the β^- decays in the ^{238}U and ^{232}Th decay chains.

Major component elements: a cosmochemical classification term for Fe, Ni, Mg, & Si, with half-mass condensation temperatures (T_c) between 1355 and 1250 K. These elements condense from a cooling nebular gas into silicates (first olivine, then pyroxene) and Fe, Ni alloys and together with oxygen represents $\geq$93% of terrestrial planet's mass (McDonough & Yoshizaki, 2021).

Earth and its parts: the Earth is chemically differentiated. It has a metallic core surrounded by the BSE (aka Primitive Mantle), which initially included the mantle, oceanic and continental crust, and the hydrosphere and atmosphere; the Primitive Mantle is the undegassed and undifferentiated Earth minus the core. The present-day silicate Earth, less the hydrosphere and atmosphere, is made up of the mantle, including its bottom thermally conductive boundary layer (D"), and the top lithosphere; the latter composed of the crust an underlying lithospheric mantle. The lithosphere is the mechanically stiff (i.e., $>10^5$ higher viscosity than the underlying asthenospheric mantle) thermally conductive boundary layer. In the continents, the zone above the Moho (a seismically defined boundary between the crust and mantle) is the continental crust and below the continental lithospheric mantle (CLM). Masses and thicknesses of these domains are listed in Table 1.1.

Moderately volatile elements: a cosmochemical classification term for elements with half-mass condensation temperatures (T_c) between 1250 and 600 K. These elements include the alkali metals (lithophile), some transition metals, all the other metals, less Al, and the pnictogens and chalcogens, less N and O.

Primordial energy: the energy in the Earth from accretion and core formation. Accretion kinetic energy is $\sim$10^{32} J, assuming an Earth mass (5.97 × 10^{24} kg) and 10 km/s as an average velocity of accreting particles. The gravitational energy of core formation, which translates to heating energy, is $\approx$10^{30} J, depending on the assumed settling velocity of a core-forming metal in the growing Earth.

Radiogenic energy: energy of a nuclear reaction (Q) resulting from radioactive decay, given in units of MeV (1 MeV = 10^6 eV) or pJ (1 pJ = 10^{-12} J), where 1MeV = 0.1602 pJ. For β decays Q (MeV) = (mass$_p$ − mass$_d$) × 931.494, with mass$_p$ (mass of parent isotope), mass$_d$ (mass of daughter isotope) in atomic mass units (1 amu = 1.660539 × 10^{-27} kg = 931.494 MeV), and for α decay, Q (MeV) = (mass$_p$ − mass$_d$ − mass$_\alpha$) × 931.494, where $E = eV/c^2$, or 1 amu = 0.931494 GeV/c^2.

Refractory elements: a cosmochemical classification term for elements with half-mass condensation

Table 1.1 Mass of the Earth and its parts

Domain/reservoir	Thickness km (±)	Mass kg (±)	Citation[a]
Earth	6,371 ($^{+7}_{-20}$)	5.97218 (60) × 10^{24}	[1]
Bulk Silicate Earth (BSE)	2,895 (5)	4.036 (6) × 10^{24}	[2]
Modern mantle (DM^b + EM^c domains)	2,867 (20)[d]	4.002 (20) × 10^{24}	[2]
Oceanic crust[e]	10.5 (4.3)	0.92 (11) × 10^{22}	[3]
Continental crust[e]	40 (9)	2.22 (26) × 10^{22}	[3]
Continental lithospheric mantle (CLM)	115 (80)	6.3 (0.8) × 10^{22}	[3]

[a] cited source: 1 = Chambat et al. (2010), 2 = Dziewonski and Anderson (1981), 3 = Wipperfurth et al. (2020)
[b] DM = Depleted Mantle, the chemically depleted source of MORB (mid-ocean ridge basalts), which is viewed as the chemical complement to the continental crust.
[c] EM = Enriched Mantle, a smaller (e.g., $\frac{1}{5}$ mass), deeper, and more chemically enriched source of OIB (ocean island basalts).
[d] From PREM, assuming a uniform surface crust of 24 km.
[e] Using a LITHO1.0 model, see Table 1 in Wipperfurth et al. (2020).

temperatures (T_c) > 1355 K; they condense at the earliest stage of the cooling of high-temperature nebular gas. These elements are in equal relative proportion (±15%) in chondritic meteorites. In terrestrial planets, many of these elements are classified as lithophile (dominantly coupling with oxygen and hosted in the crust and mantle), or siderophile (dominantly metallically bonded and hosted in the core). The refractory elements include: Be, Al, Ca, Ti, Sc, V, Sr, Y, Zr, Nb, Mo, Rh, Ru, Ba, REE, Hf, Ta, W, Re, Os, Ir, Pt, Th, and U. The core contains ≥90% of the Earth's budget of Mo, Rh, Ru, W, Re, Os, Ir, and Pt, about half of its V, and potentially a minor fraction of its Nb.

Surface heat flux: the total surface heat flux from the Earth's interior is reported as 46 ± 3 TW (Jaupart et al., 2015) or 47 ± 2 TW (Davies, 2013). On average the surface heat flux is about 86 mW/m², with that for the continents being 65 mW/m² and for the oceans being 96 mW/m² (Davies, 2013). Energy contributions to this surface flux come from the core (primordial, plus a minor [~1% of the surface total] amount due to inner core crystallization), mantle (a combination of primordial and radiogenic), and crust (radiogenic). Other contributions include negligible additions from tidal heating and crust–mantle differentiation.

1.3. DETECTORS AND DETECTION TECHNOLOGY

Electron antineutrinos ($\bar{v}_e$) come mostly from the radioactive decays of ^{40}K, ^{232}Th, ^{235}U, and ^{238}U (i.e., geoneutrinos; Krauss et al., 1984), plus contributions from local anthropogenic sources (i.e., nuclear reactor plants). The Earth emits some 10^{25} s^{-1} geoneutrinos (McDonough et al., 2020), with 65% coming from β^- decays of ^{40}K (Fig. 1.1).

The detection of an electron antineutrino uses the IBD reaction: $\bar{v}_e + p \rightarrow e^+ + n$, which has a reaction threshold energy of $E^{thr}_{\bar{v}_e} = 1.806$ MeV.

$$E^{thr}_{\bar{v}_e} = \frac{(M_n + m_e)^2 - M_p^2}{2M_p} = 1.806 [MeV] \quad (1.3)$$

assuming the laboratory frame (i.e., stationary target) and where M_p, M_n, and m_e are the masses of the proton (938.272 MeV), neutron (939.565 MeV), and electron (0.5110 MeV). The neutrino mass is unknown and contributes negligibly to this reaction. Although its upper limit is <0.8 eV/c^2 (Aker et al., 2022), estimates of the neutrino's mass is of the order of 10s–100s of meV (de Salas et al., 2018). This energy threshold restricts the detectable antineutrinos to β^- decays from the ^{238}U and ^{232}Th decay chains (Fig. 1.1).

1.3.1. Technical Details for Detecting Geoneutrinos

Here, we highlight some relevant aspects of the detection scheme.

Detection occurs when an antineutrino interacts with a free proton, transforming it to a neutron plus a positron, which then causes two flashes of light close in space and time. Each flash of light occurring in these large liquid scintillation spectrometers, which are sited

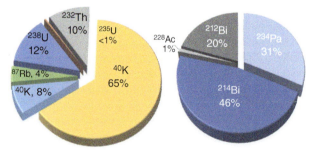

Figure 1.1 Relative proportions of the Earth's present-day flux of naturally occurring geoneutrinos (left) and the detectable geoneutrinos (right) from the ^{232}Th decay chain (gray; ^{228}Ac and ^{212}Bi) and ^{238}U decay chain (blue; ^{234}Pa and ^{214}Bi, not shown is the negligible contribution from ^{212}Tl) (see also Table 1.2). ^{40}K has two branches: β^- to ^{40}Ca (65%) and EC to ^{40}Ar (8%).

Table 1.2 Detectable $\bar{\nu}_e$ events

β^- decay events	$\bar{\nu}_e$ Max E[a] (MeV)	Branching fraction	Max IBD cross–section[b] σ_{IBD} (10^{-43} cm^2)	% of total[c] $\bar{\nu}_e$ signal
		^{232}Th decay chain		
^{228}Ac → ^{228}Th	2.134	1.00	4.3	1
^{212}Bi → ^{212}Po	2.252	0.64	4.8	20
		^{238}U decay chain		
^{234}Pa → ^{234}U	2.197	1.00	4.6	31
^{214}Bi → ^{214}Po	3.270	1.00	33	46
^{212}Tl → ^{212}Po	4.391	0.0002	90	<<1%

[a] There is a spectrum of energies for each antineutrino generated during a β^- decay, where typically the $\bar{\nu}_e$ takes about $\frac{2}{3}$ and the e^- about $\frac{1}{3}$ of the reaction energy (Q).
[b] As the energy of the antineutrino decreases so does its interaction cross-section.
[c] Numbers from Tables 3 and 5 in Fiorentini et al. (2007).

1–2 km underground to shield them from descending, atmospherically produced muons, are detected by the thousands of photomultiplier tubes covering the inner walls, each facing the detector's central volume.

Energy conservation requires that $E_{\bar{\nu}_e} + M_p = T_{e^+} + m_e + M_n + T_n$, with T_{e^+} and T_n being the kinetic energy of the positron and neutron. The prompt event involves the positron being annihilated in picoseconds by an electron, with the signal being the sum of the reaction releasing a 1.022 MeV energy flash (the sum of the masses of these two leptons) plus the characteristic kinetic energy inherited by the positron from the antineutrino ($E_{prompt} = (E_{\bar{\nu}_e} + M_p - m_e - M_n - T_n) + 2m_e$, or $= E_{\bar{\nu}_e} - T_n - 0.782$ [MeV]). The accompanying emitted neutron undergoes a cascade of collisions (thermalizing events, loss of energy to its surroundings as it goes toward thermal equilibrium) over about 200 μs and approximately 15 cm from the initial interaction point. This neutron is ultimately captured by a second free proton creating ^{2}H, resulting in a 2.22 MeV (binding energy) flash. This rare event sequence is eminently detectable because of its characteristics: two flashes of light in space and time, with the second flash having a specific energy. This reaction chain eliminates most background and improves the signal to noise ratio by a million-fold.

The organic liquid scintillator is mostly a long chain, aromatic ring hydrocarbon with approximately an H:C proportion of ~2. A wavelength shifting fluor compound is added to the scintillator to set the fluorescence peak emission at ~350–400 nm in order to reach the maximum quantum efficiency of the photomultiplier tubes. The photon yield for the liquid scintillator is typically a light yield of 200–400 photons/MeV.

The interaction cross-section of antineutrinos (and neutrinos) scales with their energy. For each β^- decay, there is a spectrum of $\bar{\nu}_e$ emitted energies, which in turn means a spectrum of interaction cross-sections. For IBD detection (i.e., starting at 1.806 MeV), the probability of a geoneutrino detection is low (order $1/10^{19}$), given that their cross-sections are between 10^{-48} and 10^{-46} m^2 (Vogel & Beacom, 1999). The overall emission above the energy threshold level are 0.40 U $\bar{\nu}_e$ per decay and 0.156 Th $\bar{\nu}_e$ per decay. There are four decay chains which produce detectable antineutrinos: two from the ^{232}Th and two from the ^{238}U decay chains (Table 1.2). The BSE has a Th/U mass ratio of 3.77 (or molar Th/U = 3.90) (Wipperfurth et al., 2018). However, despite Th being four times more abundant than U, attributes of the IBD detection method (i.e., $E_{\bar{\nu}_e}$, branching fraction, and σ_{IBD}) make U much easier to detect.

1.3.2. Detectors: Existing, Being Built, Being Planned

Currently, there are two detectors (Fig. 1.2) measuring the Earth's geoneutrino flux: KamLAND (1 kt), in Kamioka, Japan (Araki et al., 2005; Watanabe, 2019), and SNO+ (1 kt) in Sudbury, Ontario, Canada (Andringa et al., 2016), experiments. The Borexino detector (0.3 kt), in Gran Sasso, Italy (Agostini et al., 2020), has finished. The JUNO (20 kt) experiment in Jiangmen, Guangdong province, China (An et al., 2016), is currently being built. The Jinping (~4 kt) experiment in Jinping Mountains, Sichuan province, China (Beacom et al., 2017), is in development, with prototype detectors onsite testing future detector materials and technologies. Detectors in the proposal stage include Baksan in the Caucasus mountains in Russia (Domogatsky et al., 2005), Andes in Agua Negra tunnels linking the borders of Chile and Argentina (Dib et al., 2015), and a proposed ocean bottom detector

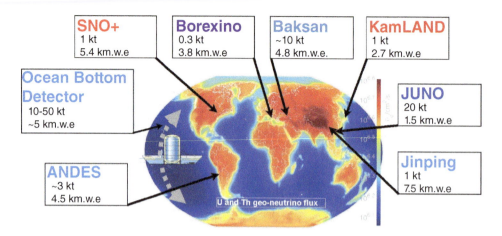

Figure 1.2 Present-day global distribution of detectors counting geoneutrino (red), have counted (purple), and are in the development and/or planning stage (blue). JUNO (bold blue) is in the construction phase. Background figure is the calculated global geoneutrino flux (order 10^6 $\bar{v}_e$ cm^{-2} s^{-1}) (Usman et al., 2015); the relatively high flux density seen in the Himalayas is directly correlated with its greater crustal thickness.

(OBD). Significantly, the Andes detector is the only proposed detector to be sited in the Southern Hemisphere.

There are ongoing developments for a ocean-going detector. A team of particle physicists from the University of Hawaii put forth a proposal more than 10 years ago for an OBD called Hanohano (Learned et al., 2007). A team of Japanese particle physicists and engineers and Earth scientists from JAMSTEC (Japan Agency for Marine-Earth Science and Technology) are currently moving forward with a project to deploy and test a mobile prototype detectors ("Ocean Bottom Detector" (OBD) scale is not yet set, but envisaged to up to 1 ton) off the coast of Japan. A mobile ocean-going detector offers a complementary measurement to land-based experiments. By sitting in the middle of the Pacific Ocean, 3,000 km from South America, 3,000 km from Australia, and ~3,000 km from the core-mantle boundary, such a detector gets a "mostly-mantle" signal.

1.4. LATEST RESULTS FROM THE PHYSICS EXPERIMENTS

Results from the physics experiments follow counting statistics, with increasing exposure (time spent counting) uncertainties reduce. These experiments are attentive to systematic and statistical uncertainties and addressed these issues in great detail in their publications.

The measured geoneutrino flux is reported in $\bar{v}_e$ cm^{-2} µs^{-1} for the KamLAND experiment and in TNU (Terrestrial Neutrino Unit) for the Borexino experiments. Mantovani et al. (2004) introduced TNU as a way to normalize the differences between detectors. A 1 TNU signal represents the detection of one event in a 1 kiloton liquid scintillation detector over a one-year exposure with a 100% detection efficiency. A 1 kiloton liquid scintillation detector has $\sim 10^{32}$ free protons (the detection target). Each detector has its own efficiency relative to a 1 kiloton fiducial volume detector, which accounts for the differences in the size of detector, its photomultiplier coverage, its response efficiency of the scintillator, and other factors. The conversion factor between signal in TNU and flux in $\bar{v}_e$ cm^{-2} µs^{-1} depends on the Th/U ratio and has a value of 0.11 $\bar{v}_e$ cm^{-2} µs^{-1} TNU^{-1} for Earth's Th/U$_{molar}$ of 3.9.

The most recent results for the KamLAND and Borexino experiments are 32.1 ± 5.0 (Watanabe, 2019) and $47.0^{+8.4}_{-7.7}$ (Agostini et al., 2020) TNU, respectively. [The SNO+ detector began counting in 2020 with a partial filled volume (with delays due to the covid-19 pandemic) and is yet to report their data.] The conversion between TNU and TW depends on the geological model assumed for the distribution of Th and U. Using geological models developed for both experiments, the radiogenic heating of the Earth ranges from 14 to 25 TW for KamLAND and 19 to 40 TW for Borexino (McDonough et al., 2020; Wipperfurth et al., 2020). The combined KamLAND and Borexino results mildly favors an Earth model with 20 TW present-day total power. Development of these and others geological models for each experiment are discussed later.

1.5. COMPOSITIONAL MODELS FOR THE EARTH

Earth scientists estimate the planet's radiogenic power within the bounds of 20 ± 10 TW (10^{12} watts), with many favoring estimates between 15 and 25 TW. In contrast, estimates from particle physics experiments (including a 19% contribution from K, not measured) range from

$15.3^{+4.9}_{-4.9}$ TW (KamLAND, Japan) (Watanabe, 2019) to $38.2^{+13.6}_{-12.7}$ TW (Borexino, Italy) (Agostini et al., 2020). The latter estimate exceeds most predictions by geologists and so one might dismiss such finding from this emerging field of science. The rub here, however, is that this broad range of TW estimates from particle physicists is mostly due to geological uncertainties associated with characterizing the surrounding ~500 km of lithosphere upon which these detectors are sited. These local lithologies contribute ~ 40% of the signal seen at a detector (Araki et al., 2005; Wipperfurth et al., 2020).

Once a geoneutrino flux is measured at a detector, a geological model for the immediately surrounding ~500 km of continental crust is applied. This model is coupled with a model for the global lithosphere and mantle contributions (see Huang et al., 2013, for method details). Given that a detector's sensitivity scales with 1/distance^2, each experiment is most sensitive to the local flux. By taking the total measured signal and subtracting the contributions from the lithosphere (with all of its unknowns), the mantle's contribution to the planetary radiogenic power is identified. Because of its low abundances of Th and U relative to the continental crust (i.e., ng/g vs. µg/g), the mantle's $\bar{v}_e$ contribution can be approximated as relatively homogeneous globally, with mantle heterogeneities in the signal being of the order 10–15% of its total signal (Šrámek et al., 2013).

There are three major models that predict the composition of the BSE, and thus the bulk Earth. In general, these models agree that the core has a negligible role in hosting the HPE. Although the Earth's core is often invoked as a host for radioactive HPE, mostly to power the geodynamo, there is a little support for such speculations. Recently, Wipperfurth et al. (2018) show that a maximum of <0.5% of the Earth's budget of Th and U could be hosted in the core. Petrologists have identified K-bearing sulfides that might have been extracted into the Earth's core. However, this evidence does not demonstrate the existence of potassium in the core, it only allows for its possibility. Such plausibility arguments need to be coupled with corroborating paragenetic evidence that is also free of negating geochemical consequences. In all instances, the extraction of a K-bearing phase into the core is not supported by other geochemical observations (e.g., no evidence for a range of alkali metals and refractory elements [Ca, REE] in the core). Finally, particle physics experiments have focused on the question of a nuclear reactor in the center of, or surrounding, the Earth's core. Conclusions of these studies limit the power of a geological reactor to <3.7 TW (Gando et al., 2013) and <2.4 TW (Agostini et al., 2020) at the 95% confidence limit.

The three major models can be characterized by their relative heat production (H): *low H*, *medium H*, and *high H*. Heat production models range from 10 to 38 TW, assumes Th/U$_{molar}$ = 3.90 and K/U = 14,000 (Arevalo et al., 2009; Wipperfurth et al., 2018), and have relative heat contributions of ~20% from K, ~40% from Th, and ~40% from U (McDonough et al., 2020). Given these ratios there is a simple multiplier for these model compositions: a 20 TW model has 20 ng/g U (a 10 TW model has 10 ng/g U), 77 ng/g Th, and 280 µg/g K. Both Th and U are refractory lithophile elements, like Al, Ca, Ti, and the REE. Also, the Earth has been demonstrated multiple times through elemental and isotopic data to be chondritic (McDonough & Sun, 1995; Willig et al., 2020). Therefore, if we know the absolute abundance of one of these elements (e.g., U) in the bulk Earth, then we know the abundances of all 36 (e.g., assuming 20 ng/g U in the BSE and (Al/U)$_{chondritic}$ = 1.08 × 10^6, then Al$_{BSE}$ = 21.7 mg/g).

The *low H* models typically have heat production levels of about 10 TW. Models like Javoy et al. (2010) and Faure et al. (2020) have 11 TW of radiogenic power, and their refractory lithophile elements abundances in the BSE are enriched by 1.5 times CI chondrite. This enrichment factor is equivalent to core separation, meaning their bulk Earth model has a CI chondrite refractory element composition. Many of the *low H* models were constructed to explain a putative ^{142}Nd isotope anomaly for the Earth (Boyet & Carlson, 2005). Other *low H to intermediate H* models for the Earth invoke non-chondritic refractory element abundances due to collisional erosion processes, which involved losing a substantial fraction of the early Earth's crust (Campbell & O'Neill, 2012; Jackson & Jellinek, 2013; O'Neill & Palme, 2008). Advocates for these models have largely fallen silent, as a simpler explanation for the ^{142}Nd isotope anomaly became obvious (Bouvier & Boyet, 2016; Burkhardt et al., 2016). Finally, a consequence of the *low H* model is that the mantle has very little radiogenic power. Assuming a BSE with about 10 TW of radiogenic power and the continental crust contains $7.1^{+2.1}_{-1.6}$ TW (Table 1.3), then the mantle has only ~3 TW of radiogenic power and thus the dominant driver of the major geodynamic processes is primordial energy.

The *high H* models typically have heat production levels of ≥30 TW. Compositional models proposed by Turcotte et al. (2001) and Agostini et al. (2020) conclude the BSE as having 30–38 TW of radiogenic power. These models have no parallels in the cosmochemistry of meteorites. They predict that the bulk Earth is enriched in refractory lithophile elements by a factor of 2.5–3.2 × CI chondrite. A survey of chondritic meteorites documents enrichment factors for refractory elements ranging from 1 to 2.2 × CI (Alexander, 2019a, 2019b). In opposition to *low H* models, the *high H* models have radiogenic energy as dominantly driving the Earth's geodynamic processes.

Table 1.3 Heat producing elements in the continental crust

Citation[a] Estimates of uncertainty	Models of the continental crust				BSE model
	1 (±30%)	2 low	2 high	3 (±35%)	4 (±10%)
U (µg/g) in bulk Cont. crust	1.3	1.09	1.33	1.27	0.020
Th (µg/g) in bulk Cont. crust	5.6	4.20	5.31	5.64	0.077
K (µg/g) in bulk Cont. crust.	15,000	11,900	18,800	11,700	280
Heat production (nW/kg)	0.325	0.253	0.333	0.312	0.005
Heat production (µW/m³)[b]	0.943	0.733	0.967	0.906	0.023
Radiogenic power (TW)[c]	7.2	5.7	7.4	7.1	20
% total U in CC	36%	30%	37%	36%	–
% total Th in CC	40%	30%	38%	40%	–
% total K in CC	29%	23%	37%	23%	–

[a] 1 = Rudnick and Gao (2014), 2 = Hacker et al. (2015), 3 = Wipperfurth et al. (2020), and 4 = McDonough and Sun (1995), Arevalo et al. (2009), Wipperfurth et al. (2018).
[b] Assumes average ρ (density) is 2,900 kg/m³ for the continental crust and 4,450 kg/m³ for the BSE.
[c] See Table 1.1 for reservoir masses.

Most *medium H* models have heat production levels of 20 ± 5 TW (McDonough & Sun, 1995; Palme & O'Neill, 2014), have enrichments in refractory elements consistent with that seen in chondritic meteorites, and are built from a residuum-melt relationship between peridotites (mantle rocks) and basalts (partial melts of the mantle). The BSE's composition comes from the least melt-depleted peridotites. This compositional model applies to the whole mantle and does not envisage any compositional layering (e.g., upper versus lower mantle domains). Moreover, this conclusion is supported by tomographic images of subducting oceanic slabs that penetrate into the lower mantle, documenting mass transfer between the upper and lower mantle and by inference whole mantle convection. In addition, the methodology used also captures into the primitive composition any potential domains that might have been created in the early Earth and since been isolated. *Medium H* models have approximately $\frac{2}{3}$ primordial and $\frac{1}{3}$ radiogenic energy driving the Earth's major geodynamic processes (i.e., mantle convection, plate tectonics, and the geodynamo). Recently, Yoshizaki and McDonough (2021) also showed that the Earth and Mars are equally enriched in refractory elements by 1.9 × CI, which might reflect the enrichment levels for all of the terrestrial planets and point to a fundamental attribute of inner solar system's building blocks.

1.6. THE GEOLOGICAL PREDICTIONS

All of our models depend upon an accurate description of the abundances and distribution of Th and U in the continents. The role of the geologist in collaboration with neutrino scientists is to determine this spatial distribution of the HPE (Huang et al., 2013). Although this task is relatively straightforward, it offers many challenges and foremost among them are the difficulties in defining the structure and composition of rocks in the crust beneath the surface, particularly in areas where detectors are sited. Global scale models predicting the composition of the continental crust are available (Hacker et al., 2015; Rudnick & Gao, 2014; Sammon & McDonough, 2021; Wipperfurth et al., 2020); however, differences in these models lead to differences in their geoneutrino signal (Enomoto et al., 2007; Huang et al., 2013; Mantovani et al., 2004).

Wipperfurth et al. (2020) showed that geophysical characterization of the continental crust, using global seismic models, CRUST 2.0, CRUST 1.0, and LITHO1.0, did not significantly contribute to the uncertainty in models describing the distribution of the HPEs in the Earth. In contrast, however, the geochemical models were identified as contributing the greatest uncertainty to the model. This finding places the onus of responsibility squarely in the camp of the geochemists and the geologists in terms of understanding the 3-D distribution of lithologies and compositions within the crust.

The total geoneutrino signal represents contributions of three components (Fig. 1.3): Near-field crust (~40%), Far-field crust (~35%, i.e., the global lithospheric signal), and Mantle signals (~25%) (Wipperfurth et al., 2020). Paramount in modeling the measured signal at a detector is understanding the contribution from the closest ~500 km of lithosphere adjacent to the detector (i.e., 250 km outward in any direction from the detector). This region is often referred to as the Near-field, local, or regional crustal signal. This part of the continental crust, and particularly the upper crust, is the region of greatest interest, because of the relative separation distance between source neutrino emitter and the detector, and because it is the brightest geoneutrino emitter due to geological processes that concentrate the HPEs upward in the continental crust.

On average the continental crust is enriched in the HPE by more than a factor of 60 over the BSE and >100 over the present-day abundances in the modern mantle. Moreover, the upper part of the crust is 10 times enriched in the HPE as compared to the lower crust (Hacker et al., 2015; Rudnick & Gao, 2014; Sammon & McDonough, 2021;

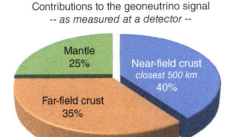

Figure 1.3 The relative contributions to the total geoneutrino signal: Near-field crust (out to 250 km in all directions), Far-field crust (the rest of the crust around the globe), and Mantle. The local and global Continental Lithospheric Mantle (CLM) contributes ~1 TNU or ~2% to the total geoneutrino signal at most detector sites.

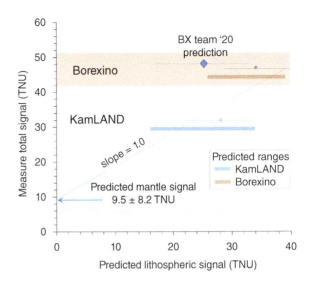

Figure 1.4 Model predictions for the geoneutrino signal at KamLAND and Borexino and predicted mantle flux (i.e., ~12 TW). See Table 1.4 for data sources and references. KamLAND's measured flux from Watanabe (2019) (32.1 ± 5.0 TNU, uncertainty shown as a light blue, horizontal band; y-axis only constraint). Borexino's measured flux from Agostini et al. (2020) ($47.0^{+8.4}_{-7.7}$ TNU, uncertainty shown as a light orange, horizontal band; y-axis only constraint). The slope = 1 is because its a two-component system (crust + mantle). Estimated signals at future detector locations (Table 1.4) will better constrain the intercept (mantle prediction).

Wipperfurth et al., 2020). Bulk compositional estimates of the continental crust report that it hosts about 30–40% of the HPEs in the BSE (Table 1.3).

The estimated model compositions of the continental crust (Table 1.3) come with a considerable amount of uncertainty in their predicted abundances. Importantly, all of these models consider the upper crust to represent about the top third of the continent's mass and assume a shared upper crustal compositional model (Rudnick & Gao, 2014), which dominates the geoneutrino flux. Using the global geophysical models that describe the continental crust (e.g., LITHO1.0) and a global model for the composition of the continents (Table 1.3), the geoneutrino signal for the Far-field crust (i.e., the crust minus the local contribution at the detector) ranges from ~8 to 19 TNU, depending on the local setting (Wipperfurth et al., 2020); locations like KamLAND and JUNO, near the coast, fall on the lower end of the scale, whereas locations like Jinping and SNO+, surrounded by a significant amount of continental crust, fall on the high end of the range. Wipperfurth et al. (2020) showed that when combining the data for KamLAND and Borexino and using their geological models found a mantle signal of 9.2 ± 8.5 TNU and found the Earth's global radiogenic power to be 21.5 ± 10.4 TW.

The lithospheric signal can be treated globally, but the local lithospheric signal is best done with a detail analysis of the geological, geochemical, and geophysical data of the region. Using this combination of data, one can build a 3-D model of the chemical and physical attributes of the local lithosphere. It is noteworthy, however, that such attempts have been conducted at both the KamLAND and Borexino experiment locations. Figure 1.4 and Table 1.4 present the range of predicted results for the geoneutrino flux at KamLAND, Borexino, and other locations.

1.7. DETERMINING THE RADIOACTIVE POWER IN THE MANTLE

The particle physics experiment, at its simplest level, reports the geoneutrino flux at a detector. With this number in hand, along with a model for the local and global lithospheric contribution, the physicist then determines the planetary geoneutrino flux and the flux from the mantle. This idea was first laid out by Krauss et al. (1984). Later, Raghavan et al. (1998), in a comparative analyses of signals from detectors in continental and oceanic settings, specifically developed a scheme to determine the mantle flux.

Surface heat flux observations constrain the Earth to radiating 46 ± 3 TW (i.e., 0.09 W/m²). Based on a compositional model for the continental crust (Table 1.3), it contains about $7.1^{+2.1}_{-1.6}$ TW of radiogenic power, leaving ~40 TW as a mantle flux contribution. This latter flux contains contributions from the core (basal heating), which is estimated to be order 10 ± 5 TW, and mantle (i.e., order 30 ± 5 TW, which includes radiogenic and primordial additions). A combined analysis of the KamLAND and Borexino yields a radiogenic mantle signal of 13 ± 12 TW (Wipperfurth et al., 2020). Consequently,

Table 1.4 Geological estimates of the signal contributions at various antineutrino detectors

Detector	Reference	Total signal	Global crustal contribution	FFC	NFC	Mantle contribution[a]
KamLAND	(Enomoto et al., 2007)	38.5	28.2	10.5	17.7	10.3
KamLAND	(Huang et al., 2013)	30.7	20.6	7.3	13.3	8.8
KamLAND	(Fiorentini et al., 2012)	–	26.5	8.8	17.7	–
KamLAND	(Wipperfurth et al., 2020)	37.9	27.0	8.8	18.2	9.4
Borexino	(Huang et al., 2013)	43.5	29.0	13.7	15.3	8.7
Borexino	(Coltorti et al., 2011)	43.5	26.2	16.0	10.2	9.9
Borexino	(Fiorentini et al., 2012)	–	25.3	15.7	9.7	–
Borexino	(Agostini et al., 2020)	47.0	25.5	16.3	9.2	20.6
Borexino	(Wipperfurth et al., 2020)	43.9	32.5	14.8	18.2	9.4
SNO+	(Huang et al., 2014)	40.0	30.7	15.1	15.6	7.0
SNO+	(Strati et al., 2017)	43.1	30.5	15.2	15.3	6.9
SNO+	(Wipperfurth et al., 2020)	46.8	34.3	14.7	19.6	9.1
JUNO	(Strati et al., 2015)	39.7	28.2	13.4	17.4	8.8
JUNO	(Gao et al., 2020)	49.1	38.3	9.8	28.5[b]	8.7
JUNO	(Wipperfurth et al., 2020)	40.5	29.8	12.7	17.1	9.5
Jinping	(Šrámek et al., 2016)	58.5	50.3	18.7	31.6	8.2
Jinping	(Wan et al., 2017)	59.4	49.0	–	–	10.4
Jinping	(Wipperfurth et al., 2020)	60.0	49.0	18.7	30.3	9.3
Hanohano	(Huang et al., 2013)	12.0	2.6	2.6	–	9.0

Notes: All numbers are reported in units of TNU (see text for details).
[a] Reported or calculated flux. NFF = Near-field Crust contribution, FFC = Far-field crust contribution.
[b] Authors defined the JUNO NFC as 10° × 10°. The flux contribution from the CLM (Continental Lithospheric Mantle) is not included, as it is not always reported by authors; its contribution is typically on the order of 1 to 2 TNU.

a large range of compositional models are acceptable for the Earth's radiogenic power (e.g., Table 1.4).

An important step toward reducing the uncertainties on the estimated mantle radiogenic power would be either (1) determine the mantle flux in oceanic setting and/or (2) improve the accuracy and precision of the local signal (Near-field). In the next section, we address measuring the signal with an OBD. Here, we consider differences in estimates of the local signal.

Figure 1.4 and Table 1.4 illustrate the problem. The graph is designed to extract the mantle signal from the combined analysis of data from different detectors. The signal is made up of crust plus mantle; it is a two-component only system (i.e., slope = 1). The mantle signal (the ordinate intercept) can be considered, at this scale, homogeneous. Note, Šrámek et al. (2013) found that even when modeling a mantle with two large, approximately antipodal structures (e.g., LLSVP: Large low shear velocity provinces), which might contain a factor of five or more difference in K, Th, and U abundances, the global mantle signal varied by only ±10%. Thus, the mantle flux estimate highly depends on the accuracy of the crustal predictions for each detector.

Figure 1.4 also reveals the challenge with different model predictions for the geoneutrino flux from the bulk lithosphere; each model and their uncertainties yield different geoneutrino flux estimates, and in turn this influences the prediction for the mantle signal. It is unclear which model provides a more accurate representation of the crust. In 2020, the Borexino particle physics team reported an updated prediction for the local lithospheric signal and mantle (Agostini et al., 2020), which leads to their distinctly higher prediction (and large uncertainty) for the Earth's total radiogenic power (i.e., $38^{+13.6}_{-12.7}$ TW). Thus, a clear challenge for the geological community is to independently construct a 3-D model for the local lithosphere and from this predict a geoneutrino flux. By doing so, it would test this hypothesized lithospheric model constructed by the particle physics team and their markedly radiogenic prediction for the Earth's HPE abundances.

Future geoneutrino detectors are being built at locations in Guangzhou, China (JUNO), and the eastern slope of the Tibetan plateau (Jinping). Each location has already had a geological model prediction of their geoneutrino signal, which critically depends upon the modeling of the local lithospheric signal (i.e., crust plus lithospheric mantle). The challenge faced by the geological community is how to treat differences in the production of the local lithospheric signal. Recently, Strati et al. (2015) and Gao et al. (2020) reported their predicted both crust signal for the JUNO detector as $28.2^{+5.2}_{-4.5}$ TNU and 38.3 ± 4.8 TNU, respectively. These numbers do not overlap with their combined uncertainties and challenge us to resolve this difference. Consequently, the physics experiments are bringing further insights into our understanding of the HPE in the Earth, which are forcing geologists to improve their 3-D physical and chemical descriptions of the Earth.

The question remains – how to critically assess the accuracy of these competing predictions? Some insights will come by adding more detectors and comparing multiple signals in order to predict a mantle signal. The SNO+ detector has begun counting and so we look forward to adding that data point to Figure 1.4 in the near future.

1.8. FUTURE PROSPECTS

The signal at all of these detectors depends critically upon the signal from the local lithosphere. Given two unknowns parameters (crust and mantle), then the problem is reduced to identifying exclusively the mantle signal. The best way to determine the mantle signal is to detect it far away from continental sources. This means determining the mantle geoneutrino flux from the deep ocean basin likely somewhere in the central Pacific Ocean. An ideal location in the central southern Pacific that is approximately a core – mantle distance (~3,000 km) away from continental masses rimming the Pacific.

The idea of an ocean-going detector that could isolate the mantle signal was introduced by Raghavan et al. (1998). In 2007, the Hanohano project (Learned et al., 2007) provided a detailed technical report showing that an ocean-going neutrino detector could be deployed in the deep ocean and serve multiple applications. Šrámek et al. (2013) identified a series of target locations in a north-south transit through the Pacific that could map out potential mantle structures and identify compositional heterogeneities in the deep mantle.

The concept envisages an ocean-going geoneutrino detector that is deployed on a yearly basis, anchored just above the ocean floor, recovered, serviced, and redeployed. Such a detector would be able to measure the mantle geoneutrino flux over multiple deployments. Given a small lithospheric signal from the Far-field continents and the oceanic crust, it is estimated that 75% of the signal measured by an OBD would be from the mantle.

Table 1.5 Estimated one-year signals[a] and backgrounds for 1.5 kt OBD by detector simulation[a]

Signals		Backgrounds		
			Rate[b]	
Source	Rate[b] (mantle)	Source	<8.5 MeV	<2.6 MeV (geo $\bar{\nu}_e$ region)
U	7.4 (5.5)	Reactor $\bar{\nu}_e$	4.5	1.7
Th	1.8 (1.3)	Accidental	1.8	1.8
Total	9.2 (6.8)	^{9}Li, ^{8}He	6.2	0.6
		^{13}C(α,n)^{16}O	3.6	2.6
		Total	16.1	6.7

[a] Simulation assumes the detector was deployed off the coast of Hawaii at 2.7 km depth. To reduce the radioactive background relative to the detector surface, a fiducial cut was applied (72 cm).
[b] Rate is in units of counted events/year.
Geoneutrino and reactor neutrino spectrum from Dye and Barna (2015).

Table 1.5 and Figure 1.5 summarize estimated signals and backgrounds for 1.5 kt OBD by detector simulation. The detector was assumed to be deployed off the coast of Hawaii at the depth of 2.7 km. Radioactive contamination in the detector components, such as liquid scintillator, acrylic vessel, photomultiplier tube, and its water pressure resistant vessel, can be the sources of the backgrounds (e.g., accidental and ^{13}C(α,n)^{16}O). Detector cleanness controls how much scintillator mass we can use as the fiducial volume. Seawater acts as a shield for cosmic-ray muons, which induces radioactive isotopes (e.g., ^{9}Li, ^{8}He). The deeper a detector can be anchored just above the seafloor, the greater the reduction in the muon-related background. A 1.5 kt OBD has sensitivity to measure mantle geoneutrino at 3.4σ level with three-year exposure measurement. A simple comparison between OBD and KamLAND (Watanabe, 2019), with both as 1.5 kt detectors, yields a geoneutrino flux of 7 [5] and 9 [24] events/year for mantle and non mantle (including the crust and other sources listed in Table 1.5) signals, respectively.

Such an instrument (OBD) is capable of accurately determining the mantle geoneutrino flux, which in turn can be used to unravel the geological signals measured at land-based experiments. Given this, we would have the capability to interpret the local lithospheric signal from all of the existing and future detectors.

Neutrino Geoscience compares geoneutrino flux measurements from particle physics experiments with flux predictions derived from integrating data from geology, geophysics, and geochemistry. The particle physics

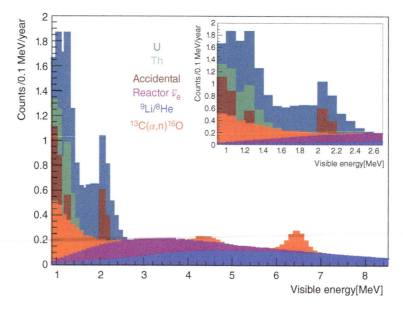

Figure 1.5 Expected energy spectrum of 1.5 kt OBD estimated by detector simulation. Right top panel focuses on geoneutrino energy range.

community are focusing on improving their measurement precision and detector technology. Efforts in geosciences are focused on improving our models for the abundance and distribution of the HPEs. Improvements in crustal studies are of greatest need. Global and regional geochemical studies, including studies of crustal terrains and xenoliths, can help constrain the composition of the deep crust.

Holistic approaches are essential for developing improved models of the crust. For example, considerable promise comes from studies that incorporate thermodynamic modeling (e.g., Perple_X; (Connolly, 2005)) with geochemical and seismic data, in combination with constraints from crustal geotherms (e.g., surface heat flux studies, curie depth maps, reflection data estimates on Moho temperatures, etc.), all in an effort to determine crustal compositional trends from the surface to the Moho (Sammon et al., 2020; Takeuchi et al., 2019). The other geological priority needing improvements is reducing or resolving the up to 30% spread in crustal predictions of the geoneutrino flux for all but the SNO+ experiment (Table 1.4).

ACKNOWLEDGMENTS

W. F. McDonough gratefully acknowledges NSF grant support (EAR1650365 and EAR2050374). H. Watanabe gratefully acknowledges grants (JP15H05833 and JP20H01909) from the Japan Society for the Promotion of Science. W. F. McDonough and H. Watanabe appreciate review comments from many colleagues and are very thankful to the two anonymous reviewers and the editor handling our chapter. There are no financial or other conflicts of interest with this work.

Both authors discussed the concepts introduced in the chapter. The writing of the main text was principally conducted by WFM; both authors were involved in the calculations and editing process. The authors have read and approved this manuscript.

REFERENCES

Agostini, M., Altenmüller, K., Appel, S., Atroshchenko, V., Bagdasarian, Z., Basilico, D., et al. (2020). Comprehensive geoneutrino analysis with Borexino. *Physical Review D*, *101*(1), 012009. doi: 10.1103/PhysRevD.101.012009

Aker, M., Beglarian, A., Behrens, J., Berlev, A., Besserer, U., Bieringer, B., et al. (2021). First direct neutrino-mass measurement with sub-eV sensitivity. *arXiv preprint*, arXiv:2105.08533.

Alexander, C. M. O. (2019a). Quantitative models for the elemental and isotopic fractionations in chondrites: The carbonaceous chondrites. *Geochimica et Cosmochimica Acta*, *254*, 277–309. doi: 10.1016/j.gca.2019.02.008

Alexander, C. M. O. (2019b). Quantitative models for the elemental and isotopic fractionations in the chondrites: The non-carbonaceous chondrites. *Geochimica et Cosmochimica Acta*, *254*, 246–276. doi: 10.1016/j.gca.2019.01.026

An, F., An, G., An, Q., Antonelli, V., Baussan, E., Beacom, J., et al. (2016). Neutrino Physics with JUNO. *Journal of*

Physics G: Nuclear and Particle Physics, 43(3), 030401. doi: 10.1088/0954-3899/43/3/030401

Andringa, S., Arushanova, E., Asahi, S., Askins, M., Auty, D. J., Back, A. R., et al. (2016). Current status and future prospects of the SNO+ experiment. *Advances in High Energy Physics*, 2016, 6194250. doi: 10.1155/2016/6194250

Araki, T., Enomoto, S., Furuno, K., Gando, Y., Ichimura, K., Ikeda, H., et al. (2005). Experimental investigation of geologically produced antineutrinos with KamLAND. *Nature*, 436(7050), 499–503. doi: 10.1038/nature03980

Arevalo, R., McDonough, W. F., & Luong, M. (2009). The K/U ratio of the silicate Earth: Insights into mantle composition, structure and thermal evolution. *Earth and Planetary Science Letters*, 278(3–4), 361–369. doi: 10.1016/j.epsl.2008.12.023

Beacom, J. F., Chen, S., Cheng, J., Doustimotlagh, S. N., Gao, Y., Gong, G., et al. (2017). Physics prospects of the jinping neutrino experiment. *Chinese Physics C 41*(2), 023002. doi: 10.1088/1674-1137/41/2/023002

Bouvier, A., & Boyet, M. (2016). Primitive Solar System materials and Earth share a common initial ^{142}Nd abundance. *Nature*, 537(7620), 399–402. doi: 10.1038/nature19351

Boyet, M., & Carlson, R. W. (2005). ^{142}Nd evidence for early (>4.53 Ga) global differentiation of the silicate Earth. *Science*, 309(5734), 576–581. doi: 10.1126/science.1113634

Burkhardt, C., Borg, L. E., Brennecka, G. A., Shollenberger, Q. R., Dauphas, N., & Kleine, T. (2016). A nucleosynthetic origin for the Earth's anomalous ^{142}Nd composition. *Nature*, 537(7620), 394–398. doi: 10.1038/nature18956

Campbell, I. H., & O'Neill, H. S. C. (2012). Evidence against a chondritic Earth. *Nature*, 483(7391), 553–558. doi: 10.1038/nature10901

Chambat, F., Ricard, Y., & Valette, B. (2010). Flattening of the Earth: Further from hydrostaticity than previously estimated: Hydrostatic flattening. *Geophysical Journal International*, 183(2), 727–732. doi: 10.1111/j.1365-246X.2010.04771.x

Coltorti, M., Boraso, R., Mantovani, F., Morsilli, M., Fiorentini, G., Riva, A., et al. (2011). U and Th content in the Central Apennines continental crust: A contribution to the determination of the geo-neutrinos flux at LNGS. *Geochimica et Cosmochimica Acta 75*(9), 2271–2294.

Connolly, J. A. D. (2005). Computation of phase equilibria by linear programming: A tool for geodynamic modeling and its application to subduction zone decarbonation. *Earth and Planetary Science Letters*, 236(1), 524–541. doi: 10.1016/j.epsl.2005.04.033

Davies, J. H. (2013). Global map of solid earth surface heat flow. *Geochemistry, Geophysics, Geosystems 14*(10), 4608–4622. doi: 10.1002/ggge.20271

de Salas, P. F., Gariazzo, S., Mena, O., Ternes, C. A., & Tórtola, M. (2018). Neutrino mass ordering from oscillations and beyond: 2018 status and future prospects. *Frontiers in Astronomy and Space Sciences*, 5, 36. doi: 10.3389/fspas.2018.00036

Dib, C. O. (2015). ANDES: An underground laboratory in South America. *Physics Procedia*, 61, 534–541. doi: 10.1016/j.phpro.2014.12.118

Domogatsky, G., Kopeikin, V., Mikaelyan, L., & Sinev, V. (2005). Neutrino geophysics at Baksan I: Possible detection of georeactor antineutrinos. *Physics of Atomic Nuclei*, 68(1), 69–72. doi: 10.1134/1.1858559

Dye, S., & Barna, A. (2015). *Global antineutrino modeling for a web application.* (arXiv:1510.05633)

Dziewonski, A. M., & Anderson, D. L. (1981). Preliminary reference Earth model. *Physics of the Earth and Planetary Interiors*, 25(4), 297–356. doi: 10.1016/0031-9201(81)90046-7

Enomoto, S., Ohtani, E., Inoue, K., & Suzuki, A. (2007). Neutrino geophysics with KamLAND and future prospects. *Earth and Planetary Science Letters*, 258(1–2), 147–159. doi: 10.1016/j.epsl.2007.03.038

Faure, P., Boyet, M., Bouhifd, M., Manthilake, G., Hammouda, T., & Devidal, J.-L. (2020). Determination of the refractory enrichment factor of the bulk silicate Earth from metal-silicate experiments on rare earth elements. *Earth and Planetary Science Letters*, 116644. doi: 10.1016/j.epsl.2020.116644

Fiorentini, G., Fogli, G., Lisi, E., Mantovani, F., & Rotunno, A. (2012). Mantle geoneutrinos in KamLAND and Borexino. *Physical Review D*, 86(3), 033004. doi: 10.1103/PhysRevD.86.033004

Fiorentini, G., Lissia, M., & Mantovani, F. (2007). Geoneutrinos and earth's interior. *Physics Reports 453*(5–6), 117–172. doi: 10.1016/j.physrep.2007.09.001

Gando, A., Gando, Y., Hanakago, H., Ikeda, H., Inoue, K., Ishidoshiro, K., et al. (2013). Reactor on-off antineutrino measurement with KamLAND. *Physical Review D*, 88, 033001. doi: 10.1103/PhysRevD.88.033001

Gao, R., Li, Z., Han, R., Wang, A., Li, Y., Xi, Y., et al. (2020). JULOC: A local 3-D high-resolution crustal model in South China for forecasting geoneutrino measurements at JUNO. *Physics of the Earth and Planetary Interiors*, 299, 106409.

Hacker, B. R., Kelemen, P. B., & Behn, M. D. (2015). Continental Lower Crust. *Annual Review of Earth and Planetary Sciences*, 43(1), 167–205. doi: 10.1146/annurev-earth-050212-124117

Huang, Y., Chubakov, V., Mantovani, F., Rudnick, R. L., & McDonough, W. F. (2013). A reference Earth model for the heat-producing elements and associated geoneutrino flux. *Geochemistry, Geophysics, Geosystems*, 14(6), 2003–2029. doi: 10.1002/ggge.20129

Huang, Y., Strati, V., Mantovani, F., Shirey, S. B., & McDonough, W. F. (2014). Regional study of the Archean to Proterozoic crust at the Sudbury Neutrino Observatory (SNO+), Ontario: Predicting the geoneutrino flux. *Geochemistry, Geophysics, Geosystems*, 15(10), 3925–3944. doi: 10.1002/2014GC005397

Jackson, M. G., & Jellinek, A. M. (2013). Major and trace element composition of the high ^{3}He/^{4}He mantle: Implications for the composition of a nonchonditic earth. *Geochemistry, Geophysics, Geosystems*, 14(8), 2954–2976. doi: 10.1002/ggge.20188

Jaupart, C., Labrosse, S., Lucazeau, F., & Mareschal, J.-C. (2015). Temperatures, heat, and energy in the mantle of the Earth. In D. Bercovici (Ed.), *Mantle Dynamics* (Vol 7, p. 223–270). Oxford Elsevier. Editor-in-chief G. Schubert doi: 10.1016/B978-0-444-53802-4.00126-3

Javoy, M., Kaminski, E., Guyot, F., Andrault, D., Sanloup, C., Moreira, M., et al. (2010). The chemical composition of the Earth: Enstatite chondrite models. *Earth and Planetary Science Letters 293*(3), 259–268. doi: 10.1016/j.epsl.2010.02.033

Krauss, L. M., Glashow, S. L., & Schramm, D. N. (1984). Anti-neutrinos astronomy and geophysics. *Nature, 310*, 191–198. doi: 10.1038/310191a0

Learned, J. G., Dye, S. T., & Pakvasa, S. (2007). Hanohano: A deep ocean anti-neutrino detector for unique neutrino physics and geophysics studies. In *Neutrino Telescopes. Proceedings, 12th International Workshop, Venice, Italy*, March 6–9, 2007 (p. 235–269). (arXiv:0810.4975)

Mantovani, F., Carmignani, L., Fiorentini, G., & Lissia, M. (2004). Antineutrinos from Earth: A reference model and its uncertainties. *Physical Review D, 69*(1), 013001. doi: 10.1103/PhysRevD.69.013001

McDonough, W. F., & Sun, S. S. (1995). The composition of the Earth. *Chemical Geology, 120*(3–4), 223–253. doi: 10.1016/0009-2541(94)00140-4

McDonough, W. F., Šrámek, O., & Wipperfurth, S. A. (2020). Radiogenic power and geoneutrino luminosity of the Earth and other terrestrial bodies through time. *Geochemistry, Geophysics, Geosystems, 21*(7), e2019GC008865. doi: 10.1029/2019GC008865

McDonough, W. F., & Yoshizaki, T. (2021). Terrestrial planet compositions controlled by accretion disk magnetic field. *Progress in Earth and Planetary Science, 8*, 39. doi: 10.1186/s40645-021-00429-4

O'Neill, H. S. C., & Palme, H. (2008). Collisional erosion and the non-chondritic composition of the terrestrial planets. *Philosophical Transactions of the Royal Society of London A: Mathematical, Physical and Engineering Sciences, 366*(1883), 4205–4238. doi: 10.1098/rsta.2008.0111

Palme, H., & O'Neill, H. S. C. (2014). Cosmochemical estimates of mantle composition. In R. W. Carlson (Ed.), *The Mantle and Core* (Vol. 3 of *Treatise on Geochemistry* (Second Edition), pp. 1–39). Oxford: Elsevier. (Editors-in-chief H. D. Holland and K. K. Turekian) doi: 10.1016/B978-0-08-095975-7.00201-1

Raghavan, R. S., Schoenert, S., Enomoto, S., Shirai, J., Suekane, F., & Suzuki, A. (1998). Measuring the global radioactivity in the Earth by multidetector antineutrino spectroscopy. *Physical Review Letters, 80*(3), 635. doi: 10.1103/PhysRevLett.80.635

Rudnick, R. L., & Gao, S. (2014). Composition of the Continental Crust. In R. Rudnick (Ed.), *The Crust* (Vol. 4 of *Treatise on Geochemistry (Second Edition)*, p. 1–51). Elsevier. (Editors-in-chief H. D. Holland and K. K. Turekian) doi: 10.1016/B978-0-08-095975-7.00301-6

Sammon, L., Gao, C., & McDonough, W. (2020). Lower crustal composition in the southwestern United States. *Journal of Geophysical Research: Solid Earth, 125*(3), e2019JB019011. doi: 10.1029/2019JB019011

Sammon, L. G., & McDonough, W. F. (2021). A geochemical review of amphibolite, granulite, and eclogite facies lithologies: Perspectives on the deep continental crust. *Journal of Geophysical Research: Solid Earth, 126*, e2021JB022791. doi: 10.1029/2021JB022791

Šrámek, O., McDonough, W. F., Kite, E. S., Lekić, V., Dye, S. T., & Zhong, S. (2013). Geophysical and geochemical constraints on geoneutrino fluxes from Earth's mantle. *Earth and Planetary Science Letters, 361*, 356–366. doi: 10.1016/j.epsl.2012.11.001

Šrámek, O., Roskovec, B., Wipperfurth, S. A., Xi, Y., & McDonough, W. F. (2016). Revealing the Earth's mantle from the tallest mountains using the Jinping Neutrino Experiment. *Scientific Reports, 6*(1), 33034. doi: 10.1038/srep33034

Strati, V., Baldoncini, M., Callegari, I., Mantovani, F., McDonough, W. F., Ricci, B., et al. (2015). Expected geoneutrino signal at JUNO. *Progress in Earth and Planetary Science, 2*(1)1–7. doi: 10.1186/s40645-015-0037-6

Strati, V., Wipperfurth, S. A., Baldoncini, M., McDonough, W. F., & Mantovani, F. (2017). Perceiving the crust in 3-D: A model integrating geological, geochemical, and geophysical data. *Geochemistry, Geophysics, Geosystems, 18*(12), 4326–4341. doi: 10.1002/2017GC007067

Takeuchi, N., Ueki, K., Iizuka, T., Nagao, J., Tanaka, A., Enomoto, S., et al. (2019). Stochastic modeling of 3-D compositional distribution in the crust with Bayesian inference and application to geoneutrino observation in Japan. *Physics of the Earth and Planetary Interiors, 288*, 37–57. doi: 10.1016/j.pepi.2019.01.002

Turcotte, D. L., Paul, D., & White, W. M. (2001). Thorium-uranium systematics require layered mantle convection. *Journal of Geophysical Research: Solid Earth, 106*(B3), 4265–4276. doi: 10.1029/2000JB900409

Usman, S. M., Jocher, G. R., Dye, S. T., McDonough, W. F., & Learned, J. G. (2015). AGM2015: Antineutrino Global Map 2015. *Scientific Reports* (51)13945.doi: 10.1038/srep13945

Vogel, P., & Beacom, J. F. (1999). Angular distribution of neutron inverse beta decay, $\bar{v}_e + p \rightarrow e^+ + n$. *Physical Review D, 60*(5), 053003. doi: 10.1103/PhysRevD.60.053003

Wan, L., Hussain, G., Wang, Z., & Chen, S. (2017). Geoneutrinos at Jinping: Flux prediction and oscillation analysis. *Physical Review D, 95*, 053001. doi: 10.1103/PhysRevD.95.053001

Watanabe, H. (2019). *Geo-neutrino measurement with KamLAND*. Presentation at "Neutrino Geoscience 2019 Prague", Czech Republic, October 21–23. Retrieved from https://indico.cern.ch/event/825708/contributions/3552210/

Willig, M., Stracke, A., Beier, C., & Salters, V. J. M. (2020). Constraints on mantle evolution from Ce-Nd-Hf isotope systematics. *Geochimica et Cosmochimica Acta, 272*, 36–53. doi: 10.1016/j.gca.2019.12.029

Wipperfurth, S. A., Guo, M., Šrámek, O., & McDonough, W. F. (2018). Earth's chondritic Th/U: Negligible fractionation during accretion, core formation, and crust–mantle differentiation. *Earth and Planetary Science Letters, 498*, 196–202. doi: 10.1016/j.epsl.2018.06.029

Wipperfurth, S. A., Šrámek, O., & McDonough, W. F. (2020). Reference models for lithospheric geoneutrino signal. *Journal of Geophysical Research, 125*(2), e2019JB018433. doi: 10.1029/2019JB018433

Yoshizaki, T., & McDonough, W. F. (2021). Earth and Mars–distinct inner solar system products. *Geochemistry 81*, 125746. doi: 10.1016/j.chemer.2021.125746

2

Trace Element Abundance Modeling with Gamma Distribution for Quantitative Balance Calculations

Sanshiro Enomoto[1,2], Kenta Ueki[3], Tsuyoshi Iizuka[4], Nozomu Takeuchi[5], Akiko Tanaka[6], Hiroko Watanabe[7], and Satoru Haraguchi[3,5]

ABSTRACT

Rocks have highly variable trace element abundances, even those with similar major element compositions. To estimate the trace element compositions of geochemical reservoirs containing various rocks, such as the continental crust, trace element distributions need to be modeled. Normal and log-normal distributions have been widely used to model trace element concentrations of rocks and the median has been widely used as a representative value of trace element abundance. However, such modeling or treatment does not preserve the mean value of the distribution, leading to biased results from mass balance calculations. On the other hand, it is often not possible to directly use the mean of the samples because of sample imperfections due to analytical detection limits, outliers, and other reasons. Here, we present a method to model unknown true distributions without causing bias on the means, combined with proper treatment of sample cutoffs. We show that our new method based on the gamma distribution model preserves the mean values of ideal samples of any distributions and is robust against actual sample imperfections. Although this method was primarily developed for geoneutrino analysis in Japan, for which mass and energy balance calculations among the crust, mantle, and the bulk Earth are essential, it is applicable to many balance calculations that require the conservation of mean values.

2.1. INTRODUCTION

Earth is chemically differentiated due to magmatic and sedimentary processes over geologic history. Viewing Earth as a system consisting of uniform reservoirs with distinct chemical compositions has proven to be a simple but useful approach to understanding its evolution. For example, assuming that the bulk Earth has chondritic relative abundances of refractory elements (i.e., the chondritic Earth model), the mass ratios of these elements between the silicate Earth and core have been used to estimate the conditions and timing of core-mantle differentiation (e.g., Jacobsen, 2005; Wood et al., 2006). In the context of the chondritic Earth model, considering that incompatible lithophile elements are concentrated in the continental crust and, in a complementary fashion, depleted in the mantle source for mid-ocean ridge basalts, elemental mass balance calculations have provided constraints on the mass fraction of the depleted mantle and growth of the continental crust (e.g., O'Nions et al., 1979; McCulloch and Bennett, 1994; Hofmann, 1997; Campbell, 2003). Furthermore, the mass ratios

[1]Kavli Institute for the Physics and Mathematics of the Universe, The University of Tokyo, Kashiwa, Japan
[2]Department of Physics, University of Washington, Seattle, Washington, USA
[3]Research Institute for Marine Geodynamics, Japan Agency for Marine-Earth Science and Technology, Yokosuka, Japan
[4]Department of Earth and Planetary Sciences, The University of Tokyo, Tokyo, Japan
[5]Earthquake Research Institute, The University of Tokyo, Tokyo, Japan
[6]Geological Survey of Japan, National Institute of Advanced Industrial Science and Technology, Tsukuba, Japan
[7]Research Center for Neutrino Science, Tohoku University, Sendai, Japan

Core-Mantle Co-Evolution: An Interdisciplinary Approach, Geophysical Monograph 276, First Edition.
Edited by Takashi Nakagawa, Taku Tsuchiya, Madhusoodhan Satish-Kumar, and George Helffrich.
© 2023 American Geophysical Union. Published 2023 by John Wiley & Sons, Inc.
DOI: 10.1002/9781119526919.ch02

of refractory lithophile elements in accessible silicate reservoirs have been used to infer the presence of a hidden reservoir in the deep mantle (Rudnick et al., 2000; Boyet and Carlson, 2005). Accurate and precise geochemical compositions of the accessible reservoirs are essential for such geochemical mass balance models. However, robust determination of the geochemical compositions of reservoirs remains difficult because they are heterogeneous. For example, the continental crust is vertically stratified in terms of its lithological and geochemical compositions, with the lower crust dominated by mafic rocks and the upper crust by granitic and sedimentary rocks (Christensen and Mooney, 1995; Rudnick and Gao, 2003). Moreover, the rocks constituting the continental crust display highly variable trace elemental abundances. Thus, estimating the chemical composition of the continental crust and other geochemical reservoirs requires adequate combination of the variable element abundances of the constituent rocks. However, combination of elemental abundance data has not always been done for mass balance calculations in mind, and it is quite common that published representative values for reservoirs or rock types are not suitable for addition or subtraction among these values as an operation to combine or compare them (hereafter referred to as "not additive," for quantities with which no meaningful addition operation is defined; see section 2.7.1). A typical example is the use of the median for the representation of the elemental concentration of a reservoir (e.g., crust) or rock type (e.g., granite). For most rock types, the median estimate of the elemental concentration is lower than its arithmetic mean and, therefore, using the median for a balance calculation will result in underestimation, which in some cases can be significant.

We encountered this problem when constructing a stochastic crustal model for geoneutrino flux calculations (Takeuchi et al., 2019a, 2019b), where the element distribution of each rock type is multiplied by probabilistic volumes and then summed to obtain the total amounts of U and Th in a defined region. We noticed that numerous geochemical studies used nonadditive values, such as the median, as representative estimations. While this might be suitable for some purposes, such as classifications and descriptions where proximity to the most populated region would make sense, those numbers are not suitable for our geophysical applications. We also noticed, on the other hand, making additive representative estimations is not straightforward, which is why the median or other values for similar purposes, such as the geometric mean, have commonly been used.

In this study, we developed a method to model elemental abundances in rocks that can be used for balance calculations. The procedure itself has been described in Takeuchi et al. (2019a). Here, we present the theoretical background, validation, and comparison with other methods, with emphasis on the problems, which have not been widely recognized.

We start by describing the problems and common practices used in section 2.2. The development of our method is presented in section 2.3. This method was then applied to a dataset prepared for a geoneutrino investigation. The validation of the method is described in section 2.4, with a discussion of the method in section 2.5. The conclusions are summarized in section 2.6.

2.2. PROBLEMS

One common approach to construct a compositional model is to collect samples and extract a measure of central tendency as a representative value of the sample distribution. In balance calculations, such representative values must be unbiased means, as otherwise the values obtained after addition, integration, and/or subtraction among the representative values will be biased. Other commonly used measures of central tendency, such as the median, mode, and geometric mean, are not additive in general and thus cannot be used for balance calculations. This can be shown by providing one simple example, given in section 2.7.1.

In many cases, the median is preferred over the arithmetic mean of samples, which is presumably due to imperfection of samples. Most importantly, detection limits are unavoidable in sample analysis. Reported values in low-concentration regions are sometimes not reliable. Outliers from unintended inclusion of exotic samples in an influential high-concentration region are also problematic. A common practice to improve the robustness of the mean value estimation, rather than using the median, is to remove samples in the tail regions of the distribution. However, such quantile cuts simply pull the estimated mean values toward the median value by an arbitrary amount. To correct for sample imperfections, assumptions regarding the imperfections must be made, leading to model-based approaches. Note that applying a quantile cut is already based on a model of the imperfection, explicitly or not.

The most common model of concentration distributions, which is often used implicitly or without much attention, is the normal distribution model. Reporting only the mean and standard deviation of samples is based on this model, or at least a meaningful approximation to a normal distribution is implied. An obvious problem with the normal distribution model is the tail in the negative-concentration region. The estimated mean value is biased if the nonzero probability of negative concentration is simply neglected. This problem is more serious for rocks depleted in trace elements, where the mean value is close to zero and the distribution is often highly skewed. Figure 2.1 shows histograms of the U and Th concentrations of gabbros (GAB), amphibolites (AMP), tonalites/tonalitic gneisses (TNL), granites/granitic gneisses (GRA), and shales (SHL) from the Japan Arc. Mafic rocks (GAB and AMP) have relatively low U and Th concentrations with highly skewed distributions, whereas evolved rocks exhibit symmetrical

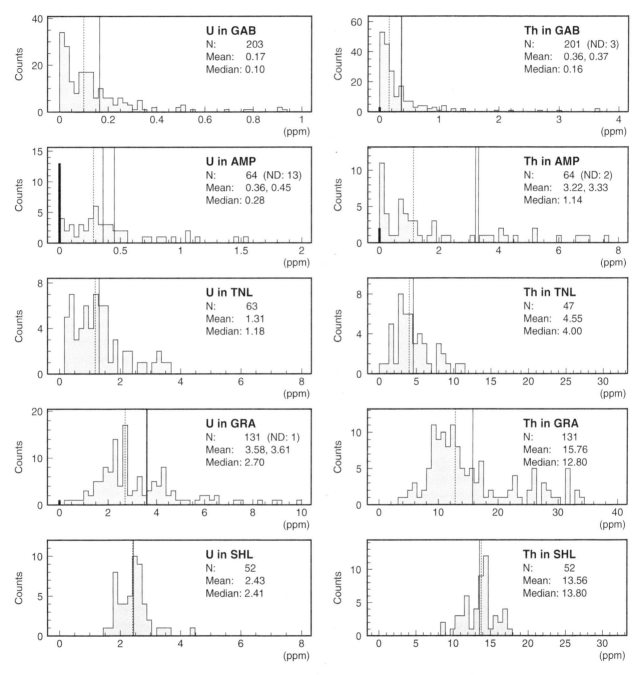

Figure 2.1 Distributions of the U and Th concentrations in several rock types from the Japan Arc (unit: ppm). Data sources are listed in Takeuchi et al. (2019a), although the dataset used in this study has been slightly updated. For distributions that include samples reported as not-detected (ND), two arithmetic mean values were calculated: one excluding the ND samples and the other by replacing the ND values with zero concentrations. The median was calculated by assuming that all the ND samples were below the calculated median value. The vertical thin solid and dotted lines are the arithmetic means and medians, respectively, and the thick solid lines show the number of ND samples.

distributions. Accordingly, the difference between the mean and median becomes evident for the mafic rocks.

Another common model that is applied when the deficiency of the normal distribution model is recognized, in particular for trace elements in depleted rocks, is the log-normal distribution model. This model was proposed by Ahrens (1954) for rock composition distributions. Despite criticism by Chayes (1954) who noted the nonuniqueness of the log-normal model for describing skewed distributions, it has been extensively used, including a widely used crustal compositional models for geoneutrino flux predictions.

The log-normal model assumes the logarithm of the concentration, log(x), follows a normal distribution, that is

$$\log(x) \sim \mathcal{N}(\mu, \sigma) = \frac{1}{\sqrt{2\pi\sigma^2}} e^{-\frac{(x-\mu)^2}{2\sigma^2}} \qquad (2.1)$$

where the tilde operator denotes that the left-hand side quantity follows the right-hand side distribution, and μ and $\sigma(>0)$ are the parameters of the normal distribution, $\mathcal{N}$. Except for cases in which the true element concentration distribution is really log-normal, applying this model can lead to biased mean estimates. This can be demonstrated by providing a simple example, as described in section 2.7.2. Actual elemental concentrations are often close to a log-normal distribution and biases are often small, but this is not always the case. Moreover, the bias is sometimes largely inflated by small fluctuations of sample values, especially for trace elements in depleted rocks. Several examples are presented in section 2.4. This instability can be explained by the interaction between the heavy tail on the high-concentration side and susceptibility to low-concentration fluctuations; the symmetrical shape of the logarithmic values (i.e., the normal distribution) mirrors small fluctuations in the low-concentration region to the high-concentration region during model fitting and the nonlinear structure of the logarithm converts the small fluctuations to expansion of the high-concentration side tail exponentially, while the low concentration side is bound above zero. Although such instabilities may be rare, encountering such cases motivated us to seek a mathematical guarantee for unbiased estimation of mean values (section 2.3).

When correlations among elements in composed bodies (e.g., rocks or reservoirs) are of interest, the compositional closure constraint must be taken into account (Aitchison, 1982, 1986). Here, the sum of all elemental concentrations in a body must be 100% and, under such constraints, changes in one element concentration affect the concentrations of all other elements, leading to spurious correlations. Aitchison (1982, 1986) noted that elemental ratios are more fundamental in compositional analysis and suggested using log-ratios to transform compositional data in a simplex of D elements (in a simplex each concentration value varies between 0 and 1, with the total being 1) into $\mathbb{R}^{D-1}$ space (where the constraint of the total being 1 reduces the number of free parameters by 1), in which the typical multivariate normal models may be applied. Among several variations, one common form of the log-ratio transforms, which is the additive logistic-normal model (e.g., Chapter 6 in Aitchison, 1986), is

$$\log\left(\frac{\mathbf{x}_{-D}}{x_D}\right) \sim \mathcal{N}^{D-1}(\boldsymbol{\mu}, \boldsymbol{\Sigma}), \qquad (2.2)$$

where $\mathbf{x}_{-D}$ is a vector of element concentrations excluding the Dth element, and x_D is the concentration of the Dth element. $\mathcal{N}^{D-1}(\boldsymbol{\mu}, \boldsymbol{\Sigma})$ is the $D-1$ dimensional multivariate normal distribution with a location vector $\boldsymbol{\mu}$ and covariance matrix $\boldsymbol{\Sigma}$. Although the formulation depends on the choice of the Dth element, many common statistical properties derived from the covariance matrix are invariant under permutations of the elements (e.g., Chapters 5 and 7 in Aitchison, 1986).

While the compositional closure plays a critical role in correlation analysis of major elements, it has negligible influence on trace element abundance modeling. First, the concentrations of trace elements (typically <0.01%) are often lower or comparable to the errors on the major element concentration analyses, making the closure constraint noninfluential to trace elements. Second, if a major element is chosen for the Dth element in equation (2.2), the distribution shape of the logistic-normal model for a trace element becomes very similar to that of the log-normal model, because variations in major element concentrations (typically tens of percent at most) are much smaller than variations in trace element concentrations (typically several factors or even orders of magnitude). Although compositional analysis with log-ratios is presumably not primarily intended to estimate absolute abundances of trace elements, rather than correlations among elements, if it is used for abundance modeling, then the mean value of the model does not conserve the arithmetic mean of the samples, just in the same way as the log-normal model. In section 2.7.3, one such example is presented.

In geochemical studies, the median has been widely used (e.g., Reimann and Filzmoser, 1999). Unless it is used for balance calculations, it may be a good measure to describe the element abundance because of its robustness against sample imperfections and its proximity to the region where samples are more populated. Despite criticisms, the log-normal distribution model has also been widely used. It is at least better than the normal distribution model in terms of logical consistency of not having negative values, and because of the description of skewed distributions (e.g., Chayes, 1954; Shaw, 1961). In some cases, log-normal distributions are derived from geological process models, often based on the partitioning behavior of trace elements where ratios in abundance play the primary role (e.g., Shaw, 1961; Allègre and Lewin, 1995; Albarède, 2005). Due to the multiplicative nature of log-normal distributions, the distributions are characterized by the geometric mean, in contrast to the arithmetic mean for normal distributions, and the geometric mean is sometimes adopted for better description of rocks (e.g., for mid-ocean ridge basalts, Albarède, 2005). If the distribution is log-normal, then the median coincides with the geometric mean. These successful applications of the log-normal model, geometric mean, and median lead us to speculate that the geometric mean may have some relevance to the underlying geological processes, although the geometric mean cannot be directly used for balance calculations.

2.3. METHOD

Our goal is to develop a method to estimate the unbiased mean from a sample distribution that could be distorted due to detection limits, outliers, and other reasons.

Cutoffs based on quantiles, leverages, or any other criteria, such as data accuracy, are unavoidable, at least with respect to detection limits, and probably necessary for other reasons such as for removing influential outliers. If cutoffs are applied without appropriate care for possible side effects, then these will results in a bias, or worse, arbitrary control to the results. However, if the shape of the distribution is given (e.g., log-normal), the fraction of samples outside the cutoffs can be predicted, allowing us to make stable estimations.

We take a systematic bottom-up approach for the distribution models. We limit our scope to two-parameter distributions, until we find it insufficient. We cannot expect that the modeled distribution is the exact true element distribution, as by a common aphorism in statistics, "all models are wrong," but we require as the first constraint of model construction that whatever the true distribution is, the mean value of the modeled distribution equals the mean value of the samples, as long as the samples are ideal. In view of the successful application of the log-normal model and median in previous geochemical studies, we take the geometric mean as our second critical constraint. As mentioned before, note that the geometric mean, a characteristic parameter of a log-normal distribution, approaches the median of the distribution if the distribution is close to log-normal.

We now have two constraints to determine a two-parameter distribution. Here, we do not assume anything further. This means that the modeled distribution must have a maximum entropy (i.e., minimum prior knowledge) under these two constraints. For a distribution $f(x)$ defined for positive real numbers x, its entropy H is given by

$$H[f(x)] = -\int_0^\infty f(x) \log f(x) dx \qquad (2.3)$$

and the two constraints, arithmetic mean m_a and the geometric mean m_g of $f(x)$, are given by

$$m_a[f(x)] = \int_0^\infty x f(x) \, dx \qquad (2.4)$$

and

$$m_g[f(x)] = \exp\left[\int_0^\infty (\log x) f(x) \, dx\right], \qquad (2.5)$$

respectively. Our model distribution to find is the distribution $f(x)$ that maximizes H under a given set of constraint values of m_a and m_g. This is in a form of a typical problem in the "calculus of variations," and can be solved with the method of Lagrange multipliers, yielding $f(x)$ being the gamma distribution (Lisman and van Zuylen, 1972).

The gamma distribution is a continuous two-parameter probability distribution defined for positive real numbers x. In a parameterization with a shape parameter k (>0) and scale parameter θ (>0), its probability density function (PDF) is defined as

$$f(x;k,\theta) = \text{Gamma}(x;k,\theta) = \frac{x^{k-1}e^{-\frac{x}{\theta}}}{\Gamma(k)\theta^k}, \qquad (2.6)$$

where $\Gamma(k)$ is the gamma function. This might look complicated, but it is simply a product of a power function and exponential function of scaled x, x/θ, as $\text{Gamma}(x;k,\theta) \propto (x/\theta)^{k-1} \cdot e^{-x/\theta}$, and the rest ($\frac{1}{\Gamma(k)\theta}$) is only for normalization to make the integration (total probability) equal to one. The mean value (expected value) of the distribution is given by $k\theta$. The skewness is given by $2/\sqrt{k}$, allowing the distribution to be highly skewed with a small k value to a Gaussian-like quasi-symmetric distribution with a large k value. The exponential distribution is a special case of the gamma distribution with $k = 1$. The shape of the gamma distribution approaches an inverse proportional ($1/x$) for $k \to 0$. The scale parameter, θ, as the name implies, scales the entire shape of the distribution. Unlike a normal distribution, this distribution does not have a location parameter (a parameter shifting the location of the distribution, directly controlling the mean value).

We use the maximum likelihood method to determine the model parameters. Note that the method of moments cannot be used because the true mean value is unknown.

The likelihood consists of two parts: one is to describe the shape of the distribution for the samples that are not removed by cutoff, and the other is to describe the fraction of the samples removed by the cutoff.

If no cutoff is applied to the samples, the PDF of the distribution model can be directly used for the likelihood function. Let $f(x; p)$ be the distribution model (not necessarily the gamma distribution), where x is the random variable for sample concentration observations and p is the model parameters to estimate. For a given observation of N rock samples, $a = \{a_1, a_2, \ldots, a_N\}$, the negative log likelihood function (with a factor of -2 for conventional reasons) is given by

$$-2 \log \mathcal{L}(p; a) = -2 \sum_{j=1}^{N} \log f(a_j; p). \quad (2.7)$$

Minimizing the negative log-likelihood function with respect to p for the given a yields the best-fit estimation of the parameters, $\hat{p}$. In the case of the gamma distribution model, the best-fit PDF, $f(x; \hat{p}) = \text{Gamma}(x; \hat{k}, \hat{\theta})$, where $\hat{k}$ and $\hat{\theta}$ are the best-fit parameters determined by the minimization, has an expected value exactly equal to the arithmetic mean of the samples, regardless of the actual distribution of the samples (section 2.7.4), that is

$$E(x) = \hat{k}\hat{\theta} \equiv \frac{1}{N} \sum_{j=1}^{N} a_j. \quad (2.8)$$

This mathematically assures that fitting a gamma model preserves the sample mean of any underlying distribution.

With a cutoff for a lower bound for detection limits or possibly inaccurate concentration analysis, the shape likelihood is modified to account for the renormalization due to the partial removal of samples. With $B (> 0)$ as a lower cutoff boundary, this is given by

$$-2 \log \mathcal{L}_{\text{shape}}(p; a) = -2 \sum_{j=0}^{N} \Theta(a_j - B) \log \left[\frac{f(a_j; p)}{\int_B^\infty f(\alpha; p) d\alpha} \right] \quad (2.9)$$

where $\Theta(a)$ is the Heaviside step function.

For samples below the cutoff, we simply count the number of such samples, as the concentration values are unknown or unreliable. The probability of having N_0 samples below the cutoff is described by a binomial distribution, leading to a negative log-likelihood of

$$-2 \log \mathcal{L}_{\text{count}}(p; a) = -2 \log \left[\frac{N!}{N_0!(N - N_0)!} \right.$$
$$\left. \times \left(\int_0^B f(\alpha; p) d\alpha \right)^{N_0} \left(\int_B^\infty f(\alpha; p) d\alpha \right)^{N-N_0} \right]. \quad (2.10)$$

The final log-likelihood is a combination of these two:

$$-2 \log \mathcal{L}(p; a) = -2 \log \mathcal{L}_{\text{shape}}(p; a) - 2 \log \mathcal{L}_{\text{count}}(p; a). \quad (2.11)$$

If a cutoff for an upper bound is applied in addition to one for the lower bound, the likelihood functions are modified for three segments by replacing the binomial distribution with a multinomial distribution. With $B_{\text{lower}} (> 0)$ and $B_{\text{upper}} (> B_{\text{lower}})$ as lower and upper cutoff boundaries, respectively, the log-likelihood functions are given by

$$-2 \log \mathcal{L}_{\text{shape}}(p; a) = -2 \sum_{j=0}^{N} \Theta(a_j - B_{\text{lower}}) \Theta(B_{\text{upper}} - a_j)$$
$$\times \log \left[\frac{f(a_j; p)}{\int_{B_{\text{lower}}}^{B_{\text{upper}}} f(\alpha; p) d\alpha} \right] \quad (2.12)$$

and

$$-2 \log \mathcal{L}_{\text{count}}(p; a) = -2 \log \left[\frac{N!}{N_0!(N - N_0 - N_1)!N_1!} \right.$$
$$\times \left(\int_0^{B_{\text{lower}}} f(\alpha; p) d\alpha \right)^{N_0}$$
$$\left. \times \left(\int_{B_{\text{lower}}}^{B_{\text{upper}}} f(\alpha; p) d\alpha \right)^{N-N_0-N_1} \left(\int_{B_{\text{upper}}}^\infty f(\alpha; p) d\alpha \right)^{N_1} \right] \quad (2.13)$$

where N_1 is the number of samples above the upper cutoff.

If the model extends to negative values (e.g., the normal distribution model), another segment for the negative-concentration region is added to the binomial/multinomial distribution. However, setting the number of samples with negative concentrations to zero reduces it to the same equation as above; with N_z being the number of negative concentration samples, the count likelihood for the lower-cutoff-only case is extended to

$$-2 \log \mathcal{L}_{\text{count}}(p; a) = -2 \log \left[\frac{N!}{N_z!N_0!(N - N_z - N_0)!} \right.$$
$$\times \left(\int_{-\infty}^0 f(\alpha; p) d\alpha \right)^{N_z}$$
$$\left. \times \left(\int_0^B f(\alpha; p) d\alpha \right)^{N_0} \left(\int_B^\infty f(\alpha; p) d\alpha \right)^{N-N_z-N_0} \right], \quad (2.14)$$

in which setting $N_z = 0$ reduces the equation to its original form, equation (2.10). Note that the model still has a nonzero probability in the negative concentration region despite the zero observation and thus the model can never be valid. The probability of negative concentrations becomes significantly large for depleted rocks.

2.4. EVALUATION OF THE METHOD

The method was applied to a dataset prepared for crustal modeling for geoneutrino observations in Kamioka, Japan, presented in Takeuchi et al. (2019a, 2019b). Minor updates and simplifications on combining the rock types were made to better suit the purpose of this study. Following Takeuchi et al. (2019a), a conservative common cutoff at 0.2 ppm for the lower bound was adopted for all rock types, based on the observed increased dispersion of the Th/U ratio below U = 0.2 ppm, possibly indicating inaccurate composition analysis in the dataset. This conservative cutoff threshold was also meant in this study to demonstrate the ability of the method to deal with the cutoff. A cutoff at an upper boundary was not applied.

Table 2.1 summarizes the results of fitting the gamma, log-normal, and normal distribution models to the dataset. Errors of the model parameter estimates due to statistical fluctuations of the samples are also provided, which were calculated from the shape of the likelihood function around the best-fit point in a common way; specifically, the Hessian matrix, $\mathcal{H}$, of the log-likelihood function at the best-fit point was constructed by $\mathcal{H}_{i,j} = \left.\frac{\partial^2 (\log \mathcal{L}(p;a))}{\partial p_i \partial p_j}\right|_{p=\hat{p}}$, where the inverse of the negative of the Hessian matrix becomes the variance-covariance matrix. The errors in Table 2.1 are the square roots of the diagonal elements of the variance-covariance matrix. Here, the variance-covariance matrix was constructed for the quantities shown in Table 2.1 columns, instead of the explicit parameters in the PDFs (e.g., used parameters of $k\theta$ and k instead of θ and k for the gamma model). As the first column of each model is its expected value, and parameter errors from a variance-covariance matrix are symmetrical by construction, the errors on the expected values are all symmetrical.

Additional information regarding the dataset and fitting results can be found in Takeuchi et al. (2019a, Tables 4, 5 and Figure 8). Although the datasets are not identical, the features of the method discussed here still apply. An important point in the results is the exact agreement between the arithmetic mean of the samples and the expected value of the fitted gamma distribution model PDF, given there are no samples below the cutoff threshold. This is not the case for the log-normal distribution models, as predicted by construction.

If a part of the samples is below the cutoff threshold, the mean values of the samples and models cannot be compared. The influence of the cutoff was investigated by examining the dependence of the estimated model means on the choice of the cutoff. Figure 2.2 shows that the mean concentrations (abundances) estimated with the gamma model are stable irrespective of the choice of cutoff, unless a large fraction of the samples falls below the cutoff. In the case of our dataset, even if half the samples are put below the cutoff, the estimates only deviate by 3%–4% for depleted rocks such as GAB.

Takeuchi et al. (2019a) used a conservative common cutoff value of 0.2 ppm for both U and Th for all rock

Table 2.1 Model fitting results (in ppm). The model PDF parameterizations of gamma, log-normal, and normal distribution models are $\frac{1}{\Gamma(k)\theta^k} x^{k-1} e^{-\frac{x}{\theta}}$, $\frac{1}{\sqrt{2\pi\sigma^2}x} e^{-\frac{(\log x - \mu)^2}{2\sigma^2}}$, and $\frac{1}{\sqrt{2\pi\sigma^2}} e^{-\frac{(x-\mu)^2}{2\sigma^2}}$, respectively

Element	Rock	Sample arithmetic mean	Gamma model Mean ($k\theta$)	k	Log-normal model Mean ($e^{\mu+\frac{\sigma^2}{2}}$)	σ	Normal model Mean (μ)	σ
U	GAB	0.17	0.15 ± 0.02	0.35 ± 0.08	0.17 ± 0.02	1.16 ± 0.14	0.19 ± 0.02	0.26 ± 0.01
	AMP	0.45	0.38 ± 0.05	0.96 ± 0.22	0.40 ± 0.06	0.94 ± 0.12	0.39 ± 0.05	0.40 ± 0.04
	TNL	1.31	1.31 ± 0.11	2.31 ± 0.39	1.35 ± 0.14	0.72 ± 0.06	1.31 ± 0.11	0.85 ± 0.08
	GRA	3.61	3.58 ± 0.19	2.62 ± 0.31	3.65 ± 0.23	0.66 ± 0.04	3.58 ± 0.21	2.41 ± 0.15
	SHL*	2.43	2.43 ± 0.07	24.39 ± 4.75	2.43 ± 0.07	0.20 ± 0.02	2.43 ± 0.07	0.51 ± 0.05
Th	GAB	0.37	0.35 ± 0.04	0.44 ± 0.06	0.36 ± 0.04	1.26 ± 0.11	0.37 ± 0.04	0.62 ± 0.03
	AMP	3.33	3.22 ± 0.60	0.46 ± 0.08	4.80 ± 1.77	1.75 ± 0.19	3.23 ± 0.63	5.07 ± 0.45
	TNL*	4.55	4.55 ± 0.37	3.19 ± 0.63	4.66 ± 0.45	0.61 ± 0.06	4.55 ± 0.36	2.49 ± 0.26
	GRA*	15.76	15.76 ± 0.66	4.40 ± 0.52	15.72 ± 0.70	0.48 ± 0.03	15.76 ± 0.72	8.22 ± 0.51
	SHL*	13.56	13.56 ± 0.30	40.16 ± 7.84	13.57 ± 0.31	0.16 ± 0.02	13.56 ± 0.29	2.07 ± 0.20

Notes: The "Mean" columns list the expected values of the fitted PDF, calculated from the best-fit parameters with the equations given in the columns. The asterisk indicates that all samples are above the cutoff threshold of 0.2 ppm. The errors represent the statistical fluctuations of the samples (68.3% confidence interval).

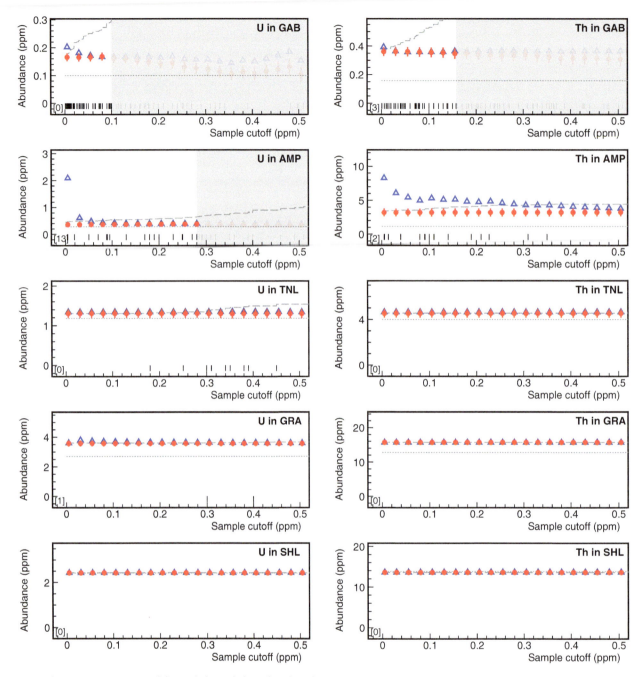

Figure 2.2 Estimates of the U (left) and Th (right) abundances for several rock types for various lower cutoff values. Filled circles (red) and open triangles (blue) represent estimates based on the gamma and log-normal distribution models, respectively. Error bars are only shown for the gamma distribution model. The gray shading indicates the region in which more than half of the samples fall below the cutoff. In addition, the arithmetic mean of the samples above the cutoff (dashed lines), median of all samples (dotted lines), individual sample values in the plotting range (short vertical lines in the bottom; for all sample values, see Fig. 2.1), and number of samples below the detection limits (square brackets) are shown.

types. The results in Figure 2.2 show that the estimated mean values of the gamma model would insignificantly change for a different choice of the cutoff value. This is not the case for the log-normal model. Based on the actual rock composition dataset, Figure 2.2 also shows that: (1) the median of the samples is an underestimation of the mean concentration (abundance) if the distribution is positively skewed, which is a common feature for almost all rock types and particularly large for GAB due to the large skewness (Fig. 2.1); (2) the arithmetic mean of the samples is an overestimation of the abundance if a part of the samples is below the detection limit and neglected; and (3) if all samples are above the cutoff, then the expected value of the gamma model PDF agrees with the arithmetic mean of the samples, whereas that of the log-normal model can have offset.

The Akaike information criterion (AIC) for each model was provided by Takeuchi et al. (2019a) for comparison of the models. In addition, we also compared the models and data. For this purpose, two cumulative distributions were constructed: one from the data and one from the modeled distribution. The shapes were then compared with each other using the Kolmogorov-Smirnov test (Smirnov, 1948). To construct cumulative distributions with a cutoff, the first point of the data accumulation was set to the first sample above the cutoff, and all samples below the cutoff were included in the accumulation of the first point. This treatment degrades the statistical power of the test, but such a degradation is a natural consequence of applying a cutoff to remove the shape information below the cutoff.

Figure 2.3 shows the results of the Kolmogorov-Smirnov test against the null hypotheses of the true distribution being gamma, log-normal, or normal. Within the statistical power, both the gamma and log-normal models cannot be rejected at the 99% confidence level for all rock types.

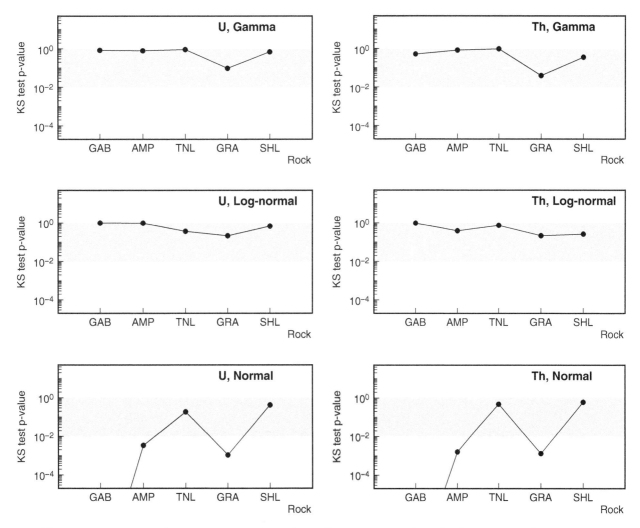

Figure 2.3 Kolmogorov-Smirnov test (KS test) results for the comparison of the distribution shapes between data and models. The p-values obtained from the KS test are shown for each rock type for the null hypothesis of the true distribution being gamma (top), log-normal (middle), or normal (bottom), for U (left) and Th (right). Note that the p-value distributes uniformly between 0 and 1 if the null hypothesis is true. The shaded band indicates the region with a p-value above 1%.

2.5. DISCUSSION

One constraint on the model construction was that the model mean is an unbiased estimation of the mean of the unknown true distribution, for whatever the true distribution is. Conservation of the mean is demonstrated if all samples are above the cutoff and, even if some samples are below the cutoff, the estimation is stable against the choice of the cutoff. In many cases, the differences between the gamma and log-normal models are small, but the log-normal model does not mathematically assure the conservation of the mean value. In some cases, the mean value is significantly biased (often overestimated; e.g., AMP in Fig. 2.2). If the geometric mean parameter of the log-normal model is used instead of the expected value of the distribution PDF, as in previous studies, including a widely used geoneutrino flux prediction model, the estimate is similar to the median (often underestimated, in some cases severely; e.g., GAB in Fig. 2.2).

Although we do not fully trust the sample values below our nominal cutoff (0.2 ppm), the flat structure of the estimated abundance down to a zero concentration is encouraging. As is evident from the estimated U and Th abundances of GAB and AMP in Figure 2.2, the estimates are stable, even after removing half of the samples, while the log-normal model estimates deviate by factors. This stability might be due to: (1) the gamma model shape being in good agreement with the true distribution shape, and (2) the method being robust against fluctuations of a small fraction of samples in the low-concentration regions. Note that the sample distribution below the highest detection limit is deformed if datasets with different detection limits are merged, and thus model fittings including this region suffer from inaccuracy. The gamma model is robust against this effect, presumably because the fitting results are dominated by the majority of the samples and a small number of inaccurate low-concentration samples have limited influence. In contrast, the log-normal model tends to produce overestimates because of the susceptibility in the low-concentration region (section 2.3); in order to make the log-normal model robust and stable against the small inaccuracies of low-concentration analysis, a model of analysis errors must be constructed and included in the fit PDF.

On model construction, our critical requirement for balance calculations was conservation of the arithmetic mean, and for our another freedom in modeling we adopted the geometric mean with a hope that it would bring a good fit of the model to actual elemental distributions of real rocks, which are controlled by geological processes. The resultant good fit for all different rock types is not an obvious outcome. As shown in Figure 2.1, among rock types, there exist quite large variations in their means, variances, and skewness of the concentration data, whereas the gamma model has only two free parameters to adjust, and if the mean and variance are determined the skewness will be automatically fixed, but the model skewness agrees with the data well for all the rock types, from mafic and felsic igneous rocks to sedimentary rocks. This removes the need to artificially switch between models depending on rock types [e.g., between the log-normal and normal models based on the variance-to-mean ratios as proposed by Bes de Berc (1954) and adopted by Shaw (1961)]. Although done with a hope, this agreement beyond the freedom of the model leads us to think about some connection to the underlying geological processes, as discussed below.

One interesting feature of the gamma distribution is its reproductive property. With this property, if two independent samples are taken from a gamma distribution, for example two rock pieces from gabbro, and mixed, the distribution of the mixed samples will also follow a gamma distribution, with an increased k parameter value (reduced skewness of $2/\sqrt{k}$). If the mixing is repeated, the distribution shape becomes more symmetrical (i.e., Gaussian-like) but remains a gamma distribution. If this reproductive structure is an approximate description of the geological mixing processes, this would explain why all observed elemental concentration distributions can be fitted with the gamma distribution. Elemental fractionation processes under an enrichment or partition factor can be approximately described by a change in the scaling parameter θ. Interestingly, as shown in Table 2.1, the shape parameter k, which increases monotonically with the reproductive mixing, basically traces the rock evolution chain from ultramafic rocks, mafic (e.g., GAB and AMP), felsic (GRA and TNL), to sedimentary rocks (SHL).

It is often experienced that samples collected from a small region exhibit large compositional variations. For a similar line, it is commonly advised to collect samples of a sufficiently large size (i.e., the dimensions of each sample) for stable compositional analysis. By mixing multiple samples from one location or collecting large samples, the compositional variation is reduced by averaging, and the concentration distribution is narrowed. The reproductive property of the gamma distribution naturally describes this narrowing, which assures the model is independent of sample size scales. On the other hand, distribution models without this reproductive property, such as the log-normal model, need a specific sample size (i.e., if the log-normal model is valid for a specific sample size, then the model is no longer valid for a different sample size), although such scale dependence is typically not accounted for.

2.6. CONCLUSIONS

Geophysical and geochemical balance calculations using rock compositional models, such as the mass balance of elements among reservoirs, energy balance involving radiogenic heating, and geoneutrino flux calculations, require unbiased estimations of the mean element concentrations (abundances). We have shown that practically all commonly used methods cause bias in such estimations. This includes the median, arithmetic mean of samples (even with quantile cuts to remove outliers), and geometric mean of samples, as well as elemental distribution modeling based on the normal, log-normal, and logistic-normal distributions.

We constructed a bias-free model based on the gamma distribution, in combination with a treatment for unavoidable cutoffs to deal with the detection limits, inaccurate low-concentration analysis, and influential outliers. The gamma distribution model mathematically assures that the expected value (mean) of the modeled distribution is equal to the arithmetic mean of the samples, irrespective of the true distribution, as long as the samples are ideal. For imperfections, the gamma distribution model is robust against inaccurate sample analyses, whereas the log-normal model has susceptivity to small fluctuations in the low-concentration region. The treatment of samples outside the cutoff boundaries with a binomial or multinomial probability distribution enables us to apply conservative cutoff thresholds. In the case of our dataset, up to half the samples can be removed from the shape fitting. In combination with a good strategy for setting cutoff thresholds, such as data accuracy analysis for the low-concentration region and leverage analysis for the high-concentration region, our method eliminates the need to use the median or geometric means, which are commonly used for robustness and/or the description of skewed distributions.

We emphasize again that the conservation of the mean is a key requirement for balance calculations. Among the discussed models and methods, only the gamma distribution model satisfies this requirement for trace element abundance modeling, making it the only choice for this application. This does not necessarily mean that the gamma distribution model is better than other models for other purposes or imply that the gamma distribution model agrees with the actual elemental distributions better than the other models. Nevertheless, at least for the investigated dataset, the gamma distribution model describes the actual elemental concentration distributions consistently well for all rock types, from depleted igneous rocks to surface sedimentary rocks. This is presumably due to the hybrid nature of the gamma distribution model, namely, having both the conservation of the geometric mean and the reproductive property, where the log-normal model has the former with possible relevance to the partitioning processes (section 2.2), and the normal model has the latter with direct correspondence to the rock mixing processes (section 2.5).

Trace element concentrations in depleted rocks generally have highly skewed distributions, and median estimations can lead to several factors of underestimations in balance calculations. Applying the element abundance modeling method proposed in this study to crustal rocks, combined with a stochastic method to infer the 3-D crustal lithology (e.g., Takeuchi et al., 2019a, 2019b), would enable accurate determination of the trace element abundances of the continental crust, leading to a better understanding of the continental growth history. Furthermore, the combination of crustal U and Th abundance quantification with geoneutrino observation would allow us to evaluate the validity of the chondritic Earth model which is the cornerstone for our current understanding of the chemical and thermal evolution of the Earth's core-mantle system.

2.7. SUPPORTING EXAMPLES AND PROOF

2.7.1. An Example Showing that the Median is Not Additive

Here, we show that the median cannot be used as a measure of central tendency for balance calculations, by providing one simple example.

Consider two same-sized sets of samples, $A : \{10,10,40\}$ and $B : \{30,30,30\}$. The mean and median representations, including weighting by size denoted by $\times$, are $A : \{20 \times 3\}$ and $B : \{30 \times 3\}$, and $A : \{10 \times 3\}$ and $B : \{30 \times 3\}$, respectively. If the two sets of samples are merged, denoted by $\oplus$, it will become $A \oplus B : \{10,10,30,30,30,40\}$, with a mean and median of $A \oplus B : \{25 \times 6\}$ and $A \oplus B : \{30 \times 6\}$, respectively. While a balance calculation using the mean $(20 \times 3 + 30 \times 3$ for a sum in this example) produces the mean of the merged set, 25×6, which is assured because the mean is obtained from a weighted sum, a balance calculation using the median $(10 \times 3 + 30 \times 3)$ produces neither median nor mean estimates of the merged set (i.e., neither 25×6 nor 30×6).

Using the same example above, it can be shown that many other common measures of central tendency, such as the geometric mean and mode, are also not generally additive.

2.7.2. An Example Showing that Log-normal Models have Bias on the Mean Value

Here, we show that fitting a log-normal model with the maximum likelihood method does not always conserve the mean value of a sample distribution, even if the sample distribution is not affected by a cutoff or any other artifacts, by providing one example.

Consider a set of rock samples, $\{1, e, e^2\}$. The maximum likelihood fitting of the log-normal distribution model, $f(x; \mu, \sigma^2) = \frac{1}{\sqrt{2\pi\sigma^2}x} e^{-\frac{(\log x - \mu)^2}{2\sigma^2}}$, results in a best-fit estimation of $\hat{\mu} = 1$ and $\hat{\sigma} = \sqrt{2/3}$. Therefore, the expected value of the best-fit model PDF is $\hat{E}(x) = e^{\hat{\mu} + \frac{\hat{\sigma}^2}{2}} = e^{\frac{4}{3}} \approx 3.79$, which differs from the arithmetic mean of the samples, $\frac{1 + e + e^2}{3} \approx 3.70$.

2.7.3. An Example Showing that Logistic-normal Models have Bias on the Mean Value

Here, we show that fitting a logistic-normal model with the maximum likelihood method does not always conserve the mean value of a sample distribution, just like the log-normal model case, using the same example as for the log-normal model.

Consider a set of rock samples with concentrations of an element of interest of $\{1, e, e^2\}$ and concentrations of another element for x_D in equation (2.2) of $\{1, 1, 1\}$. The maximum likelihood fitting of the logistic-normal distribution model results in identical numbers as the log-normal model, and thus the expected value of the modeled distribution differs from the arithmetic mean of the samples.

2.7.4. Proof of the Conservation of the Mean Value in the Gamma Model

Here, we show that if a gamma model is fitted to samples using the maximum likelihood method, the expected value of the fit model PDF is always equal to the mean value of the samples. In the parameterization with a shape parameter $k > 0$ and a scale parameter $\theta > 0$, the PDF of the gamma distribution is

$$f(x; k, \theta) = \frac{x^{k-1} e^{-\frac{x}{\theta}}}{\Gamma(k)\theta^k}. \tag{2.15}$$

Its expected value is given by

$$E(x) = k\theta. \tag{2.16}$$

For the samples $\{a_1, a_1, \ldots, a_N\}$, the log-likelihood is

$$\log \mathcal{L}(k, \theta) = \sum_{j=1}^{N} \log[f(a_j; k, \theta)]$$
$$= (k-1) \sum_{j=1}^{N} \log a_j - \frac{1}{\theta} \sum_{j=1}^{N} a_j$$
$$- N[\Gamma(k) + k \log \theta]. \tag{2.17}$$

To find the maximum with respect to θ, take the derivative:

$$\frac{\partial \log \mathcal{L}}{\partial \theta} = \frac{1}{\theta^2} \sum_{j=1}^{N} a_j - Nk\frac{1}{\theta}. \tag{2.18}$$

The maximum likelihood estimators of θ and k, $\hat{\theta}$ and $\hat{k}$, are given by the values that make this derivative equal to zero, yielding:

$$\hat{k}\hat{\theta} = \frac{1}{N} \sum_{j=1}^{N} a_j. \tag{2.19}$$

Therefore, the best-fit gamma distribution PDF, $f(x; \hat{k}, \hat{\theta})$, has an expected value equal to the mean value of the samples:

$$\hat{E}(x) = \hat{k}\hat{\theta} = \frac{1}{N} \sum_{j=1}^{N} a_j. \tag{2.20}$$

ACKNOWLEDGMENTS

We sincerely thank an anonymous reviewer who noted that we had overlooked an important aspect in an earlier version of our manuscript, and two other anonymous reviewers for extraordinarily careful and critical readings that significantly improved our manuscript. This research was supported by the Japan Society for the Promotion of Science (Grants JP15H05833, JP19K04026, and JP20H01909) and the Earthquake Research Institute of the University of Tokyo (Grants 2020-B-06 and 2020-G-06). S. H. was supported by the Japan Science and Technology Agency CREST (Grant JPMJCR1761).

REFERENCES

Ahrens, L. H. (1954). The lognormal distribution of the elements (a fundamental law of geochemistry and its subsidiary). *Geochimica et Cosmochimica Acta, 5*, 49–73.

Aitchison, J. (1982). The Statistical Analysis of Compositional Data. *Journal of the Royal Statistical Society: Series B, 44*, 139–177.

Aitchison, J. (1986). *The Statistical Analysis of Compositional Data*. London: Chapman and Hall.

Allègre, C. J., & Lewin, E. (1995). Scaling laws and geochemical distributions. *Earth and Planetary Science Letters, 132*, 1–13.

Albarède, F. (2005). The survival of mantle geochemical heterogeneities. In R. D. Van Der Hilst, J. D. Bass, J. Matas, & J. Trampert (Eds.), *Earth's Deep Mantle: Structure, Composition, and Evolution. Geophysical Monograph Series*, (Vol. 160, pp. 27–46). Washington, DC: American Geophysical Union.

Bes de Berc, O. (1954). Influence sur la précision des calculs du rendement poids et du rendement métal, des erreurs commises sur les teneurs. *Revue de l'Industrie Minérale A, 35*, 366–383.

Boyet, M., & Carlson, R. W. (2005). ^{142}Nd evidence for early (>4.53 Ga) global differentiation of the silicate Earth. *Science, 309*, 576–581.

Campbell, I. H. (2003). Constraints on continental growth models from Nb/U ratios in the 3.5 Ga Barberton and other Archaean basalt-komatiite suites. *American Journal of Science, 303*, 319–351.

Chayes, F. (1954). The lognormal distribution of the elements: A discussion. *Geochimica et Cosmochimica Acta, 6*, 119–120.

Christensen N. I., & Mooney, W. D. (1995). Seismic velocity structure and composition of the continental crust: A global view. *Journal of Geophysical Research, 100*, 9761–9788.

Hofmann, A. W. (1997). Mantle geochemistry: The message from oceanic volcanism. *Nature, 385*, 219–229.

Jacobsen, S. B. (2005). The Hf–W isotopic system and the origin of the Earth and Moon. *The Annual Review of Earth and Planetary Sciences, 33*, 531–570.

Lisman J. H. C., & Av van Zuylen, M. C. (1972). Note on the generation of most probable frequency distributions. *Statistica Neerlandica, 26*, 19–23.

McCulloch, M. T., & Bennett, V. C. (1994). Progressive growth of the Earth's continental crust and depleted mantle: Geochemical constraints. *Geochimica et Cosmochimica Acta, 58*, 4717–4738.

O'Nions, R. K., Evensen, N. M., & Hamilton, P. J. (1979). Geochemical modeling of mantle differentiation and crustal growth. *Journal of Geophysical Research, 84*, 6091–6101.

Reimann, C. & Filzmoser, P. (1999). Normal and lognormal data distribution in geochemistry: Death of a myth. Consequences for the statistical treatment of geochemical and environmental data. *Environmental Geology, 39*, 1001–1014.

Rudnick, R. L., Barth, M., Horn, I., & McDonough, W. F. (2000). Rutile-bearing refractory eclogites: Missing link between continents and depleted mantle. *Science, 287*, 278–281.

Rudnick, R. L., & Gao, S. (2003). Composition of the continental crust. In R. L. Rudnick, J. D. Holland, & K. K. Turekian (Eds.), *The Crust, Treatise on Geochemistry*. Oxford: Elsevier. pp. 1–64.

Shaw D. M. (1961). Element distribution laws in geochemistry. *Geochimica et Cosmochimica Acta, 23*, 116–134.

Smirnov, N. (1948). Table for estimating the goodness of fit of empirical distributions. *Annals of Mathematical Statistics, 19*, 279–281.

Takeuchi N., Ueki, K., Iizuka, T., Nagao, J., Tanaka, A., Enomoto, S., et al. (2019a). Stochastic modeling of 3-D compositional distribution in the crust with Bayesian inference and application to geoneutrino observation in Japan. *Physics of the Earth and Planetary Interiors, 288*, 37–57.

Takeuchi N., Ueki, K., Iizuka, T., Nagao, J., Tanaka, A., Enomoto, S., et al. (2019b). Numerical Data of Probabilistic 3-D Lithological Map of the Japanese Crust. *Data in Brief, 26*, 104497.

Wood, B. J., Walter, M. J. & Wade, J. (2006). Accretion of the Earth and segregation of its core. *Nature, 441*, 825–833.

3

Seismological Studies of Deep Earth Structure Using Seismic Arrays in East, South, and Southeast Asia, and Oceania

Satoru Tanaka[1] and Toshiki Ohtaki[2]

ABSTRACT

We review studies that focus on the seismic structure of the lower mantle and outer and inner cores and that used seismic arrays and networks throughout East, South, and Southeast Asia, and Oceania. Studies using seismic arrays in these areas have contributed to the development of many important seismic images, leading to new discoveries. For example, (1) small-scale low-velocity regions several hundred kilometers wide exist in the lowermost mantle beneath the western Pacific; (2) the inner-core boundary beneath East Asia has a very complex structure; (3) the ultralow-velocity zones and scatterers in the lowermost mantle appear to be randomly distributed with no relationship to the large-scale heterogeneity; and (4) the structures of the upper inner core beneath the both polar regions show the characteristics of a quasi-Western Hemisphere. However, the tomographic images of the mantle beneath the South Pacific, the eastern area of the Pacific large low shear velocity province, have not resolved whether a large low-velocity body occupies the entire lower mantle or whether it is populated with individual, vertical, low-velocity columns. Furthermore, the inferred boundaries of the inner core hemispheres in the northern area do not coincide. To address these and other issues, such as azimuthal anisotropy and the structure of seismic attenuation in the lower mantle, improving ray coverage in the studied areas is crucial. Thus, we should continue seismic array observations not only on land but also in the oceanic areas.

3.1. INTRODUCTION

The study of Earth's deep interior by seismology has progressed through the complementary research targets of both global structures (e.g., 1-D seismic velocity structure of the average Earth, 3-D global seismic tomography) and small-scale structures that are examined by local or regional observations. In the early days, when the concept of lateral heterogeneity was not widespread, studying a representative 1-D structure of the Earth at a specific depth range by using regional observation was adequate. For example, seismic velocity anomalies in the lowermost mantle were reported by observations in a limited area and supported by corresponding observations from other areas (e.g., Cleary, 1969; Bolt et al., 1970). Later, as more observations were made, regional variations became vaguely apparent (e.g., Niazi, 1973; Mondt, 1977). Finally, global tomographic studies of the lowermost mantle revealed approximations of a bigger picture, which led to its geophysical discussion and geological interpretation (e.g., Castle et al., 2000; Kuo et al., 2000). When whole mantle tomography revealed regions that did not have sufficient spatial resolution or interesting regions (e.g., Dziewonski, 1984; Tanimoto, 1990), reviewing data from existing observation networks or conducting new temporary observations to target these regions sometimes led to new discoveries (e.g., Ritsema et al., 1997; Wysession et al., 1999).

[1] Research Institute for Marine Geodynamics, Japan Agency for Marine-Earth Science and Technology, Yokosuka, Japan
[2] Geological Survey of Japan, National Institute of Advanced Industrial Science and Technology, Tsukuba, Japan

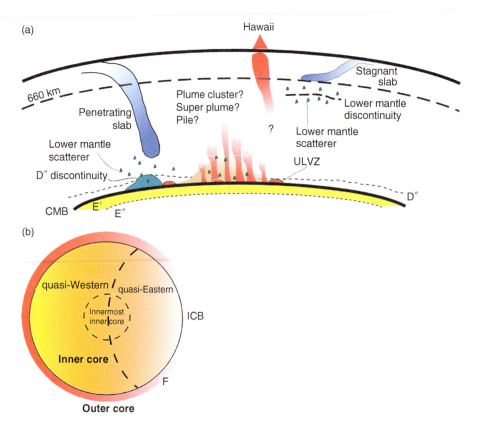

Figure 3.1 Schematic images of the cross-sections of the Earth's deep interior, with several hypotheses for (a) the lower mantle and outermost core and (b) the base of the outer core and inner core. See the text for more explanations of the abbreviations and terms used in this figure; the core-mantle boundary (CMB), the D″ layer (the base of the lower mantle), the D″ discontinuity, ultralow-velocity zone (ULVZ), E′ and E″ layers (the top and rest layers of the outer core), F layer (the bottom of the outer core), inner core boundary (ICB), quasi-Eastern and quasi-Western Hemispheres of the inner core, and innermost inner core. A plume cluster or super plume or plies are discussed as an interpretation of the large low shear wave velocity provinces (LLSVPs). A color figure is available in the online version.

An example of a current concept for the deep Earth structure based on the knowledge accumulated in this way over the years is shown in Figure 3.1. For details on each item, please refer to the references cited below. First, Figure 3.1a shows the structure of the lower mantle and core-mantle boundary (CMB). Subducting slabs show diversity in morphology, depending on their locations, that stagnates in the mantle transition zone or upper part of the lower mantle or subduct deeply into the lower mantle (e.g., see review by Fukao et al., 2001, 2009). Seismic reflectors (or scatterers) in the lower (or mid-) mantle are distributed around stagnant slabs and upwellings (Waszek et al., 2018). In the lowermost mantle (the D″ layer), the D″ discontinuities (e.g., see review by Wysession et al., 1998) and seismic velocity anisotropy (Lay et al., 1998; Nowacki et al., 2011; Creasy et al., 2019) have been observed, and these seismological features are explained by a phase change from bridgmanite to post-perovskite in the lowermost mantle (Hirose, 2007). On a larger scale, large low shear wave velocity provinces (LLSVPs) have been found in the lowermost mantle beneath the Pacific Ocean and Africa. However, whether LLSVPs are massive mantle plumes, aggregates of plumes, or piles of material swept up by mantle convection is still unclear (Garnero et al., 2016; McNamara, 2019). The relationship between LLSVPs and ultralow-velocity zones (ULVZs) is recognized as a key to understanding mantle dynamics (Yu and Garnero, 2018; McNamara, 2019). It has been argued that the D″ layer beneath a subduction region is thicker at lower temperatures because of the slab and thinner at higher temperatures in LLSVPs (Lay et al., 2008). The uppermost outer core (the E′ layer) has a larger seismic velocity gradient than the underlying E″ layer, and the possibility of stratification in the outermost core has been discussed (Souriau and Calvet, 2015). Second, Figure 3.1b shows a conceptual diagram of the F layer (a basal layer of the outer core) and the inner core. In the lowermost outer core (the F layer), heterogeneities can be formed through the accumulation of heavy fluids and release of lighter elements

(Tkalčić, 2015, 2017). In the upper part of the inner core, seismic velocity and intensity of seismic anisotropy differ between the quasi-Eastern and quasi-Western Hemispheres, and the presence of an innermost inner core with a different anisotropic nature is suggested (Souriau and Calvet, 2015).

In this chapter, from the viewpoint of how array observations have contributed to the formation of this kind of concept for the Earth's deep interior, we review the findings of previous studies using seismic array observations that were conducted in parts of Asia and Oceania, which were the target areas of the planned research related to seismic observation in the MEXT Grant-in-Aid for Scientific Research on Innovative Areas "Interaction and Coevolution of the Core and Mantle – Toward Integrated Deep Earth Science." Finally, we summarize the achievements to date and briefly discuss the future perspectives.

3.2. SEISMIC ARRAYS IN EAST, SOUTH, AND SOUTHEAST ASIA, AND OCEANIA THAT CONTRIBUTE TO DEEP EARTH STUDIES

This section provides an overview of the seismic arrays in the East, South, and Southeast Asia and Oceania regions that are used for the study of deep Earth structures. Table 3.1 lists the representative arrays introduced in this chapter. Most of the information for these arrays is available through the Incorporated Research Institutions for Seismology (IRIS), GEOFON, the Earthquake Research Institute (ERI) of the University of Tokyo, the National Research Institute for Earth Science and Disaster Resilience (NIED), the Japan Agency for Marine-Earth Science and Technology (JAMSTEC), etc. Although many other temporary arrays are shown on the IRIS and GEOFON websites, we focus on arrays with records used for deep Earth research. The seismic

Table 3.1 Summary of seismic arrays in Asia and Oceania for deep Earth studies described in the text

Arrays	Country/Region	Instrument	Existing period
Matsushiro (MSAS, MJAR)	Japan	SP	1983–present
J-array (1st gen.)	Japan	SP	1991–1996.3
J-array (2nd gen.)	Japan	SP	1996.6–present
Hi-net	Japan	SP	2000–present
F-net	Japan	BB	1995–present
CNSN	China	BB	–present
INDEPTH-II	Southwest China	BB, SP	1994.6–1994.12
INDEPTH III	Southwest China	BB	1997.7–1999.6
INDEPTH IV	Southwest China	BB	2007.4–2009.6
Hi-CLIMB	Southwest China	BB	2002.10–2005.8
Namche Barwa	Southwest China	BB, SP	2003.7–2004.10
2003MIT-China	Southwest China	BB	2003.9–2004.9
GHENGIS	West China (+ Kyrgyzstan)	BB	1997.9–2000.8
NCISP1	North China	BB	2000.1–2001.12
NCISP2	North China	BB	2001.1–2003.12
NCISP3	North China	BB	2003.6–2003.7
NCISP4	North China	BB	2005.9–2006.9
NCISP5	North China	BB	2007.1–2008.5
NCISP6	North China	BB	2007.9–2008.9
NCISP7	North China	BB	2008.6–2009.8
NECESSArray	Northeastern China	BB	2009.8–2011.8
TTSN	Taiwan	SP	1970'–1990'
CWBSN	Taiwan	SP(+BB)	1990'–present
BATS	Taiwan	BB	1990'–present
TAIGER	Taiwan	SP+BB	2006.4–2009.6
KSAR	Korea	SP	–present
CMAR	Thailand	SP	–present
TSAR	Thailand	BB	2016.2–2019.1
GBA	India	SP	1966.2–?
WRA	North Australia	SP	1966.3–1998
WRA	North Australia	BB	1998–present
ASAR	Central Australia	SP	1956–present
SKIPPY	Australia	BB	1993.1–1996.12
WOMBAT EAL2	East Australia	SP	2010.5–2010.12
JISNET	Indonesia	BB	1997–2010
CDPapua	Papua New Guineas	BB	2010.3–2011.12
SPANET	Fiji, Tonga	BB	1997–2003.4
SPaSE	Fiji, Tonga	BB	1993.11–1995.12
North Vietnam	North Vietnam	BB	2005.12–?
ERI-Vietnam	Vietnam	BB	2000.2–2005.10
PLUME (FP)	French Polynesia	BB	2001.10–2005.9
FP-BBOBS	French Polynesia	BB	2003.1–2004.1 2004.8–2005.6
TIARE	French Polynesia	BB	2009.2–2010.7
OJP(BBOBS)	Ontong Java Plateau	BB	2014.9–2017.2
PLUME(Hawaii)	Hawaii	BB	2006.4–2007.12 (OBS) 2004.8–2007.12 (Island)

Notes: BB: Broadband seismograph; BBOBS: Broadband Ocean Bottom Seismograph; SP: Short-period seismograph.

Table 3.2 Summary of targets, analyzed phases, and epicentral distances

Targets	Analyzed phases	Epicentral distances (°)
Lower mantle	P	30–90
	SdP	30–90
	Precursors to PKKP	0–60
	Precursors to PKP	130–140
	ScS–S	30–85
	ScS, PcP	60–80
	SKS–S	85–130
D″	PdP, SdS	70–80
	ScP	30–50
	SdKS	100–130
	Pdiff, Sdiff	100–180
Outer core		
Top (E′)	SmKS (m≥2)	90–180
	P4KP	40–70
Whole body (E″)	P4KP, P7KP	40–70
	P5KP	140–160
Bottom (F)	PKP(CD)–PKP(DF)	110–140
	PKP(CD)–PKP(BC)	145–150
	PKP(Cdiff)	155–160
	PKP(BC) dispersion	150–155
ICB	Precritical PKiKP	10–70
Inner core	PKP(CD)	120–152
	PKP(DF)	120–180
	PKP(CD)–PKP(DF)	110–140
	PKP(BC)–PKP(DF)	145–155
	PKP(Cdiff)–PKP(DF)	155–160
	PKP(AB)–PKP(DF)	150–180
	PKIIKP–PKIKP(PKP(DF))	170–180
	PKJKP	70–180

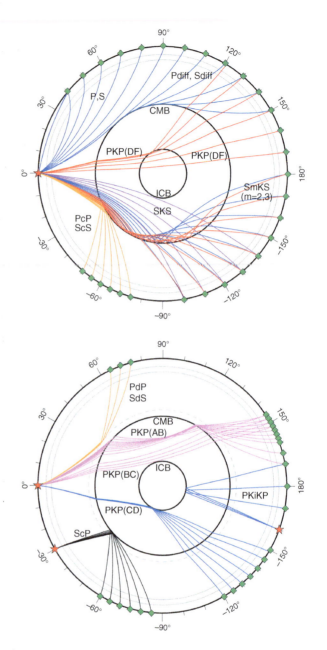

Figure 3.2 Representative seismic ray paths used for studies of the Earth's deep interior. Stars and diamonds represent hypocenters and stations, respectively. A color figure is available in the online version.

phases described in this chapter and their observational conditions are listed in Table 3.2. Figure 3.2 shows the ray paths of the representative phases. Because some studies used more than one array, the first description will be slightly more detailed, and the second and subsequent descriptions will be briefly mentioned.

3.2.1. Japan

The Japanese islands are located at an epicentral distance of approximately 70° from the Tonga-Fiji Islands, where deep earthquakes occur, and are suitable for observing ScS and PcP waves reflected from the outer core surface as well as SdS and PdP waves reflected from the surface of the D″ discontinuities (e.g., Weber, 1993). It is also located at distances greater than 30° from the seismogenic zone along the western Pacific rim, making it suitable for studying the structure of the lower mantle beneath the western Pacific Ocean. Japan is also located at epicentral distances of 140°–160° from South America, where deep earthquakes are relatively frequent, making it suitable for core phase observations.

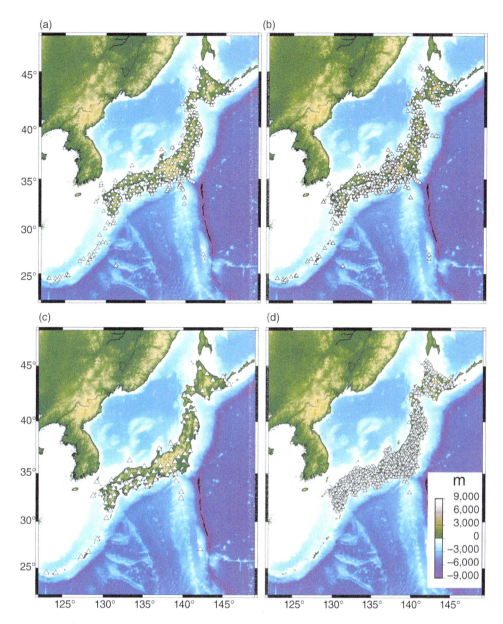

Figure 3.3 Geographical distribution of seismic stations as of December 2021: (a) The Japan Meteorological Agency (JMA) (254 stations), (b) New J-array (431 stations), (c) F-net (73 stations), and (d) Hi-net (781 stations). ETOPO1 was used to draw the topography and bathymetry. A color figure is available in the online version.

Figure 3.3 shows the current distribution of the seismic stations of representative networks in Japan including the Japan Meteorological Agency (JMA), J-array, F-net, and Hi-net.

Matsushiro Seismic Array System (MSAS) and Seismic Observation Networks of the Japan Meteorological Agency and Japanese Universities

Matsushiro is one of the JMA's representative seismic stations. It has contributed to seismic research by being at the forefront of seismic observation system technology, such as the Benioff seismograph, in cooperation with the International Geophysical Year in the 1950s as beginning to participate in international seismic observations. This was followed by the World-Wide Standardized Seismograph Network (WWSSN), Abbreviated Seismic Research Observatory (ASRO), and IRIS, as the seismograms at Matsushiro indicated the lowest background noise level in Japan. In addition, the JMA began operating the Matsushiro Seismic Array System (MSAS) in 1983, as shown in Figure 3.4 (Osada et al., 1984). This seismic array observation system has also recently contributed to the Comprehensive Nuclear-Test-Ban Treaty Organization (CTBTO) as the Matsushiro seismic

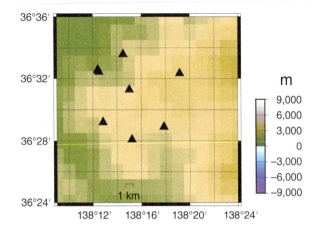

Figure 3.4 The station distribution of the Matsushiro Seismic Array System (MSAS). A color figure is available in the online version.

array (MJAR), which belongs to the International Monitoring System (IMS).

The seismic observation network covering Japan was initiated by the Central Meteorological Observatory in the 1890s and is currently operated by the JMA (Ohtake and Ishikawa, 1995) (Fig. 3.3a). Microearthquake observations by Japanese national universities and research institutes have been developed since 1965 based on the (Japanese) National Program for Earthquake Prediction (Ohtake and Ishikawa, 1995). It was not until the latter half of the 1980s that such observation networks began to be used as large-aperture arrays for the study of the Earth's deep structure.

Morita (1987) used the MSAS and Umetsu (1990) used the microearthquake observation network of Tohoku University to discuss the short-wavelength topography of the inner core boundary (ICB) by using the waveforms and spectra of the PKP(Cdiff) coda. Nakanishi (1990) interpreted the origin of the PKP(Cdiff) coda as scattered waves in the upper mantle using the slowness inferred from the microearthquake observation networks of Hokkaido, Hirosaki, and Tohoku universities. Morita (1991) investigated the difference in the Q-values of the inner core estimated from the amplitude and spectral ratios of PKP(DF) and PKP(AB) using the MSAS and found that the scatter of the Q-values estimated from the amplitude ratios was smaller than that estimated from the spectral ratios and also varied with the epicentral distance. He inferred the depth variation of Q-values in the inner core from the PKP(DF)/PKP(AB) amplitude ratio using seismograms recorded by the JMA network (Morita, 1991). Yamada and Nakanishi (1996) applied a double array analysis (Kruger et al., 1993) to the seismograms from the microearthquake observation networks of Hokkaido University to detect PdP with improved signal-to-noise ratio and inferred a velocity jump of 4% at the D″ discontinuity within a small, local area. Subsequently, Yamada and Nakanishi (1998) extended the area of their analysis by adding seismograms from the microearthquake observation networks of the University of Tokyo and Kyoto University to that of Hokkaido University and found regional variations in the D″ reflection surface in the southwestern Pacific. Kaneshima et al. (1994) estimated the velocity structure from the lowest part of the outer core to the upper inner core beneath the northeast Pacific using the traveltime differences of PKP(BC)–PKP(DF) and PKP(CD)–PKP(DF) and the amplitude ratios of PKP(DF)/PKP(BC) measured on the seismograms from Kyoto University and other Japanese stations. Frost et al. (2017) collected precursors to PKKP phases observed at the MJAR (same as the MSAS), other IMS and other arrays and networks and detected scatterers in the lowermost few hundred kilometers in the mantle.

J-Array

In 1990, the Grants-in-Aid for Scientific Research on Priority Area "Central Core of the Earth" was launched in Japan. On this occasion, a large-scale array consisting of 218 short-period seismic stations (J-array) was organized by integrating seismic data obtained from microearthquake observation networks operated by several universities and research institutes (J-Array Group, 1993). The characteristic seismic waves observed by the J-array were described in detail by Morita (1996). Since 1996, the data acquisition system has been updated, and the array has been operated as a new J-array (Fig. 3.3b) (http://jarray.eri.u-tokyo.ac.jp/jarray/). The following subsections are organized by research targets to review the results to date.

The Lower Mantle Using SdP waves (converted from S to P at a depth "d") observed by the J-array, Kawakatsu and Niu (1994) detected a new seismic discontinuity at a depth of 920 km beneath the Tonga subduction zone. Vinnik et al. (2001) detected seismic discontinuities at depths of approximately 900, 1,200, and 1,700 km beneath the Kermadec-Fiji-Tonga region using relatively long-period seismograms (T ~ 10 s) provided by some broadband seismic stations belonging to the J-array and others. Vanacore et al. (2006) detected a lower mantle discontinuity (970 km deep) beneath the source region in Indonesia using short-period seismograms from the J-array and other arrays. Using a similar method, Kaneshima (2013) found scatterers in the upper part of the lower mantle, 50–100 km below the Tonga-Fiji plate, and interpreted them as transition zone material entrained by slab subduction. As a study using the J-array as supplementary data, Zhao (2004) constructed a 3-D

P-wave global tomography model by adding P-wave traveltime measurements from observation networks around the world, including the J-array, to the traveltime catalog issued by the International Seismological Center (ISC).

The D″ Region Shibutani et al. (1993) searched for short-period reflected PdP waves using seismograms provided from the first year of J-array operation, but failed to find a significant signal above noise levels that corresponded to the D″ discontinuities with velocity jumps greater than 1.6%. After collecting more data, Shibutani et al. (1995) succeeded in detecting PdP waves by improving the signal-to-noise ratio using deconvolution of the source time functions and were able to estimate the velocity jump of the D″ discontinuity to be 1–2%. Kito and Kruger (2001) and Kito et al. (2004) applied migration analysis to seismograms of Tonga-Fiji events and detected a very weak scatterer in the D″ region just below the western Pacific Ocean. Kito et al. (2003) applied migration analysis to the seismograms of a nuclear test at Lake Lop Nor and detected scatterers in the lowermost mantle (200–375 km above the CMB) below northeast China. Idehara (2011) used ScP waveforms to determine the distribution of the ULVZ (e.g., Garnero and Helmberger, 1995) on the CMB just below the Philippines.

The Outer Core Tanaka and Hamaguchi (1996) discussed the Q-value in the outer core and the possibility of its frequency dependence on the P4KP/PcP spectral amplitude ratios. Helffrich and Kaneshima (2004) showed that the P4KP waveform did not support the existence of a 3–12 km thick layer at the top of the outer core. Tanaka (2010) estimated the even-order topography of the CMB using P4KP–PcP differential traveltimes collected from the J-array and other stations worldwide. Helffrich and Kaneshima (2010) and Kaneshima and Helffrich (2013) applied the tau-p method, and Kaneshima and Matsuzawa (2015) applied both the tau-p method and genetic algorithms to SmKS traveltimes measured on seismograms from seismic arrays and networks around the world, including the J-array. They found that the velocity gradient in the uppermost 300–450 km of the outer core was slightly steeper than that of the preliminary reference Earth model (PREM) (Dziewonski and Anderson, 1981).

The Inner Core Yamada and Nakanishi (1993) detected precritical PKiKP on the seismograms of the 1992 nuclear test near Lake Lop Nor, China, and showed by the PKiKP/PcP amplitude ratio that the density jump at the ICB was 0.6 g/cm^3, which was smaller than previous estimates using the same method. Kaneshima et al. (1994) also used seismograms from the J-array to estimate the velocity structure from the lowermost outer core to the uppermost inner core in the northeastern Pacific. Niu and Wen (2001) detected a hemispherical structure in the uppermost 100 km of the inner core by measuring the PKP(CD)–PKP(DF) differential traveltimes from worldwide arrays, including the J-array. Cao and Romanowicz (2004) estimated the hemispherical variation in seismic wave attenuation in the uppermost 100 km of the inner core using seismograms from the J-array and others. Tkalčić et al. (2010) showed that PKiKP and PcP were rarely observed simultaneously at the same station and that the large variation in the PKiKP/PcP amplitude ratio was likely due to not only the CMB but also the heterogeneity in the upper mantle. Mattesini et al. (2010) measured the traveltime differences between PKP(CD) and PKP(DF) recorded on global observation networks and arrays, including the J-array, and claimed that the iron constituting the inner core was a body-centered cubic lattice based on the characteristics of anisotropy in the Eastern and Western Hemispheres of the inner core. Krasnoshchekov et al. (2019) analyzed the PKiKP–PcP traveltime differences and PKiKP/PcP amplitude ratios from the J-array and other Japanese, Far East Russian, and South American stations, and estimated that the outer core was 1–3 km thinner beneath Northeast Asia than in South America, and that the density jump in the ICB was 0.3 g/cm^3 beneath Northeast Asia.

Hi-net

Since 2000, Hi-net, a high-sensitivity seismic network consisting of approximately 700 short-period seismic stations installed at the bottom of boreholes, has been in operation to improve the detection threshold of small earthquakes in Japan (Fig. 3.3d) (Okada et al., 2004). It has become a powerful tool for studying the deep Earth structure. In addition, the tiltmeters installed in the Hi-net can be used as the horizontal components of a broadband seismometer for analyses. Subsections for each research target are presented below to introduce the results to date.

The Lower Mantle Kaneshima (2018b) detected scatterers in the upper part of the lower mantle below the Tonga-Fiji hypocentral regions. Kaneshima (2019) also detected those below Indonesia and the Philippines. Vanacore et al. (2006) detected a lower mantle discontinuity (approximately 1,000 km in depth) beneath Indonesia.

The D″ Region Kito et al. (2004) detected a very weak scatterer in the D″ region just below the western Pacific Ocean. Idehara et al. (2007) and Idehara (2011) analyzed the pre- and postcursors to PcP and ScP from earthquakes near Indonesia and detected ULVZs just below the Philippines. Pachhai et al. (2014) performed automated waveform modeling of ScP waves based on Bayesian inference and revealed the detailed structure of the ULVZ with uncertainty beneath off the east coast of the Philippines. Shen et al. (2016b) showed that the variation of PcP/ScP amplitude ratio could be explained by the bump or tilt of the CMB beneath eastern Sakhalin. Kawai et al. (2009) applied waveform inversion to the Hi-net tiltmeter seismograms of European earthquakes and detected a high-velocity region in the D″ region beneath northern Asia. Takeuchi and Obara (2010) found an anticorrelation between the height of the D″ discontinuity detected by stacking tiltmeter records and the residuals of ScS–S differential traveltimes.

The Outer Core Kaneshima and Matsuzawa (2015) and Kaneshima (2018a) analyzed SmKS waves recorded by Hi-net tiltmeters. Ohtaki and Kaneshima (2015) estimated the velocity structure of the lowermost outer core under the northeast Pacific Ocean using the frequency dependence of the PKP(BC)–PKP(DF) traveltime differences and the PKP(CD)–PKP(BC) traveltime differences observed in the seismograms of earthquakes in South America. Ohtaki et al. (2018) conducted a similar analysis on those in the South Sandwich Islands and detected lateral heterogeneity in the seismic velocity in the lowermost outer core. As a cause of lateral variation, the concentration of oxygen in the lowermost outer core beneath Australia was estimated using *ab initio* calculations of elastic properties.

The Inner Core Kawakatsu (2006) used PKiKP/PcP amplitude ratios from earthquakes under the Mariana Islands to estimate the sharpness of the ICB and showed a transparent inner core based on the small amplitudes of PKiKP codas. Dai et al. (2012) used PKiKP/PcP and PKiKP–PcP from earthquakes in Indonesia to estimate ICB topography. Jiang and Zhao (2012) estimated the fine velocity structure near the ICB using PKiKP and its pre- and postcursors. Tanaka and Tkalčić (2015) characterized the ICB topography and complexity using the spectral amplitude ratios of PKiKP/PcP from earthquakes in the western Pacific. Tian and Wen (2017) used PKiKP waveform modeling to show that sharp and mushy areas were scattered on the ICB beneath the vicinity of Japan. Kazama et al. (2008) used PKP waves from earthquakes in South America to estimate seismic velocity and attenuation structures in the upper part of the inner core beneath the East Pacific. Furthermore, Iritani et al. (2010) determined Q in the upper inner core beneath the northeastern Pacific through the separation of superimposed PKP(DF), PKP(BC), and PKP(AB) branches. Based on variations in the differential traveltimes of PKP(DF)–PKP(BC) from earthquakes in South America and the South Sandwich Islands, Yee et al. (2014) revealed the heterogeneity of seismic velocity in the upper inner core with spatial scales of tens and hundreds of kilometers. Wookey and Helffrich (2008) detected PKJKP from an earthquake in Mozambique in southern Africa and pointed out that anisotropy might cause the splitting of PKJKP.

F-net

The broadband seismic array in Japan was established by the FREESIA project (Fukuyama et al., 1996), which started in the late 1990s and has been continuously operated as the F-net (Okada et al., 2004). The F-net consists of approximately 70 stations equipped with broadband seismometers, such as STS-1 and STS-2 (Fig. 3.3c). The research results are presented in subsections for each research target, as described above.

The Lower Mantle Vinnik et al. (2001) detected lower mantle discontinuities beneath the Kermadec-Fiji-Tonga region using broadband seismograms of the F-net (described as FREESIA in their paper) and others. Obayashi et al. (2013) collected seismograms from seismic stations in East Asia, including the F-net as part of the dataset, measured P-wave traveltimes at different frequencies, and performed finite-frequency tomography to construct a global P-wave velocity model GAP_P4. Bagley et al. (2013) probably used the F-net to find discontinuities in the lower mantle (750–1,600 km depth) that distributed from northeast China to the northwest Pacific Ocean using mantle reflection analysis with ScS reverberations. However, the use of the F-net was not specified in the text, and this was inferred from the distribution of stations in the figure (Bagley et al., 2013). Based on an S-wave splitting analysis of deep earthquakes, Kaneshima (2014) showed that there was no significant deformation in the uppermost part of the lower mantle near the Tongan slab.

The D″ Region He and Wen (2011) revealed a seismic velocity anomaly in the lowermost mantle from China to Southeast Asia by analyzing the residuals of PcP–P and ScS–S differential traveltimes from the F-net and other permanent and temporary observation networks in East Asia; they showed that the D″ discontinuity beneath northern Vietnam to southern China corresponded to a high-velocity anomaly region. Konishi et al. (2014) conducted waveform inversion using S to ScS waves from the Tonga-Fiji earthquakes and detected several low-velocity regions in the lowermost mantle below the western Pacific Ocean. Konishi et al. (2020) simultaneously estimated the local 3-D S-wave velocity and attenuation structure beneath the western end of the Pacific LLSVP (e.g., Garnero et al., 2016) using finite-frequency tomography and detected low-velocity and low-Q regions. Takeuchi

(2012) verified the existence of a ridge-like LLSVP beneath the Pacific Ocean that was revealed using his global tomography model and the residuals of the ScS–S differential traveltimes obtained from the F-net. Yao and Wen (2014) estimated the S-wave velocity structure in the lowermost mantle beneath Southeast Asia from the ScS–S and ScP–P traveltime differences by combining F-net data with those from Chinese seismic observation networks.

The Outer Core Tanaka (2007) performed waveform modeling of stacked SmKS waveforms collected from seismic stations around the world, including the F-net, with theoretical waveforms and speculated that the uppermost 90–140 km of the outer core had a low-velocity layer. Helffrich and Kaneshima (2010), Kaneshima and Helffrich (2013), Kaneshima and Matsuzawa (2015), and Kaneshima (2018a) also analyzed the traveltime differences of SmKS waves recorded in arrays around the world, including the F-net. Zou et al. (2008) analyzed the traveltime differences of PKP(Cdiff)–PKP(DF) and amplitude ratios of PKP(Cdiff)/PKP(DF) at a rather long period of 3 seconds from several broadband seismic arrays, including the F-net, and concluded from the traveltime that the velocity in the lowermost outer core was slower than those in the PREM, but the amplitude ratio matched those expected from the PREM. This discrepancy suggested that either that the PKP(Cdiff) was scattered by the ICB or that the Q-value in the lowermost outer core was small. Adam and Romanowicz (2015) detected the M phase that follows PKP(Cdiff) in seismograms from around the world, including the F-net, and proposed a slightly higher velocity region from 400 to 50 km above the ICB and a lower velocity region within 50 km of the ICB. Adam et al. (2018) further suggested a low-velocity layer in the lowermost 100 km of the outer core and a low-Q region (Q = 600) in the lowermost 600 km.

The Inner Core Helffrich et al. (2002) collected the ray paths of the Tonga-Fiji source to the UK SPICeD array and those of the South American source to the F-net and measured PKP(DF)–PKP(BC) traveltime differences as the ray paths crossed under the northeast Pacific Ocean. They discussed azimuthal anisotropy of the seismic velocity in the uppermost inner core and reported that the difference in traveltime was only 0.19 s, which was slightly above the detection limit. Based on the amplitude ratio of PKP(DF)/PKP(BC), the Q-value in the uppermost 340 km of the inner core was estimated to be approximately 130. Yu et al. (2005) inferred the velocity structure from the lowermost outer core to the uppermost inner core using worldwide array data, including the F-net. Tanaka (2012) collected PKP(Cdiff)–PKP(DF) traveltime differences from several broadband seismic arrays, including the F-net, and discussed the depth extent of the east-west hemisphere difference in the inner core. Iritani et al. (2014a, 2014b) used array records to separate PKP(DF), PKP(BC), and PKP(AB) branches that overlap to form a package of PKP waveforms at epicentral distances of approximately 145° to estimate the seismic velocity and attenuation structure of the uppermost inner core. In the collected data, the seismograms of the South American and South Sandwich Islands earthquakes recorded by the F-net were included, which helped reveal the structures in the Western and Eastern Hemispheres. Ivan and He (2017) discussed the location of the boundary between the Eastern and Western Hemispheres by measuring PKP(DF)–PKP(Bdiff) and PKP(DF)–PKP(Cdiff). Ivan et al. (2018) extracted the non-hemispheric component of heterogeneity in the uppermost approximately 100 km of the inner core from global observations, including the F-net, of the differential traveltimes of PKP(CD)–PKP(DF).

3.2.2. China and Adjacent Areas

China covers epicentral distances of 70°–130° from the Tonga-Fiji Islands seismic zone, which is suitable for studying the lowermost mantle to the outermost core just below the western Pacific Ocean. It is also suitable for core phase observations to study the inner core structure beneath the region from the northern Atlantic Ocean to North America, with epicentral distances from 140° to nearly 180° from the deep seismic zones in South America.

China Digital Seismograph Network (CDSN), China National Seismic Network (CNSN), and Chinese Regional Seismic Network (CRSN)

According to Zheng et al. (2010), there are more than 1,000 permanent stations covering China (China National Seismic Network; CNSN), which are called the China Digital Seismograph Network (CDSN) or China Regional Seismic Network (CRSN) in some papers. However, few studies have used all of them to study the deep Earth interior, and many of researchers have only limited access to several tens of the open stations through the IRIS or the International Federation of Digital Seismograph Networks (FDSN). Niu and Chen (2008) observed both major and minor arcs of PKIKP (PKP(DF)) and PKIIKP waves from the Argentinian earthquake at the CDSN stations in east-central China and were able to discuss the anisotropy of the innermost inner core. Xia et al. (2016) applied noise and seismic coda correlation methods to extract PKP wave branches between the Chinese stations belonging to the CRSN and CNSN and the Brazilian stations with angular distances greater than 145°. Wu et al. (2018) applied the coda correlation method to the stations of the CRSN and CNSN as an array and compared the theoretical and observed waveforms of the phases corresponding to PKIKP (PKP(DF)) and PKIKP2 (reciprocating PKIKP) and to PKIIKP and PKIIKP2 (reciprocating PKIIKP), which were sensitive to the Earth's center.

INDEPTH, Hi-CLIMB, Namche Barwa, 2003MIT-China, and GHENGIS

Several seismic projects have been conducted to reveal the tectonics of the Indo-Himalayan collision zone as main objectives. Here, we introduce five array projects that have been utilized to study the deep Earth structure.

One representative project is the InterNational Deep Profiling of Tibet and the Himalaya (INDEPTH) I–IV (http://www.geo.cornell.edu/geology/indepth/indepth.html) (see also James, 2015). Since INDEPTH-II, passive seismic surveys using broadband seismometers have been conducted, and data that are useful for the study of deep Earth structures have been accumulated. The next is the study of the Himalayan-Tibetan Continental Lithosphere during Mountain Building (Hi-CLIMB) (Nábelek et al., 2009), which was conducted through an international collaboration by Nepal, China, the United States, France, and Germany. The third and fourth are the Namche Barwa (Meltzer et al., 2003) and 2003MIT-China (Lev et al., 2006) projects, which were conducted by Lehigh University and the Massachusetts Institute of Technology (MIT), respectively, in cooperation with Chinese institutions. The fifth is the Multidisciplinary Investigation of the Mountain Building in the Tien Shan (GHENGIS), a collaborative study by Kyrgyzstan, China, and the United States (Roecker, 2001). The station distributions are shown in Figure 3.5.

These arrays are located at epicentral distances of 30°–90° from deep earthquake regions distributed from Japan to the Aleutian Islands or Indonesia to the New Hebrides and are suitable for performing structural analysis of the lower mantle. From Tonga-Fiji, it is at a distance of 90° or more, which is suitable for CMB studies using SKS and SmKS and core diffracted waves (Pdiff and Sdiff). It is also located approximately 150° from South America, making it suitable for the study of the inner core.

The D″ Region Vanacore et al. (2010) found a particularly coherent phase in precursors to PKP recorded by the Namche Barwa array and inferred a reflection surface in the lowermost mantle just below Central America, which they interpreted as the boundary of the ULVZ. Using INDETPH and other arrays worldwide, Ivan and Wang (2013) pointed out that the P4KP/P(S)cP and P5KP/PKP(AB) amplitude ratio anomalies could

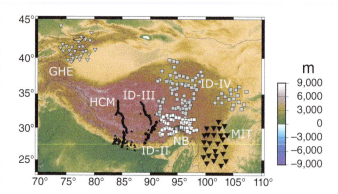

Figure 3.5 The arrays in Tibet, southwest China, west China. Shown are the arrays for INDEPTH-II, -III, and -IV (ID-II: black circle, -III: black star, -IV: gray square); Hi-CLIMB (HCM: black triangle); Namche Barwa (NB: open square); 2003MIT-China (MIT: black inverted triangle); and GHENGIS (GHE: gray inverted triangle). ETOPO1 was used to draw the topography and bathymetry. A color figure is available in the online version.

be due to focusing or scattering caused by structural anomalies at the CMB. Waszek et al. (2015) analyzed precursors to PKP using array data deployed in Tibet (e.g., INDEPTH, Hi-CLIMB, and 2003MIT-China) and other arrays worldwide to show that the distribution of scatterers in the CMB was not related to long-wavelength heterogeneity in the lowermost mantle. Ma and Thomas (2020) showed that the distribution of scatterers in the lowermost mantle was a mixture of strong and weak scattered areas and that there was no clear relationship with the large-scale structure seen in global tomography. Chaloner et al. (2009) used GHENGIS and other arrays to detect faint D″ reflections beneath northern Vietnam. Kito et al. (2007) developed a slowness-back azimuth weighted migration and applied it to the seismograms from GHENGIS and other arrays to obtain a clearer scatterer image. Cobden and Thomas (2013) used GHENGIS and other arrays to detect a D″ reflector under northern Vietnam (Southeast Asia) with P-wave velocity jumps of −1% to −2% and S-wave velocity jumps of +1% to 3% from earthquakes in Indonesia (Banda Sea). Although not specified in the paper by Rao and Kumar (2014), we judged that they used the seismograms of the Tibetan stations based on the station distribution shown in their figure, as well as seismograms from arrays in India, Africa, and elsewhere. Rao and Kumar (2014) showed that the anomalies in the traveltime differences of ScS–S and amplitude ratios of ScS/S corresponded to the low geoid region extending from India to the Indian Ocean. They concluded that there was a high-velocity anomaly region in an approximately 1,000 km thick layer above the CMB below the Indian Ocean, which was interpreted as slab graveyards. Furthermore, Rao et al. (2017) extracted high-quality waveform data suitable for

ScS splitting analysis. The ScS splitting parameters in the lowermost mantle beneath off the east coast of the Philippines, corresponding to the high geoid region, were obtained from ScS waves observed at the Namche Barwa array. Suzuki et al. (2020) applied waveform inversion to S-to-ScS waveform data from stations in Asia and Oceania, including the arrays in this region, to obtain the 3-D S-wave velocity structure of the lowermost mantle in the western Pacific. They found several high-velocity regions that appeared to be past subducting plates and several vertically extending low-velocity regions that appeared to be plumes.

The Core Tanaka (2007) used GHENGIS data to analyze SmKS waveforms. Zou et al. (2008) used INDEPTH data to analyze the traveltimes and amplitudes of PKP(Cdiff). Tanaka (2012) discussed the depth extent of inner core hemispheric structures from PKP(Cdiff)–PKP(DF) traveltimes, including INDEPTH and Hi-CLIMB data. Yu and Wen (2007) collected PKP(CD)–PKP(DF) data passing under Africa from GHENGIS and observation networks including China, Europe, and South America, revealing complex aspects such as lateral heterogeneity and anisotropy of the seismic velocity and attenuation structure in the uppermost 80 km of the inner core. Wang et al. (2015) and Wang and Song (2018) applied coda interferometry to the waveforms from arrays such as INDEPTH and detected antipodal waveforms to discuss the characteristics of anisotropy in the innermost inner core. Ivan and He (2017) used the array data in this region to collect PKP(DF)–PKP(Cdiff) based on the station distribution plotted in their figure.

North China Craton

Many temporary broadband seismic surveys have been conducted over the years to reveal the structure of the North China Craton (Zhu et al., 2012). Among them, the station distribution of the North China Interior Structure Project (NCISP) arrays that are available is plotted in Figure 3.6. These data are suitable for studies of the lower mantle to the lowermost mantle using earthquakes in the southwest Pacific. For South American earthquakes, they are suitable for studies of the deep inner core, as the epicentral distances often exceed 160°.

He and Wen (2009) discussed the structure of the LLSVP under the Pacific Ocean using a number of S, Sdiff, ScS, SKS, and SKKS traveltimes, including those from the F-net and China's temporary and permanent observation networks. They used S, SdS, and ScS waveform data recorded by one of the NCISP arrays to examine the structure of the large-scale low-velocity region in the lowermost mantle beneath the western Pacific Ocean by comparing the observed waveforms with theoretical waveforms that could account for the 2-D

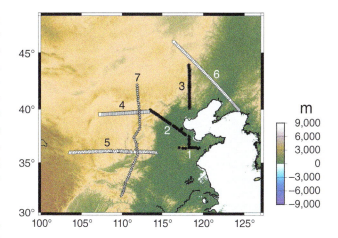

Figure 3.6 The North China Interior Structure Project (NCISP; 1–7) arrays accessible from the Incorporated Research Institutions for Seismology (IRIS) database. The numbers correspond to the order in which the arrays were implemented. According to Zhu et al. (2012), there are more linear arrays. ETOPO1 was used to draw the topography and bathymetry. A color figure is available in the online version. Source: Adapted from Zhu et al. (2012).

structure. Shen et al. (2016a) detected PKiKP phases with reflection points at the ICB beneath East Asia (from the east coast of China to the Sea of Japan, Yellow Sea, East China Sea, and South China Sea) from the seismograms of the NCISP arrays. They concluded the mosaic-like structure of the ICB from the PKiKP–PcP traveltime differences and the PKiKP/PcP amplitude ratios, and found that the ICB was concave and had large density jumps beneath the Yellow Sea. Suzuki et al. (2020) used waveform data from this region to estimate the S-wave velocity structure in the lowermost mantle.

NECESSArray

The NECESSArray (https://www.eri.u-tokyo.ac.jp/people/hitosi/NECESSArray/; Tanaka et al., 2006) was a temporary broadband seismic network consisting of 129 stations deployed in northeastern China (Fig. 3.7) and was supported by a joint project by Japan, China, and the United States. The primary purpose of this project was to elucidate the seismological structure related to the geological background of the Sino-Korean Craton, Northeast China Shield, and Inner Mongolia folds, as well as the actual features of the subducting plate under the Chinese continent. As a research achievement in the field of deep Earth structure, it has been used in the following studies. Tang et al. (2014) conducted S- and P-wave tomography and found that the magma plumbing system under the Changbaishan volcano originated below the 660 km discontinuity, the uppermost part of the lower mantle, and was continuous through a gap in the stagnant slab. Iritani et al. (2014a, 2014b) used this

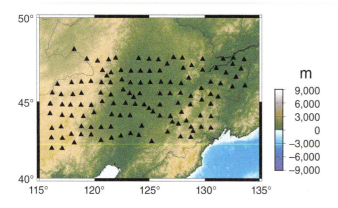

Figure 3.7 Station distribution of the NECESSArray. ETOPO1 was used to draw the topography and bathymetry. A color figure is available in the online version.

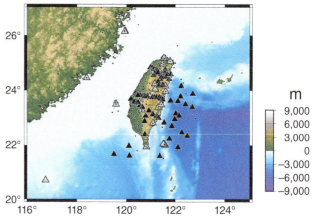

Figure 3.8 Seismic networks in Taiwan as of December 2021. Broadband Array in Taiwan for Seismology (BATS): gray triangles, Taiwan Integrated Geodynamic Research (TAIGER): closed triangles, Central Weather Bureau Seismic Network (CWBSN): dots. ETOPO1 was used to draw the topography and bathymetry. A color figure is available in the online version.

array in their inner core study. Ivan and He (2017) also used this array to analyze PKP(Cdiff)–PKP(DF) and PKP(Bdiff)–PKP(DF). Tanaka et al. (2015) analyzed the relative P-wave traveltimes and PcP waveforms and found structural changes in the lowermost mantle along the western margin of the Pacific LLSVP with the scale of hundreds of kilometers. Ma and Thomas (2020) used the NECESSArray to reveal the distribution of scatterers in the lowermost mantle by analyzing PKP precursors. Kaneshima (2019) used the NECESSArray to detect lower mantle scatterers just below the epicentral regions of Indonesia, Mindanao, and Mariana.

3.2.3. Taiwan

Taiwan is located at epicentral distances of 15°–40° from the Kuril, Japan, and Mariana deep seismic zone; 40°–80° from the Solomon Islands, New Hebrides, and Tonga-Fiji; and less than 90° from Italy, making it suitable for the study of the lower mantle and lowermost mantle structures. It is also located at epicentral distances of 150°–180° from South America and 140° from the South Sandwich Islands and is suitable for the observation of PKP phases, which are used to study the inner core and the lowermost mantle.

The Taiwan Telemetered Seismographic Network (TTSN) is a short-period seismograph network constructed in the early 1970s (Wang, 1989). This network was later incorporated into the Central Weather Bureau Seismic Network (CWBSN), which started in the early 1990s and now consists of 75 stations (https://scweb.cwb.gov.tw/en-US/page/ObservationNetwork/206) (Fig. 3.8). A broadband seismic network was also established in the early 1990s, which was later named the Broadband Array in Taiwan for Seismology (BATS) and consists of 28 permanent stations, as shown in Figure 3.8 (Kao et al., 1998). Huang et al. (2009) stated that more than 50 portable broadband seismic instruments were operated as part of the BATS. The Taiwan Integrated Geodynamic Research (TAIGER) project also conducted seismic observations, including broadband seismographs (Fig. 3.8) (Wu et al., 2007).

Huang (1996) used the TTSN to compare PKP(Cdiff)–PKP(DF) traveltime differences with those calculated using existing models and demonstrated that VMOI (Kaneshima et al., 1994) and ak135 (Kennett et al., 1995) explained the observations relatively well. Tseng and Huang (2001) used the CWBSN and BATS with the German Regional Seismic Network to compare the Q-structure of the uppermost inner core from the observations of the PKP(DF)/PKP(BC) spectral ratios between South America and Taiwan and between Tonga-Fiji and Germany. They found that the Q-structure of the uppermost inner core beneath Siberia was smaller than that of the western North America. Bagley et al. (2013) analyzed mantle discontinuity using multiple ScS waves also using the observation network in Taiwan. Wang et al. (2015) and Wang and Song (2018) discussed the anisotropy of the innermost inner core from the coda interferometry using BATS and TAIGER as well. Helffrich and Kaneshima (2010) used broadband stations in Taiwan to analyze SmKS. Suzuki et al. (2020) used BATS for waveform inversion to estimate the S-wave velocity structure in the lowermost mantle beneath the western Pacific.

3.2.4. Korea

Korea is located at epicentral distances of approximately 70°–90° from the Tonga-Fiji Islands and is suitable for studies of the lowermost mantle. It is also more than 30° from the seismogenic zone distributed along the western Pacific periphery, making it suitable for studies of the lower mantle beneath the western Pacific. It is also located at epicentral distances of 140°–160° from South America, making it suitable for core phase observations.

The Korea Institute of Geoscience and Mineral Resources maintains and operates a seismic network in Korea and operates the Korea Seismic Research Station (KSRS) array consisting of 26 borehole stations (Fig. 3.9) in Wonju, Korea (Lee et al., 2000). The KSRS array contributes to the detection of nuclear tests under the CTBTO as the South Korean Seismic Array (KSAR) of the IMS.

Idehara et al. (2007) studied the distribution of the ULVZ beneath the western Pacific Ocean from ScP and PcP waveforms using the KSAR. Koper et al. (2003) used IMS arrays, including the KSAR, to measure PKiKP–PcP traveltime differences and estimated the thickness of the outer core. Koper and Pyle (2004) also used IMS arrays, including the KSAR, to estimate density jumps, and P- and S-wave velocity jumps at the ICB using PKiKP/PcP amplitude ratios. Koper et al. (2004) discussed small-scale heterogeneity in the inner core using PKiKP recorded in IMS arrays, including the KSAR. Tanaka (2005) showed that the origin of PKP(Cdiff) coda waves recorded at IMS arrays, including the KSAR, was predominantly the CMB. Using the KSAR and other IMS arrays, Leyton et al. (2005) reported that sharp discontinuities with impedance ratios greater than 3% and a thicknesses of less than 3–4 km were not detected in the inner core. Leyton and Koper (2007), using the KSAR and other IMS arrays, revealed a regional pattern of small coda-Q values (large coda wave amplitudes) in the inner core of the hemisphere from the central Pacific to East Asia. Frost et al. (2017) used the KSAR to analyze the PKKP precursors.

3.2.5. Vietnam

Vietnam is located at epicentral distances of 70°–80° from Tonga-Fiji and is suitable for investigating the lowermost mantle beneath the vicinity of New Guinea Island. It is also located at epicentral distances of 70°–80° from Alaska and is suitable for investigating the lowermost mantle beneath northern Japan. As Huang et al. (2009) have already pointed out, its epicentral distance from South America is greater than 160°, making it suitable for studying the center of the Earth.

A Taiwanese team conducted temporary observations around North Vietnam (Huang et al., 2009), and a team from the ERI deployed six broadband seismic stations across Vietnam. Using data from the ERI array, Takeuchi et al. (2008) observed the traveltime differences of ScS–S from earthquakes that occurred in Tonga-Fiji and Vanuatu and found that the western edge of the Pacific LLSVP was located under the center of New Guinea Island. Idehara et al. (2013) added the ScS–S traveltimes from earthquakes near Japan observed in Australia to the data of Takeuchi et al. (2008), confirmed the western edge of the LLSVP, and suggested that a high-velocity anomaly region lies adjacent to the western edge of the LLSVP. Suzuki et al. (2020) used the data from the ERI array for waveform inversion.

3.2.6. India

India is located at epicentral distances of 30°–90° from the northwest Pacific to the West Pacific and 90°–120° from New Hebrides to Tonga-Fiji, making it suitable for studying the lower mantle, CMB, and outermost core. In addition, the South American earthquakes are aligned in a broad azimuth range with an epicentral distance of approximately 150°, which makes earthquakes useful for studies of the inner core beneath the entire African continent.

Roy et al. (2014) analyzed the SKS and SKKS splitting parameters recorded by the stations of the Indian Meteorological Department (IMD), National Geophysical Research Institute (NGRI), and additional temporary observation networks and showed the existence of seismic velocity anisotropy in the lower mantle (probably the lowermost mantle) below the Indochina Peninsula. Rao and Kumar (2014) investigated the heterogeneity in the lowermost mantle beneath the Indian Ocean and its surrounding regions by collecting ScS–S differential traveltimes and ScS/S amplitude ratios, using seismograms from the IMD and NGRI stations. Roy et al.

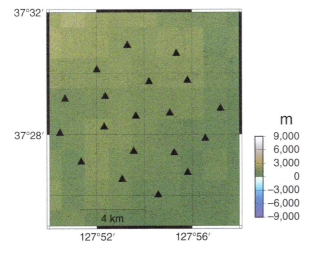

Figure 3.9 The station distribution of the South Korean Seismic Array (KSAR). ETOPO1 was used to draw the topography and bathymetry. A color figure is available in the online version.

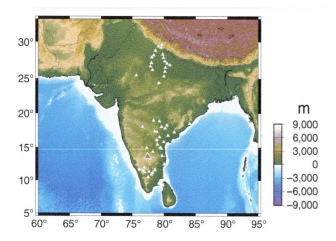

Figure 3.10 The station configuration (open triangles) used by Roy et al. (2019) and the Gauribidanur Array (GBA) (a gray triangle). ETOPO1 was used to draw the topography and bathymetry. A color figure is available in the online version. Source: Adapted from Roy et al. (2019).

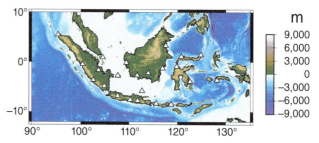

Figure 3.11 The station configuration of the Japan-Indonesia Seismic Network (JISNET) in the late 1990s. ETOPO1 was used to draw the topography and bathymetry. A color figure is available in the online version.

(2019) used Sdiff and its multipaths recorded at temporary and permanent stations to capture the Pacific LLSVP extending more westward than previously estimated. The station configuration used by Roy et al. (2019) is shown in Figure 3.10.

The Gauribidanur Array (GBA) is located in southern India (Fig. 3.10) and is an L-shaped seismic array of approximately 20 km on one side, established in 1966 by the UK Atomic Energy Administration (UKAEA) (Bache et al., 1985; Rost and Earle, 2010). Using the GBA, Datt and Varghese (1972) observed PKP phases from multiple underground nuclear explosions at the Nevada Test Site and the San Fernando earthquake and reported that no PKP precursors were detected. Rost and Earle (2010) used the GBA to detect PKKP precursors originating from the lowermost mantle beneath the Caribbean Sea and Patagonia, suggesting the presence of seismic wave scatterers due to subducted crustal material. Frost et al. (2017) also used the GBA to analyze PKKP precursors.

3.2.7. Indonesia

Indonesia is located at epicentral distances of 30°–70° from the Japan-Mariana arc, where deep earthquakes frequently occur, and 30°–90° from the Solomon, New Hebrides, and Tonga-Fiji Islands, making them suitable for analysis of the lower mantle. It is also located 140°–170° from South America and is suitable for the observation of core phases passing beneath the Antarctic region.

The Japan-Indonesia Seismic Network (JISNET), a temporary observation network consisting of 23 broadband seismometers (Fig. 3.11), was started in 1997 through an international collaboration by Japan and Indonesia under the Superplume Project (Ishida et al., 1999) to study Earth's interior (Ohtaki et al., 2000; Ohtaki et al., 2002). This network was transferred to the NIED in 2001 (Miyakawa et al., 2007). The network was then transferred to the Meteorological, Climatological, and Geophysical Agency of Indonesia in 2010 (Inoue, H., personal communication, 2011). The observation network was initially offline but was partially brought online after the 2004 Sumatran earthquake (Miyakawa et al., 2007; Yamashina et al., 2007). This network was used to study Earth's deep interior, as described below.

Tanaka (2007) used the JISNET for SmKS waveform analysis, and Ohtaki et al. (2012) used PKP waves from South American earthquakes to infer the seismic velocity structure of the inner core beneath Antarctica, showing that the region had a characteristic feature of the western hemispherical inner core.

3.2.8. Thailand

Thailand is located at epicentral distances of 30°–80° from the Japan-Mariana arc, and Aleutian-Alaska arc, and are suitable for studying the lower mantle structure beneath the North Pacific Ocean. The epicentral distance between Thailand and New Hebrides, Solomon Islands, and Tonga-Fiji are 60°–90°, which makes it suitable for studying the structure of the lower mantle to the lowermost mantle beneath New Guinea Island and its immediate surroundings. The epicentral distances are greater than 160° from South America, and it is therefore suitable for observing PKP(DF) waves passing through the deep inner core.

The Chiang Mai Seismic Array (CMAR) is a seismic array constructed near Chiang Mai by the IMS (Fig. 3.12a) and managed by the Hydrographic Department of the Royal Thai Navy. Idehara et al. (2007) used ScP and PcP waveforms to study the distribution of the ULVZ in the CMB beneath the region between Indonesia

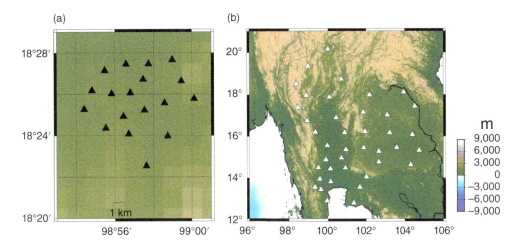

Figure 3.12 The station configurations of (a) the Chiang Mai Seismic Array (CMAR) and (b) Thai Seismic Array (TSAR). ETOPO1 was used to draw the topography and bathymetry with the TSAR. A color figure is available in the online version.

and the South China Sea. Frost et al. (2017) used the CMAR to analyze PKKP precursors.

The Thai Seismic Array (TSAR) is a mobile broadband seismic observation network supported by the MEXT Grant-in-Aid for Scientific Research on Innovative Areas "Interaction and Coevolution of the Core and Mantle" (Tanaka et al., 2019). A total of 40 stations was deployed across Thailand and spaced approximately 100 km apart (Fig. 3.12b), except for the Malay Peninsula in southern Thailand.

Suzuki et al. (2020) applied waveform inversion to TSAR data to estimate the structure of the lowermost S-wave velocities beneath the western Pacific, while Ohtaki et al. (2021) used PKP(DF)–PKP(BC) differential traveltimes from Central America observed by the TSAR and compared them with those of the F-net. They showed that the structure in the uppermost inner core beneath the Arctic region had the characteristics of the western hemispherical inner core.

3.2.9. Australia

Australia is an important region in the Southern Hemisphere for seismic studies of Earth's deep interior, but its contribution to deep Earth research has not been sufficient considering its potential. Japan, Tonga-Fiji, and northwestern India, where many deep earthquakes occur, are located at epicentral distances of 30°–90°, making it suitable for studying the mantle structure beneath the region from the western and southwestern Pacific to Southeast Asia. South America is located 120°–150° from Australia; therefore, Australia is well situated for studies of the structure of the core beneath the South Pacific and Antarctic regions.

Warramunga and Alice Springs Arrays

The Warramunga Array (WRA), located in northern Australia, is an L-shaped seismic array approximately 22 km on one side and was established by the UKAEA in 1965 (Cleary et al., 1968; Rost and Thomas, 2002). It currently consists of 24 broadband seismic stations (Fig. 3.13a), is operated by the Australian National University, and contributes to the CTBTO as an IMS seismic array.

The Alice Springs Array (ASAR) was first established in 1954, when the United States built a four-station linear array at Alice Springs in central Australia to monitor nuclear testing (Australian Earthquake Engineering Society, Newsletter, https://aees.org.au/wp-content/uploads/2013/11/AEES_2010_3.pdf). Today, the array consists of 19 short-period borehole seismic stations (Fig. 3.13b) and contributes to the CTBTO as an IMS seismic array.

Wright (1968) measured P-wave slowness at the WRA for earthquakes in Mariana and found a low-velocity zone at a depth of approximately 800 km below the northern part of New Guinea Island. Using the WRA, Wright (1970) analyzed the P-wave slowness for earthquakes with epicentral distances of 28°–99° and found anomalous velocity gradients at several depths in the lower mantle. Datt and Muirhead (1976) estimated a seismic discontinuity at a depth of approximately 770 km, with a velocity jump of 2%, based on the P-wave slowness from the Philippines observed at the WRA. Wright and Muirhead (1969) detected PKiKP and other waves in seismograms after the Novaya Zemlya nuclear test, and King et al. (1973, 1974) conducted an array analysis of PKP precursors from South American earthquakes and showed their origin to be the CMB. Wright et al. (1985), after measuring P-wave slowness using the WRA and two temporal seismic arrays, found a velocity anomaly in the

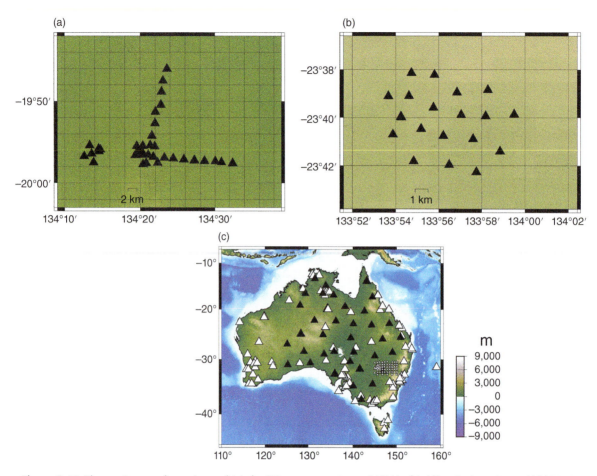

Figure 3.13 The station configurations of (a) the Warramunga Array (WRA), (b) Alice Springs Array (ASAR), (c) AU (open triangles), SKIPPY (closed triangles), and WOMBAT-EAL2 (open stars). ETOPO1 was used to draw the topography and bathymetry. A color figure is available in the online version.

lower mantle deeper than 1,700 km depth and proposed the existence of a 2.5–3% velocity discontinuity 200 km above the CMB. Souriau and Souriau (1989) added measurements of PKiKP–PcP traveltime differences and PKiKP/PcP amplitude ratios using WRA data to the previous measurements and examined the density jumps at the ICB and the ellipticity of the inner core surface. Rost and Revenaugh (2001, 2003), Rost et al. (2005), and Rost et al. (2010) used ScP waveforms to reveal the distribution of the ULVZ south of New Caledonia and off northwest Australia and reported the detection of the core rigidity zone. Frost et al. (2017) analyzed the PKKP precursors using the WRA. Poupinet and Kennett (2004) observed short-range PKiKP with the WRA.

Using ScP and PcP waveforms observed with the WRA and ASAR, Rost et al. (2006) investigated the distribution of the ULVZ off the eastern coast of Australia, and Idehara et al. (2007) investigated the distribution of the ULVZ beneath the region from the eastern coast of Australia to Indonesia. Ivan and Wang (2013) discussed the structure of the CMB from P4KP/P(S)cP and P5KP/PKP(AB) amplitude ratio anomalies detected with the WRA and ASAR. Brown et al. (2015) determined the structure of the ULVZ beneath the Coral Sea from ScP waveforms observed by the ASAR. Waszek et al. (2015) also used the WRA and ASAR to analyze PKP precursors. Ma et al. (2019) observed PKP precursors from South America using the WRA and ASAR and noted that the detected scatterers at the eastern margin of the Pacific LLSVP were interpreted as a small-scale cluster of ULVZs. Ma and Thomas (2020) revealed the distribution of scatterers in the lowermost mantle by analyzing PKP precursors with arrays around the world and seismic networks, including the WRA and ASAR. Koper et al. (2003), Koper et al. (2004), Koper and Pyle (2004), and Leyton and Koper (2007) also used WRA and ASAR records to analyze PKiKP.

SKIPPY and Others

SKIPPY was a large-scale seismic array observation program that was implemented to cover the Australian continent (Fig. 3.13c). Eight to twelve broadband seismic stations were deployed at a time, spaced approximately 400 km apart, and moved repeatedly every 3 months across the continent (van der Hilst et al., 1994). Tono and Yomogida (1997) interpreted the Pdiff coda of the 1994 Bolivian deep earthquake observed by the SKIPPY and New Zealand observational networks as scattered waves at the CMB beneath the South Pacific. Poupinet and Kennett (2004) also used SKIPPY for PKiKP observations. Pachhai et al. (2015) used PcP and ScP waves recorded by the EAL2 subarray (Fig. 3.13c) of WOMBAT (Rawlinson et al., 2008) in addition to the Hi-net data used by Pachhai et al. (2014) to analyze the ULVZ structure under the Tasman Sea. Idehara et al. (2013) read the ScS–S traveltime differences from the permanent observation network in the Australian continent, abbreviated as the AU (Fig. 3.13c). Kaneshima and Matsuzawa (2015) and Kaneshima (2018a) analyzed SmKS traveltimes and slowness using permanent observation networks on the Australian continent as an array.

3.2.10. Micronesia, Melanesia, and Polynesia

Temporary observations in these areas are essential because the density of permanent seismic stations is particularly low. To determine the 3-D mantle structure down to at least the uppermost lower mantle with teleseismic tomography, we should configure a large array with an aperture greater than approximately 1,500 km. Otherwise, teleseismic phases should be treated with traveltime and amplitude analyses and/or waveform modeling, as discussed in the previous subsections.

Using the southwest Pacific Seismic Experiment (SPaSE) (Wiens et al., 1995) and South Pacific broadband seismic network (SPANET) (Ishida et al., 1999), as shown in Figure 3.14, Suetsugu (2001) detected a low Q anomaly in the lower mantle beneath the South Pacific by analyzing multiple ScS waves. From the differential traveltimes of Sdiff–SKS and SKKS–SKS observed by the SPaSE and SPANET, Tanaka (2002) detected a very low velocity zone in the lowermost mantle beneath the south of the Society (Tahiti) hotspot in the South Pacific. Tanaka et al. (2009a, 2009b), Suetsugu et al. (2009), and Obayashi et al. (2016) used broadband seismographs installed by a French team on oceanic islands in French Polynesia (Barruol et al., 2002) and the broadband ocean bottom seismograph (BBOBS) (Suetsugu and Shiobara, 2014) array observations in the South Pacific as shown in Figure 3.14 (Suetsugu et al., 2005; Suetsugu et al., 2012). They determined the 3-D mantle structure under the South Pacific Ocean and revealed a super plume or large dome in the lower part of the lower mantle that extends branches to each hotspot in the South Pacific from the upper mantle.

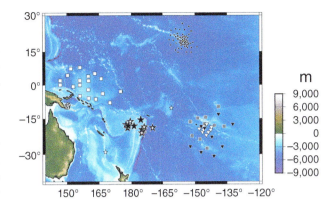

Figure 3.14 The station configurations of the southwest Pacific Seismic Experiment (SPaSE; closed stars), South Pacific broadband seismic network (SPANET; open stars), Hawaiian PLUME (dots), CDPapua (open triangles), PLUME (gray squares), French Polynesia Broadband Ocean Bottom Seismograph Project (FP-BBOBS; closed inverted triangles), Tomographic Investigation by seafloor Array Experiment for the Society hotspot (TIARES; open inverted triangles), and Ontong Java Plateau (OJP; open squares). ETOPO1 was used to draw the topography and bathymetry. A color figure is available in the online version.

Many seismic stations in Hawaii monitor volcanic activity. Garnero and Vidale (1999) and Reasoner and Revenaugh (2000) treated them as an array and applied stack processing to show the anomalies in ScP waveforms and the existence of the ULVZ beneath the southwestern Pacific Ocean. Using data from the PLUME project (Laske et al., 2009) (Fig. 3.14), which deployed a broadband submarine seismic network around Hawaii, Wolfe et al. (2009, 2011) estimated the 3-D mantle structures of P and S waves down to a depth of approximately 1,000 km and showed that the root of the mantle plume forming the Hawaii hotspot was located at least in the upper part of the lower mantle. Thomas and Laske (2015) detected PdP waves using a Hawaii PLUME array. Regarding other arrays, Waszek et al. (2015) also used the Papua New Guinea array (CDPapua) (Fig. 3.14) to analyze PKP precursors.

Using broadband seismograms from the Ontong Java Plateau (OJP) array observations (Fig. 3.14) (Suetsugu et al., 2018), Suetsugu et al. (2019) analyzed the traveltimes and amplitudes of multiple ScS phases to obtain large positive traveltime residuals and high Q values for the entire mantle beneath the OJP. More recently, Obayashi et al. (2021) collected P-wave traveltimes and applied finite-frequency tomography to determine a seismic high-velocity region that can be interpreted as a stagnant slab in the mantle transition zone and a sheet-like low-velocity region interpreted as upwelling from the lower mantle beneath the Caroline volcanic chain.

3.3. DISCUSSION

3.3.1. Summary of the Achievements and Continuing Issues

As we have seen above, array-based research methods are characterized by (1) the detection of unknown phases that reflect fine structures by the stack of seismograms with time shifts that adjust the slowness and incident direction; (2) the detection of the anomalies of seismic velocity and velocity gradient by measuring slowness; and (3) tomography that incorporates data from multiple seismic arrays and networks or uses a single large-aperture array. Regarding the resultants of such analyses, the previous studies have revealed radial heterogeneity of seismic velocity structures in the deep Earth (discontinuities in the lower mantle, velocity gradients and discontinuities at the bottom of the mantle, and velocity gradient anomalies at the top and bottom of the outer core), small-scale lateral heterogeneity (high-, low-, and ultralow-velocity zones), and the presence of scatterers in the lower mantle and at the CMB. However, many regions have not yet been sampled. Figure 3.15 shows an overview of the regions sampled in the lowermost mantle by the array studies discussed in this chapter using PcP, PdP, ScS, SdS, ScP, P (Pdiff), S (Sdiff), SKS, and SPdKS. The areas just below the vicinity of the subduction zone in the West to southwest Pacific Ocean have been well investigated. When deep earthquakes in Tonga-Fiji, Indonesia, and the Philippines are observed in East Asia and Australia, the sampled regions inevitably become similar. However, the density of the data may be coarse or dense depending on the size of the array and the number of elements in the array. In exceptional cases, oceanic seismic observations in and around oceanic islands such as Hawaii sample the area beneath the center of the Pacific Ocean.

The studies mentioned in this chapter have contributed to revealing the Pacific LLSVP. These studies suggest that the western area of the Pacific LLSVP is fragmented on the CMB, that its eastern area is occupied by the large body of the low-velocity region in the lower mantle beneath French Polynesia, and that the low-velocity body in the lower mantle is branched into several hotspots in the upper mantle. However, the deep lower mantle beneath French Polynesia has not been fully examined using previous array observations. Interestingly, French and Romanowicz (2015) indicated that there were vertical low-velocity columns in the entire mantle beneath Samoa, Tahiti, MacDonald, and Pitcairn hotspots in their global 3-D tomographic model of SEMUCB-WM1, in conflict with the results of Tanaka et al. (2009a, 2009b), Suetsugu et al. (2009), and Obayashi et al. (2016). This discrepancy is caused by insufficient ray sampling of the deep mantle beneath these hotspots and, thus, should be addressed by future dense observations. Furthermore, the relationship between the ULVZ and dynamics of the deep Earth should be reconsidered in future studies. Although the distribution of the scatterers and small-scale ULVZs in the lowermost mantle appears to be random, the large ULVZs examined with diffracted waves (e.g., SPdKS and Sdiff) may be located near distinct hotspots, such as Hawaii, Samoa, and Iceland (Yu and Garnero, 2018). Thus, further analysis of the large ULVZs should be conducted to reveal a fine structure with new observations and fill the unexplored areas of the ULVZ.

The studies mentioned in this chapter have also contributed to the elucidation of the inner core. Studies using precritical reflection of PKiKP phases covered a wide area in the East Asia and revealed a pattern formed by placing small pieces of the inner core surface with different seismological properties into a mosaic-like feature. The regions of the inner core sampled by PKP(DF) observed by the arrays discussed in this chapter include the northeast Pacific (earthquakes in South America and observatories in Japan), Africa (South America and China), the Arctic region (Central America and Thailand), the Antarctic region (South America and Indonesia), and Australia (South Sandwich and Japan). The most significant observation is that the volume of the inner core of the quasi-Eastern Hemisphere is much smaller than that of the quasi-Western Hemisphere (Ohtaki et al., 2012; Ohtaki et al., 2021). However, the boundary locations of the inner core hemisphere in the Arctic region do not coincide (Iritani et al., 2019; Ohtaki et al., 2021). In addition, the hemispherical pattern observed in the seismic velocity perturbation in

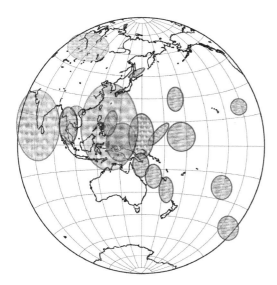

Figure 3.15 A summary map for the investigated areas in the lowermost mantle using the seismic arrays and networks mentioned in this chapter for seismic phases of PcP (PdP), ScS (SdS), ScP, P (Pdiff), S (Sdiff), SKS, and SPdKS.

the upper inner core does not correspond to that of the seismic attenuation (Attanayake et al., 2014; Iritani et al., 2014a). Because the mechanism for seismic attenuation in the inner core is still unknown, theoretical consideration will be required as well as further observations over a wide frequency range.

Although there are some examples of studies of anisotropy just beneath the slab and the upper part of the inner core, there are still only a few examples of studies of anisotropy and horizontal isotropy in the lowermost mantle using arrays in this region. To address azimuthal anisotropy, further array observations are necessary to obtain multidirectional ray paths, and we need to consider not only deep earthquakes that generate relatively simple waveforms but also shallow earthquakes with more complex source processes and the crustal structure near the hypocenter. Although the deconvolution of the source pulse has been applied (e.g., Shibutani et al., 1995; Eaton and Kendall, 2006), more advanced techniques are required to resolve such complexities. In addition, although the attenuation structure of the inner core has been addressed, few challenges to the attenuation structure of the lower mantle and lowermost mantle exist.

3.3.2. Future Perspective with Consideration for Land and Sea Floor Observations

In describing future prospects, we would like to start the discussion with observations on land. First, we can point out seismic phases and hypocentral regions that have not been analyzed by past and existing arrays and observation networks to date. The NECESSArray has not yet been used for the analyses of ScS and ScP phases. Additionally, no studies of the lower mantle using earthquakes in the southwest Pacific observed by the Indonesian network have been conducted. The Thai and Vietnamese seismic networks have not been used for the analysis of the lower mantle beneath the northwestern Pacific using earthquakes in the northern Pacific, the Aleutian Islands, and Alaska. Furthermore, although some studies collected crossing paths or multidirectional paths to study the lowermost mantle beneath the western Pacific, as introduced in section 3.2 (Idehara et al., 2013; Suzuki et al., 2020), these studies did not examine seismic anisotropy. If we consider further collection of the S and ScS phases at epicentral distances of 40°–60°, earthquakes in the regions of Indonesia, Papua New Guinea, the Izu-Bonin-Mariana arc, and the China-Russia border can be observed by the seismic arrays or networks of the F-net, NECESSArray, and NCISP, and those in Thailand, Myanmar, and Indonesia. These seismic ray paths can intensively sample the lower mantle beneath the Philippines with various azimuths. The lowermost mantle beneath the Philippines is interesting because detection of ULVZs has been reported even though they are somewhat distant from the margin of the Pacific LLSVP (e.g., Idehara et al., 2007; Idehara, 2011). With respect to the inner core structure, the NCISP has not been used for the analysis of the core phase from South America, the ray paths of which cover north of the North American continent to the Arctic region. In addition, the observation of earthquakes in South America by seismic networks in India and Sri Lanka will play an important role in studying wide region beneath the African continent in the east-west direction. Furthermore, if we consider the multidirectional path passing through the inner core, we should use seismic arrays that were not discussed in this chapter. For example, the combination of Asian and European seismic arrays has already been analyzed, as introduced in section 3.2 (Helffrich et al., 2002). Although we did not mention Isse and Nakanishi (2002) in section 3.2 because they used individual permanent stations, they studied seismic anisotropy in the inner core beneath Australia using stations in Asia and Antarctica. Currently, seismic arrays and networks in these areas have significantly improved. Therefore, we can extend their analysis further by using modern array data. Before starting the analysis, the researchers should consider their own scientific objectives and issues that should be addressed as well as the background of the deep Earth dynamics.

Next, we gathered information about new permanent seismic networks. In Vietnam, the Institute of Geophysics (IGP) of the Vietnam Academy of Science and Technology (VAST) has completed the deployment of 31 broadband seismic stations in 2018, as shown in Figure 3.16 (Nguyen, 2018, personal communication). This network is very promising for observing deep earthquakes in South America, where it is located at the antipodal point of the country to examine the innermost inner core. In Myanmar, the Earth Observatory of Singapore (EOS) has established and is operating a broadband seismic network (https://earthobservatory.sg/project/myanmar-seismology; Wang et al., 2019). A huge seismic array can be formed across Southeast and South Asia through the collaboration of the new seismic networks in Vietnam and Myanmar, with the networks existing in India, Thailand, and Indonesia. Several broadband seismic observations have been carried out in Australia, and the AusArray project has begun since 2016, with 135 seismic stations arranged in a grid at intervals of approximately 55 km as a partial array. The partial arrays will be moved after a certain period of observation and eventually cover Australia (https://www.ga.gov.au/eftf/minerals/nawa/ausarray). In addition, the ChinArray (e.g., Xiao et al., 2020) can be a useful huge tool for the deep Earth studies.

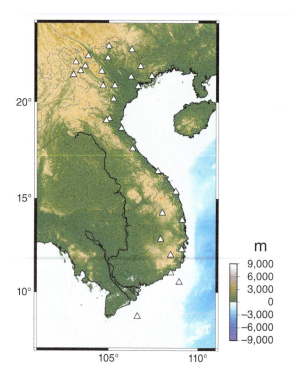

Figure 3.16 Geographical map of new Vietnamese broadband seismographic stations (open triangles). ETOPO1 was used to draw the topography and bathymetry. A color figure is available in the online version.

However, the global availability of some arrays and networks is still limited. We know that array construction is hard work and agree that a priority period should be allowed for researchers involved. We also know that maintaining a website for waveform distribution is also difficult. Nevertheless, as global seismologists, we hope that the seismograms from these arrays will be made publicly available, at least after a moratorium period.

Finally, if we plan new observations for 1–2 years, we should construct a promising strategy that considers epicentral distances from the hypocentral regions of the Tonga-Fiji Islands because of its very high seismic activity. As shown in Table 3.2, each seismic phase that is suitable for deep structure studies has a different observable epicentral distance range and corresponding depth suitable for the study. For the observation of ScP waves, epicentral distances of 30°–50° from the Tonga-Fiji Islands cover the Papua New Guinea Island and the central and eastern parts of the Australian continent. The islands of the Philippines and Borneo are suitable for observations of ScS and PcP waves. West China, Mongolia, and India are expected to be suitable locations for observations of SKS+SPdKS, SmKS, Sdiff, and Pdiff. However, the observation environment in regions where no seismologist has ever been present will be quite severe owing to many difficulties, such as lack of accessibility, rough terrain, transportation problems, flooding, lightening, wild animals, insects, deserts, and jungles. Thus, researchers should first look for reliable local collaborators who know the regions of concern well. Additionally, it is important to consider the benefits of collaborators and construct good relationships.

Although we discussed land observations, the limited distribution of earthquakes and seismic stations on land alone limits the areas and depths that can be analyzed. The prospects for future observation planning will likely require that we expand our focus to oceanic areas. Kawakatsu (2016) noted that international cooperation is important for the Pacific Array, which is currently investigating the structure of the oceanic lithosphere, to play a full role in deep Earth research. For the regions covered in this chapter, the Pacific LLSVP would be one of the most interesting observation targets, as discussed above. To capture the entire image, we need to conduct observations in oceanic areas, such as the Pacific Array, to cover the entire Pacific Ocean. The Mobile Earthquake Recording in Marine Areas by Independent Divers (MERMAID) is one of the most promising observation systems and collects hydro-acoustic signals using a hydrophone floating at mid-column water depths (Simons et al., 2009). A project using MERMAID has recently elucidated the seismic structure beneath the Galapagos hotspot (Nolet et al., 2019), and MERMAID is currently being used to study the South Pacific (Simon et al., 2022). Unfortunately, while MERMAID is a good system for observing teleseismic P waves, it is not suited for observations of S waves. To understand the Pacific LLSVP more deeply, obtaining the S-wave structure is also important. Thus, seafloor seismic observations, such as by the BBOBS, are required to obtain seismic signals. More than 20 years have passed since BBOBS began to be fully operational in the late 1990s (Suetsugu and Shiobara, 2014). Because self-pop-up ocean bottom seismographs sit only on the seafloor, they result in much larger horizontal components of noise than the vertical component because of the effects of ocean currents (Suetsugu and Shiobara, 2014). However, it is possible to observe S to Sdiff and SKS in the long-period component, as shown in Figure 3.17 (Suetsugu et al., 2010).

Here, we present an example of a practical idea for project planning. To collect data as soon as possible to reconcile with funding-oriented scientific purposes for the structure of the crust and upper mantle, as the first step, we need to put medium-sized arrays with an aperture of ~500 km through multiple observations and finally form a long zonal array across the South Pacific to view the Pacific LLSVP in two dimensions. Because we already have broadband seismograms from the OJP array (Suetsugu et al., 2018) and the projects in French Polynesia (Barruol et al., 2002; Suetsugu et al., 2005;

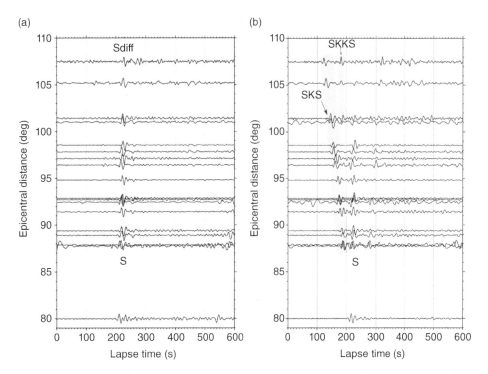

Figure 3.17 (a) Transverse and (b) radial components of seismograms of the Philippines deep earthquake on 26 May 2003, recorded by the broadband ocean bottom seismographs (BBOBS) and broadband seismic stations in French Polynesia. S and Sdiff phases are aligned on a 200 seconds timelapse. SKS and SKKS that are shown with hatches can be found before S (Sdiff) phases. A band-passed filter with periods of 10–30 seconds was applied.

Suetsugu et al., 2012), the regions around the Samoan and Easter Islands are important for filling the gap in the long zonal array. Because our ultimate goal is to obtain the 3-D structure of the entire Pacific LLSVP, arrays that form a long zonal array in the north-south direction and that fill the gap between the Hawaiian PLUME array (Laske et al., 2009) and the French Polynesian arrays are also important. Because earthquakes may not occur in the epicentral distance range suitable for analytical purposes – for example, observations of S, Sdiff, ScS, and SKS to reveal the deeper lower mantle during the limited observation period (typically 1–2 years) – it may be necessary to repeat observations in the same area. To achieve the ultimate goal, patience and long-lived enthusiasm, as well as the successors maintaining these mindsets, are required.

ACKNOWLEDGMENTS

This chapter was supported by MEXT/JSPS KAKENHI 15H05832. Figures were plotted using the Generic Mapping Tools (GMT) (Wessel et al., 2013). ETOPO1 was provided by the NOAA (https://www.ncei.noaa.gov/access/metadata/landing-page/bin/iso?id=gov.noaa.ngdc.mgg.dem:316). Dr. Le-Minh Nguyen provided the information about the Vietnamese broadband seismographic network. We greatly appreciate all the people involved in the seismic arrays. The authors are grateful for the valuable comments and suggestions of two anonymous reviewers, which improved this manuscript.

REFERENCES

Adam, J. M.-C., Ibourichène, A., & Romanowicz, B. (2018). Observation of core sensitive phases: Constraints on the velocity and attenuation profile in the vicinity of the inner-core boundary. *Physics of the Earth and Planetary Interiors*, *275*, 19–31. Doi:10.1016/j.pepi.2017.12.008.

Adam, J. M.-C., & Romanowicz, B. (2015). Global scale observations of scattered energy near the inner-core boundary: Seismic constraints on the base of the outer-core. *Physics of the Earth and Planetary Interiors*, *245*, 103–116. Doi:10.1016/j.pepi.2015.06.005.

Attanayake, J., Cormier, V. F., & de Silva, S. M. (2014). Uppermost inner core seismic structure – new insights from body waveform inversion. *Earth and Planetary Science Letters*, *385*, 49–58. Doi:10.1016/j.epsl.2013.10.025.

Bache, T. C., Marshall, P. D., & Bache, L. B. (1985). Q for teleseicmic P waves from central Asia. *Journal of Geophysical Research*, *90*, 3575–3587.

Bagley, B., Courtier, A. M., & Revenaugh, J. (2013). Seismic shear wave reflectivity structure of the mantle beneath northeast China and the northwest Pacific. *Journal of Geophysical*

Research, Solid Earth, 118, 5417–5427. Doi:10.1002/jgrb.50385.

Barruol, G., Bosch, D., Clouard, V., Debayle, E., Doin, P., Fontaine, F.R., et al. (2002). PLUME investigates South Pacific Superswell. *EOS Transactions, American Geophysical Union, 83,* 511–514.

Bolt, B. A., Niazi, M., & Somerville, M. R. (1970). Diffracted ScS and the shear velocity at the core boundary. *Geophysical Journal of the Royal Astronomical Society, 19,* 299–305.

Brown, S. P., Thorne, M. S., Miyagi, L., & Rost, S. (2015). A compositional origin to ultralow-velocity zones. *Geophysical Research Letters, 42.* Doi:10.1002/2014GL062097.

Cao, A., & Romanowicz, B. (2004). Hemispherical transition of seismic attenuation at the top of the earth's inner core. *Earth and Planetary Science Letters, 228*(3–4), 243–253. Doi:10.1016/j.epsl.2004.09.032.

Castle, J. C., Creager, K. C., Wincheser, J. P., & van der Hilst, R. D. (2000). Shear wave speeds at the base of the mantle. *Journal of Geophysical Research, 105,* 21543–21557.

Chaloner, J. W., Thomas, C., & Rietbrock, A. (2009). P- and S-wave reflectors in D″ beneath southeast Asia. *Geophysical Journal International, 179*(2), 1080–1092. Doi:10.1111/j.1365-246X.2009.04328.x.

Cleary, J. R. (1969). The S velocity at the core-mantle boundary, from observations of diffracted S. *Bulletin of the Seismological Society of America, 59,* 1399–1405.

Cleary, J. R., Wright, C., & Muirhead, K. J. (1968). The Effects of Local Structure upon Measurements of the Travel Time Gradient at the Warramunga Seismic Array. *Geophysical Journal of the Royal Astronomical Society, 16,* 21–29.

Cobden, L., & Thomas, C. (2013). The origin of D reflections: A systematic study of seismic array data sets. *Geophysical Journal International, 194,* 1091–1118. Doi:10.1093/gji/ggt152.

Creasy, N., Pisconti, A., Long, M. D., Thomas, C., & Wookey, J. (2019). Constraining lowermost mantle anisotropy with body waves: A synthetic modelling study. *Geophysical Journal International, 217,* 766–783. Doi:10.1093/gji/ggz049.

Dai, Z., Wang, W., & Wen, L. (2012). Irregular topography at the Earth's inner core boundary. *Proceedings of the National Academy of Sciences, 109*(20), 7654–7658. Doi:10.1073/pnas.1116342109.

Datt, R., & Muirhead, K. J. (1976). Evidence for a sharp increase in P-wave velocity at about 770 km depth. *Physics of the Earth and Planetary Interiors, 13,* 37–46.

Datt, R., & Varghese, T. G. (1972). Array detection and location of core shadow events. *Bulletin of the Seismological Society of America, 62,* 231–245.

Dziewonski, A. M. (1984). Mapping the lower mantle: Determination of lateral heterogeneity in P velocity up to degree and order 6. *Journal of Geophysical Research, 89,* 5929–5952.

Dziewonski, A. M., & Anderson, D. L. (1981). Preliminary reference Earth model. *Physics of the Earth and Planetary Interiors, 25*(4), 297–356.

Eaton, D. W., & Kendall, J. M. (2006). Improving seismic resolution of outermost core structure by multichannel analysis and deconvolution of broadband SmKS phases. *Physics of the Earth and Planetary Interiors, 155,* 104–119.

French, S. W., & Romanowicz, B. (2015). Broad plumes rooted at the base of the Earth's mantle beneath major hotspots. *Nature, 525,* 95–99.

Frost, D. A., Rost, S., Garnero, E. J., & Li, M. M. (2017). Seismic evidence for Earth's crusty deep mantle. *Earth and Planetary Science Letters, 470,* 54–63. Doi:10.1016/j.epsl.2017.04.036.

Fukao, Y., Obayashi, M., & Nakakuki, T. (2009). Stagnant Slab: A Review. *Annual Review of Earth and Planetary Sciences, 37*(1), 19–46. Doi:10.1146/annurev.earth.36.031207.124224.

Fukao, Y., Widiyantoro, S., & Obayashi, M. (2001). Stagnant slabs in the upper and lower mantle transition region. *Reviews of Geophysics, 39,* 291–323.

Fukuyama, E., Ishida, M., Hori, S., Sekiguchi, S., & Watada, S. (1996). Broadband seismic observation conducted under the FREESIA project. *Report of the National Research Institute for Earth Science and Disaster Resilience, 57,* 23–31 (in Japanese with English abstract).

Garnero, E. J., & Helmberger, D. V. (1995). A very slow basal layer underlying large-scale low-velocity anomalies in the lower mantle beneath the Pacific – Evidence from core phases. *Physics of the Earth and Planetary Interiors, 91*(1–3), 161–176.

Garnero, E. J., McNamara, A. K., & Shim, S. H. (2016). Continent-sized anomalous zones with low seismic velocity at the base of Earth's mantle. *Nature Geoscience, 9*(7), 481–489. Doi:10.1038/ngeo2733.

Garnero, E. J., & Vidale, J. E. (1999). ScP; A probe of ultralow velocity zones at the base of the mantle. *Geophysical Research Letters, 26*(3), 377–380.

He, Y., & Wen, L. (2009). Structural features and shear-velocity structure of the "Pacific Anomaly." *Journal of Geophysical Research, 114,* B02309. Doi:10.1029/2008JB005814.

He, Y., & Wen, L. (2011). Seismic velocity structures and detailed features of the D″ discontinuity near the core-mantle boundary beneath eastern Eurasia. *Physics of the Earth and Planetary Interiors, 189,* 176–184. Doi:10.1016/j.pepi.2011.09.002.

Helffrich, G., & Kaneshima, S. (2004). Seismological Constraints on Core Composition from Fe-O-S Liquid Immiscibility. *Science, 306,* 2239–2242.

Helffrich, G., & Kaneshima, S. (2010). Outer-core compositional stratification from observed core wave speed profiles. *Nature, 468*(7325), 807–810. Doi:10.1038/nature09636.

Helffrich, G., Kaneshima, S., & Kendall, J. M. (2002). A local, crossing-path study of attenuation and anisotropy of the inner core. *Geophysical Research Letters, 29,* 9-1–9-4. doi:10.1029/2001GL014059.

Hirose, K. (2007). Discovery of post-perovskite phase transition and the nature of D″ layer. In K. Hirose, J. Brodholt, T. Lay, & D. Yuen (Eds.), *Post-Perovskite: The Last Mantle Phase Transition.* Washington: American Geophysical Union. doi:10.1029/174GM04.

Huang, B. S. (1996). Investigation of the inner-outer core boundary structure from the seismograms of a deep earthquake recorded by a regional seismic array. *Geophysical Research Letters, 23*(3), 209–212. doi:10.1029/96gl00087.

Huang, B. S., Le, T. S., Liu, C. C. i., Toan, D. V., Huang, W. G., Wu, Y. M., et al. (2009). Portable broadband seismic network in Vietnam for investigating tectonic deformation, the Earth's interior, and early-warning systems for earthquakes and tsunamis. *The Journal of Asian Earth Sciences*, *36*(1), 110–118. doi:10.1016/j.jseaes.2009.02.012.

Idehara, K. (2011). Structural heterogeneity of an ultra-low-velocity zone beneath the Philippine Islands: Implications for core-mantle chemical interactions induced by massive partial melting at the bottom of the mantle. *Physics of the Earth and Planetary Interiors*, *184*(1–2), 80–90. doi:10.1016/j.pepi.2010.10.014.

Idehara, K., Tanaka, S., & Takeuchi, N. (2013). High-velocity anomaly adjacent to the western edge of the Pacific low-velocity province. *Geophysical Journal International*, *192*(1), 1–6. doi:10.1093/gji/ggs002.

Idehara, K., Yamada, A., & Zhao, D. (2007). Seismological constraints on the ultralow velocity zones in the lowermost mantle from core-reflected waves. *Physics of the Earth and Planetary Interiors*, *165*, 25–46.

Inoue, H. (2011). *Personal Communication*.

Iritani, R., Kawakatsu, H., & Takeuchi, N. (2019). Sharpness of the hemispherical boundary in the inner core beneath the northern Pacific. *Earth and Planetary Science Letters*, *527*, 115796. doi:10.1016/j.epsl.2019.115796.

Iritani, R., Takeuchi, N., & Kawakatsu, H. (2010). Seismic attenuation structure of the top half of the inner core beneath the northeastern Pacific. *Geophysical Research Letters*, *37*, L19303. doi:10.1029/2010gl044053.

Iritani, R., Takeuchi, N., & Kawakatsu, H. (2014a). Intricate heterogeneous structures of the top 300 km of the Earth's inner core inferred from global array data: II. Frequency dependence of inner core attenuation and its implication. *Earth and Planetary Science Letters*, *405*, 231–243. doi:10.1016/j.epsl.2014.08.038.

Iritani, R., Takeuchi, N., & Kawakatsu, H. (2014b). Intricate heterogeneous structures of the top 300 km of the Earth's inner core inferred from global array data: I. Regional 1D attenuation and velocity profiles. *Physics of the Earth and Planetary Interiors*, *230*, 15–27. doi:10.1016/j.pepi.2014.02.002.

Ishida, M., Maruyama, S., Suetsugu, D., Matsuzaka, S., & Eguchi, T. (1999). Superplume Project: Towards a new view of whole Earth dynamics. *Earth Planets Space*, *51*, i–v.

Isse, T., & Nakanishi, I. (2002). Inner-core anisotropy beneath Australia and differential rotation. *Geophysical Journal International*, *151*(1), 255–263.

Ivan, M., & He, X. (2017). Uppermost inner core heterogeneity from differential travel times of PKIKP vs. PKP-Bdiff and PKP-Cdiff phases. *Pure and Applied Geophysics*, *174*, 249–259. doi:10.1007/s00024-016-1391-z.

Ivan, M., & Wang, R. (2013). Anomalous high amplitude ratios of P5KP/PKPab and P4KP/P(S)cP observed globally around 1 Hz. *Journal of Seismology*, *17*, 453–464. doi:10.1007/s10950-012-9330-7.

Ivan, M., Wang, R., & Hofstetter, R. (2018). Non quasi-hemispherical seismological pattern of the Earth's uppermost inner core. *Scientific Reports*, *8*, 2270. doi:10.1038/s41598-018-20657-x.

J-Array Group (1993). The J-Array program: System and present status. *Journal of geomagnetism and geoelectricity*, *45*, 1265–1274.

James, D. E. (2015). Crust and lithospheric structure – Natural source portable array studies of the continental lithosphere. In B. Romanowicz, & A. M. Dziewonski (Eds.), *Treatise on Geophysics*, 2nd edn, pp. 513–555. Amsterdam: Elsevier. doi:10.1016/B978-0-444-53802-4.00014-2.

Jiang, G., & Zhao, D. (2012). Observation of high-frequency PKiKP in Japan: Insight into fine structure of inner core boundary. *The Journal of Asian Earth Sciences*, *59*(SI), 167–184.

Kaneshima, S. (2013). Lower mantle seismic scatterers below the subducting Tonga slab: Evidence for entrainment of transition zone material. *Physics of the Earth and Planetary Interiors*, *222*, 35–46.

Kaneshima, S. (2014). Upper bounds of seismic anisotropy in the Tonga slab near deep earthquake foci and in the lower mantle. *Geophysical Journal International*, *197*, 351–368. doi:10.1093/gji/ggt494.

Kaneshima, S. (2018a). Array analyses of SmKS waves and the stratification of Earth's outermost core. *Physics of the Earth and Planetary Interiors*, *276*, 234–246. doi:10.1016/j.pepi.2017.03.006.

Kaneshima, S. (2018b). Seismic scatterers in the mid-lower mantle beneath Tonga-Fiji. *Physics of the Earth and Planetary Interiors*, *274*, 1–13. doi:10.1016/j.pepi.2017.09.007.

Kaneshima, S. (2019). Seismic scatterers in the lower mantle near subduction zones. *Geophysical Journal International*, *219*, S2–S20. doi:10.1093/gji/ggz241.

Kaneshima, S., & Helffrich, G. (2013). Vp structure of the outermost core derived from analysing large-scale array data of SmKS waves. *Geophysical Journal International*, *193*(3), 1537–1555. doi:10.1093/gji/ggt042.

Kaneshima, S., Hirahara, K., Ohtaki, T. & Yoshida, Y. (1994). Seismic structure near the inner-core outer core boundary. *Geophysical Research Letters*, *21*(2), 157–160.

Kaneshima, S., & Matsuzawa, T. (2015). Stratification of earth's outermost core inferred from SmKS array data. *Progress in Earth and Planetary Sciences*, *2*, 15. 10.1186/s40645-40015-40046-40645, doi:10.1186/s40645-015-0046-5.

Kao, H., Jian, P. R., Ma, K. F., Huang, B.-S., & Liu, C. C. (1998). Moment-tensor inversion for offshore earthquakes east of Taiwan and their implications to regional collision. *Geophysical Research Letters*, *25*, 3619–3622. doi:3619–3622.

Kawai, K., Sekine, S., Fuji, N., & Geller, R. J. (2009). Waveform inversion for D″ structure beneath northern Asia using Hi-net tiltmeter data. *Geophysical Research Letters*, *36*, L20314. doi:10.1029/2009gl039651.

Kawakatsu, H. (2006). Sharp and seismically transparent inner core boundary region revealed by an entire network observation of near-vertical PKiKP. *Earth Planets Space*, *58*, 855–863.

Kawakatsu, H. (2016). Pacific Array of, by and for Global Deep Earth Research. AGU Fall Meeting, DI23C-01.

Kawakatsu, H., & Niu, F. (1994). Seismic evidence for a 920-km discontinuity in the mantle. *Nature*, *371*, 301–395.

Kazama, T., Kawakatsu, H., & Takeuchi, N. (2008). Depth-dependent attenuation structure of the inner core inferred from short-period Hi-net data. *Physics of the Earth and Planetary Interiors*, *167*, 155–160.

Kennett, B. L. N., Engdahl, E. R., & Buland, R. (1995). Constraints on seismic velocities in the Earth from travel-times. *Geophysical Journal International*, *122*(1), 108–124.

King, D. W., Haddon, R. A. W., & Cleary, J. R. (1973). Evidence for seismic scattering in the D" layer. *Earth and Planetary Science Letters*, *20*, 353–356.

King, D. W., Haddon, R. A. W., & Cleary, J. R. (1974). Array analysis of precursors to PKIKP in the distance range 128ε to 142ε. *Geophysical Journal of the Royal Astronomical Society*, *37*, 157–173.

Kito, T., & Kruger, F. (2001). Heterogeneities in D" beneath the southwestern Pacific inferred from scattered and reflected P-waves. *Geophysical Research Letters*, *28*, 2545–2548.

Kito, T., Krüger, F., & Negishi, H. (2004). Seismic heterogeneous struture in the lowermost mantle beneath the southwestern Pacific. *Journal of Geophysical Research*, *109*, B09304. doi:10.1029/2003JB002677.

Kito, T., Rietbrock, A., & Thomas, C. (2007). Slowness-backazimuth weighted migration: A new array approach to a high-resolution image. *Geophysical Journal International*, *169*, 1201–1209. doi:10.1111/j.1365-246X.2007.03379.x.

Kito, T., Shibutani, T., & Hirahara, K. (2003). Scattering objects in the lower mantle beneath northeastern China observed with a short-period seismic array. *Physics of the Earth and Planetary Interiors*, *138*, 55–69.

Konishi, K., Fuji, N., & Deschamps, F. (2020). Three-dimensional elastic and anelastic structure of the lowermost mantle beneath the western Pacific from finite-frequency tomography. *Journal of Geophysical Research: Solid Earth*, *125*, e2019JB018089. doi:10.1029/2019JB018089.

Konishi, K., Kawai, K., Geller, R. J., & Fuji, N. (2014). Waveform inversion for localized three-dimensional seismic velocity structure in the lowermost mantle beneath the Western Pacific. *Geophysical Journal International*, *199*(2), 1245–1267. doi:10.1093/gji/ggu288.

Koper, K. D., Franks, J. M. & Dombrovskaya, M. (2004). Evidence for small-scale heterogeneity in Earth's innre core from a global study of PKiKP coda waves. *Earth and Planetary Science Letters*, *228*, 227–241.

Koper, K. D., & Pyle, M. L. (2004). Observations of PKiKP/PcP amplitude ratios and implications for Earth structure at the boundaries of the liquid core. *Journal of Geophysical Research*, *109*, B03301. doi:10.1029/2003JB002750.

Koper, K. D., Pyle, M. L., & Franks, J. M. (2003). Constraints on aspherical core structure from PKiKP-PcP differential travel times. *Journal of Geophysical Research*, *108*(B3), 2168. doi:2110.1029/2002JB001995.

Krasnoshchekov, D., Ovtchinnikov, V., & Valentin Polishchuk, V. (2019). Dissimilarity of the Earth's inner core surface under south America and northeastern Asia revealed by core reflected phases. *Journal of Geophysical Research: Solid Earth*, *124*, 4862–4878. doi:10.1029/2019JB017408.

Kruger, F., Weber, M., Scherbaum, F. & Schlittenhardt, J. (1993). Double-beam analysis of anomalies in the core-mantle boundary region. *Geophysical Research Letters*, *20*(14), 1475–1478. doi:10.1029/93gl01311.

Kuo, B. Y., Garnero, E. J., & Lay, T. (2000). Tomographic inversion of S-SKS times for shear velocity heterogeneity in D": Degree 12 and hybrid models. *Journal of Geophysical Research*, *105*(B12), 28139–28157.

Laske, G., Collins, J. A., Wolfe, C. J., Solomon, S. C., Detrick, R. S., Orcutt, J. A., et al. (2009). Probing the hawaiian hot spot with new ocean bottom instruments. *EOS, Transactions, American Geophysical Union*, *90*, 362–363.

Lay, T., HernLund, J., & Buffett, B. A. (2008). Core-mantle boundary heat flow. *Nature geoscience*, *1*, 25–32. doi:10.1038/ngeo.2007.44.

Lay, T., Williams, Q., Garnero, E. J., Kellogg, L. H., & Wysession, M. E. (1998). Seismic wave anisotropy in the D" region and its implications. In M. Gurnis, M. E. Wysession, E. Knittle, & B. A. Buffett (Eds.), *The Core-Mantle Boundary Region*. pp. 299–318. Washington: AGU.

Lee, H.-I., Chi, H.-C., Park, J.-H., Cho, C.-S., Kim, G.-Y., & Lim, I.-S. (2000). *Seismic research network in KIGAM*. Paper Presented at Proceedings of the Earthquake Engineering Society of Korea Conference.

Lev, E., Long, M., & van der Hilst, R. D. (2006). Seismic anisotropy in Eastern Tibet from shear-wave splitting reveals changes in lithosphere deformation. *Earth and Planetary Science Letters*, *251*, 293–304. doi:10.1016/j.epsl.2006.09.018.

Leyton, F., & Koper, K. D. (2007). Using PKiKP coda to determine inner core structure: 2. Determination of Q_C. *Journal of Geophysical Research*, *112*, B05317. doi:10.1029/2006JB004370.

Leyton, F., Koper, K. D., Zhu, L., & Dombrovskaya, M. (2005). On the lack of seismic discontinuities within the inner core. *Geophysical Journal International*, *162*, 779–786.

Ma, X., Sun, X., & Thomas, C. (2019). Localized ultra-low velocity zones at the eastern boundary of Pacific LLSVP. *Earth and Planetary Science Letters*, *507*, 40–49. doi:10.1016/j.epsl.2018.11.037.

Ma, X., & Thomas, C. (2020). Small-Scale Scattering Heterogeneities in the Lowermost Mantle From a Global Analysis of PKP Precursors. *Journal of Geophysical Research: Solid Earth*, *125*, e2019JB018736. doi:10.1029/2019JB018736.

Mattesini, M., Belonoshko, A. B., Buforn, E., Ramírez, M., Simak, S. I., Udías, A., et al. (2010). Hemispherical anisotropic patterns of the Earth's inner core. *Proceedings of the National Academy of Sciences*, *107*, 9507–9512. doi:10.1073/pnas.1004856107.

McNamara, A. (2019). A review of large low shear velocity provinces and ultra low velocity zones. *Tectonophysics*, *760*, 199–220. doi:10.1016/j.tecto.2018.04.015.

Meltzer, A., Sol, S., Zurek, B., Xuanyang, Z., Jianlong, Z., & Wenqing, T. (2003). The Eastern Syntaxis Seismic Experiment. *EOS Transactions, American Geophysical Union*, *84*, T42B–T0292.

Miyakawa, K., Yamashina, T., Inoue, H., Ishida, M., Masturyono, Harjadi, P., et al. (2007). Deployment of Satellite Telemetered Broadband Seismic Network in Indonesia. *Technical Note of the National Research Institute for Earth Science and Disaster Resilience*, *304*, 25–40 (in Japanese with English Abstract).

Mondt, J. C. (1977). SH waves: Theory and observations for epicentral distances greater than 90 degrees. *Physics of the Earth and Planetary Interiors*, *15*, 46–59.

Morita, Y. (1987). Analysis of short-period PKP phases (1): Topography of the ICB. *Abstracts of Annual Meeting of Seismological Society of Japan, No.2*, 301 (in Japanese).

Morita, Y. (1991). The attenuation structure in the inner-core inferred from seismic body waves. *Central Core of the Earth*, *1*, 65–75 (in Japanese with English abstract).

Morita, Y. (1996). The characteristics of J-array seismograms. *Journal of Physics of the Earth*, *44*, 657–668.

Nábelek, J., Hetényi, G., Vergne, J., Sapkota, S., Kafle, B., Jiang, M., et al. (2009). Underplating in the Himalaya-Tibet collision zone revealed by the Hi-CLIMB experiment. *Science*, *325*, 1371–1374. doi:10.1126/science.1167719.

Nakanishi, I. (1990). High-frequency waves following PKP-Cdiff at distances greater than 155-degrees. *Geophysical Research Letters*, *17*(5), 639–642.

Nguyen, L.-M. (2018). *Personal communication*.

Niazi, M. (1973). SH travel times and lateral heterogeneities in the lower mantle. *Bulletin of the Seismological Society of America*, *63*, 2035–2046.

Niu, F., & Chen, Q. (2008). Seismic evidence for distinct anisotropy in the innermost inner core. *Nature Geoscience*, *1*, 692–696. doi:10.1038/ngeo314.

Niu, F. L., & Wen, L. X. (2001). Hemispherical variations in seismic velocity at the top of the Earth's inner core. *Nature*, *410*(6832), 1081–1084.

Nolet, G., Hello, Y., van der Lee, S., Bonnieux, S., Ruiz, M. C., Pazmino, N. A., et al. (2019). Imaging the Galápagos mantle plume with an unconventional application of floating seismometers. *Scientific Reports*, *9*, 1326. doi:10.1038/s41598-018-36835-w.

Nowacki, A., Wookey, J., & Kendall, J. M. (2011). New advances in using seismic anisotropy, mineral physics and geodynamics to understand deformation in the lowermost mantle. *Journal of Geodynamics*, *52*(3–4), 205–228. doi:10.1016/j.jog.2011.04.003.

Obayashi, M., Yoshimitsu, J., Nolet, G., Fukao, Y., Shiobara, H., Sugioka, H., et al. (2013). Finite frequency whole mantle P wave tomography: Improvement of subducted slab images. *Geophysical Research Letters*, *40*(21), 5652–5657. doi:10.1002/2013gl057401.

Obayashi, M., Yoshimitsu, J., Suetsugu, D., Shiobara, H., Sugioka, H., Ito, A., et al. (2021). Interrelation of the stagnant slab, Ontong Java Plateau, and intraplate volcanism as inferred from seismic tomography. *Scientific Reports*, *11*, 20966. doi:10.1038/s41598-021-99833-5.

Obayashi, M., Yoshimitsu, J., Sugioka, H., Ito, A., Isse, T., Shiobara, H., et al. (2016). Mantle plumes beneath the South Pacific superswell revealed by finite frequency P tomography using regional seafloor and island data. *Geophysical Research Letters*, *43*(22), 11628–11634. doi:10.1002/2016gl070793.

Ohtake, M., & Ishikawa, Y. (1995). Seismic observation networks in Japan. *Journal of Physics of the Earth*, *43*, 563–584.

Ohtaki, T., & Kaneshima, S. (2015). Independent estimate of velocity structure of Earth's lowermost outer core beneath the northeast Pacific from PKiKP–PKPbc differential traveltime and dispersion in PKPbc. *Journal of Geophysical Research: Solid Earth*, *120*, 7572–7586. doi:10.1002/2015JB012140.

Ohtaki, T., Kaneshima, S., Ichikawa, H., & Tsuchiya, T. (2018). Seismological Evidence for Laterally Heterogeneous Lowermost Outer Core of the Earth. *Journal of Geophysical Research: Solid Earth*, *123*(12), 10903–10917. doi:10.1029/2018jb015857.

Ohtaki, T., Kaneshima, S., & Kanjo, K. (2012). Seismic structure near the inner core boundary in the south polar region. *Journal of Geophysical Research: Solid Earth*, *117*, B03312. doi:10.1029/2011jb008717.

Ohtaki, T., Kanjo, K., Kaneshima, S., Nishimura, T., Ishihara, Y., Yoshida, Y., et al. (2000). Broadband Seismic Network in Indonesia - JISNET -. *Bulletin of the Geological Survey of Japan*, *51*, 189–203 (in Japanese with English abstract).

Ohtaki, T., Suetsugu, D., Kanjo, K., & Purwana, I. (2002). Evidence for a thick mantle transition zone beneath the Philippine Sea from multiple-ScS waves recorded by JISNET. *Geophysical Research Letters*, *29*, 24-1–24-4. doi:10.1029/2002GL014764.

Ohtaki, T., Tanaka, S., Kaneshima, S., Siripunvaraporn, W., Boonchaisuk, S., Noisagool, S., et al. (2021). Seismic velocity structure of the upper inner core in the north polar region. *Physics of the Earth and Planetary Interiors*, *311*, 106636. doi:10.1016/j.pepi.2020.106636.

Okada, Y., Kasahara, K., Hori, S., Obara, K., Sekiguchi, S., Fujiwara, H., et al. (2004). Recent progress of seismic observation networks in Japan –Hi-net, F-net, K-NET and KiK-net–. *Earth Planets Space*, *56*, xv–xxviii.

Osada, Y., Kashiwabara, S., Nagai, A., Takayama, H., Wakui, S., Morishita, I., et al. (1984). Matsushiro Seismic Array System (1) - Outline and hypocenter determination of near earthquakes. *Technical Reports of the Matsushiro Seismological Observatory, Japan Meteorological Agency*, *5*, 13–31 (in Japanese).

Pachhai, S., Tkalčić, H., & Dettmer, J. (2014). Bayesian inference for ultralow velocity zones in the Earth's lowermost mantle: Complex ULVZ beneath the east of the Philippines. *Journal of Geophysical Research: Solid Earth*, *119*, 8346–8365. doi:10.1002/2014JB011067.

Poupinet, G., & Kennett, B. L. N. (2004). On the observation of high frequency PKiKP and its coda in Australia. *Physics of the Earth and Planetary Interiors*, *146*, 497–511.

Rao, B. P., & Kumar, M. R. (2014). Seismic evidence for slab graveyards atop the Core Mantle Boundary beneath the Indian Ocean Geoid Low. *Physics of the Earth and Planetary Interiors*, *236*, 52–59. doi:10.1016/j.pepi.2014.08.005.

Rao, B. P., Kumar, M. R., & Singh, A. (2017). Anisotropy in the lowermost mantle beneath the Indian Ocean Geoid Low from ScS splitting measurements. *Geochemistry, Geophysics, Geosystems*, *18*, 558–570. doi:10.1002/2016GC006604.

Rawlinson, N., Tkalčić, H., & Kennett, B. L. N. (2008). New results from WOMBAT: An ongoing program of passive seismic array deployments in Australia. *EOS, Transactions, American Geophysical Union*, *89*(53), Fall Mett. Suppl., Abstract S22A-03.

Reasoner, C., & Revenaugh, J. (2000). ScP constraints on ultralow-velocity zone density and gradient thickness beneath the Pacific. *Journal of Geophysical Research*, *105*, 28173–28182.

Ritsema, J., Garnero, E., & Lay, T. (1997). A strongly negative shear velocity gradient and lateral variability in the lowermost mantle beneath the Pacific. *Journal of Geophysical Research*, *102*(B9), 20395–20411.

Roecker, S. (2001). Constraints on the crust and upper mantle of the Kyrgyz Tien Shan from the preliminary analysis of GHENGIS broad-band seismic data. *Geologiya I Geopizika*, *42*, 1554–1565.

Rost, S., & Earle, P. S. (2010). Identifying regions of strong scattering at the core-mantle boundary from analysis of PKKP precursor energy. *Earth and Planetary Science Letters*, *297*(3–4), 616–626. doi:10.1016/j.epsl.2010.07.014.

Rost, S., Garnero, E., & Stefan, W. (2010). Thin and intermittent ultralow-velocity zones. *Journal of Geophysical Research*, *115*, B06312. doi:10.1029/2009JB006981.

Rost, S., Garnero, E. J., & Williams, Q. (2006). Fine-scale ultralow-velocity zone structure from high-frequency seismic array data. *Journal of Geophysical Research*, *111*, B09310. doi:10.1029/2005JB004088.

Rost, S., Garnero, E. J., Williams, Q., & Manga, M. (2005). Seismological constraints on a possible plume root at the core-mantle boundary. *Nature*, *435*(7042), 666–669.

Rost, S., & Revenaugh, J. (2001). Seismic detection of rigid zones at the top of the core. *Science*, *294*, 1911–1914.

Rost, S., & Revenaugh, J. (2003). Small-scale ultralow-velocity zone structure imaged by ScP. *Journal of Geophysical Research*, *108*, 2056. doi:10.1029/2001JB001627.

Rost, S., & Thomas, C. (2002). Array seismology: Methods and applications. *Reviews of Geophysics*, *40*, 3. doi:10.1029/2000RG000100.

Roy, S. K., Kumar, M. R., & Srinagesh, D. (2014). Upper and lower mantle anisotropy inferred from comprehensive SKS and SKKS splitting measurements from India. *Earth and Planetary Science Letters*, *392*, 192–306. doi:10.1016/j.epsl.2014.02.012.

Roy, S. K., Takeuchi, N., Srinagesh, D., Kumar, M. R., & Kawakatsu, H. (2019). Topography of the western Pacific LLSVP constrained by S-wave multipathing. *Geophysical Journal International*, *218*(1), 190–199. doi:10.1093/gji/ggz149.

Shen, Z., Ai, Y. S., He, Y., & Jiang, M. (2016a). Using pre-critical PKiKP–PcP phases to constrain the regional structures of the inner core boundary beneath East Asia. *Physics of the Earth and Planetary Interiors*, *252*, 37–48. doi:10.1016/j.pepi.2016.01.001.

Shen, Z., Ni, S., Wu, W., & Sun, D. (2016b). Short period ScP phase amplitude calculations for core–mantle boundary with intermediate scale topography. *Physics of the Earth and Planetary Interiors*, *253*, 64–73. doi:10.1016/j.pepi.2016.02.002.

Shibutani, T., Hirahara, K., & Kato, M. (1995). P-wave velocity discontinuity in D" layer beneath western Pacific with J-array records. In T. Yukutake (Ed.), *The Earth's Central Part: Its Structure and Dynamics*, pp. 1–11. Tokyo: Terra Scientific Publishing Company.

Shibutani, T., Tanaka, A., Kato, M., & Hirahara, K. (1993). A study of P-wave velocity discontinuity in D" layer with J-Array records: Preliminary results. *Journal of geomagnetism and geoelectricity*, *45*, 1275–1285.

Simon, J. D., Simons, F. J., & Irving, J. C. E. (2022). Recording earthquakes for tomographic imaging of the mantle beneath the South Pacific by autonomous MERMAID floats. *Geophysical Journal International*, *228*, 147–170. doi:10.1093/gji/ggab271.

Simons, F. J., Nolet, G., Georgief, P., Babcock, J. M., Regier, L. A., & Davis, R. E. (2009). On the potential of recording earthquakes for global seismic tomography by low-cost autonomous instruments in the oceans. *Journal of Geophysical Research: Solid Earth*, *114*, B05307. doi:10.1029/2008JB006088.

Souriau, A., & Calvet, M. (2015). Deep Earth structure – The Earth's cores. In B. Romanowicz, & A. M. Dziewonski (Eds.), *Treatise on Geophysics, 2nd ed., vol. 1, Seismology and Structure of the Earth*. Amsterdam: Elsevier.

Souriau, A., & Souriau, M. (1989). Ellipticity and density at the inner core boundary from subcritical PKiKP and PcP data. *Geophysical Journal International*, *98*, 39–54.

Suetsugu, D. (2001). A low Q_{ScS} anomaly near the South Pacific superswell. *Geophysical Research Letters*, *28*(2), 391–394.

Suetsugu, D., Isse, T., Tanaka, S., Obayashi, M., Shiobara, H., Sugioka, H., et al. (2009). South Pacific mantle plumes imaged by seismic observation on islands and seafloor. *Geochemistry, Geophysics, Geosystems*, *10*, Q11014. doi:10.1029/2009GC002533.

Suetsugu, D., & Shiobara, H. (2014). Broadband Ocean-Bottom Seismology. *Annual Review of Earth and Planetary Sciences*, *42*, 27–43. doi:10.1146/annurev-earth-060313-054818.

Suetsugu, D., Shiobara, H., Sugioka, H., Ito, A., Isse, I., Kasaya, T., et al. (2012). TIARES Project—Tomographic investigation by seafloor array experiment for the Society hotspot. *Earth Planets Space*, *64*, i–iv. doi:10.5047/eps.2011.11.002.

Suetsugu, D., Shiobara, H., Sugioka, H., Ito, A., Isse, T., Ishihara, Y., et al. (2019). High Q_{ScS} beneath the Ontong Java Plateau. *Earth Planets Space*, *71*, 97. doi:10.1186/s40623-019-1077-8.

Suetsugu, D., Shiobara, H., Sugioka, H., Tada, N., Ito, A., Isse, T., et al. (2018). The OJP array: Seismological and electromagnetic observation on seafloor and islands in the Ontong Java Plateau. *Report of Research and Development*, *26*, 54–64. doi:10.5918/jamstecr.26.54.

Suetsugu, D., Sugioka, H., Isse, T., Fukao, Y., Shiobara, H., Kanazawa, T., et al. (2005). Probing South Pacific mantle plumes with ocean bottom seismographs. *EOS Transactions, American Geophysical Union*, *86*, 429–435.

Suetsugu, D., Tanaka, S., Shiobara, H., Sugioka, H., Kanazawa, T., Fukao, Y., et al. (2010). Constraints on the Large Low Shear Velocity Province beneath the Pacific Ocean from joint ocean floor and islands broadband seismic experiments in French Polynesia. *AGU Fall Meeting*, DI21B-1956.

Suzuki, Y., Kawai, K., Geller, R. J., Tanaka, S., Siripunvaraporn, W., Boonchaisuk, S., et al. (2020). High-resolution 3-D S-velocity structure in the D" region at the western margin of the Pacific LLSVP: Evidence for small-scale plumes and paleoslabs. *Physics of the Earth and Planetary Interiors*, *307*, 106544. doi:doi.org/10.1016/j.pepi.2020.106544.

Takeuchi, N. (2012). Detection of ridge-like structures in the Pacific Large Low-Shear-Velocity Province. *Earth and Planetary Science Letters*, *319–320*, 55–64.

Takeuchi, N., Morita, Y., Xuyen, N. D., & Zung, N. Q. (2008). Extent of the low-velocity region in the lowermost mantle beneath the western Pacific detected by the Vietnamese Broadband Seismograph Array. *Geophysical Research Letters*, *35*, L05307. doi:10.1029/2008GL033197.

Takeuchi, N., & Obara, K. (2010). Fine-scale topography of the D″ discontinuity and its correlation to volumetric velocity fluctuations. *Physics of the Earth and Planetary Interiors*, *183*(1–2), 126–135. doi:10.1016/j.pepi.2010.06.002.

Tanaka, S. (2002). Very low shear wave velocity at the base of the mantle under the South Pacific Superswell. *Physics of the Earth and Planetary Interiors*, *203*, 879–893.

Tanaka, S. (2005). Characteristics of PKP-Cdiff coda revealed by small-aperture seismic arrays: Implications for the study of the inner core boundary. *Physics of the Earth and Planetary Interiors*, *153*, 49–60.

Tanaka, S. (2007). Possibility of a low P-wave velocity layer in the outermost core from global SmKS waveforms. *Earth and Planetary Science Letters*, *259*, 486–499.

Tanaka, S. (2010). Constraints on the core-mantle boundary topography from P4KP–PcP differential travel times. *Journal of Geophysical Research: Solid Earth*, *115*, B04310. doi:10.1029/2009jb006563.

Tanaka, S. (2012). Depth extent of hemispherical inner core from PKP(DF) and PKP(Cdiff) for equatorial paths. *Physics of the Earth and Planetary Interiors*, *210–211*, 50–62. doi:10.1016/j.pepi.2012.08.001.

Tanaka, S., & Hamaguchi, H. (1996). Frequency-dependent Q in the Earth's outer core from short-period P4KP/PcP spectral ratio. *Journal of Physics of the Earth*, *44*(6), 745–759.

Tanaka, S., Kawakatsu, H., & Obayashi, M. (2006). NECESSArray Project: Earth dynamics viewed from the Chinese continent. *Chikyu Monthly*, *28*, 592–596 (in Japanese).

Tanaka, S., Kawakatsu, H., Obayashi, M., Chen, Y. J., Ning, J. Y., Grand, S. P., et al. (2015). Rapid lateral variation of P-wave velocity at the base of the mantle near the edge of the Large-Low Shear Velocity Province beneath the western Pacific. *Geophysical Journal International*, *200*(2), 1050–1063. doi:10.1093/gji/ggu455.

Tanaka, S., Siripunvaraporn, W., Boonchaisuk, S., Noisagool, S., Kim, T., Kawai, K., et al. (2019). Thai Seismic Array (TSAR) Project. *Bulletin of Earthquake Research Institute*, *94*, 1–11.

Tanaka, S., & Tkalčić, H. (2015). Complex inner core boundary from frequency characteristics of the reflection coefficients of PKiKP waves observed by Hi-net. *Progress in Earth and Planetary Sciences*, *2*(1). 10.1186/s40645-40015-40064-40643. doi:10.1186/s40645-015-0064-3.

Tang, Y., Obayashi, M., Niu, F., Grand, S., Chen, Y. J., Kawakatsu, H., et al. (2014). Changbaishan volcanism in northeast China linked to subduction-induced mantle upwelling. *Nature geoscience*, *7*, 470–475. doi:10.1038/NGEO2166.

Tanimoto, T. (1990). Long-wavelength S-wave velocity structure throughout the mantle. *Geophysical Journal International*, *100*, 327–336.

Thomas, C., & Laske, G. (2015). D″ observations in the Pacific from PLUME ocean bottom seismometer recordings. *Geophysical Journal International*, *200*, 851–862. doi:10.1093/gji/ggu441.

Tian, D., & Wen, L. (2017). Seismological evidence for a localized mushy zone at the Earth's inner core boundary. *Nature Communications*, *8*, 165. doi:10.1038/s41467-017-00229-9.

Tkalčić, H. (2015). Complex inner core of the Earth: The last frontier of global seismology. *Reviews of Geophysics*, *53*, 1–36. doi:10.1002/2014RG000469.

Tkalčić, H. (2017). *The Earth's inner core: Revealed by observational seismology, edited*. Cambridge: Cambridge University Press.

Tkalčić, H., Cormier, V. F., Kennett, B. L. N., & He, K. (2010). Steep reflections from the earth's core reveal small-scale heterogeneity in the upper mantle. *Physics of the Earth and Planetary Interiors*, *178*(1–2), 80–91. doi:10.1016/j.pepi.2009.08.004.

Tono, Y., & Yomogida, K. (1997). Origin of short-period signals following P-diffracted waves: A case study of the 1994 Bolivian deep earthquake. *Physics of the Earth and Planetary Interiors*, *103*, 1–16.

Tseng, T. L., & Huang, B. S. (2001). Depth-dependent attenuation in the uppermost inner core from the Taiwan short period seismic array PKP data. *Geophysical Research Letters*, *28*, 459–462.

Umetsu, I. (1990). *A study on the boundaries of the Earth's core using PKP waves*. Ms. thesis, p.101. Sendai (in Japanese): Tohoku Univ.

Vanacore, E., Niu, F., & Kawakatsu, H. (2006). Observations of the mid-mantle discontinuity beneath Indonesia from S to P converted waveforms. *Geophysical Research Letters*, *33*, L04302. doi:10.1029/2005GL025106.

Vanacore, E., Niu, F., & Ma, Y. (2010). Large angle reflection from a dipping structure recorded as a PKIKP precursor: Evidence for a low velocity zone at the core–mantle boundary beneath the Gulf of Mexico. *Earth and Planetary Science Letters*, *293*, 54–62. doi:10.1016/j.epsl.2010.02.018.

van der Hilst, R. D., Kennett, B. L. N., Christie, D., & Grant, J. (1994). Project Skippy explores lithosphere and mantle beneath Australia. *EOS Transactions, American Geophysical Union*, *75*, 177–181, DOI: 10.1029/94EO00857.

Vinnik, L., Kato, M., & Kawakatsu, H. (2001). Search for seismic discontinuities in the lower mantle. *Geophysical Journal International*, *147*, 41–56.

Wang, J. H. (1989). The Taiwan telemetered seismographic network. *Physics of the Earth and Planetary Interiors*, *58*, 9–18. doi:10.1016/0031-9201(89)90090-3.

Wang, T., & Song, X. (2018). Support for equatorial anisotropy of Earth's inner-inner core from seismic interferometry at low latitudes. *Physics of the Earth and Planetary Interiors*, *276*, 247–257. doi:10.1016/j.pepi.2017.03.004.

Wang, T., Song, X., & Xia, H. H. (2015). Equatorial anisotropy in the inner part of Earth's inner core from autocorrelation of earthquake coda. *Nature geoscience*, *9*. doi:10.1038/NGEO2354.

Wang, X., Wei, S., Wang, Y., Maung, P. M., Hubbard, J., Banerjee, P., et al. (2019). A 3-D shear wave velocity model for Myanmar region. *Journal of Geophysical Research: Solid Earth*, *124*, 504–526. doi:10.1029/2018JB016622.

Waszek, L., Schmerr, N., & Ballmer, M. D. (2018). Global observations of reflectors in the mid-mantle with implications for mantle structure and dynamics. *Nature Communications, 9,* 385. doi:10.1038/s41467-017-02709-4.

Waszek, L., Thomas, C., & Deuss, A. (2015). PKP precursors: Implications for global scatterers. *Geophysical Research Letters, 42,* 3829–3838. doi:10.1002/2015GL063869.

Weber, M. (1993). P-wave and S-wave reflections from anomalies in the lowermost mantle. *Geophysical Journal International, 115*(1), 183–210. doi:10.1111/j.1365-246X.1993.tb05598.x.

Wessel, P., Smith, W. H. F., Scharroo, R., Luis, J. F., & Wobbe, F. (2013). Generic Mapping Tools: Improved version released. *EOS Transactions, American Geophysical Union, 94,* 409–410.

Wiens, D. A., Shore, P. J., McGuire, J. J., Roth, E., Bevis, M. G., & Draunidalo, K. (1995). The Southwest Pacific Seismic Experiment. *IRIS Newsletter, 14,* 1–4.

Wolfe, C. J., Solomon, S. C., Laske, G., Collins, J. A., Detrick, R. S., Orcutt, J. A., et al. (2009). Mantle shear-wave velocity structure beneath the Hawaiian hot spot. *Science, 326*(5958), 1388–1390. doi:10.1126/science.1180165.

Wolfe, C. J., Solomon, S. C., Laske, G., Collins, J. A., Detrick, R. S., Orcutt, J. A., et al. (2011). Mantle P-wave velocity structure beneath the Hawaiian hotspot. *Earth and Planetary Science Letters, 303*(3–4), 267–280. doi:10.1016/j.epsl.2011.01.004.

Wookey, J., & Helffrich, G. (2008). Inner-core shear-wave anisotropy and texture from an observation of PKJKP waves. *Nature, 454*(7206), 873–U824. doi:10.1038/nature07131.

Wright, C. (1968). Evidence for a low velocity layer for P waves at a depth close to 800km. *Earth and Planetary Science Letters, 5,* 35–40.

Wright, C. (1970). P-wave travel time gradient measurements and lower mantle structure. *Earth and Planetary Science Letters, 8*(1), 41. doi:10.1016/0012-821x(70)90097-x.

Wright, C., & Muirhead, K. J. (1969). Longitudinal waves from the Novaya Zemlya nuclear explosion of October 27, 1966, recorded at the Warramunga seismic array. *Journal of Geophysical Research, 74,* 2034–2048.

Wright, C., Muirhead, K. J., & Dixon, A. E. (1985). The P wave velocity structure near the base of the mantle. *Journal of Geophysical Research, 90*(B1), 623–634. doi:10.1029/JB090iB01p00623.

Wu, B., Xia, H. H., Wang, T., & Shi, X. (2018). Simulation of core phases from coda interferometry. *Journal of Geophysical Research: Solid Earth, 123,* 4983–4999. doi:10.1029/2017JB015405.

Wu, F. T., Lavier, L. L., & Teams, U. T. (2007). Collision Tectonics of Taiwan and TAIGER Experiments. *AGU Fall Meeting,* T51A-0321.

Wysession, M. E., Langenhorst, A., Fouch, M. J., Fischer, K. M., Al-Eqabi, G. I., Shore, P. J., et al. (1999). Lateral variations in compressional/shear velocities at the base of the mantle. *Science, 284*(5411), 120–125. doi:10.1126/science.284.5411.120.

Wysession, M. E., Lay, T., Revenaugh, J., Williams, Q., Garnero, E. J., Jeanloz, R., et al. (1998). The D" discontinuity and its implications. In M. Gurnis, M. E. Wysession, E. Knittle, & B. A. Buffett (Eds.), *The Core-Mantle Boundary Region.* pp. 273–297. Washington: AGU.

Xia, H. H., Song, X., & Wang, T. (2016). Extraction of triplicated PKP phases from noise correlations. *Geophysical Journal International, 205,* 499–508. doi:10.1093/gji/ggw015.

Xiao, Z., Fuji, N., Iidaka, T., Gao, Y., Sun, X., & Liu, Q. (2020). Seismic structure beneath the Tibetan plateau from iterative finite-frequency tomography based on ChinArray: New insights into the Indo-Asian collision. *Journal of Geophysical Research: Solid Earth, 125,* e2019JB018344. doi:10.1029/2019JB018344.

Yamada, A., & Nakanishi, I. (1993). The density jump across the ICB and constraints on P-reflector in the D" layer from observation on the 1992 Chinese nuclear explosion. *Geophysical Research Letters, 20,* 2195–2198.

Yamada, A., & Nakanishi, I. (1996). Detection of P-wave reflector in D" beneath the south-western Pacific using double-array stacking. *Geophysical Research Letters, 23,* 1553–1556.

Yamada, A., & Nakanishi, I. (1998). Short-wavelength lateral variation of a D" P-wave reflector beneath the southwestern Pacific. *Geophysical Research Letters, 25,* 4545–4548.

Yamashina, T., Miyakawa, K., Negishi, H., Inoue, H., Ishida, M., Fauzi, et al. (2007). Construction of Data Processing System of JISNET. *Technical Note of the National Research Institute for Earth Science and Disaster Resilience, 304,* 41–47 (in Japanese with English Abstract).

Yao, J., & Wen, L. (2014). Seismic structure and ultra-low velocity zones at the base of the Earth's mantle beneath Southeast Asia. *Physics of the Earth and Planetary Interiors, 233,* 103–111.

Yee, T. G., Rhie, J., & Tkalčić, H. (2014). Regionally heterogeneous uppermost inner core observed with Hi-net array. *Journal of Geophysical Research: Solid Earth, 119*(10), 7823–7845. doi:10.1002/2014jb011341.

Yu, S. L., & Garnero, E. J. (2018). Ultralow Velocity Zone Locations: A Global Assessment. *Geochemistry, Geophysics, Geosystems, 19*(2), 396–414. doi:10.1002/2017gc007281.

Yu, W. C., Wen, L. & Niu, F. (2005). Seismic velocity structure in the Earth's outer core. *Journal of Geophysical Research, 110,* B02302. doi:10.1029/2003JB002928.

Yu, W. C., & Wen, L. X. (2007). Complex seismic anisotropy in the top of the Earth's inner core beneath Africa. *Journal of Geophysical Research: Solid Earth, 112,* B08304. doi:10.1029/2006jb004868.

Zhao, D. (2004). Global tomographic images of mantle plumes and subducting slabs: Insight into deep Earth dynamics. *Physics of the Earth and Planetary Interiors, 146*(1–2), 3–34.

Zheng, X.-F., Yao, Z.-X., Liang, J.-H., & Zheng, J. (2010). The role played and opportunities provided by IGP DMC of China National Seismic Network in Wenchuan Earthquake disaster relief and researches. *Bulletin of the Seismological Society of America, 100,* 2886–2872. doi:10.1785/0120090257.

Zhu, R., Xu, Y., Zhu, G., Zhang, H., Xia, Q., & Zheng, T. (2012). Destruction of the North China Craton. *SCIENCE CHINA Earth Sciences, 55,* 1565–1587. doi:10.1007/s11430-012-4516-y.

Zou, Z. H., Koper, K. D., & Cormier, V. F. (2008). The structure of the base of the outer core inferred from seismic waves diffracted around the inner core. *Journal of Geophysical Research, 113*(B5), B05314. 05310.01029/02007jb005316; doi:10.1029/2007jb005316.

4

Preliminary Results from the New Deformation Multi-Anvil Press at the Photon Factory: Insight into the Creep Strength of Calcium Silicate Perovskite

Andrew R. Thomson[1], Yu Nishihara[2], Daisuke Yamazaki[3], Noriyoshi Tsujino[3], Simon A. Hunt[4], Yumiko Tsubokawa[5], Kyoko Matsukage[6], Takashi Yoshino[3], Tomoaki Kubo[5], and David P. Dobson[1]

ABSTRACT

A D111 deformation multi-anvil press, which is a larger version of the Deformation T-cup, has been installed at beamline NE7A at PF-AR, KEK, Tsukuba, Japan. Using this apparatus, controlled deformation experiments can be performed by independently moving two opposite second-stage anvils in a Kawai-type 6/8 geometry. This allows both pure and simple shear experiments with sample strain rates of $\sim 10^{-6}$ s^{-1} to be conducted at pressure and temperature conditions up to at least 27 GPa and 1,700 K. Here, the capabilities of the D111 press are demonstrated using experiments investigating the creep strength of calcium silicate perovskite at mantle conditions. Experiments performed at ~13 GPa and 1150–1373 K with quantitative stress and strain rate measurements have allowed preliminary evaluation of the creep strength of calcium silicate perovskite. Observations indicate that under dry and wet conditions samples possess grain sizes of between 0.5 and ~10 μm, and undergo deformation in a diffusion creep regime with an apparent activation energy of ~364 kJ mol^{-1} and a stress exponent of 1.28. Qualitative comparison of preliminary results with the properties of mantle silicate minerals including bridgmanite, wadsleyite, and dry olivine indicates that calcium silicate perovskite is very weak, such that it may strongly influence the geodynamics of the deep mantle especially in regions of subduction.

4.1. INTRODUCTION

Compared to the other rocky planets, Earth is dynamic, with an active dynamo, volcanism, seismicity, and repeated, continuous, cycling of material between the surface and deep interior for at least the last 2.5 billion years (Krusky et al., 2001). The plate-tectonic style of convection appears to be unique to the Earth and results in regions of dynamic complexity at the surface and the core-mantle boundary, the major thermal boundary layers of the mantle system. Geophysical studies show a third region of complexity in the mid-mantle, associated with the stabilization of silicate perovskite around 660 km depth. In this region, subducted slab material shows a wide range of complex behaviors, ranging from penetration into the lower mantle, through ponding around 1,000–1,200 km depth, to stagnation in the transition zone (van der Hilst, 1995). This tomographically imaged complexity is accompanied by further evidence of basalt enrichment in the region around 660 km (Greaux et al., 2019; Thomson et al., 2019), implying that subducted crustal mid-ocean ridge basalt (MORB) can mechanically

[1]*Department of Earth Sciences, University College London, London, UK*
[2]*Geodynamics Research Center, Ehime University, Matsuyama, Japan*
[3]*Institute for Planetary Materials, Okayama University, Misasa, Japan*
[4]*Department of Materials, University of Manchester, Manchester, UK*
[5]*Department of Earth and Planetary Sciences, Kyushu University, Fukuoka, Japan*
[6]*Department of Natural and Environmental Science, Teikyo University of Science, Uenohara, Japan*

Core-Mantle Co-Evolution: An Interdisciplinary Approach, Geophysical Monograph 276, First Edition.
Edited by Takashi Nakagawa, Taku Tsuchiya, Madhusoodhan Satish-Kumar, and George Helffrich.
© 2023 American Geophysical Union. Published 2023 by John Wiley & Sons, Inc.
DOI: 10.1002/9781119526919.ch04

decouple from the depleted, ultramafic components of the slab interior. The complex behavior of the subducting slab is likely to be due to changes in density and, more importantly, viscosity of the slab and surrounding mantle as the Mg_2SiO_4-dominated upper mantle mineral assemblages transform into $MgSiO_3$-bridgmanite dominated assemblages in the lower mantle (e.g., Karato, 1997). The rheological behaviors of the crustal components of subducting slabs are further complicated in this region because of its more silica- and calcium-enriched chemistry compared with ultramafic lithologies, which results in a distinct bulk mineralogy in the transition zone and lower mantle variously containing components of stishovite, $CaSiO_3$ perovskite, NAL phase, and CF-phase (e.g., Perrillat et al., 2006).

In order to understand this complexity and to accurately model the convective behavior of the mantle, it is necessary to measure the thermoelastic and rheological properties of the constituent rocks and minerals under the appropriate conditions of pressure, temperature, and stress. There has therefore been considerable effort to develop apparatus to maintain simultaneous high pressure and temperature in macroscopic ($\sim mm^3$) samples for extended durations and, recently, to modify these presses to allow controlled deformation of samples under mantle conditions. The most successful static large-volume high-pressure devices are multi-anvil presses where steel wedges transfer load from the press ram(s) to an inner arrangement of anvils composed of tungsten carbide or other, super-hard, materials (Kawai and Endo, 1970; Kawai et al., 1973; Ito, 2007; for a review of multi-anvil history see Liebermann, 2011). These inner anvils compress a ceramic pressure cell which itself contains an electrically conductive furnace, insulation materials, means of measuring temperature, and sample capsule. There is a very large stress gradient between the high-pressure region within the ceramic cell, which can be at several tens of gigapascals, and the air gap between the inner anvils, and much effort has been invested in optimizing the materials of the cell and gaskets, and the geometry of the press. Currently, the highest pressures are attainable using an arrangement where an octahedral pressure cell is compressed by eight hard inner (second-stage) anvils, which pack together to form a cubic volume (Fig. 4.1a,b). This cubic "nest" is compressed by six (first-stage) steel wedges (the 6/8 multi-anvil arrangement) with either a threefold (111) or a fourfold (100) axis aligned parallel to the direction of compression of the primary press ram. The last two decades have seen the development of deformation multi-anvil presses where one opposing pair of anvils can be advanced (or retracted) independently of the remain-

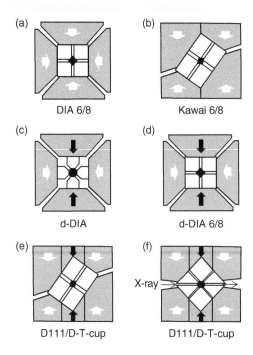

Figure 4.1 Schematic illustrations showing the geometries, hydrostatic compression directions (white arrows), and differential deformation (black arrows) of various high-pressure experimental apparatus. The (a) DIA 6/8 and (b) Kawai 6/8 (which may also be named the "111 6/8") geometries both utilize six primary anvils in combination with eight secondary truncated cubic anvils and an octahedral pressure medium to perform hydrostatic experiments, with the significant difference being the wedge-shaped primaries and rotated orientation of the secondary anvils in the Kawai 6/8 geometry. The (c) d-DIA geometry utilizes six secondary anvils and a cubic pressure medium to apply differential strain in the vertical direction, whereas (d) the d-DIA 6/8 geometry uses the same primary anvils to deform an octahedral pressure medium within eight secondary anvils, with differential strain loading the pyrophyllite gaskets. (e) and (f) show two alternative cross-sections of the D111/D-T-cup drawn (e) parallel and (f) perpendicular to the incident X-ray beam. The D111/D-T-cup geometry uses a Kawai-type geometry of primary wedges with differential strain applied onto two opposing anvil truncations along the 111 axis of the secondary anvil set.

ing anvils, applying an approximately axi-symmetric strain to the pressure medium and sample. The d-DIA geometry (Fig. 4.1c) applies strain along the fourfold axis of a cubic carbide anvil set in a single stage press (Wang et al., 2003) and has been combined with synchrotron radiation at several facilities (e.g., Nishiyama et al., 2008; Wang et al., 2009; Guignard and Crichton, 2015; Farmer et al., 2020). The single stage (six-anvil) DIA geometry is limited to about 10 GPa confining pressure, which can be increased in the 6/6 geometry using multistaging coupled with anvil guides to reach ~18 GPa (Nishiyama et al., 2008; Kawazoe et al., 2016). Alternatively, the d-DIA can be used with an internal 6/8 geometry cube-set to perform deformation experiments at conditions reaching ~25 GPa (Fig. 4.1d) (Tsujino et al., 2016). However, in the 6/8 implementation of d-DIA, the deformation axis is coaxial to the (100) axis of the octahedral pressure cell (and hence centered on the gaskets, rather than through an anvil). This results in complex sample strain possibilities due to gasket relaxations along the compression axis. An alternative deformation press geometry (Fig. 4.1e,f), the deformation-T-cup, was developed over the last decade (Hunt and Dobson, 2017; Hunt et al., 2014, 2019). This is based on the Kawai-type 6/8 geometry and the deformation axis is oriented along the (111) axis, centered on a pair of opposed anvil faces. This geometry is capable of controlled deformation at sample pressures in excess of 23 GPa and, here, we report the implementation of this deformation geometry at beamline NE7A of the Photon Factory at KEK and demonstrate its capabilities using preliminary data collected on $CaSiO_3$ perovskite samples.

4.2. THE D111 PRESS ON NE7A AT KEK

The deformation geometry employed in the D111 press is described elsewhere (Hunt et al., 2014); so, here we will concentrate on details specific to this implementation. A cutaway schematic of the press tooling installed on beamline NE7A is shown in Figure 4.2a. The tooling is designed to be a compact single unit which can be replaced in the press with alternative tooling geometries. Oil-driven secondary actuators which control the uniaxial deformation of the cell are contained within the guide blocks of the tooling so the primary pressure generation is driven using a standard single-piston four-post press. This means that, if the secondary actuators are not moved during an experiment the system behaves identically to a traditional Kawai-type hydrostatic multi-anvil press. The implementation on beamline NE7A uses the existing 700 ton MAX-III press and is interchangeable for different tooling designs. In the Kawai geometry, each second-stage anvil experiences loads of up to 175 tons from the main ram. The differential actuators have a capacity of 314 tons each but only act on a single anvil – this excess load capacity in the differential actuators ensures that frictional losses can be overcome, and deformation is possible right up to the 700 ton maximum load of the press. The two deformation anvils are advanced by pumping oil into the secondary actuators. This shortens the octahedral pressure medium along its vertical (111) symmetry axis, and the anvils in the equatorial plane dilate in response to the reduced component of the main-ram end load which they are supporting.

High-pressure cells and carbide tooling are constructed and loaded in the standard manner for 6/8 experiments, with the addition of hard components along the cell axis to transfer differential stress to the sample. This cubic nest of second-stage anvils and cell are loaded into the D111 tooling outside the press frame (Fig. 4.2b) with a (111) symmetry axis vertical. The first-stage Kawai-geometry

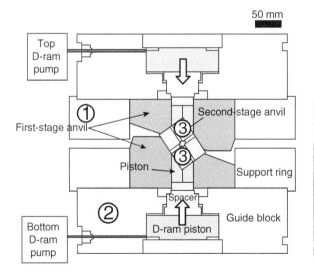

Figure 4.2 (a) A schematic of the D111 guide block at NE7A of KEK depicting the Kawai 6/8 geometry arrangement within a split confining ring and position of the differential deformation actuators. (b) A view of D111 module consisting of the split confining ring (1) and lower differential actuator (2) next to the MAX-III load frame on NE7A taken from an upstream position and showing the direction of the synchrotron X-ray beam in relation to the press.

wedges compress six of the eight second-stage anvils but the two anvils which align along the vertical (111) axis (labeled (3) in Fig. 4.2a) are compressed by a hexagonal arrangement of pistons which communicate with the deformation actuators. The original hexagonal deformation pistons of Hunt et al. (2014) are split such that the hexagonal column, which supports the deformation anvils, is composed of three identical pieces, each occupying 120° of rotation about the (111) symmetry axis (Hunt and Dobson, 2017). The front ends of these pieces are ground such that when installed they support the entire back faces of the second-stage carbide deformation anvils. This improves anvil performance over the original D-T-cup design (Hunt and Dobson, 2017). Similarly, to the deformation-DIA geometry the primary mode of deformation in the D111 geometry is axi-symmetric, resulting in a pure shear sample deformation mode. The axi-symmetric cell deformation can be transformed into simple shear deformation by placing the sample between pistons whose ends are cut at 45° (Karato and Rubie, 1997).

In the D111 implementation at beamline, NE7A X-rays traverse the second-stage anvil set perpendicular to the vertical mirror plane of the second-stage anvils and octahedral pressure cell (Fig. 4.1f) passing through two diagonally opposite gaskets. The pyrophyllite gaskets through which X-rays pass are replaced by a low-absorbing, typically B-dominated, gasket material. Careful alignment of the first-stage anvil wedges within the split confining ring is required to ensure that the X-ray cut-outs maintain the correct geometry with respect to the X-ray beam (Fig. 4.2b), which can also be adjusted by rotating the entire MAX-III load frame. The beamline delivers a photon flux of 10^8–10^9 mm^{-2} s^{-1} over a 3 mm high by 80 mm wide beam, measured at 20 m from the source. The photon energy range is 10–140 keV and the beamline can be operated in white beam or monochromatic (10–60 keV) modes. In order to perform successful rheological experiments, either X-ray transparent anvils (e.g., polycrystalline diamond [PCD], or cubic Bornon Nitride [cBN]) or carbide anvils with suitable slot/cone cut-outs are used in place of the downstream second-stage anvils to allow sufficient radiographic and diffraction data collection (Irifune et al., 1992; Dobson et al., 2012).

Stress-strain data on NE7A are collected by combined radiography and diffraction using a monochromatic incident X-ray beam and angle-dispersive diffraction. This is generated by passing the raw synchrotron beam through a two-crystal Si(111) monochromator to achieve an incident beam with monochromatic energy of 50–60 keV. The transmitted X-ray beam is typically imaged using a YAG(Ce) or GAGG(Ce) scintillator crystal combined with a CCD or CMOS camera positioned downstream of the sample on motorized linear translation stages. Highly absorbent foils (typically Au, Pt, or Re) are placed on both ends of samples as strain markers which show up in the radiographic images, allowing changes in the sample length (and strain rate) to be monitored throughout deformation experiments. Radiographic images are normally 0.5–5 seconds in duration, depending on the absorption properties of the sample assembly. Their resolution depends on the exact camera model used; however, it is typically around 1 μm which is sufficient to allow strains of 10^{-3} to be monitored in millimeter-sized samples when an appropriate cross-correlation algorithm is employed (e.g., Li et al., 2003). Diffraction patterns are collected using a Dexela 2923 area detector, whose position is calibrated using a certified standard material (CeO_2 in this study). Individual diffraction patterns are collected from an illuminated sample area defined using collimation slits placed upstream of the sample position and typically have durations of 1–4 minutes depending on sample characteristics. During experiments the lattice strain of each sample is determined from the azimuthal distortions of the Debye-Scherrer rings from the polycrystalline samples. Lattice strain is subsequently converted into a stress measurement by utilizing the elastic moduli of the sample material, which is required to be known prior to experiments. This experimental approach has been widely applied in d-DIA and 6/8 geometries at global synchrotron sources and achieves stress resolution, depending on the elastic properties of the sample, of ~100 MPa for most silicate minerals. Each data collection cycle consists of sequential X-ray radiographic images followed by X-ray diffraction pattern collections, between which both the diffraction slits and CCD/CMOS imaging camera are moved into/out of the X-ray beam path using motorized linear translation stages. The total time for each data collection cycle can vary between approximately 2 and 8 minutes depending on the required radiograph and diffraction times, and whether one or two samples are being monitored throughout deformation. A schematic of the experimental setup of the D111 on NE7A is provided in Figure 4.3.

The D111 setup on NE7A has so far been used to perform deformation experiments using various sized multi-anvil assemblies at primary ram loads up to ~500 tonf coupled with 2, 3, or 5 mm truncations. Samples investigated to date have included hcp iron, bridgmanite, ringwoodite, and $CaSiO_3$ perovskite (Ca-Pv), which have been studied at a range of pressure and temperature (*P-T*) conditions extending between 12 and 27 GPa, 650 and 1,700 K with strain rates of 1×10^{-6}–1.1×10^{-4} s^{-1} (Fig. 4.4; after Nishihara et al., 2020). In the subsequent sections of this chapter, we describe initial experiments performed to investigate the rheological properties of

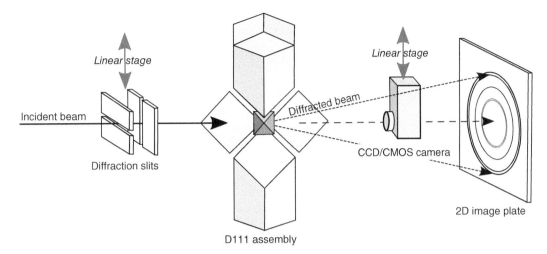

Figure 4.3 A schematic of beamline layout on NE7A at KEK when setup for monochromatic diffraction experiments using the D111 module, including the position of the X-ray collimation slits, radiographic CCD/CMOS camera and image plate detector, and the motorized linear stages used to switch between imaging and diffraction modes.

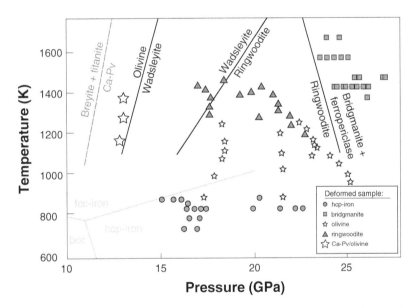

Figure 4.4 Pressure and temperature conditions of deformation experiments performed using the D111 at NE7A up until now, plotted alongside relevant phase relations. Samples studied are indicated by symbols as follows: circles = hcp-iron, large white stars = Ca-Pv/olivine (this study), small gray stars = olivine, dark gray triangles = ringwoodite, and squares = bridgmanite.

Ca-Pv. These provide a demonstration of the current capabilities and examples of data that can achieved using the existing D111 system on NE7A.

4.3. THE RHEOLOGICAL BEHAVIOR OF CA-PV

Ca-Pv is one of the major mineralogical constituents of Earth's deep mantle and is thermodynamically stable at pressures from ~10 GPa to the core-mantle boundary. After entering the phase assemblages of subducting lithologies at ~19–21 GPa (Holland et al., 2013), it constitutes ~5 to ~30 wt% of depleted ultramafic to mafic lower mantle assemblages, respectively (e.g., Kesson et al., 1998; Perrillat et al., 2006). At these abundances, especially in subducting mafic assemblages, Ca-Pv is sufficiently abundant that it might well form an interconnected network of grains (Handy, 1994). In this situation, depending on the rheological properties of Ca-Pv, it may dominate the strength of subducting mafic lithologies and potentially be important in controlling the geodynamics of subduction and slab delamination throughout the transition zone and lower mantle.

Studies of Ca-Pv are complicated by the inability to recover samples of it from synthesis to ambient conditions, as it spontaneously undergoes amorphization during decompression at room temperature (e.g., Liu and Ringwood, 1975; Wang and Weidner, 1994). This feature of Ca-Pv means that many of its physical properties, including its elastic and rheological properties, are poorly understood and can only reasonably be assessed using *in-situ* experiments where it can be synthesized and investigated in a single experimental run. Synchrotron-based large volume press experiments (e.g., Wang and Weidner, 1994; Greaux et al., 2019; Thomson et al., 2019) provide a suitable approach to achieve this, and here we present our initial experimental results to constrain the viscous strength of Ca-Pv at *P-T* conditions inside its stability field using the KEK D111 press.

4.3.1. Experimental Details

As the elastic moduli of Ca-Pv, which are required for experimental stress measurements, are poorly known, two experimental setups were used in this study. While a solitary sample of Ca-Pv was studied in one experiment, two further experiments utilized a deformation column containing stacked samples of olivine and Ca-Pv. Starting materials for all deformation experiments consisted of sintered cylinders of synthetic olivine and wollastonite. Cylinders of fine-grained powder, ground to ~5 μm grainsize using an agate mortar, were fabricated by pressing using a pellet die prior to sintering in a 1-atmosphere furnace. In two experiments, 17.5 wt% of the wollastonite sample was replaced with a 1:1 mixture of $Ca(OH)_2$ + SiO_2 glass, in order to provide a small quantity of water (~2 wt% H_2O in the bulk composition) at experimental conditions to promote the kinetics of Ca-Pv formation (e.g., Gasparik et al., 1994). In all experiments, the samples were deformed between two fully dense Al_2O_3 pistons which had 10 μm thick Pt marker foils placed at either end of each sample to allow sample strain to be monitored. Deformation columns were housed within multi-anvil assemblies consisting of a Co-doped MgO octahedron of 7 mm edge length, a TiB_2:BN ceramic furnace, and MgO inner parts. The pyrophyllite gaskets and ceramic components within the experimental assembly along the X-ray beam path were replaced with rods of low-density boron epoxy to maximize transmission for radiographic imaging and diffraction. Type-D thermocouples with junctions placed adjacent to the samples were used to monitor experimental temperatures.

Experimental assemblies were compressed using secondary anvils with 3 mm truncations for pressure generation. 6 × 26 mm TF05 carbide anvils were combined with two X-ray transparent composite anvils placed downstream of the sample to allow collection of angle dispersive diffraction data. These composite anvils consisted of 14 mm PCD cubes backed by tungsten carbide spacers which bring the composite cube size up to the 26 mm size of the remaining second-stage anvils. The backing spacers are chamfered to provide a solid cone angle of 20° from the sample position permitting 360° azimuthal diffraction to a 2θ angle of ~10° (Fig. 4.5). Pressure was applied by gradually increasing the primary ram's oil pressure to 190 tonf over 2–3 hours, corresponding to ~15–16 GPa as determined using the shift of X-ray diffraction peaks from the Pt marker foils (Matsui et al., 2009). The sample was then gradually heated by applying power to the TiB_2:BN furnace at a rate of between 20 and 100 K/min until the target temperature (1150–1373 K) was achieved. During this heating cycle, the wollastonite sample was observed to undergo transformation to Ca-Pv at ~1,000 K, indicated by an abrupt shortening of the sample (due to large volume change from wollastonite to perovskite structure) and the identification of diffraction rings corresponding to Ca-Pv in subsequent diffraction patterns. After heating all experiments were observed to be at a pressure of 12–13 GPa, which was maintained throughout deformation.

After initial heating, the target *P-T* conditions were maintained for a minimum of 30 minutes to anneal the samples, and until no diffraction evidence of remnant low pressure $CaSiO_3$ phases was present. Prior to deformation the differential actuators were preloaded to an oil pressure of 3 MPa. Sample deformation was commenced by driving both differential actuators, which are controlled by linear displacement transducers, at a fixed rate of 0.5–10 μm min^{-1} such that the sample was shortened along an axis parallel to the furnace and the stress-strain data collection loop was commenced. We have observed that after commencing differential ram motion that their preload increases to approximately 25% of the main ram's load before motion and sample deformation commences (Nishihara et al., 2020). For these experiments, radiographic images of 0.5–2 seconds duration with resolutions of ~1 μm were coupled with 180 seconds sample diffraction patterns using 200 × 200 μm illumination slits. Diffraction counting times were reduced to 60 seconds when differential actuator velocities of 10 μm min^{-1} were being used to ensure sufficient data were collected at high strain rates. Data collections at constant actuator velocities were repeated for approximately 1 hour to ensure both sample strain rate and stress had reached steady state equilibrium. Multiple strain rates, controlled by three or four differential actuator velocities, were studied in each experiment. All experiments were performed at a single sample temperature of

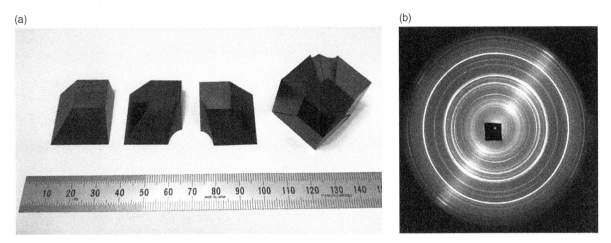

Figure 4.5 (a) WC spacers for monochromatic diffraction, consisting of three shaped trapezoidal WC spacers with cone cut-out for diffraction, adjacent to a fully assembled composite anvil consisting of three spacers combined with a 14 mm PCD cube. (b) An example X-ray diffraction pattern collected using two downstream composite anvils collected from a Ca-Pv sample at ~13 GPa and 1273 K. The direct beam can be observed as a small spot in the center of the diffraction rings within the Pb beamstop that was attached to the image plate detector, and Debye-Scherrer rings extend to 10° 2θ (equivalent to a d-spacing of ~1.17 Å using monochromatic X-rays of 0.2066 Å). The more intense stripe inclined from the lower left to upper right is due to the increased X-ray transmission through the anvil gap.

1,150–1,373 K. Following completion of data collection at the final experimental strain rate, the sample temperature was quenched by turning off the furnace power and the differential actuators were stopped. The sample was then gradually decompressed at room temperature while the actuators were also decompressed over the same time duration in attempt to minimize breakage of the PCD anvils.

4.3.2. Stress-Strain Data Processing

X-ray radiographs were sequentially processed using correlative image processing to determine the relative length change throughout each experiment (e.g., Hunt et al., 2019). Changes in sample length were subsequently converted into the strain ($\varepsilon = \frac{l-l_0}{l_0}$) experienced by each sample as a function of experimental time. Examples of strain measurement data are provided in Figure 4.6. Strain rates ($\dot{\varepsilon}$ in s^{-1}) were calculated using these strain measurements after steady state was achieved at each subsequent differential actuator velocity. Observed strain rates range between 3×10^{-6} and 2×10^{-4} s^{-1}.

X-ray diffraction data were initially processed using a combination of IPAnalyzer and PDIndexer (Seto, 2010), as available at NE7A, to allow evaluation of experiments in real time. Latterly, the diffraction datasets were processed using the newly developed *Continuous Peak Fit* package (Chen et al., 2021; Hunt and Fenech, *in prep*). This package implements a new approach to fitting 2-D diffraction (or any other continuous 2-D data) by assuming that each diffraction peak, which is normally described in 1D data as a peak function (e.g., a pseudo-Voigt or Gaussian), varies smoothly as a function of azimuth. This description is achieved by fitting each peak parameter (d-spacing, amplitude, half-width, etc.) as a Fourier series in azimuthal space. In this case, the d-spacing of a single Debye-Scherrer ring was described as follows:

$$d^{hkl} = a_0^{hkl} + a_1^{hkl} \sin\psi + b_1^{hkl} \cos\psi + a_2^{hkl} \sin 2\psi + b_2^{hkl} \cos 2\psi \quad (4.1)$$

where a_0^{hkl} is the mean d-spacing of the ring, the two first-order terms (a_1^{hkl} and b_1^{hkl}) represent the displacement of the center of the diffraction ring relative to the calibration center and the second-order terms (a_2^{hkl} and b_2^{hkl}) represent the ellipticity of the diffraction ring and contain the information corresponding to the sample's differential strain. This model is fitted sequentially to each of the Debye-Scherrer rings of interest that are not significantly overlapped by peaks from other cell components, in the 2-D diffraction pattern. In this case, the 110, 200, 211, and 220 peaks of Ca-Pv and the 131, 222, 240, 130, 021, 122, and 140 peaks of olivine samples were fitted using continuous pseudo-Voigt functions, although not all fitted peaks were used for stress evaluations for every experiment. *Continuous Peak Fit* relies on knowledge of the sample-detector geometry as calibrated using an ambient diffraction collection from a CeO$_2$ standard processed using *Dioptas* (Prescher and Prakapenka, 2015). The fitting procedure directly provides the lattice strain

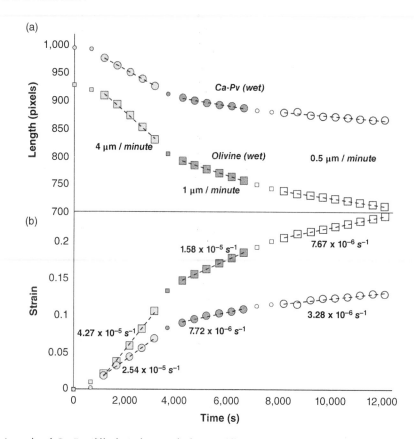

Figure 4.6 (a) Length of Ca-Pv (filled circles) and olivine (filled squares) at 13 GPa and 1273 K under hydrous conditions as a function of time throughout deformation, as derived from radiographic images, and colored by the differential ram velocity (as annotated). (b) The strain of Ca-Pv and olivine samples as a function of time throughout the same experiment. Small symbols were not used to derive strain rates (annotated and indicated by dashed lines) as these had not yet reached steady state after change in ram velocities.

indicated by each peak accompanied by an uncertainty based on fitting statistics from input of the unprocessed diffraction images in one computational process. A typical example of a fitted Deybe-Scherrer ring for the Ca-Pv 200 at ~13 GPa, 1373 K, is given in Figure 4.7.

The output models from *Continuous Peak Fit* provide the deviatoric lattice strain and the minimum/maximum d-spacings for each reflection as a function of azimuth, which are evaluated from the second-order a_2^{hkl} and b_2^{hkl} Fourier coefficients and trigonometric relations. These parameters can then be used to calculate the differential stress following the approach of Singh et al. (1998), where the relationship between axial stress, lattice strain, and azimuthal variations in the peak position is as follows:

$$d_{hkl}(\psi) = d_{hkl}^0 \left[1 + (1 - 3\cos^2\psi) \frac{\sigma}{6\langle G_{hkl} \rangle} \right] \quad (4.2)$$

In equation (4.2), d_{hkl} is the *d-spacing* measured as a function of azimuthal angle ψ, d_{hkl}^0 is the *d-spacing* under hydrostatic pressure, G_{hkl} is the appropriate shear modulus for a given *hkl* orientation, and σ is the axial differential stress for that *hkl*.

An approximate evaluation of the differential stress in each sample has been calculated using the maximum and minimum *d-spacing* from fitted models of each diffraction peak combined with either (1) an isotropic shear modulus for olivine or (2) an appropriately adjusted elastic tensor for Ca-Pv (following the relations in Singh et al., 1998). This approach neglects the effects of crystallographic orientations and elastic anisotropy in olivine samples, but sufficient diffraction peaks are observed that this should not add significant uncertainty to stress estimates. As there are fewer diffraction peaks from Ca-Pv available, and because Ca-Pv has large elastic anisotropy (Kawai and Tsuchiya, 2015), orientation adjusted elastic moduli were used in equation (4.2). Throughout stress estimations performed in this chapter, it was assumed that the isotropic shear modulus of olivine at run conditions (~13 GPa and 1150–1373 K) was 90 ± 10 GPa, based on recent experimental measurements (Mao et al., 2015). An elastic tensor for Ca-Pv providing values of c_{11}, c_{12}, and c_{44} (of 347.9, 158.7, and 179.9 GPa, respectively) from the molecular dynamics *ab initio* calculations of Kawai and Tsuchiya (2015) was used to derive differential

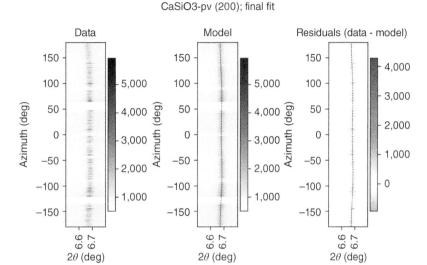

Figure 4.7 Example of the fitted output from *Continuous Peak Fit* for wet Ca-Pv 200 at 13 GPa and 1373 K. The left-hand panel plots the raw data as a function of diffraction azimuth and 2θ (both in degrees). Regions of the data that are white are masked due to large intensity changes in the anvil gap, which hampers successful fitting of the diffraction rings. The middle panel plots the calculated model of the fitted peak, with the black dashed line indicating the model peak position. The right-hand panel plots the residual misfit between the raw data and the fitted peak model. All panels are colored by intensity, indicated by the adjacent color bars (color figure provided as supplementary figure).

stresses in perovskite samples. We note that the isotropic adiabatic shear modulus of Ca-Pv reported by Kawai and Tsuchiya (2015) of ~155 GPa is significantly larger than those determined in recent ultrasonic interferometry experiments by Greaux et al. (2019) and Thomson et al. (2019), which are ~130 and ~105 GPa, respectively. Thus, the true adiabatic shear modulus of Ca-Pv remains unclear, and the results from Kawai and Tsuchiya (2015) are preferred here because this is the only study to report full c_{ij}'s required to calculate orientation adjusted values of the shear modulus. However, due to these discrepancies, we caution that the differential stresses derived in this study are based on the largest recent estimate of Ca-Pv's shear modulus and may overestimate the true values of differential stress in these experiments by more than 25%. Uncertainties caused by errors in G_{hkl} of Ca-Pv were assumed to be ±25%, such that estimated stresses incorporate this source of uncertainty; however, in several cases, the fitting uncertainties dominate. After calculating the apparent stress from each fitted diffraction ring, an average differential stress for the olivine and/or Ca-Pv samples was calculated as a weighted mean of all observed stresses. The reported differential stress at each strain rate was calculated as the mean of these weighted averages from each data collection cycle, with uncertainty estimated as two standard deviations. Examples of differential stress measurements are provided in Figure 4.8.

4.4. RESULTS

A summary of conditions, starting materials, measured strain rates, and estimated differential stresses from all experiments is provided in Table 4.1. In all experiments, the wollastonite sample material was observed to very rapidly transform into Ca-Pv structure. In the dry experiment, diffraction peaks corresponding to Ca-Pv possessed sharp, complete, and non-spotty diffraction rings, suggesting the Ca-Pv sample had an average grainsize that was large enough to reduce peak broadening, but still small enough that a very large number of grain orientations are being sampled during diffraction. Although this cannot be verified upon recovery, it is assumed samples possess a grainsize between 0.5 and 5 μm, somewhat smaller than those of olivine starting material which were 10–15 μm. In wet experiments, diffraction rings from the Ca-Pv were pseudo-continuous but have a slightly spotty texture implying a larger sample grainsize. Again this cannot be rigorously quantified, but we suggest a similar grainsize to the olivine samples (which have similarly spotty diffraction rings) of ~10 μm.

In the single experiment performed on nominally dry starting materials, strain was observed to preferentially partition into the Ca-Pv sample relative to olivine, with strain rates in the dry Ca-Pv sample 1.9–5.4 times higher than those of the dry olivine sample. Additionally, despite being simultaneously deformed within the same column, the differential stress appears to be significantly lower in

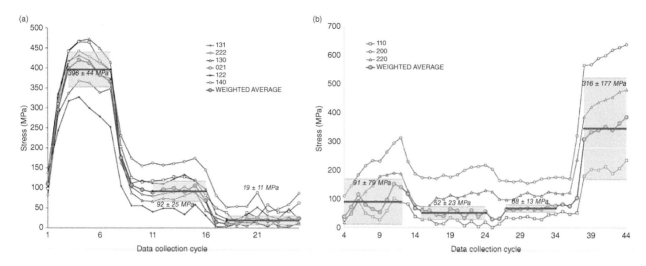

Figure 4.8 Examples of differential stress data collected throughout (a) experiment B for hydrous olivine and (b) experiment C for hydrous Ca-Pv.

Table 4.1 Summary of experimental run conditions and results from deformation

Expt.	P (GPa)	T (K)	Starting materials	Sample hkls used for stress	Deformation velocity (μm min^{-1})	$\dot{\varepsilon}$ (s^{-1})	±	σ (MPa)	±
A (dry)	13	1150	CaSiO$_3$	Ca-Pv 200, 110	4	1.91×10^{-5}	4.48×10^{-5}	116	24
					1	2.51×10^{-5}	1.83×10^{-5}	109	27
					10	1.83×10^{-4}	8.12×10^{-5}	260	99
			(Mg$_{1.8}$Fe$_{0.2}$)SiO$_4$	olivine 021, 122, 130, 140, 222	4	1.00×10^{-5}	1.18×10^{-5}	1,480	400
					1	4.65×10^{-6}	5.8×10^{-6}	1,300	70
					10	4.87×10^{-5}	1.93×10^{-5}	1,760	240
B (wet)	13	1273	CaSiO$_3$ + 2wt% H$_2$O	Ca-Pv 200, 110, 211	4	2.54×10^{-5}	3.1×10^{-6}	430	129
					1	7.72×10^{-6}	2.4×10^{-6}	164	23
					0.5	3.28×10^{-6}	4.3×10^{-6}	132	25
			Mg$_2$SiO$_4$	olivine 131, 222, 130, 021, 122, 140	4	4.27×10^{-5}	1.2×10^{-6}	396	44
					1	1.58×10^{-5}	1.1×10^{-6}	92	25
					0.5	7.67×10^{-6}	8.6×10^{-7}	19	11
C (wet)	13	1373	CaSiO$_3$ + 2wt% H$_2$O	Ca-Pv 200, 110, 220	4	6.42×10^{-5}	9.2×10^{-6}	91	79
					1	2.04×10^{-5}	1.7×10^{-6}	52	23
					0.5	9.98×10^{-6}	1.3×10^{-6}	68	13
					10	1.73×10^{-4}	1.02×10^{-5}	316	177

the Ca-Pv than olivine in this experiment. As observed in previous studies, dry olivine exhibits a creep strength of 1–2 GPa at strain rates of approximately $1-2 \times 10^{-5}$ s^{-1} (e.g., Kawazoe et al., 2009; Li et al., 2006). In contrast, the simultaneously deforming Ca-Pv sample in our experiments has an apparent creep strength that appears to be an order of magnitude smaller at similar strain rates. This is unusual, as it is normally expected that all samples in a single deforming column should experience the same differential stress, which is clearly not the case in this experiment. In the case where two samples being simultaneously deformed have very different creep strengths, it is plausible that the flow in the weaker phase is sufficiently rapid that deformation of the surrounding pressure medium becomes rate limiting for this sample. That would then cause the differential stress in the weaker sample to be smaller than that in the stronger sample as the pressure medium would be confining the weak sample more strongly. Assuming this is indeed the situation, this experiment suggests Ca-Pv is significantly weaker than dry olivine, as indicated by observations of higher strain rate at lower differential stress within the Ca-Pv sample. We note that the differential stress for Ca-Pv in this experiment is derived using the weighted mean stress from two Debye-Scherrer rings, as opposed to five for olivine samples. In combination with the large uncertainty for Ca-Pv's shear modulus, the differential stress estimates for Ca-Pv are relatively larger than those for olivine. Even accounting for this, however, the Ca-Pv appears very significantly weaker than olivine under dry conditions.

Two further experiments were performed using Ca-Pv starting materials containing 2 wt% H_2O provided by a component of the starting mix being replaced by a mixture of $Ca(OH)_2$ and SiO_2 glass. This means that these latter deformation experiments are effectively performed in a hydrous environment. Samples of Ca-Pv and olivine simultaneously deformed under hydrous conditions at ~13 GPa and 1273 K have more comparable strengths. Strain rates in the hydrous olivine sample were observed to be 1.7–2.3 times higher than those for Ca-Pv, and differential stress appears to be similar in both samples at strain rates of $2–4 \times 10^{-5}$ s^{-1}, with olivine becoming relatively weaker at lower strain rates. The final experiment was performed on a solitary sample of Ca-Pv at a temperature of 1373 K and under hydrous conditions. This solitary Ca-Pv experiment was performed as the strain rate of Ca-Pv samples in previous experiments had been high, and including only Ca-Pv permitted a longer initial sample length. At equivalent differential stress, increasing temperature weakens the Ca-Pv sample, which was observed to strain faster than at 1273 K (Fig. 4.9).

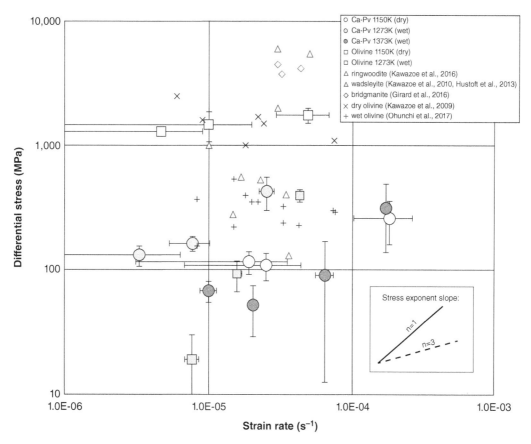

Figure 4.9 Summary of the creep strength of Ca-Pv (large circles) and olivine samples (large squares), plotted as strain rate versus differential stress, as observed in experiments performed using the D111 at KEK in this study. Uncertainties in observed strain rates and differential stresses, as reported in Table 4.1, are plotted and in some cases are smaller than the symbol size. These are plotted alongside creep strength data for other silicate minerals from similar experiments using d-DIA or Rotational Drickamer apparatuses for comparison; white triangles = creep strength data for ringwoodite (Kawazoe et al., 2016), grey triangles = wadsleyite (Kawazoe et al., 2010; Hustoft et al., 2013), grey diamonds = bridgmanite (Girard et al., 2016), plus symbols = wet olivine (Ohuchi et al., 2017), and cross symbols = dry olivine (Kawazoe et al., 2009).

4.5. DISCUSSION

The presented experiments and data analysis approach using the *Continuous Peak Fit* package, as described above, demonstrate the capabilities of the D111 at KEK NE7A for deformation experiments at mantle conditions. Experiments are reported with strain rates as low as ~3–7 × 10^{-6} s^{-1} with differential stresses of ~20 ± 10 MPa in hydrous olivine samples. Equally, anhydrous olivine samples deformed at similar strain rates were observed to support differential stress of 1,300 ± 70 MPa at a strain rate of ~5 × 10^{-6} s^{-1}. These observations for both hydrous and anhydrous samples overlap published rheological properties for wet and dry olivine, respectively (e.g., Kawazoe et al., 2009; Ohuchi et al., 2017), verifying the capabilities of the D111 to perform rheological experiments similar to other devices available (e.g., d-DIA and RDA) at synchrotron beamlines worldwide. The apparent slope of data for strain rate versus differential stress suggests that hydrous samples of olivine may be deforming in the diffusion creep regime (n < 1), while anhydrous samples are deforming in dislocation creep (n > 3) in the current study (Fig. 4.9, equation (4.3)).

Observations on Ca-Pv samples, although preliminary, allow some initial observations of its rheological properties at mantle *P-T* conditions. Across all three experiments, in both dry and hydrous sample conditions, the maximum average differential stress observed in Ca-Pv samples was 430 ± 129 MPa. While these stress estimates are subject to large errors in part due to the uncertainty in the elastic tensor of Ca-Pv, it is unambiguous that they are low compared with other high-pressure silicate phases at equivalent strain rates. In the anhydrous experiment, it appeared, comparing both differential stresses and observed sample strain rates, that Ca-Pv may be more than 1 order of magnitude weaker than coexisting olivine. Taken absolutely at face value, these observations point to Ca-Pv being substantially weaker than dry olivine (Fig. 4.9). However, it is noted that Ca-Pv samples are synthesized *in-situ* and their grainsize is not known accurately. While the diffraction characteristics described above imply Ca-Pv samples probably have a grainsize of 0.5–5 μm, which is smaller than that of the olivine starting material, and is potentially sufficiently small to make it anomalously weak. Since the spontaneous amorphization of Ca-Pv during decompression prevents a rigorous grainsize measurement, this could not be verified, and it was in order to try and study Ca-Pv samples with larger average grainsizes that subsequent experiments were performed under hydrous conditions. While the slight spottiness of Ca-Pv diffraction rings in hydrous conditions suggested the success of this approach in creating coarser perovskite samples, it also induced a significant hydrolytic weakening in coexisting olivine (e.g., Mei and Kohlstedt, 2002a, 2002b). However, the presence of hydrogen did not appear to weaken Ca-Pv in a similar manner as hydrous Ca-Pv deforming at 1273 K (123 K above the dry experiment) appeared to be stronger than that under dry conditions. We assume this increase in strength is the expression of increased sample grainsize. The strength of Ca-Pv under hydrous conditions did reduce with temperature increasing from 1273 to 1373 K, as is expected. A brief inspection of the slope of data for all Ca-Pv sample in Figure 4.9 suggests, in all three experiments Ca-Pv was deforming in a diffusion creep regime at run conditions.

In order to consider this further, the strain rate versus stress data for Ca-Pv were fitted to a thermally activated creep flow law of the form:

$$\dot{\varepsilon} = A \frac{\sigma^n}{d^m} exp\left(-\frac{Q}{RT}\right) \qquad (4.3)$$

where A is the constant of proportionality, n is the dimensionless stress dependence, d is the grainsize and m is the dimensionless grainsize dependence, Q is the activation energy, R is the universal gas constant, and T is the temperature. By utilizing a flow law of this form, we are inherently assuming that the presence (and concentration) of water has no effect on the creep strength of Ca-Pv. While this is a significant, and unjustified, assumption, our qualitative observations all suggest Ca-Pv is relatively weak compared with other silicates irrespective of the presence or lack of water in our experiments; thus, it is arguably not the most significant component in controlling Ca-Pv's creep strength. Additionally, by making this assumption estimates for other flow law parameters can be obtained. Finally, in order to fit the data for Ca-Pv, we assumed that wet experiments had a grainsize of 10 μm and that the grainsize dependence of Ca-Pv's creep is fixed at $m = 2$, that is, we assume Nabarro-Herring creep. Making these assumptions and performing a weighted regression, we obtain values of Q and n of 364 kJ mol^{-1} and 1.28, respectively, with an inferred grainsize in the dry experiment of 0.6 μm. If we assume Coble creep ($m = 3$), the dry grainsize changes to 0.7 μm. This predicted activation energy is entirely dependent on data from the two wet experiments and is not strongly influenced by our assumptions about the effect of water on creep strength or the exact grainsize in the experiments. Additionally, an activation energy of 364 kJ mol^{-1} is very similar to the activation energy for Si diffusion in bridgmanite (~330–350 kJ mol^{-1}; Dobson et al., 2008; Yamazaki et al., 2000) and as such appears both reasonable and fairly robust. This implies the fitted value for n, of 1.28, and the relative grainsizes of wet/dry experiments are also likely to be reliable since the exact grainsizes are effectively an adjustable parameter, such that we conclude Ca-Pv in these experiments is deforming in a diffusion creep regime.

Irrespective of Ca-Pv's deformation regime in experiments here, comparison with stress-strain data from similar experiments allows a direct comparison of its strength with additional mantle silicates. Ringwoodite samples deformed using the d-DIA at ~17 GPa and 1,300–1,700 K sustain similar differential stress (130–560 MPa) at strain rates of 10^{-5} s^{-1} (Fig. 4.9; Kawazoe et al., 2016). This suggests that Ca-Pv and ringwoodite possess broadly similar rheological properties. Comparisons with data for wadsleyite (Hustoft et al., 2013; Kawazoe et al., 2010) and/or bridgmanite (Girard et al., 2016) samples suggest Ca-Pv is more than an order of magnitude weaker than either of these materials at the strain rates probed in this study (Fig. 4.9). Such comparisons only provide a qualitative understanding of relative strengths of mantle phases, but suggest that Ca-Pv may be capable of inducing rheological contrasts at deep Earth conditions. During subduction, the similar properties of Ca-Pv and ringwoodite might suggest that basaltic and harzburgitic slab assemblages retain similar strengths throughout subduction within the upper mantle and transition zone. However, after entry to the lower mantle, the rheology of these lithologies may significantly diverge. The rheological behavior of subducting basalts, which contain up to 30 vol.% Ca-Pv (Perrillat et al., 2006), may be controlled by interconnected grains of this phase, while harzburgite assemblages contain >90% bridgmanite (Ishii et al., 2019). In this situation, the strength contrast of Ca-Pv and bridgmanite may cause downwelling slabs to delaminate and could promote slab stagnation in the upper/lower mantle boundary region (e.g., Fukao and Obayashi, 2013). Additionally, in more fertile ultramafic assemblage throughout the lower mantle, both Ca-Pv and MgO may act as weak phases coexisting with bridgmanite and together may constitute >20 vol.% of phase assemblages (Perrillat et al., 2006). If these grains of Ca-Pv and MgO are sufficient to generate an interconnected network within a bridgmanite matrix, this will allow strain partitioning into these phases and cause an overall weakening of peridotitic assemblages in the lower mantle. Such rheological contrasts between depleted and fertile components of an ultramafic lower mantle assemblage may allow generation of 3-dimensionally distributed domains of rigid materials (e.g., bridgmanite-enriched ancient mantle structures [BEAMS], Ballmer et al., 2017). Additional studies of the rheological properties of Ca-Pv in relation to those of lower mantle materials are required to further investigate these possibilities.

4.6. CONCLUSIONS

A D111 deformation multi-anvil press has been successfully installed at beamline NE7A of KEK and utilized to performed deformation experiments at controlled strain rates of 10^{-6}–10^{-4} s^{-1} at mantle pressure and temperature conditions. In this study, we have demonstrated the capabilities of the D111 geometry using controlled strain rate experiments of Ca-Pv and olivine samples under both wet and dry conditions with sample strain rates between 3×10^{-6} s^{-1} and 2×10^{-4} s^{-1} occurring in response to differential stresses of ~20–1,760 MPa. Preliminary data suggest that Ca-Pv is up to 1 order of magnitude weaker compared with other deep mantle silicate minerals including olivine, wadsleyite, and bridgmanite and may strongly influence the geodynamics of the Earth's deep mantle.

The stability of the Kawai 6/8 geometry, which forms the basis of the D111 deformation press, means that P-T conditions at controlled strain rates may be extended significantly beyond those achievable in other existing deformation geometries, such as the d-DIA and Rotational Drickamer. To date, experiments have been performed at conditions up to ~27 GPa and 1,700 K, to investigate the creep strength of Earth-forming materials including hcp-structured iron, ringwoodite, and bridgmanite (Nishihara et al., 2020). Further developments, including use of harder anvil materials (e.g., Yamazaki et al., 2019) and new X-ray transparent ceramic components (e.g., Xie et al., 2020a, 2020b; Xu et al., 2020), should allow additional expansion of the accessible P-T conditions using this deformation geometry.

ACKNOWLEDGMENTS

We thank two anonymous reviews for their comments and Taku Tsuchiya for his editorial handling of this chapter. We acknowledge the support of UKRI grant NE/P017657/1 and grants 15H05827, 15H03749, 19H00723, and 18H05232 from the Japan Society for Promotion of Science.

REFERENCES

Ballmer, M. D., Houser, C., Hernlund, J. W., Wentzcovitch, R. M., & Hirose, K. (2017). Persistence of strong silica-enriched domains in the Earth's lower mantle. *Nature Geoscience*, 10(3), 236–240. http://doi.org/10.1038/ngeo2898

Chen, Y., Clark, S. J., Collins, D. M., Marussi, S., Hunt, S. A., Fenech, D. M., et al. (2021). Correlative synchrotron X-ray imaging and diffraction of directed energy deposition additive manufacturing. *Acta Materialia*, 209, 116777.

Dobson, D. P., Dohmen, R., & Wiedenbeck, M. (2008). Self-diffusion of oxygen and silicon in MgSiO$_3$ perovskite. *Earth and Planetary Science Letters*, 270(1–2), 125–129.

Dobson, D. P., Hunt, S. A., & Müller, H. J. (2012). Slotted Carbide Anvils: Improved X-ray Access for Synchrotron Multi-Anvil Experiments. *High Pressure Research*, *32*, 532–536.

Farmer, N., Rushmer, T., Wykes, J., & Mallmann, G. (2020). The Macquarie Deformation-DIA facility at the Australian Synchrotron: A tool for high-pressure, high-temperature experiments with synchrotron radiation. *Review of Scientific Instruments*, *91*(11), 114501.

Fukao, Y., & Obayashi, M. (2013). Subducted slabs stagnant above, penetrating through, and trapped below the 660 km discontinuity. *Journal of Geophysical Research B: Solid Earth*, *118*(11), 5920–5938. http://doi.org/10.1002/2013JB010466

Gasparik, T., Wolf, K., & Smith, C. M. (1994). Experimental determination of phase relations in the $CaSiO_3$ system from 8 to 15 GPa. *American Mineralogist 79*(11–12), 1219–1222.

Girard, J., Amulele, G., Farla, R., Mohiuddin, A., & Karato, S. -I. (2016). Shear deformation of bridgmanite and magnesiowüstite aggregates at lower mantle conditions. *Science*, *351*(6269), 144–147. http://doi.org/10.1126/science.aad3113

Gréaux, S., Irifune, T., Higo, Y., Tange, Y., Arimoto, T., Liu, Z., et al. (2019). Sound velocity of $CaSiO_3$ perovskite suggests the presence of basaltic crust in the Earth's lower mantle. *Nature*, *565*(7738), 218–221. http://doi.org/10.1038/s41586-018-0816-5

Guignard, J., & Crichton, W. A. (2015). The large volume press facility at ID06 beamline of the European synchrotron radiation facility as a High Pressure-High Temperature deformation apparatus. *Review of Scientific Instruments*, *86*(8), 085112.

Handy, M. R. (1994). Flow laws for rocks containing 2 nonlinear viscous phases. *Journal of Structural Geology*, *16*(12), 1727. http://doi.org/10.1029/2002JB001833/full

van der Hilst, R. (1995). Complex morphology of subducted lithosphere in the mantle beneath the Tonga trench. *Nature*, *374*, 154–157.

Holland, T. J., Hudson, N. F., Powell, R., & Harte, B. (2013). New thermodynamic models and calculated phase equilibria in NCFMAS for basic and ultrabasic compositions through the transition zone into the uppermost lower mantle. *Journal of Petrology*, *54*(9), 1901–1920.

Hunt, S. A., Dobson, D. P., Li, L., McCormack, R. J., Vaughan, M. T., Weidner, D. J., et al. (2014). Deformation T-Cup: A new apparatus for high temperature, controlled strain-rate deformation experiments at pressures in excess of 18 GPa. *Review of Scientific Instruments*, *85*, 085103.

Hunt, S. A., & Dobson, D. P. (2017). Modified anvil design for improved reliability in DT-cup experiments. *Review of Scientific Instruments*, *88*, 126106.

Hunt, S. A., Whitaker, M. L., Bailey, E., Mariani, E., Stan, C. V., & Dobson, D. P. (2019). An experimental investigation of the relative strength of the silica polymorphs quartz, coesite, and stishovite. *Geochemistry, Geophysics and Geosystems*, *20*(4), 1975–1989.

Hustoft, J., Amulele, G., Ando, J. I., Otsuka, K., Du, Z., Jing, Z., et al. (2013). Plastic deformation experiments to high strain on mantle transition zone minerals wadsleyite and ringwoodite in the rotational Drickamer apparatus. *Earth and Planetary Science Letters*, *361*, 7–15.

Irifune, T., Utsumi, W., & Yagi. T. (1992). Use of a new diamond composite for multianvil high-pressure apparatus. *Proceedings of the Japan Academy*, *68*, 161–166 DOI: 10.2183/pjab.68.161

Ishii, T., Kojitani, H., & Akaogi, M. (2019). Phase relations of Harzburgite and MORB up to the uppermost lower mantle conditions: Precise comparison with Pyrolite by multisample cell high-pressure experiments with implication to dynamics of subducted slabs. *Journal of Geophysical Research: Solid Earth*, *124*(4), 3491–3507.

Ito, E. (2007). Theory and practice – Multianvil cells and high-pressure experimental methods. *Treatise on Geophysics*, *2*, 197–230.

Kawai, N., & Endo, S. (1970). The generation of ultrahigh hydrostatic pressure by a split sphere apparatus. *Reviews of Scientific Instruments*, *41*, 425–428.

Kawai, N., Togaya, M., & Onodera, A. (1973). New device for pressure-vessels. *Proceedings of the Japan Academy*, *49*(8), 623–626.

Kawai, K., & Tsuchiya, T. (2015). Small shear modulus of cubic $CaSiO_3$ perovskite. *Geophysical Research Letters*, *42*(8), 2718–2726. http://doi.org/10.1002/2015GL063446

Kawazoe, T., Karato, S.-I., Otsuka, K., Jing, Z., & Mookherjee, M. (2009). Shear deformation of dry polycrystalline olivine under deep upper mantle conditions using a rotational Drickamer apparatus (RDA). *Physics of the Earth and Planetary Interiors*, *174*(1–4), 128–137. http://doi.org/10.1016/j.pepi.2008.06.027

Kawazoe, T., Karato, S.-I., Ando, J. I., Jing, Z., Otsuka, K., & Hustoft, J. W. (2010). Shear deformation of polycrystalline wadsleyite up to 2,100 K at 14–17 GPa using a rotational Drickamer apparatus (RDA). *Journal of Geophysical Research B: Solid Earth*, *115*(8), B08208. http://doi.org/10.1029/2009JB007096

Kawazoe, T., Nishihara, Y., Ohuchi, T., Miyajima, N., Maruyama, G., Higo, Y., et al. (2016). Creep strength of ringwoodite measured at pressure–temperature conditions of the lower part of the mantle transition zone using a deformation–DIA apparatus. *Earth and Planetary Science Letters*, *454*, 10–19. http://doi.org/10.1016/j.epsl.2016.08.011

Karato, S.-i. (1997). On the separation of crustal component from subducted oceanic lithosphere near the 660 km discontinuity. *Physics of the Earth and Planetary Interiors*, *99*, 103–111.

Karato, S. –I., & Rubie, D. C. (1997). Toward an experimental study of deep mantle rheology: A new multianvil sample assembly for deformation studies under high pressures and temperatures. *Journal of Geophysical Research*, *102*, 20111–20122.

Kesson, S. E., Fitz Gerald, J. D., & Shelley, J. M. (1998). Mineralogy and dynamics of a pyrolite lower mantle. *Nature*, *393*(6682), 252–255. http://doi.org/10.1038/30466

Krusky T. M., Li, J. –H., & Tucker, R. D. (2001).The Archean Dongwanzi Ophiolite Complex, North China Craton: 2.505-Billion-Year-Old Oceanic Crust and Mantle. *Science*, *292*, 1142–1145. DOI: 10.1126/science.1059426

Li, L. I., Weidner, D., Raterron, P., Chen, J., Vaughan, M., Mei, S., et al. (2006). Deformation of olivine at mantle pressure using the D-DIA. *European Journal of Mineralogy*, *18*(1), 7–19.

Li, L., Raterron, P., Weidner, D. J., & Chen, J. (2003). Olivine flow mechanisms at 8 GPa. *Physics of the Earth and Planetary Interiors*, *138*, 113–129.

Liebermann, R. C. (2011). Multi-anvil, high pressure apparatus: A half-century of development and progress. *High Pressure Research*, *31*, 493–532.

Liu, L. -G., & Ringwood, A. E. (1975). Synthesis of a perovskite-type polymorph of $CaSiO_3$. *Earth and Planetary Science Letters*, *28*(2), 209–211. http://doi.org/10.1016/0012-821X(75)90229-0

Mao, Z., Fan, D., Lin, J. -F., Yang, J., Tkachev, S. N., Zhuravlev, K., et al. (2015). Elasticity of single-crystal olivine at high pressures and temperatures. *Earth and Planetary Science Letters*, *426*, 204–215.

Matsui, M., Ito, E., Katsura, T., Yamazaki, D., Yoshino, T., Yokoyama, A., et al. (2009). The temperature-pressure-volume equation of state of platinum. *Journal of Applied Physics*, *105*(1), 013505.

Mei, S., & Kohlstedt, D. L. (2000a). Influence of water on plastic deformation of olivine aggregates 1. Diffusion creep regime. *Journal of Geophysical Research B: Solid Earth*, *105*(B9), 21457–21469.

Mei, S., & Kohlstedt, D. L. (2000b). Influence of water on plastic deformation of olivine aggregates 2. Dislocation creep regime. *Journal of Geophysical Research B: Solid Earth*, *105*(B9), 21471–21481.

Nishihara, Y., Yamazaki, D., Tsujino, N., Doi, S., Kubo, T., Imamura, M., et al. (2020). Studies of deep earth rheology based on high-pressure deformation experiments using d111-type apparatus. *Review of High Pressure Science and Technology/Koatsuryoku No Kagaku To Gijutsu*, *30*(2), 78–84.

Nishiyama, N., Wang, Y., Sanehira, T., Irifune, T., & Rivers, M. L. (2008). Development of the Multi-anvil Assembly 6–6 for DIA and D-DIA type high-pressure apparatuses. *High Pressure Research*, *28*(3), 307–314. DOI: 10.1080/08957950802250607

Ohuchi, T., Kawazoe, T., Higo, Y., & Suzuki, A. (2017). Flow behavior and microstructures of hydrous olivine aggregates at upper mantle pressures and temperatures. *Contributions to Mineralogy and Petrology*, *172*(8), 1–26.

Perrillat, J. P., Ricolleau, A., Daniel, I., Fiquet, G., Mezouar, M., Guignot, N., et al. (2006). Phase transformations of subducted basaltic crust in the upmost lower mantle. *Physics of the Earth and Planetary Interiors*, *157*(1–2), 139–149.

Prescher, C., & Prakapenka, V. B. (2015). DIOPTAS: A program for reduction of two-dimensional X-ray diffraction data and data exploration. *High Pressure Research*, *35*(3), 223–230.

Seto, Y. (2010). Development of a software suite on X-ray diffraction experiments. *The Review of High Pressure Science and Technology*, *20*, 269–276.

Singh, A. K., Balasingh, C., Mao, H. -K., Hemley, R. J., & Shu, J. (1998). Analysis of lattice strains measured under nonhydrostatic pressure. *Journal of Applied Physics*, *83*(12), 7567–7575.

Thomson, A. R., Crichton, W. A., Brodholt, J. P., Wood, I. G., Siersch, N. C., Muir, J., et al. (2019). Seismic velocities of $CaSiO_3$ perovskite can explain LLSVPs in Earth's lower mantle. *Nature*, *572*, 643–647.

Tsujino, N., Nishihara, Y., Yamazaki, D., Seto, Y., Higo, Y., & Takahashi, T. (2016). Mantle dynamics inferred from the crystallographic preferred orientation of bridgmanite. *Nature*, *539*, 81–84. https://doi.org/10.1038/nature19777

Wang, Y., & Weidner, D. J. (1994). Thermoelasticity of $CaSiO_3$ perovskite and implications for the lower mantle. *Geophysical Research Letters*, *21*(10), 895–898. http://doi.org/10.1029/94GL00976

Wang, Y., Durham, W. B., Getting, I. C., & Weidner, D. J. (2003). The deformation-DIA: A new apparatus for high temperature triaxial deformation to pressures up to 15 GPa. *Review of Scientific Instruments*, *74*(6), 3002–3011.

Wang, Y., Rivers, M., Sutton, S., Nishiyama, N., Uchida, T., & Sanehira, T. (2009). The large-volume high-pressure facility at GSECARS: A "Swiss-army-knife" approach to synchrotron-based experimental studies. *Physics of the Earth and Planetary Interiors*, *174*(1–4), 270–281.

Xie, L., Yoneda, A., Xu, F., Higo, Y., Wang, C., Tange, Y., et al. (2020a). Boron–MgO composite as an X-ray transparent pressure medium in the multi-anvil apparatus. *Review of Scientific Instruments*, *91*(4), 043903.

Xie, L., Yoneda, A., Liu, Z., Nishida, K., & Katsura, T. (2020b). Boron-doped diamond synthesized by chemical vapor deposition as a heating element in a multi-anvil apparatus. *High Pressure Research*, *40*(3), 369–378.

Xu, F., Xie, L., Yoneda, A., Guignot, N., King, A., Morard, G., et al. (2020). TiC-MgO composite: An X-ray transparent and machinable heating element in a multi-anvil high pressure apparatus. *High Pressure Research*, *40*(2), 257–266.

Yamazaki, D., Ito, E., Yoshino, T., Tsujino, N., Yoneda, A., Gomi, H., et al. (2019). High-pressure generation in the Kawai-type multianvil apparatus equipped with tungsten-carbide anvils and sintered-diamond anvils, and X-ray observation on $CaSnO_3$ and $(Mg,Fe)SiO_3$. *Comptes Rendus Geoscience*, *351*(2–3), 253–259. http://doi.org/10.1016/j.crte.2018.07.004

Yamazaki, D., Kato, T., Yurimoto, H., Ohtani, E., & Toriumi, M. (2000). Silicon self-diffusion in $MgSiO_3$ perovskite at 25 GPa. *Physics of the Earth and Planetary Interiors*, *119*(3–4), 299–309.

5

Deciphering Deep Mantle Processes from Isotopic and Highly Siderophile Element Compositions of Mantle-Derived Rocks: Prospects and Limitations

Katsuhiko Suzuki[1], Gen Shimoda[2], Akira Ishikawa[3], Tetsu Kogiso[4], and Norikatsu Akizawa[5]

ABSTRACT

Geochemical and geophysical studies on the core and mantle have allowed us to place constraints on the physical properties and the geochemical and isotopic compositions of the bulk Earth. To advance our understanding of the geochemical evolution of the Earth, we review recently published data on the geochemistry of mantle-derived rocks and data from high-pressure experiments in an attempt to reinterpret the current hypotheses regarding the chemical heterogeneity of the deep mantle, interactions between the core and mantle, and the evolution of early Earth. Our investigation suggests that the chemical heterogeneity of the deep mantle observed as Sr-Nd-Pb isotopic diversities in ocean island basalts (OIBs) can be mostly attributed to the recycling of subducted crustal materials. Deficits of ^{182}W found in OIBs and other mantle-derived rocks suggest that lower mantle compositions are influenced by core-mantle interactions, as the core is expected to have a lower $^{182}W/^{184}W$ ratio than the mantle. However, it remains unclear whether or not the core signatures of W isotopes imprinted on the deep mantle are affected by other processes such as silicate crystallization in the early Earth magma ocean and recycling of crustal material in the mantle, and if they are, to what extent. Abundances of highly siderophile elements (HSEs) in the mantle, which have been estimated to be much higher than what are expected from core-mantle equilibrium, are the key to understand the core-mantle interactions and the Earth's accretionary history. However, the estimated HSE concentrations in the primitive mantle entail large uncertainties resulting from shallow mantle processes such as melt extraction and metasomatism. Deciphering the true HSE composition of the primitive mantle is further hindered by the discrepancies between the HSE data from partitioning experiments and those from analyses of mantle peridotites, which require them to have undergone not only melt extraction but also metasomatism.

[1]Submarine Resources Research Center, Japan Agency for Marine-Earth Science and Technology, Yokosuka, Japan
[2]Geological Survey of Japan, National Institute of Advanced Industrial Science and Technology, Tsukuba, Japan
[3]Department of Earth and Planetary Sciences, Tokyo Institute of Technology, Tokyo, Japan
[4]Graduate School of Human and Environmental Studies, Kyoto University, Kyoto, Japan
[5]Atmosphere and Ocean Research Institute, The University of Tokyo, Kashiwa, Japan

5.1. INTRODUCTION

Although research on the structure and mineral composition of the Earth's deep interior has achieved remarkable observational and experimental progress in recent years, interactions between the mantle and core have remained the subject of debate for over 60 years because materials from the deep Earth cannot be sampled directly. Since the core and mantle account for 20% and 80% of the Earth

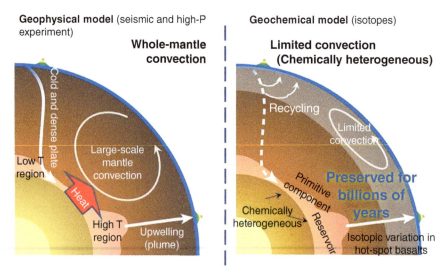

Figure 5.1 Comparison between the geophysical and geochemical models of Earth mantle convection. Source: Grant-In-Aid from the Japan Society for the Promotion of Science "Core-Mantle Interaction and its Co-Evolution" (Website: http://core-mantle.jp/).

by volume, respectively, understanding the interactions between them is crucial for reconstructing the chemical evolution of the Earth.

Ever since the formation of the Earth, mantle convection has acted as a heat engine that releases thermal energy stored in the Earth's interior as well as the thermal energy from the decay of radionuclides. As shown in Figure 5.1 [modified from the figure in the Core-Mantle Coevolution project (http://core-mantle.jp/)], seismic mantle tomography (Fukao et al., 2003, 2009) and simulation models (Yuen et al., 2007) support whole-mantle circulation. By contrast, results of high-pressure experiments suggest that the bulk composition of the lower mantle is different from that of the upper mantle (Ricolleau et al., 2009; Murakami et al., 2012). Geochemically, isotopic data from mid-ocean ridge basalts (MORBs) and ocean island basalts (OIBs) suggest that the mantle comprises several chemically distinct domains: a depleted MORB mantle (DMM), enriched mantle (EM1 and EM2), high-μ (HIMU) mantle, and primitive mantle (PM) (Hofmann, 2014). Such chemical heterogeneity within the mantle is inconsistent with whole-mantle convection and suggests that the mantle has a more complex structure than those inferred from geophysical studies.

Except for the DMM, which is generally agreed to correspond to the upper mantle, residences of the other chemically distinct reservoirs are not so well constrained. Several models have been proposed to explain the chemical heterogeneity of the mantle, such as the layered mantle model in which the upper and lower mantle convect separately (DePaolo and Wasserburg, 1976; Jacobsen and Wasserburg, 1979; O'Nions et al., 1979), the plum-pudding mantle model with lumps of chemically heterogeneous materials scattered throughout the mantle (Morris and Hart, 1983; Zindler et al., 1984), and the marble-cake mantle model with layered heterogeneities (Allègre and Turcotte, 1986). Another model referred to as the statistical upper mantle assemblage (SUMA) model (Meibom and Anderson, 2004) is rather unique and assumes a heterogeneous distribution of depleted residues and enriched subducted oceanic crust in the upper mantle, each acting as the source for MORBs or OIBs, respectively. The last model does not require any distinct reservoirs to exist in the lower mantle. Because none of the models have yet to be verified (Tackley, 2000), it remains uncertain if some of the geochemical reservoirs are silicate mantle chemically influenced by the core, or if the metallic core itself functions as one of the reservoirs. A key to solving these uncertainties is to elucidate how, where, and when these geochemical reservoirs have been generated in the mantle.

Core-mantle chemical interactions have been suggested to be taking place at the bottom of the mantle based on geophysical and geochemical observations. Tanaka (2007) identified a 30 km thick layer located at the bottom of the mantle exhibiting a 10% decrease in S-wave velocity using a global dataset of seismic waves, which is now referred to as the ultralow-velocity zone (ULVZ). Garnero and McNamara (2008) employed seismic tomography to simulate the structure and dynamics of the lower mantle and suggested that core-mantle interactions are responsible for generating this ultralow-velocity pocket, which most likely is the Earth's deepest magma chamber. Meanwhile,

Helffrich and Kaneshima (2010) also reported that a decrease in seismic wave velocity was observed in the lowermost part of the mantle.

Core-mantle interactions transfer core components into the mantle, and elements supplied from the core are likely to be incorporated into the grain boundaries of the mineral phases in the mantle and transported together with the minerals in the upward flow of the mantle. Hayden and Watson (2007) observed that the diffusion rates of Au, W, and Rh, which are faster than those of Os and Re, are fast enough to allow them to migrate several tens of kilometers via diffusion at the grain boundaries of silicate minerals in the mantle. Yoshino et al. (2020) conducted a high-pressure experiment and observed that the grain boundary diffusion of W is faster than those of the other siderophile elements and strongly dependent on temperature. In addition, they claimed that if subducted slabs were to oxidize the outer core at the core-mantle boundary, a large W flux with low $^{182}W/^{184}W$ ratios would be generated and penetrate into the lowermost mantle. These results demonstrate that isotopic and elemental concentration anomalies caused by core-mantle interactions could be transferred to magma source regions in the deep mantle and be recorded in the chemical and isotopic compositions of OIBs and other mantle-derived rocks.

In particular, the fractionation between siderophile elements and lithophile elements may provide clues for discovering traces of core-mantle interactions. Such characteristics will be reflected in Re-Os and Hf-W isotope systematics of mantle-derived rocks, and therefore geologists are now attempting to detect these isotopic anomalies derived from the core in Os and in the above-mentioned W isotopes in OIBs, komatiites, and kimberlites.

Another major challenge is clarifying the early evolution of the Earth, which is poorly constrained to date. Figure 5.2 displays one of the models proposed as a working hypothesis in the Core-Mantle Coevolution project (http://core-mantle.jp/). After the formation of proto-Earth, a giant impact with a planetary embryo is believed to have occurred, resulting in the formation of the Moon and the fusion of the core materials of the Earth and the planetary embryo. After the fusion and separation of the core, chondritic materials were supplied to the surface by subsequent impacts, in what is termed as the late veneer event or late accretion. These processes have been proposed to explain the geochemical data, such as the high abundance of highly siderophile elements (HSEs), including platinum group elements (PGEs), and the chondritic evolution curve of the Os isotopic composition of the mantle. Although the events that occurred after the formation of the Earth are important for constraining the chemical and isotopic composition of the Earth, their timing, magnitude, and whether or not some events such as the late veneer actually occurred are highly controversial. To solve these problems, precise estimation of Os and other radiogenic isotopes in the mantle will be required.

In this chapter, we review the recent geochemical studies on the chemical characteristics of the mantle that can be linked to the core-mantle interactions and deep mantle evolution. First, we review the current understandings on the origin of the geochemical heterogeneity of Sr-Nd-Pb isotopic ratios of OIB and then examine whether the Sr-Nd-Pb isotope heterogeneity of the mantle necessitates a contribution from deep mantle processes using model calculations (section 5.2). In the latter sections, we review the latest data on Os and W isotopic ratios and HSE abundances in mantle-derived rocks and discuss the relevance of the Os, W, and HSE signatures of the mantle to the core-mantle interactions (sections 5.3 and 5.4).

5.2. ISOTOPIC COMPOSITIONS OF OIBS

The Sr-Nd-Pb isotopic variations of OIB, which are the basis for identifying the geochemical reservoirs in the mantle, are notable in that the Pb-Nd-Sr (-Hf) isotopic compositions of OIBs converge to a small area in Pb-Nd-Sr isotope space (Fig. 5.3). Several definitions have been proposed for this convergent area: the Focus Zone (FOZO; Hart et al., 1992), the common component (C; Hanan and Graham, 1996), and prevalent mantle (PREMA; Zindler and Hart, 1986) (Fig. 5.3). Although the proposed isotopic components exhibit considerably

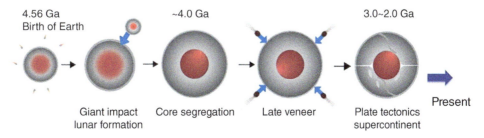

Figure 5.2 Model of the early evolution of the Earth. Source: Grant-In-Aid from the Japan Society for the Promotion of Science "Core-Mantle Interaction and its Co-Evolution" (Website: http://core-mantle.jp/).

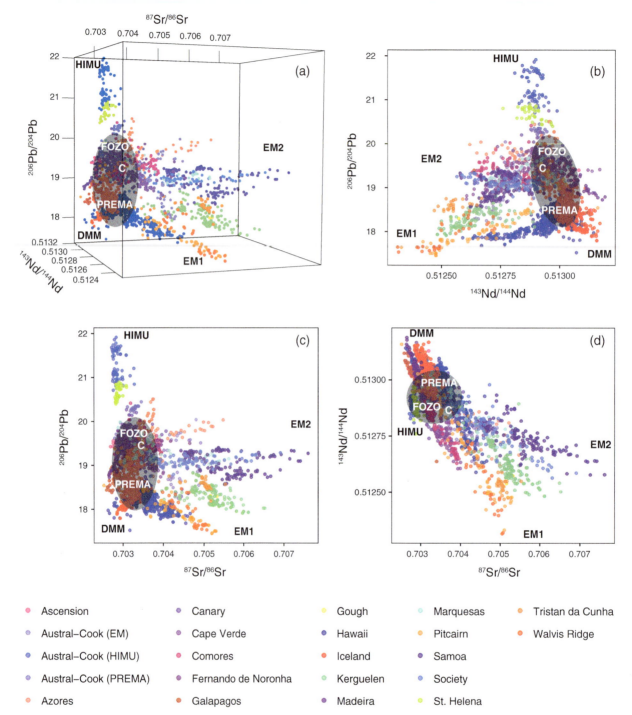

Figure 5.3 Relationships between $^{87}Sr/^{86}Sr$, $^{143}Nd/^{144}Nd$, and $^{206}Pb/^{204}Pb$ in ocean island basalts (OIBs). The plotted data were compiled by Stracke (2012). The names of the hotspots of the OIB data are given in the legend. OIB (colored dots), representative mantle end components (black labels), intermediate components (white labels), and isotopic convergence areas (gray fields) are (a) plotted in three dimensions, and projected onto (b) $^{206}Pb/^{204}Pb$ versus $^{143}Nd/^{144}Nd$, (c) $^{206}Pb/^{204}Pb$ versus $^{87}Sr/^{86}Sr$, and (d) $^{143}Nd/^{144}Nd$ versus $^{87}Sr/^{86}Sr$ diagrams. Abbreviations are C, common component; DMM, depleted MORB mantle; EM1, enriched mantle 1; EM2, enriched mantle 2; FOZO, Focus Zone; HIMU, high-μ; PREMA, prevalent mantle.

similar isotopic compositions, the isotopic compositions of OIBs appear not converge to a point but rather to a range of isotopic compositions, as described by Stracke (2012). As this convergence is one of the most significant features of the isotopic characteristics of OIBs, it is important to determine whether deep mantle processes could contribute to the phenomenon. Hereafter, we will use the term "FOZO" to represent the isotopic convergent area of OIBs (corresponding to the gray field in Fig. 5.3), which will include PREMA, C, and the original FOZO.

While the isotopic convergent area occupies only a small area in Nd-Sr isotopic space, it exhibits considerable variation in Pb isotopic ratios and follows a vertical trend in the 3-D Nd-Sr-Pb isotopic space (Fig. 5.3a). This area encompasses the PREMA, C, FOZO, and all other convergent points that have been defined and extends to the HIMU on the enriched side and to the DMM on the depleted side. Most of the isotopic arrays of OIB suites extend from this vertical trend to the enriched side of the Nd-Sr isotopic space in varying directions and are thus referred to as the enriched mantle (EM) trends (Fig. 5.3) (Willbold and Stracke, 2006, 2010). Their essential feature is that each EM trend displays relatively low Pb isotopic variations, resulting in a nearly horizontal array on the Nd-Sr isotopic plane in the 3-D-isotopic plot (Fig. 5.3a).

In this section, we discuss the cause of the isotopic convergence to a limited area (vertical trend in Fig. 5.3a) based on the model proposed by Shimoda and Kogiso (2019). We calculated the Pb-Nd-Sr-Hf isotopic compositions of recycled oceanic crusts for approximately a million different subduction conditions to prove that a sufficiently large number of different subduction conditions are capable of producing the converging isotopic distribution of OIBs. For the calculations, we varied recycling age, which is defined as the time interval from the onset of subduction modification of the oceanic crust to the initiation of melting events of the source materials in OIB, as the interval is the major controlling factor for the isotopic diversity of the recycled crust.

5.2.1. Previous Works on the Origin of Isotopic Convergent Areas and Their Issues

It has been inferred that the OIBs plotted in the isotopically convergent areas could be derived from the lower mantle, where primitive or less differentiated materials reside (Zindler and Hart, 1986; Hart et al., 1992; Hanan and Graham, 1996). The bases for considering the lower mantle origin of OIBs are as follows: (1) the ubiquitous occurrence of OIBs with FOZO-like isotopic signatures and (2) the presence of FOZO-like OIBs exhibiting high ^{3}He/^{4}He isotopic ratios, suggesting their derivation from primitive or less degassed materials that probably reside in the lowermost mantle. However, this theory has been disputed as the isotopic compositions of FOZO are not consistent with that of bulk silicate Earth. In addition, the ^{3}He/^{4}He ratios of OIBs with FOZO-like isotopic signatures are generally not high, indicating that the primitive He isotopic signature is an exception and not the norm for most of the FOZO-like OIBs.

A contrasting theory on the origin of FOZO is that the source material of FOZO is the recycled oceanic crust (Stracke et al., 2005; Shimoda and Kogiso, 2019). According to their models, the isotopic composition of FOZO can be explained by subduction modification of the oceanic crust. However, subduction modification, which is often synonymous with dehydration of the oceanic crust (e.g., Kogiso et al., 1997; Stracke et al., 2003), is usually invoked to explain the generation of the HIMU source, whose isotopic characteristics are distinct from those of FOZO.

Several researchers have attributed this difference in isotopic characteristics between FOZO and HIMU to differences in recycling ages, with FOZO having younger age than HIMU (more than 2.0 Ga; Thirlwall, 1997; Kimura et al., 2016). However, this conflicts with the fact that FOZO exhibits a higher ^{207}Pb/^{206}Pb ratio than HIMU, which should be lower if it were younger than HIMU. In addition, a combination of forward modeling of isotopic ingrowth and Independent Component Analysis inferred that the recycling age of OIB, including HIMU, to be much younger than 2.0 Ga (Iwamori et al., 2010).

The predominant issue in all existing geochemical models is their choice of a dataset of partition coefficients of elements during dehydration of oceanic crusts, as most geochemical models are forced to employ partition coefficients determined at temperatures higher than the dehydration temperatures of oceanic crusts. This is simply because there are no available dataset of partition coefficients that have been determined at around 500°C at which the dehydration of the oceanic crusts is presumed to occur. As the solubility of elements in fluids is strongly dependent on temperature, models that utilize partition coefficients determined at high temperatures risk overestimating the effect of dehydration reactions (Green and Adam, 2003). This concern has been validated by the results of studies of natural rocks from high-pressure and ultrahigh-pressure terrains, which have shown that aqueous fluids cannot modify the chemical composition of subducted oceanic crusts to any significant degree (Bebout et al., 1999; Hermann et al., 2006). Thus, the results of high-pressure experiments and studies on natural rock systems are inconsistent with the results of existing geochemical models because their choice of partition coefficients results in the assumed temperature conditions being too high.

Recently, Shimoda and Kogiso (2019) proposed a geochemical model that considers the temperature

Table 5.1 Chemical compositions of oceanic crust and serpentinite

Oceanic crust composition (ppm)	Enriched					ALL MORB (average)	Depleted					Serpentinite[a]
	+2.0 SD	+1.5 SD	+1.0 SD	+0.5 SD	+0.25 SD		−0.25 SD	−0.5 SD	−1.5 SD	−1.0 SD	−2.0 SD	
Rb	3.32	3.21	3.10	2.99	2.94	**2.88**	2.83	2.77	2.66	2.55	2.44	0.0102
Sr	133	132	131	130	129.5	**129**	128.5	128	127	126	125	0.0081
Nd	12.81	12.62	12.42	12.23	12.13	**12.03**	11.9	11.84	11.64	11.45	11.3	0.039
Sm	3.97	3.93	3.90	3.86	3.84	**3.82**	3.80	3.78	3.745	3.71	3.67	0.14
Lu	0.55	0.55	0.54	0.54	0.53	**0.53**	0.528	0.53	0.52	0.52	0.510	4.39
Hf	2.94	2.90	2.87	2.83	2.81	**2.79**	2.77	2.75	2.715	2.68	2.64	0.0793
Pb	0.60	0.59	0.59	0.58	0.574	**0.57**	0.6	0.56	0.555	0.55	0.54	0.00922
Th	0.485	0.465	0.445	0.424	0.414	**0.404**	0.39	0.384	0.364	0.343	0.32	0.00410
U	0.132	0.129	0.126	0.122	0.12	**0.119**	0.117	0.116	0.113	0.109	0.106	0.00599

[a] Serpentinite composition is the log-normal average of abyssal serpentinites of which protoliths were harzburgite (Deschamps et al., 2013).

dependence of the solubility of elements. In this model, major chemical fractionation of the oceanic crust occurs through fluid-rock reaction as a result of dehydration of the serpentinite underlying the subducting oceanic crust. As the fluid-rock reaction occurs at a higher temperature than the dehydration reaction of the oceanic crust, it can induce more extensive chemical fractionation in the subducted oceanic crust. This model also suggests that subduction modifications via dehydration and fluid-rock interactions can produce a region within the mantle exhibiting FOZO-like chemical composition. In other words, the isotopic convergence of OIBs can be simply explained by recycling of the oceanic crust, which is a major reservoir of incompatible elements. The variety of isotopic arrays observed in the OIBs can be reproduced by varying the proportion of the added sediment, thereby forming the nearly horizontal arrays in the Pb-Nd-Sr isotopic space.

In the next section, we will provide an outline of the model of Shimoda and Kogiso (2019). We also provide the results of newly calculated isotopic compositions of subduction-modified oceanic crust, which take into account the variations in recycling ages, degrees of dehydration, fluid-rock reactions, and chemical compositions of oceanic crusts. The purpose of our model calculations was to test whether a sufficiently large number of combinations of these factors can produce the converging isotopic distribution to FOZO.

5.2.2. Modeling Parameters

Chemical Compositions of the Oceanic Crust

For the chemical composition of the oceanic crust in our model, we utilized the global average composition of MORBs determined by Gale et al. (2013) (denoted as ALL MORB), instead of the more commonly used N-MORB composition determined by Sun and McDonough (1989). Although the N-MORB composition is appropriate for representing a melt directly derived from the upper mantle, it was not utilized here because most of the measured chemical compositions of MORBs are different from N-MORB compositions (Gale et al., 2013). The chemical variation within the oceanic crust was approximated using standard deviations (SDs) determined for ALL MORBs, that is, ALL MORB ± 0.25 SD, ± 0.5 SD, ± 1.0 SD, ± 1.5 SD, and ±2.0 SD (11 compositions, Table 5.1). These compositions were utilized as the original oceanic crust compositions modified by subduction processes.

Recycling Ages

Debate is still ongoing regarding the recycling ages of oceanic crusts. Rudge (2006) estimated the mean recycling age of oceanic crusts to be 0.5 Ga. By contrast, Kimura et al. (2016) assessed recycling ages from 0.5 to 2.5 Ga, in which recycling ages of FOZO, HIMU, and EM were estimated to be ca. 1.0, 2.0, and 2.5 Ga, respectively. Because it is difficult to select recycling ages of oceanic crust and their distribution with accuracy, we adopted 200 recycling ages utilizing a Gaussian distribution with a mean age of 1.0 Ga and employed two standard deviations of ±0.2 Ga. In total, the Pb-Nd-Sr-Hf isotopic compositions of 2,200 oceanic crusts were calculated (11 chemical compositions × 200 recycling ages). Note that the calculated isotopic compositions depend on the assumed ages and that the results are subject to change for different mean ages and age distributions.

Dehydration Conditions of Oceanic Crust

The water content of oceanic crusts varies widely due to the difference in the degree of alteration (Staudigel et al., 1995, 1996; Shipboard Scientific Party, 2000; Bach et al., 2001, 2003). The practical upper limit of water content in oceanic crusts is approximately 6% (Poli and Schmidt, 1995; Schmidt and Poli, 1998; Hacker et al., 2003; Hacker, 2008). Thus, we tested the chemical effect of dehydration reactions with dehydration degrees varied between 0.5% and 6% at increments of 0.5%. In total, 26,400 sets of Pb-Nd-Sr-Hf isotopic compositions of the modeled dehydrated oceanic crust were calculated (11 × 200 × 12 degrees of dehydration).

The temperature conditions of the dehydration reactions were estimated based on the P-T paths of descending slabs estimated by Syracuse et al. (2010) and the stability fields of hydrous minerals reported by Hacker et al. (2003). Based on their results, the major dehydration reactions were assumed to occur at approximately 500°C and around 3 GPa, irrespective of the slab age. The relatively constant dehydration temperature results from the temperature dependency of the major dehydration reactions. As partition coefficients have not been determined at such low temperatures, we utilized partition coefficients determined at the lowest investigated pressure and temperature conditions of dehydration (650°C and 3 GPa) (Green and Adam, 2003). Note that the solubility of elements in fluids depends on temperature (Nakamura and Kushiro, 1974; Manning, 2004; Plank et al., 2009) and that the selected partition coefficients were determined at a higher temperature than the dehydration temperature of around 500°C. Thus, the maximum chemical effects of dehydration on the oceanic crust are estimated.

In this study, the effects of dehydration were estimated by implementing the batch melting equation (Shaw, 1970). The choice of which fractional melting equation to use in the calculations does not affect the argument below because difference in parent-to-daughter elemental ratios determined with the different equations is similar due to the low solubility of the elements in the fluid.

Dehydration Conditions of Serpentinites

There are a few constraints regarding the extent of serpentinization in subducted slabs (van Keken et al., 2011). We assumed that the maximum thickness of the serpentinized layer was 40 km, which was estimated for the Pacific plate off the northeast Japan arc (Garth and Rietbrock, 2014). The maximum degree of serpentinization was assumed to be 10%, which corresponds to the upper limit for maintaining the negative buoyancy of the slab (Schmidt and Poli, 1998). The degree of serpentinization was varied from 0.5% to the maximum value at increments of 0.5%.

The amount of fluid released by the dehydration of serpentinite depends on subduction conditions. Specifically, serpentinite can release 13% of its mass as water in hot subduction zones (Bose and Ganguly, 1995; Ulmer and Trommsdorff, 1995; Peacock and Wang, 1999; Iwamori, 2000, 2004; Hacker, 2008). Conversely, serpentinite can release only 9% of its mass as water in cold subduction zones (Peacock, 2001; Iwamori, 2000, 2004; Komabayashi et al., 2005). Thus, we assumed two conditions for the degree of water release during the dehydration, namely, 13% and 9% for hot and cold subduction zones, respectively. The maximum water-rock ratios were set to 0.07:1 in hot subduction zones and 0.05:1 in cold subduction zones, given the average thickness of 7 km of oceanic crust (White et al., 1992; Mutter and Mutter, 1993).

The compositions of serpentinite-derived fluids were determined using the average serpentinite composition based on Deschamps et al. (2013), the batch melting equation, and the partition coefficients listed in Table 5.2. The estimated chemical composition of the oceanic crust after fluid-rock reactions was taken from Shimoda and Kogiso (2019).

Based on these parameters, we estimated 1,056,000 sets of Pb-Nd-Sr isotopic compositions for recycled oceanic crusts (11 chemical compositions × 200 recycling ages × 12 degrees of dehydration × 20 degrees of serpentinization × 2 degrees of water release) (Fig. 5.4).

5.2.3. Results of Modeling: Possible Isotopic Range of Recycled Oceanic Crusts

As shown in Figure 5.4, the results of calculations for approximately 1 million different combinations reveal that the isotopic compositions of oceanic crusts modified by subduction processes plot in the vicinity of FOZO. The testing conditions of subduction modification vary widely in terms of the chemical compositions of the oceanic crust, degree of dehydration, fluid-rock ratios, and recycling ages. Nevertheless, the calculated isotopic variations of the recycled oceanic crust are relatively small compared to the entire range of OIBs but are on a similar level to those of FOZO.

It is notable that dehydration of the oceanic crust, which is commonly considered to be one of the major elemental fractionation processes during subduction, produces only limited isotopic variations (Fig. 5.4). Pb, Nd, and Hf isotopic compositions of dehydrated and dry (without subduction modification) oceanic crusts plot close each other, revealing their nearly identical nature (Fig. 5.4). Only the Sr isotopic compositions of the dry and dehydrated oceanic crust exhibit notable variations. This is because Rb is soluble in fluid at the dehydration

Table 5.2 Partition coefficients used in the dehydration and fluid-rock reaction calculations

	Dehydration[a] 650 °C	Fluid-rock reaction (hot subduction zone)[b]					Fluid-rock reaction (cold subduction zone)[c]						Serpentinite[d]
		700 °C	800 °C	900 °C	a[e]	b[e]	800 °C	900 °C	1000 °C	1200 °C	a[e]	b[e]	
Rb	0.0751	0.011	0.019	0.013	0.026	−0.47	0.00419	0.00323	0.015	0.0102	0.0688	−2.08	0.0102
Sr	0.22	2.9	0.526	0.0474	1.84×10^{-6}	10.0	0.035	0.0313	0.012	0.0081	6.27×10^{-4}	3.28	0.0081
Nd	37.0	18	6.85	0.661	0.00151	6.59	0.525	0.325	0.0511	0.039	4.11×10^{-4}	5.75	0.039
Sm	65.0	39	15	2.70	0.00631	6.12	1.5	1.1	0.27	0.14	0.00403	4.78	0.14
Lu	386	238	195	107	12.3	2.09	42	60	13	4.39	1.10	3.07	4.39
Hf	126	47	16	1.9	0.00224	6.97	6.41	1.78	0.19	0.0793	2.13×10^{-5}	10.1	0.0793
Pb	6.62	0.31	0.0526	0.0318	6.25×10^{-7}	9.18	0.0402	0.0392	0.021	0.00922	0.00155	2.68	0.00922
Th	8.47	8.37	2.3	0.16	7.8×10^{-5}	8.11	0.17	0.0408	0.016	0.00410	5.57×10^{-7}	10.1	0.00410
U	8.44	6.98	0.977	0.17	9.13×10^{-7}	11.1	0.479	0.24	0.0400	0.00599	8.7×10^{-5}	6.91	0.00599

[a] Bulk partition coefficients during dehydration of oceanic crust were calculated based on reported data determined at 3 GPa and phase modes determined by Schmidt et al. (2004): 47.3% garnet, 38.9% clinopyroxene, 12.4% quartz/coesite and 1.4% rutile, after recalculation to 100% total. Partition coefficients are from Green and Adam (2003) for garnet and clinopyroxene, from Ayers (1998) for rutile, and zero is assumed for quartz/coesite. Partition coefficients of Hf and Lu into rutile were assumed to be identical to those of Zr and Tm from Ayers (1998). Bulk partition coefficients of Nd into garnet and rutile were estimated based on the equation $Kd_{Nd} = (Kd_{Sm} + Kd_{Ce})/2$.
[b] Partition coefficients are from Kessel et al. (2005) determined at 4 GPa for aqueous fluid.
[c] Partition coefficients are from Kessel et al. (2005) determined at 6 GPa for supercritical fluid.
[d] Bulk partition coefficients of serpentinite were estimated based on the assumption that serpentinite becomes harzburgite after dehydration, consisting of 70% olivine and 30% orthopyroxene. Partition coefficients of olivine and pyroxene are from Adam et al. (2014), Ayers (1998), and Brenan et al. (1995). The partition coefficient of Pb into orthopyroxene was assumed to be identical to that for olivine. The partition coefficient of Hf into olivine is assumed to be identical to that of Zr.
[e] Constants obtained from the regression lines (see text).

temperature, whereas the other elements cannot dissolve in fluids at such low temperatures. Thus, no apparent differences are produced in the isotopic compositions of dry and dehydrated oceanic crusts.

On the other hand, the fluid-rock reaction produces a relatively wide variation in Pb isotopic ratios (purple and red dots in Fig. 5.4), similar to that observed in FOZO. This implies that the major cause of the isotopic variation of FOZO is fluid-rock reactions in the oceanic crust induced by serpentinite dehydration and not the dehydration reaction of the oceanic crust itself. Thus, the isotopic diversity of the recycled oceanic crust is limited in Sr, Nd, and Hf isotopic compositions, resulting in the vertical isotopic trends observed in the 3-D isotopic plot (Fig. 5.3a).

It follows then that the OIB isotopic variations converging to FOZO are manifestations of the isotopic variations of recycled oceanic crusts created under variable subduction conditions and recycling age conditions and that the presence of primitive or less degassed components in deep mantle is not required to produce the variations. The addition of small amounts of continental crustal materials to the oceanic crust via sediment subduction or tectonic erosion can extend the isotopic array away from FOZO to EM1/EM2 without accompanying any substantial changes in Pb isotopic ratios, thereby producing the nearly horizontal arrays on the Nd-Sr isotopic plane in the 3-D isotopic plot as shown in Figure 5.3a. We thus conclude that the Sr-Nd-Pb isotopic heterogeneity in the mantle can be attributed entirely to the recycling of subducted crustal materials, that is, oceanic crusts of various ages modified under variable subduction conditions (FOZO and HIMU) with variable degrees of addition of continental crustal materials.

5.2.4. FOZO as a Candidate for the Detection of Core-Mantle Interactions

The most appropriate geochemical tracer for detecting core-mantle interactions may be the $^{182}W/^{184}W$ ratios of OIBs (see later section). Hence, we will discuss the potential link between Pb-Nd-Sr and W isotope systematics of OIB.

As discussed in the previous section, OIB sources contain recycled oceanic crust materials affected by variable degrees of addition of continental crustal materials, which could have much higher W concentrations than mantle materials (Jackson et al., 2020; Taylor and Mclennan,

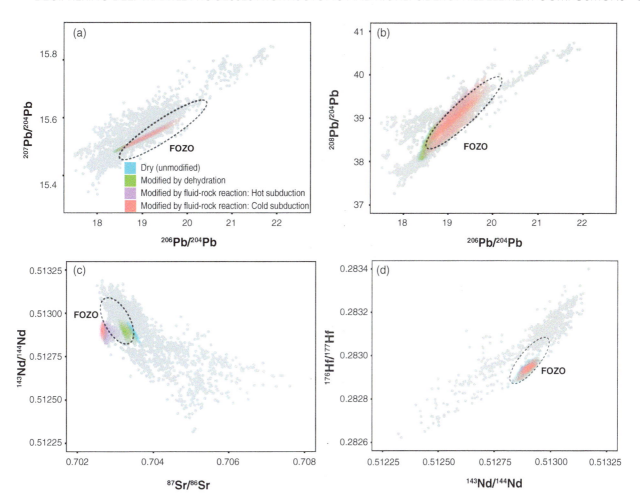

Figure 5.4 Calculated isotopic compositions of subduction modified oceanic crusts at various recycling ages, degree of dehydration, fluid-rock ratios, and chemical compositions of oceanic crusts expressed using density estimator method. Note that the estimated densities depend solely on calculation assumptions, and therefore do not predict the isotopic distributions of ocean island basalts (OIBs). (a) $^{207}Pb/^{204}Pb$ versus $^{206}Pb/^{204}Pb$, (b) $^{208}Pb/^{204}Pb$ versus $^{206}Pb/^{204}Pb$, (c) $^{143}Nd/^{144}Nd$ versus $^{87}Sr/^{86}Sr$, and (d) $^{176}Hf/^{177}Hf$ versus $^{143}Nd/^{144}Nd$. The initial chemical variation in the oceanic crust was estimated using the average chemical composition of mid-ocean ridge basalts (MORBs) and their standard deviation estimated by Gale et al. (2013). As the recycling ages of the actual oceanic crusts are unclear, we assumed 200 recycling ages normally distributed with 1.0 ± 0.2 Ga (mean age and two standard deviations). Initial isotopic ratios of MORBs are assumed to be identical to those of depleted MORB mantle (DMM) estimated by Stracke et al. (2005). Recycled oceanic crusts without subduction modification (dry oceanic crust) are shown as light blue areas. Green areas indicate recycled oceanic crusts, for which the chemical compositions were modified by dehydration reaction. Pb, Nd, and Hf isotopic ratios of dry and dehydrated oceanic crusts were nearly identical, and thus overlap in (a), (b), and (d). The isotopic compositions of the recycled oceanic crusts that experienced fluid-rock reactions at hot and cold subduction zones are shown as purple and coral areas, respectively.

1995). Hence, FOZO and HIMU, which contain minimal continental crustal materials, are considered to be the most appropriate types of OIB for investigating the W isotope signature of the deep mantle. W is likely to be removed from oceanic crusts by seafloor alteration and subduction-related processes (König et al., 2008, 2010, 2011; Bali et al., 2012; Reifenröther et al., 2021). Therefore, oceanic crusts that have been subjected to intensive fluid-rock reactions should have low W concentrations. Given that the intensity of fluid-rock reaction is proportional to the $^{206}Pb/^{204}Pb$ ratios of recycled oceanic crust, OIBs with high $^{206}Pb/^{204}Pb$ ratios can be regarded to be the most representative samples of the deep mantle in terms of W isotopes.

W and U are highly incompatible during mantle melting (Hill et al., 2000; McDade et al., 2003). In addition, W is more mobile than U during seafloor alteration at ocean basins and during dehydration processes beneath subduction zones (König et al., 2008; Reifenröther et al., 2021). Thus, the W concentration of recycled oceanic crust can be inferred from estimated U concentrations of recycled oceanic crust.

According to Shimoda and Kogiso (2019), a subduction zone with moderate temperature conditions has the potential to produce the lowest U concentration in a recycled oceanic crust after fluid-rock reactions, making it the most likely candidate involved in the production of both HIMU and FOZO-like OIBs. The higher U concentration of HIMU source compared to that of FOZO source could be explained by assuming the formation of the oceanic crusts for HIMU sources at mid-ocean ridges at low degrees of partial melting (Shimoda and Kogiso, 2019). If this is the case, then the recycled oceanic crusts with FOZO isotopic signature should have the lowest W concentrations. Thus, FOZO-like OIBs exhibiting the highest $^{206}Pb/^{204}Pb$ should provide the most appropriate material for detecting $^{182}W/^{184}W$ anomalies resulting from core-mantle interactions.

5.3. OS AND ^{182}W ISOTOPE SYSTEMATICS OF ROCKS DERIVED FROM THE DEEP MANTLE

One attempt to decode the core signature is the application of Os and W isotope systematics to deep mantle-derived rocks. We will first briefly introduce the present status of Os isotope systematics for deep-rooted volcanic rocks. Then, we will present W isotope variations observed in OIBs and kimberlites, taking into consideration the possible influence from core-mantle interactions.

5.3.1. Os Isotope Systematics of Deep Mantle-Derived Rocks

The higher Pt/Os and Re/Os ratios of the outer core compared to those of the inner core result in the elevated $^{186}Os/^{188}Os$ and $^{187}Os/^{188}Os$ ratios of the outer core. Thus, the positive correlation between $^{186}Os/^{188}Os$ and $^{187}Os/^{188}Os$ observed in the Hawaiian picrites, Gorgona komatiites, and the Siberian Trap flood basalts was considered to reflect the involvement of core materials in the chemical composition of the lowermost mantle (Brandon et al., 1998, 2003, 2005). However, Luguet et al. (2008) claimed that metasomatic sulfides derived from pyroxenite or peridotite melts in the mantle source may also be responsible for the ^{186}Os-^{187}Os signatures of OIBs, implying that core-mantle interactions are not necessarily required to generate the Os isotope diversity in basalt source regions. In addition, high-pressure experiments do not support the view that Pt/Os and Re/Os ratios of the outer core are higher than those of the inner core (Van Orman et al., 2008). Ireland et al. (2011) examined both the core-mantle interaction and the metasomatic sulfide contribution models and claimed that the dominant cause of the ^{186}Os-^{187}Os variations cannot be conclusively determined.

5.3.2. W Isotopes as a Tracer of Deep Mantle Processes

It is generally believed that the core separated from the magma ocean soon after the formation of the Earth. During this separation process, HSEs and other siderophile elements, such as W, were largely removed from the mantle, which resulted in extensive fractionation of lithophile and siderophile elements in the magma ocean. As for W isotopes, important results have been obtained with the improvement of analytical accuracy in recent years, as will be described in section 5.3.3.

^{182}Hf undergoes beta-decay to ^{182}W with a half-life of only 8.9 million years (Vockenhuber et al., 2004), and so ^{182}Hf is now extinct. W is a siderophile element and partitioned into the core, whereas Hf is strongly lithophile and remains in the mantle. Therefore, if the core separated during the first ~50 million years after the Earth's formation, Hf/W fractionation during core formation would have resulted in a difference in the W isotopic composition between the mantle and core observed today. Therefore, if chondrites and the bulk Earth have equal $^{182}W/^{184}W$ isotope ratios, the mantle and core should have higher and lower $^{182}W/^{184}W$ ratios than the chondrites, respectively (as conceptually shown in Fig. 5.5).

An alternative model for generating the W isotope variations assumed that the Hf/W fractionation occurred during early silicate differentiation before ^{182}Hf became extinct (Touboul et al., 2012; Rizo et al., 2016b). Recently, Hf partitioning between bridgmanite and silicate melt under lower mantle conditions was investigated by first-principle calculations (Deng and Stixrude, 2021). The larger Hf partition coefficient of bridgmanite at greater depths may have caused a fractionation of Hf/W between the crystallized mantle and residual melt in the magma ocean. This fractionation may account for the variations in $^{182}W/^{184}W$ observed in ancient rocks, such as komatiites and modern OIBs, as will be described later.

The typical W isotope ratio of the present upper mantle is approximately 200 ppm (ppm = 10^{-6}) higher than that of chondrites (Kleine et al., 2002; Yin et al., 2002). $^{182}W/^{184}W$ of iron meteorites, which are considered to represent samples of planetary cores, are lower than those of chondrites (e.g., Kruijer et al., 2014). During the late veneer, meteoritic materials with low $^{182}W/^{184}W$ possibly fell onto the surface of the Earth and gradually penetrated into the mantle (Willbold et al., 2011; Touboul et al., 2012,

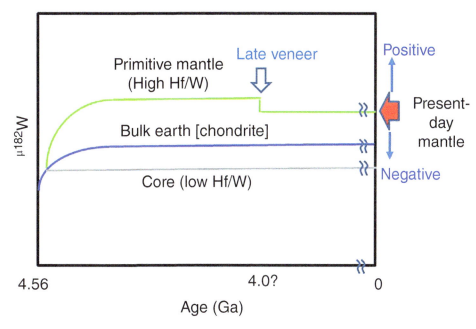

Figure 5.5 The ^{182}W/^{184}W isotopic evolution of the mantle, bulk silicate Earth, and presumed core.

2015; Kruijer et al., 2015). Due to such later influx of W, ^{182}W/^{184}W could have gradually decreased to the present value of the mantle (Fig. 5.5).

5.3.3. ^{182}W Isotope Variations of OIBs and Kimberlites

The ^{182}W/^{184}W ratios of several OIBs and volcanic rocks from the large igneous provinces (LIPs) were analyzed to investigate the W isotope signature of the deep mantle. Before 2010, the analytical uncertainties of 20–30 ppm did not allow the detection of W isotope anomalies in terrestrial rocks, except for a study by Schoenberg et al. (2002). And so, at the time, it was thought that crustal rocks do not contain materials that have undergone core-mantle interaction (Takamasa et al., 2009; Iizuka et al., 2010; Moynier et al., 2010). However, subsequent technical developments significantly improved the analytical precision and enabled the detection of small W isotope anomalies in terrestrial rocks. These data are summarized in Figure 5.6 (Rizo et al., 2019). In general, Archean rocks such as komatiites possess 5–15 ppm positive ^{182}W anomalies, whereas some OIBs exhibit negative anomalies (Touboul and Walker, 2012; Willbold et al., 2011, 2015; Rizo et al., 2016a).

Moreover, the W isotope variations in OIBs are negatively correlated with ^{3}He/^{4}He in modern OIBs, such as Hawaiian Loihi Island and Samoa (Mundl et al., 2017). Mundl-Petermeier et al. (2019, 2020) observed at least three trends in the He-W isotope space for the OIBs of Hawaii, Samoa, Iceland, Juan Fernandez, Heard and Galapagos, Moorea, Azores, Pitcairn, Caroline, Discovery, Mangaia, La Palma, and MacDonald (Fig. 5.7). They claimed that these trends were derived from mixing of the ambient mantle with mantle components affected by core-mantle interactions to varying degrees (Fig. 5.7). Rizo et al. (2019) observed low ratios for the OIBs of the Reunion and Kerguelen archipelagos and estimated the contribution of the core to be 0.3%. They also found a negative correlation between μ^{182}W and Fe/Mn for Hawaii, Reunion, Iceland, and Azores basalts and interpreted this to be evidence of a core-mantle interaction process because Fe/Mn can serve as a tracer for core-mantle interactions (Humayun et al., 2004).

The source region for some OIBs with negative μ^{182}W values is likely to correspond to the seismically identified ULVZs that may possess core μ^{182}W signatures as a result of core-mantle interactions (Mundl-Petermeier et al., 2020). Specifically, a strong negative μ^{182}W was found from Samoa Island (Mundl et al., 2017; Mundl-Petermeier et al., 2020), where the ULVZ can be clearly identified (Tanaka, 2002; Thorne et al., 2013). This supports the hypothesis that the ULVZ facilitates the transfer of core materials into the lowermost mantle (Jackson et al., 2020; Mundl-Petermeier et al., 2020).

The isotope data of kimberlites may also reveal the chemical signature of the deep mantle. Nakanishi et al. (2021) presented the homogeneous and moderately negative μ^{182}W (−5.9 ± 3.6 ppm, 2SD, n = 13) for kimberlites with ages ranging from 89 to 1153 Ma. They claimed that these kimberlites were derived from a long-lived

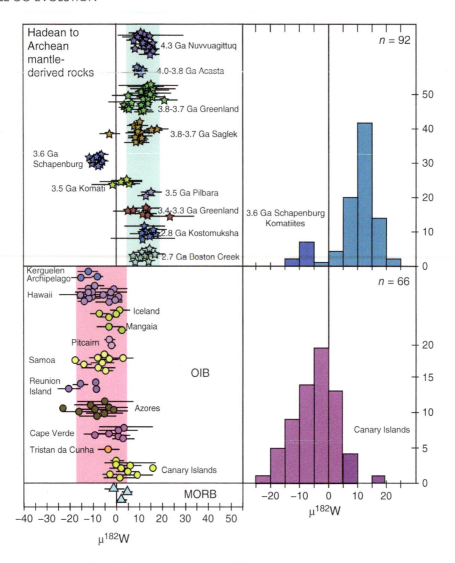

Figure 5.6 Compilation of $^{182}W/^{184}W$ data shown as $\mu^{182}W$ values of the rocks of Hadean and Archean mantle-derived rocks and modern OIBs. Shaded areas show the average $\mu^{182}W$ values ±2 standard deviations for the Hadean-Archean and OIB samples. OIB: ocean island basalts; MORB: mid-ocean ridge basalts. Data Sources: Dale et al. (2017), Kruijer and Kleine (2018), Liu et al. (2016), Mei et al. (2018), Mundl et al. (2017), Puchtel et al. (2016, 2018), Reimink et al. (2018), Rizo et al. (2016a, 2016b, 2019), Touboul et al. (2014, 2015), and Willbold et al. (2011, 2015). Source: Reproduced from Rizo et al. (2019; *Geochemical Perspective Letters*).

mantle reservoir with a negative ^{182}W anomaly, which may have been generated by core-mantle interaction, early differentiation in the mantle, and/or heterogeneous distribution of late veneer materials. On the other hand, younger kimberlites possess $\mu^{182}W$ indistinguishable from those of the modern upper mantle (Tappe et al., 2020) and may have resulted from the mixing of the kimberlite source with recycled crustal components (Nakanishi et al., 2021).

Core formation could also fractionate I-Pu especially if it occurred at high pressures where I becomes highly siderophile (Jackson et al., 2018). The fractionation of I/Pu in the mantle of the early Earth leads to Xe isotope anomalies (^{129}Xe and ^{136}Xe), which is the decay products of short-lived ^{129}I and short-lived ^{244}Pu, respectively. Such anomalies have been observed in mantle-derived rocks (Mukhopadhyay, 2012; Peto et al., 2013; Caracausi et al., 2016). Thus, it is possible that a reservoir with both W and Xe isotope anomalies in the mantle generated during core formation may have been preserved and may be contributing to the source of certain OIBs.

Meanwhile, extremely high $^{182}W/^{184}W$ ratios have been reported in modern volcanic rocks, such as the Ontong Java Plateau (eruption date 120 million years ago) and Baffin Bay (61 million years ago) (Rizo

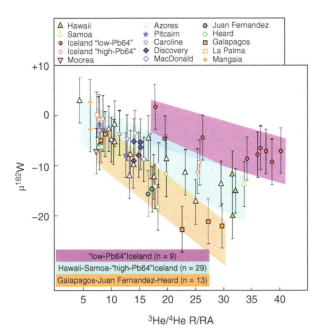

Figure 5.7 Plot of ^{3}He/^{4}He versus μ^{182}W. The purple trend represents the "low-Pb64" samples from Mundl-Petermeier et al. (2019). The light blue trend corresponds to samples from Hawaii, Samoa, and the "high-Pb64" are samples from Iceland (Mundl et al., 2017; Mundl-Petermeier et al., 2019, 2020). Samples from Juan Fernandez, Heard, and Galapagos comprise the orange trend (Mundl-Petermeier et al., 2020). Source: Reproduced from Mundl-Petermeier et al. (2020) / With permission of Elsevier.

et al., 2016a). The rocks from both these areas have chondritic Os isotopic compositions (Tejada et al., 2013 and Dale et al., 2009, for Ontong Java Plateau and Baffin Bay, respectively). The Baffin Bay volcanic rocks, in particular, exhibit high ^{3}He/^{4}He ratios. This may reflect the long-term preservation of an early formed mantle domain, which may be either an early depleted reservoir developed during solidification of the magma ocean or a primordial reservoir unaffected by the late veneer impacts. However, Kruijer and Kleine (2018) claimed that the high ^{182}W/^{184}W ratios of the basalts from these areas may be analytical artifacts most likely caused by the nuclear field shift effect during sample processing (Cook and Schönbächler, 2016; Kruijer and Kleine, 2018; Takamasa et al., 2020).

As described in the previous section, ^{182}W isotope variations observed in a small number of OIBs may imply the transfer of core materials to the lowermost mantle as a result of core-mantle interactions. However, negative μ^{182}W signatures were not observed in the lavas of LIPs. Currently, it is uncertain if the negative W isotope signature inherited from the core is masked by shallower processes during plume ascent and magma emplacement, or if contribution of the core components is absent in rocks from the LIPs. To further understand the origin of the W isotope variations in the deep mantle-derived rocks, Hf-W partitioning between metal and silicate melt and between minerals and melt under core-mantle boundary conditions needs to be better constrained.

5.3.4. Possible Effect of Crustal Recycling on ^{182}W Isotopes of OIBs

Mixing between recycled crustal materials and materials exhibiting W isotope anomalies can mask both the positive and negative W anomalies of the latter (Jackson et al., 2020; Mundl-Petermeier et al., 2020). To account for the variations displayed in W-He isotopes space of OIBs, Mundl-Petermeier et al. (2020) proposed the following three components: the Ambient Mantle representing major portions of Earth's mantle, including the DMM; the Early Formed Mantle Reservoir with high ^{3}He/^{4}He and μ^{182}W value of 0; and Core-Mantle Equilibrated Reservoir with high ^{3}He/^{4}He and negative μ^{182}W value, which was mentioned above in this section. In addition to the above, they also included the recycled subducted slab component with low ^{3}He/^{4}He, low W concentration, and positive to negative μ^{182}W (depending on the timing of formation) in their discussions. They claimed that the subducted slab could not be expected to have any notable impact on the He and W isotopic compositions of most of the OIBs, in contrast to the substantial effects the slab has produced on the Sr, Nd, and Hf isotopes.

However, Jackson et al. (2020) pointed out that among the OIBs, the highest ^{3}He/^{4}He and the largest ^{182}W anomalies are observed solely in OIBs exhibiting high ^{143}Nd/^{144}Nd and low ^{206}Pb/^{204}Pb, which are characteristics indicative of less recycled crustal materials. Another important geochemical feature pointed out by Jackson et al. (2020) is that the mantle component with high ^{3}He/^{4}He and anomalous ^{182}W possesses low W and ^{4}He concentrations compared to recycled materials and are therefore highly susceptible to overprinting by the low ^{3}He/^{4}He and normal ^{182}W characteristics of the subducted crust.

As we discussed in section 5.2, the isotope variations of lithophile elements in FOZO, HIMU, and EM clearly indicate the presence of recycled crustal components in the source of their mantle domains. As for the three trends between He-W isotopes for the OIBs reported in Mundl-Petermeier et al. (2019, 2020) (Fig. 5.6), they require another component that is mixed with the components variably affected by core-mantle interaction, which might be a domain in the deep mantle where materials from the recycled oceanic crust predominate.

5.4. HSE GEOCHEMISTRY OF THE MANTLE

HSEs consist of PGEs (Ru, Rh, Pd, Os, Ir, and Pt), Re, and Au, all of which strongly prefer metal phases to silicates as reflected by their partition coefficients between metal and silicate phases exceeding 10,000. HSEs are thus likely to have been sequestered in the Earth's core during core-mantle segregation. Despite the preferential partitioning of HSE into the core, HSE concentrations in the "primitive mantle," which is representative of the bulk mantle composition, are estimated to be much higher than those expected from metal-silicate equilibrium, and the relative abundances of HSEs are almost identical to those of chondrites (Meisel et al., 2001; Becker et al., 2006; Fischer-Gödde et al., 2011; Mann et al., 2012) (Fig. 5.8). The cause of this chondritic "overabundances" of HSEs in the primitive mantle is a key to deciphering the detailed mechanism of the core-mantle segregation and subsequent chemical evolution of the Earth's interior.

A number of hypotheses has been proposed for the cause of this HSE overabundance in the primitive mantle, including (1) the "incomplete separation hypothesis," which assumes that a part of the separated metallic phase was left behind in the mantle (Arculus and Delano, 1981; Jones and Drake, 1982); (2) the "magma ocean hypothesis," which assumes that the final separation of the metallic core was governed by partition coefficients for the high-temperature/pressure conditions at the bottom of the magma ocean (Ringwood, 1977; Murthy, 1991); (3) the "late veneer hypothesis," which assumes that the accumulation of primordial materials continued even after the formation of the metallic core and that siderophile elements were added only to the mantle (Chou, 1978; Kimura et al., 1974); and (4) the "core-mixing hypothesis," which assumes that the differentiated outer-core materials were incorporated back into the mantle (Snow and Schmidt, 1998).

To investigate the process responsible for the HSE overabundances in the mantle, accurate estimation of the HSE composition of the primitive mantle is necessary. However, this is not a straightforward process because natural peridotites, based on which the HSE composition of the primitive mantle has been estimated (Becker et al., 2006), possess a wide range of HSE concentrations, differing by more than 2 orders of magnitude. Such a wide range of HSE concentrations has been attributed to an intricate combination of factors such as melt extraction and secondary modification processes. In order to clarify the current understanding of the HSE variations in the mantle, we review major issues related to the estimation of the primitive mantle HSE composition from HSE data of natural peridotites and the experimental constraints on the fractionation of HSEs under mantle conditions.

5.4.1. HSE Composition of the Primitive Mantle

The term "primitive mantle" represents a hypothetical reservoir, which, if it were to actually exist in the Earth's interior, would be still inaccessible to us. On the other hand, natural peridotites have generally been subjected to melt extraction to considerable degrees by partial melting since the formation of the mantle. If HSE concentrations in peridotites are solely controlled by melt extraction, we can estimate the HSE composition of the primitive mantle by extrapolating the correlations of HSEs with melt extraction indices, such as Al_2O_3. In this context, peridotites with fertile compositions (i.e., enriched in components preferentially partitioned into melts such as Al_2O_3) are thought to be suitable for approximating the primitive mantle HSE composition, as they do not require extensive extrapolations. In particular, the primitive mantle HSE composition most commonly referred (Becker et al., 2006; Fischer-Gödde et al., 2011) has been estimated based on correlations between HSEs and Al_2O_3 of fertile peridotites whose compositions are assumed to have not been modified by secondary processes. It is noteworthy that these studies employed the correlations between relative concentrations of HSEs (e.g., Pd/Ir and Re/Ir) and Al_2O_3, due to the relative concentrations of HSEs exhibiting stronger correlations with Al_2O_3 than their absolute concentrations.

This methodology is based on the premise that the correlations between the relative concentrations of HSEs and Al_2O_3 of the fertile peridotites are produced by

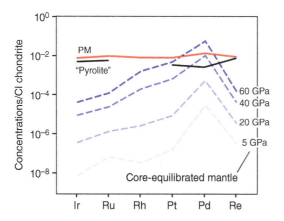

Figure 5.8 CI chondrite-normalized HSE concentrations in the primitive mantle (PM) estimated by Fischer-Gödde et al. (2011), the mantle equilibrated with the core calculated by Mann et al. (2012), and the "pyrolitic" mixture of this study. The pyrolitic mixture is the 3:1 mixture of refractory peridotite (HSE values are taken from those of the harzburgites with limited influence of metasomatism reported in Luguet and Reisberg, 2016) and MORB (data from Rehkämper et al. 1999; Peucker-Ehrenbrink et al. 2003; Bézos et al. 2005).

melt extraction. However, large gaps exist between the HSE-Al$_2$O$_3$ correlations in the fertile peridotites and the melt extraction trends expected from experimental constraints. For example, the correlation between Pd/Ir and Al$_2$O$_3$ of the fertile peridotites utilized for estimating the primitive mantle values exhibits a moderately positive slope, whereas experimental data for partitioning of Pd and Ir predict a strongly positive slope for the Pd/Ir–Al$_2$O$_3$ correlation (Figure 5D of Fischer-Gödde et al., 2011). Although this discrepancy was noted by Fischer-Gödde et al. (2011), they did not discuss the cause of the discrepancy in depth. Thus, the accuracy of the estimation of the HSE composition of the primitive mantle cannot be properly evaluated. In the section that follows, we provide an overview of the experimental constraints on the behavior of HSEs during partial melting of peridotite.

5.4.2. HSE Behavior in Partial Melting of the Mantle

Experimental Constraints on the Fractionation of HSEs During Partial Melting of Peridotite

Several studies have conducted laboratory experiments to understand HSE behavior in the Earth's interior, and a thorough review of the experimental constraints on HSE fractionation in mantle melting and magmatic evolution has been made by Brenan et al. (2016). Here, we focus on the experimental data relevant to HSE fractionation during partial melting under conditions found in the uppermost mantle, from where most natural peridotites are derived (Stagno et al., 2013).

We review the HSE behavior in sulfur-free silicate systems, as the major constituents of mantle peridotite are silicate and oxide minerals. Laboratory experiments on HSE fractionation in partial melting of sulfur-free peridotite have mostly focused on partitioning into olivine and spinel. This is because the HSE variations in natural komatiitic and basaltic lavas suggest that olivine and chromite are the major crystalline phases controlling the HSE concentrations in silicate melts (Puchtel et al., 2009). Mineral-silicate melt partitioning experiments on both olivine and spinel have demonstrated that the partition coefficients D of HSEs vary largely as a function of oxygen fugacity f_{O2} (Brenan et al., 2016 and references therein). For example, the $D^{olivine/silicate\ melt}$ of Re into olivine ranges from approximately 10^{-5}–10^{-1} within $\pm 10^4\ f_{O2}$ (Mallmann and O'Neill, 2007). Under the fayalite-magnetite-quartz buffer conditions, corresponding to the representative f_{O2} in the uppermost mantle, $D^{mineral/silicate\ melt}$ of Ir, Os, Ru (referred to collectively as iridium PGEs: IPGEs), and Rh are on the order of 10^0–10^1. However, those reported for Pt, Pd, and Re are less than 10^0. These data suggest that residual peridotite is expected to become enriched in IPGE and Rh and depleted in other HSEs with the progression of partial melting, unless discrete HSE-rich phases crystallize.

Despite the compatible behavior of a few HSEs with olivine and spinel, HSEs in natural peridotites, in which sulfide phases are commonly present, are concentrated mostly in Fe-Ni-rich base-metal sulfides (BMSs), such as monosulfide solid solution (mss) and pentlandite (Alard et al., 2000), and platinum-group minerals (PGMs), such as Ir-Os alloys and laurite-erlichmanite (Ru-Os sulfides) (Luguet et al., 2001; Kogiso et al., 2008). Several types of BMS phases are found in natural peridotites, but most sulfide phases in the actual upper mantle are expected to have mss-like stoichiometry (Aulbach et al., 2009). High-pressure melting experiments on mantle sulfides (Zhang and Hirschmann, 2016) demonstrated that the liquidus temperature of mss is lower than the geotherm of the convecting mantle. This suggests that sulfides in the uppermost mantle are commonly molten. Thus, it is expected that the HSE distribution in the uppermost mantle is mostly governed by partitioning of HSEs between the sulfide melt and silicate phases.

Although the partitioning of HSEs is influenced by both sulfur fugacity f_{S2} and f_{O2} (Brenan et al., 2016), a clear correlation of $D^{sulfide\ melt/silicate\ melt}$ with f_{O2} and f_{S2} has been demonstrated only for Re. The $D^{sulfide\ melt/silicate\ melt}$ of Re varies by 4 orders of magnitude with $\log f_{S2}$–$\log f_{O2}$ (Brenan, 2008), whereas the effects of f_{S2} and f_{O2} on other HSE partitioning behaviors are not so well constrained. For the f_{O2} and f_{S2} values that are likely found in the regions of basalt genesis in the upper mantle, the $D^{sulfide\ melt/silicate\ melt}$ of Re is on the order of 10^2–10^3 (Brenan, 2008), and those of the other HSEs are on the order of 10^4–10^5 (Mungall and Brenan, 2014). These values are much larger than the $D^{mineral/silicate\ melt}$ values of all HSEs for olivine and spinel, suggesting that HSEs are concentrated in the sulfide melt upon the partial melting of peridotite.

If sulfide melt is absent or exhausted during the partial melting of peridotite, HSEs are partitioned between the silicate melt and residual solid phases, including discrete PGM crystals. Even if a sulfide melt is present initially, it will be exhausted at a high degree of melting because the solubility of sulfur into silicate melt increases with the progression of partial melting during adiabatic upwelling (Mavrogenes and O'Neill, 1999). Before the exhaustion of the sulfide melt during partial melting, the sulfide melt reaches saturation in IPGEs due to their low solubility of approximately several tens of μg/g (Fonseca et al., 2012). Consequently, IPGE alloys will crystallize, remaining as solid phases even after the exhaustion of the sulfide melt. This is because IPGE solubilities in silicate melts are very low (Brenan et al., 2016 and references therein). Conversely, Pt will dissolve into the silicate melt upon the exhaustion of the sulfide melt due to its high solubility in sulfur-bearing silicate melts (Borisov and Palme, 1997). Pd, Re, and Au will also dissolve into silicate melts as their solubilities in silicate melts are sufficiently high (Bennett and Brenan, 2013). Thus, the sulfide melt is expected to disappear at some point as partial melting progresses in an adiabatically upwelling mantle, after which the IPGEs will remain in the residues as alloys, while the other HSEs will be partitioned into the silicate phases.

Melt Extraction Trends of HSEs in the Mantle: Experimental Constraints Versus Natural Peridotites

The aforementioned experimental constraints on HSE behavior during the partial melting of peridotite have enabled researchers to construct quantitative models for HSE fractionation during partial melting of the mantle (Fonseca et al., 2011; Mungall and Brenan, 2014; Aulbach et al., 2016). Some of the critical features of the experimental constraints are summarized in a schematic illustration in Figure 5.9a. Before the partial melting of silicate minerals, HSEs are mostly concentrated in the sulfide melt with an mss-like composition. As the partial melting of silicate minerals progresses, the amount of sulfide melt decreases with increasing degree of melting, and IPGE alloys crystallize and grow within the sulfide melt. After the sulfide melt is exhausted, IPGEs remain as alloys, Rh is partitioned into olivine, and other HSEs are partitioned into the silicate melt.

The HSE variations in the residual peridotite expected from the melting models are displayed in Figure 5.9b. As IPGEs and Rh are compatible with the alloy phases and/or residual olivine, their concentrations increase with increasing degree of melting. Conversely, as the other HSEs are incompatible with silicate minerals and alloys, they are extracted from the residues after the sulfide melt is exhausted. Thus, with the progression of partial melting, residual peridotite is expected to become enriched in IPGEs and Rh, and further depleted in other HSEs (Fig. 5.9b).

The changes expected from the experimental data in the HSE concentrations during peridotite partial melting are not consistent with those observed in natural peridotites. Many of the fertile peridotites utilized to estimate the primitive mantle HSE values (Becker et al., 2006; Fischer-Gödde et al., 2011) exhibited moderate depletions of Pt and Pd and significant depletions of Re and Au relative to IPGEs and Rh, when normalized to the primitive mantle values. This pattern remains mostly unchanged with decreasing Al_2O_3, that is, with the progression of partial melting (Fig. 5.9c). On the other

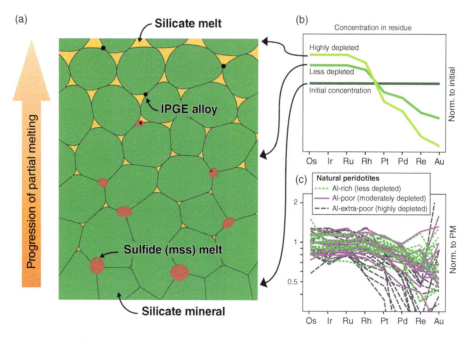

Figure 5.9 (a) Schematic illustration showing partial melting of peridotite and the fate of the phases associated with highly siderophile element (HSE) behavior. Silicate minerals at the bottom are not molten, and the degree of partial melting increases upward. (b) Expected change in HSE concentrations in residual peridotite with progression of partial melting. The vertical axis shows the values normalized to initial concentrations (not to scale). (c) HSE concentrations in the natural peridotite samples used to estimate the HSE composition of the primitive mantle (PM) (Becker et al., 2006; Fischer-Gödde et al., 2011). The vertical axis shows the values normalized to those of PM. Al-extra-poor: peridotite with Al_2O_3 less than 2.0 wt.%; Al-poor: peridotite with Al_2O_3 between 2.0 wt.% and 3.2 wt.%; Al-rich: peridotite with Al_2O_3 more than 3.2 wt.%. Source: Data from Fischer-Gödde et al. (2011).

hand, quantitative melting models (Fonseca et al., 2011; Mungall and Brenan, 2014) have succeeded in producing HSE concentration patterns in a partial melt that are consistent with those of natural mantle-derived magmas, such as MORBs. This suggests that the general trend of enrichments of IPGEs and Rh, along with depletions of other HSEs (Fig. 5.9b), may possibly be a robust feature of melt extraction from peridotite. Nevertheless, the natural fertile peridotites do not exhibit such clear enrichments of IPGEs and Rh with decreasing Al_2O_3 (Fig. 5.9c). Thus, in addition to melt extraction, there appears to be other processes that control the HSE variations in the natural fertile peridotites.

5.4.3. Influence of Metasomatism

A major process that significantly affects chemical compositions of mantle peridotites other than partial melting is "metasomatism," which refers to various (and sometimes unspecified) petrologic reactions between rock and silicate melt or aqueous/carbonate fluid phases. Mantle peridotites often undergo metasomatism during their transport from the mantle to the Earth's surface, and their HSE compositions tend to be significantly modified by metasomatism (Lorand and Luguet, 2016; Luguet and Reisberg, 2016).

A type of metasomatism often referred to as "refertilization" specifically refers to the reaction between refractory (depleted in melt components) peridotite and percolating mafic melts. Petrologic studies on natural peridotites have demonstrated that some fertile peridotites in peridotite massifs, which were utilized to estimate primitive mantle HSEs, are significantly affected by refertilization processes (Le Roux et al., 2007). This indicates that the HSE variations in the fertile peridotites utilized for the primitive mantle HSE abundance estimation were the product of not only by melt extraction but also of refertilization.

If the HSE composition of the refertilizing agent is close to that of the melt extracted from primitive mantle-like peridotite, that is, typical mantle-derived magmas like MORB, the refertilized peridotite could have HSE concentrations similar to those of the primitive mantle. In such a case, the HSE-Al_2O_3 correlations of the fertile peridotites provide a strong basis for precise estimation of HSEs in the primitive mantle, provided that the HSE-Al_2O_3 correlations are the combined effect of melt extraction and refertilization. However, the HSE compositions of the refertilizing agents are likely to differ considerably from those of typical mantle-derived melt such as MORB. For example, the fertile peridotites display a clear positive correlation between Pd and Al_2O_3 (Fig. 5.10) (Becker et al., 2006; Fischer-Gödde et al., 2011). If this correlation was produced by melt

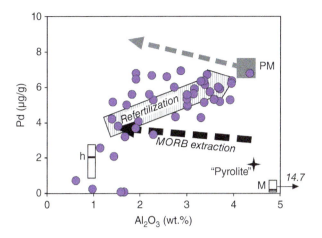

Figure 5.10 Pd-Al_2O_3 correlation of the fertile peridotites (filled circles) used for the estimation of the HSE composition of the primitive mantle (PM) (Fischer-Gödde et al., 2011). The thick hatched arrow represents a trend of refertilization of a refractory (1.5 wt.% Al_2O_3) peridotite. For the sake of clarity, the trend has been greatly simplified. The actual refertilization process would follow a more complex trend due to the complexity of melt-rock interactions. However, it is certain that the refertilizing agent has Pd concentration higher than all peridotites and PM. The black broken arrow is the trend of MORB-like melt extraction necessary to produce the refractory peridotite. The gray broken arrow is the trend of MORB-like melt extraction from PM. The composition of the MORB-like melt is assumed to be 14.7 wt.% Al_2O_3 and 0.197 µg/g Pd. Data source for MORB is the same as in Figure 5.8. Boxes show the compositional ranges (median: thick line in the middle; first and third quartiles: lower and upper sides, respectively) of harzburgite (h) and MORB (M) used for the HSE estimation of the pyrolitic mixture. Cross is the Pd concentration of the pyrolitic mixture. Source: Adapted from Fischer-Gödde et al. (2011).

extraction and subsequent refertilization, the positive Pd-Al_2O_3 correlation requires the refertilizing agent to have a higher Pd concentration than the fertile peridotites, which have much higher Pd concentrations than a MORB-like melt (Fig. 5.10). In addition, it is clear that the extraction of MORB-like melt from the primitive mantle of Fischer-Gödde et al. (2011) does not reproduce the observed trend of the fertile peridotites.

Actual refertilization processes are expected to be far more complicated than simple addition of refertilizing agents. However, there must have been a component more enriched in Pd and Al_2O_3 than the fertile peridotites that contributed to the creation of the positive Pd-Al_2O_3 correlation (Fig. 5.10). Thus, if the fertile peridotites are chemically affected by refertilization, extrapolation of the HSE-Al_2O_3 trends is not an appropriate method for estimating the HSE composition of the primitive mantle.

A key to estimating the extent of metasomatic influence on the HSE compositions of peridotite may be found in the occurrence of BMSs in peridotite samples. Petrologic and geochemical studies on BMSs in peridotite (Alard et al., 2000; Luguet and Reisberg, 2016; Luguet and Pearson, 2019) demonstrated that BMSs can be classified into several groups based on their occurrence: one (sometimes referred to as Type 1) is completely enclosed in major silicate phases in peridotite and other three (Types 2–4) are observed along the grain boundaries of other phases. Type 1 BMSs are generally enriched in IPGEs relative to other HSEs, and this IPGE enrichment is more emphasized in refractory peridotites than in fertile ones (Alard et al., 2000; Delpech et al., 2012). This signature roughly mimics the changes in whole rock HSE compositions of peridotite with increasing degree of melt extraction (Fig. 5.9b), suggesting that Type 1 BMSs preserve the HSE characteristics of residual peridotite after partial melting. Thus, the existence of Type 1 BMSs in peridotite could indicate that the peridotite sample did not suffer significant metasomatic disturbance. However, Type 1 BMSs are often closely associated with micrometer-sized sulfide inclusion arrays (Harvey et al., 2016). If these small sulfide inclusions originated from metasomatic sulfide melts filling grain boundaries, discrete sulfide inclusions that resemble Type 1 BMSs could be left after the healing of grain boundaries. Meticulous petrographic observations of BMSs in peridotite are necessary for discerning the influence of metasomatism on HSE concentrations in peridotite samples (Akizawa et al., 2017, 2020). Although the occurrence of BMSs in the fertile peridotites utilized for the estimation of primitive mantle HSE composition has been reported (Becker et al., 2006), detailed petrologic and geochemical studies on them are awaited as they would be indispensable for better estimation of the primitive mantle HSE composition.

Metasomatic influences on HSE compositions of peridotites have been further discussed concerning infiltration of host magmas into xenoliths. Peridotite xenoliths are included in the fertile peridotites utilized for primitive mantle HSE estimations (Becker et al., 2006; Fischer-Gödde et al., 2011). It has been pointed out that Os concentrations in alkali basalt-borne peridotite xenoliths are systematically lower than those in kimberlite-borne peridotite xenoliths and in rock samples taken from peridotite bodies (Handler et al., 1999; Pearson et al., 2004). Clear correlations between HSEs with other chalcophile elements observed in alkali basalt-borne xenoliths have partly been attributed to the dissolution of Os with sulfur into host alkali basaltic magmas during the transport of the xenoliths to the surface (Handler et al., 1999). Similar Os depletions have been observed in harzburgite xenoliths in alnöite (ultramafic magma similar to kimberlite) from Malaita in Solomon Islands (Ishikawa et al., 2004, 2011a, 2011b). By contrast, the majority of peridotites such as lherzolites display normal Os abundances, implying that their HSE variations may be caused by partial melting and metasomatism within the mantle rather than by interaction with the host magmas during xenolith transportation. Thus, the actual processes responsible for the decrease in Os concentration in alkali basalt-borne xenoliths have not been completely understood.

Although detailed mechanisms of the dissolution of HSEs in peridotite xenoliths into host magmas still remain to be investigated, the stability conditions of BMSs and PGMs, which are highly sensitive to oxygen fugacity, temperature, and pressure, are expected to be one of the most important factors (Liu et al., 2010). Thorough experimental studies on the behaviors of HSEs in metasomatic reactions will be essential for resolving the influence of metasomatism on HSE geochemistry in the mantle.

5.4.4. Validity of the Chondritic HSE "Overabundance" in the Mantle

As described above, our understandings of the HSE composition of the primitive mantle include considerable uncertainties, which are inherent in the methodology of the estimation of the HSE composition of the primitive mantle. Irrespective of such methodological difficulties, we attempt to constrain the probable ranges of the HSE concentrations in the primitive mantle herein.

The fertile peridotites utilized for the estimation of the primitive mantle HSE composition are likely to have been refertilized as discussed above, but refractory (or less fertile) peridotites probably have not been affected extensively by refertilization and may even retain compositions close to their pristine compositions when they were formed as residues after melt extraction in the mantle. If this is the case, we can then assume that the HSE composition of the primitive mantle is close to that of a mixture of refractory peridotite and basaltic melt, which basically follows the same line of logic as in the estimation of major element composition of "pyrolite" by Ringwood (1966).

To estimate the "pyrolitic" HSE composition, we need to determine the compositions of refractory peridotite that have not been remarkably affected by metasomatism and other processes, such as alteration at the Earth's surface. It is generally difficult to select refractory peridotites suitable for this calculation due to the scarcity of petrographic descriptions that have been reported in enough detail to assess the influence of metasomatism in existing studies on HSE geochemistry of peridotites. Nevertheless, we utilize the median value of the harzburgites (refractory peridotites) reported in Luguet and Reisberg (2016) that were interpreted to have experienced limited metasomatism, disregarding the insufficiency of the petrographic evidence.

The HSE concentrations of the pyrolitic mixture of the harzburgite and average MORB were calculated using the median concentration of each HSE in the harzburgites and average MORB (data sources are listed in the caption of Fig. 5.8). The mixing ratio of the harzburgite to the average MORB was assumed to be 3:1 so that Al_2O_3 content of the mixture falls within the range of that of the primitive mantle (4.0–4.5 wt.%; McDonough and Sun, 1995). The calculated HSE concentrations of the pyrolitic mixture are shown in Figures 5.8 and 5.10. They are all within 1 order of magnitude of their concentrations in the primitive mantle of Fischer-Gödde et al. (2011). Most are obviously higher than those expected from metal-silicate equilibrium (Mann et al., 2012) (Fig. 5.8). This suggests that if the pyrolitic composition is a rational estimation for the primitive mantle HSE composition, then the mantle can be expected to have an overabundance of HSEs.

On the other hand, the CI chondrite-normalized pattern of the pyrolitic mixture exhibits significant depletions of Pt and Pd relative to IPGEs, which are clearly inconsistent with those of the primitive mantle of Fischer-Gödde et al. (2011) (Fig. 5.8). In particular, the depletion of Pd is totally opposite to the HSE feature of the primitive mantle of Fischer-Gödde et al. (2011), which is characterized by clear enrichments of Pd and Ru relative to IPGEs (Fischer-Gödde et al., 2011). The relative enrichments of Pd and Ru are observed in some fertile peridotites, but it is not clear whether such enrichments are the general character of the mantle or were caused by metasomatic modification of the original HSE features of peridotite samples (Lorand et al., 2010; Snow and Schmidt, 1998). The clear depletion of Pd and the absence of enrichment of Ru in the pyrolitic mixture cast doubts on the view that the enrichments of Pd and Ru are the characteristics of the primitive mantle. However, the significance of our pyrolitic HSE composition should be treated tentatively until the harzburgite composition utilized in our pyrolitic mixture calculation is verified to be representative of the melt-extracted mantle composition through detailed petrographic descriptions in future studies.

The chondritic relative abundance of HSEs in the primitive mantle has also been supported by the correlations between $^{187}Os/^{188}Os$ ratios and Al_2O_3 (Meisel et al., 2001). Nearly chondritic $^{187}Os/^{188}Os$ ratios of fertile lherzolites with limited influence of metasomatism could provide strong evidence for the primitive mantle having $^{187}Os/^{188}Os$ and Re/Os ratios close to those of chondrite (Reisberg, 2021). However, note that even if this were proven to be true, it would not necessarily guarantee that the relative abundances of other HSEs in the primitive mantle are chondritic. Moreover, Re/Os ratios of the mantle could have remained unchanged from the chondritic value even with core-mantle segregation as discussed below. The partition coefficient of Ir between iron melt and silicate melt becomes close to that of Re at higher pressures (Fig. 5.8) (Mann et al., 2012), suggesting that Os, which has similar chemical features with Ir, would also exhibit partition coefficient close to that of Re. This means that even if the primitive mantle were to have chondritic Os isotopic ratio, it would not follow that the HSE composition of the primitive mantle is unrelated to those of core-mantle equilibration.

We summarize that it is probable that the HSE concentrations of the primitive mantle are mostly higher than those expected from metal-silicate equilibrium based on our current understandings of HSEs of mantle peridotites. However, whether the relative HSE abundance is closely chondritic or not is still a matter of debate. The incomplete understanding about the relative HSE abundance prevents us from making precise arguments about the origin of the HSE overabundance in the mantle. Moreover, it should be noted that the partitioning of HSEs between the core and mantle has not yet been accurately revealed. The partition coefficients of Mann et al. (2012) were determined between silicate melt and pure Fe under upper mantle conditions, but the partitioning behaviors of siderophile elements are significantly affected by the presence of light elements in the metal phases, as well as the temperature and pressure of equilibrium (Huang et al., 2021). A recent experimental study (Suer et al., 2021) demonstrated that partitioning of Pt into Fe at high pressure and temperature close to those at the core-mantle boundary is much lower than had been previously demonstrated (Mann et al., 2012). Therefore, HSE abundances in the mantle attained through metal-silicate equilibrium could be completely different from those shown in Figure 5.8, and even the "overabundance" of HSEs might be an illusion.

5.5. SUMMARY

In this chapter, we reviewed the current status of our understanding of isotopic and HSE characteristics of the mantle and their relevance to the core-mantle interactions and deep mantle evolution. The Sr-Nd-Pb isotopic data of OIB demonstrate that the isotopic diversity in the mantle can be explained by mixing between several end-member components with extreme isotopic ratios (EM1, EM2, and HIMU) and a common component (FOZO) that has a range of Pb isotopic ratios and restricted Sr-Nd isotopic ratios. Several FOZO-like OIBs exhibit primitive He isotope signatures, suggesting that FOZO might represent a primitive or less degassed component residing in the lowermost mantle. However, our calculations demonstrate that the isotopic features of FOZO can be attributed entirely to the recycling of subducted oceanic crust that has experienced fluid-rock reactions induced by serpentine dehydration in the subducting slab. Thus, deep mantle processes, such as long-term preservation of the primitive component and core-mantle interaction, are not reflected in the Sr-Nd-Pb isotopic heterogeneity of the mantle.

Negative ^{182}W anomalies in some OIBs suggest a contribution of core materials to the lower mantle as a result of core-mantle interactions. Although it is still debatable whether the low ^{182}W/^{184}W inherited from the core are masked by shallow processes during plume ascent and/or recycling of subducted crustal components, correlations of ^{182}W/^{184}W with other isotopes like ^{3}He/^{4}He appear to support that the W isotope systematics of mantle-derived rocks reflect deep mantle processes. On the other hand, ^{186}Os-^{187}Os isotopic variations in OIBs, which have sometimes been attributed to the influence of the core-mantle interaction, are likely to be affected by shallow mantle processes like metasomatism.

The chondritic "overabundance" of HSEs in the primitive mantle is the key to understanding the details of core-mantle segregation in the early Earth and subsequent chemical evolution of the mantle. However, HSE variations in natural fertile peridotites, which were utilized for estimating the HSE abundances in the primitive mantle, are not consistent with the results from mineral-silicate melt partitioning experiments under mantle conditions. This is because the natural fertile peridotites have been subjected to secondary processes that have modified their original chemical compositions, such as melt extraction and metasomatism. These observations indicate that the estimated HSE composition (i.e., the chondritic overabundance of HSEs) of the primitive mantle involves large uncertainties. Furthermore, our knowledge on the HSE partitioning between metal and silicate phases at high pressures and temperatures is still being updated. Application of the pyrolite model to mantle HSEs along with detailed petrographic characterization of refractory peridotites will be a promising approach for more precise estimation of the primitive mantle HSE composition.

As presented in this chapter, the integrated usage of multiple elements and isotope systematics is a fundamental step toward establishing a plausible scenario for the geochemical evolution of the deep Earth's mantle. In this regard, further accumulation of geochemical and experimental datasets of multiple elements and isotope systematics is strongly desired.

ACKNOWLEDGMENTS

All authors equally contributed to this paper (KS is the corresponding author). This work was supported by the Japan Society for the Promotion of Science (No. 15H05833 and 15H05831). We are grateful to Ambre Luguet for providing the HSE dataset of refractory peridotites. We wish to thank Junko Saka for editing the references, to Junko Kikuchi for preparing the figure, and to Editage (www.editage.com) and Rieko Tsurudome for English language editing. We are grateful to Tsuyoshi Iizuka, an anonymous reviewer, and Madhusoodhan Satish-Kumar (Editor) for careful reading and constructive comments, which were significantly helpful in improving the manuscript.

REFERENCES

Adam, J., Locmelis, M., Afonso, J. C., Rushmer, T., & Fiorentini, M. L. (2014). The capacity of hydrous fluids to transport and fractionate incompatible elements and metals within the Earth's mantle. *Geochemistry, Geophysics, Geosystems, 15*, 2241–2253. https://doi:10.1002/2013GC005199.

Akizawa, N., Kogiso, T., Miyake, A., Tsuchiyama, A., Igami, Y., & Uesugi, M. (2020). Formation process of submicrometer-sized metasomatic platinum-group element-bearing sulfides in Tahitian harzburgite xenolith. *Canadian Mineralogist, 58*(1), 99–114. https://doi.org/10.3749/canmin.1800082.

Akizawa, N., Miyake, A., Ishikawa, A., Tamura, A., Terada, Y., Uesugi, K., et al. (2017). Metasomatic PGE mobilization by carbonatitic melt in the mantle: Evidence from sub-μm-scale sulfide–carbonaceous glass inclusion in Tahitian harzburgite xenolith. *Chemical Geology, 475*, 87–104. https://doi.org/10.1016/j.chemgeo.2017.10.037.

Alard, O., Griffin, W. L., Lorand, J. P., Jackson, S. E., & O'Reilly, S. Y. (2000). Non-chondritic distribution of the highly siderophile elements in mantle sulphides. *Nature, 407*(6806), 891–894. https://doi.org/10.1038/35038049, PubMed: 11057664.

Allègre, C. J., & Turcotte, D. L. (1986). Implications of a two-component marble-cake mantle. *Nature, 323*(6084), 123–127. https://doi.org/10.1038/323123a0.

Arculus, R. J., & Delano, J. W. (1981). Siderophile element abundances in the upper mantle: Evidence for a sulfide signature and equilibrium with the core. *Geochimica et Cosmochimica Acta, 45*(8), 1331–1343. https://doi.org/10.1016/0016-7037(81)90226-X.

Aulbach, S., Mungall, J. E., & Pearson, D. G. (2016). Distribution and processing of highly siderophile elements in cratonic mantle lithosphere. *Reviews in Mineralogy and Geochemistry, 81*(1), 239–304. https://doi.org/10.2138/rmg.2016.81.5.

Aulbach, S., Stachel, T., Creaser, R. A., Heaman, L. M., Shirey, S. B., Muehlenbachs, K., et al. (2009). Sulphide survival and diamond genesis during formation and evolution of Archaean subcontinental lithosphere: A comparison between the Slave and Kaapvaal Cratons. *Lithos, 112*, 747–757. https://doi.org/10.1016/j.lithos.2009.03.04.

Bach, W., Alt, J. C., Niu, Y., Humphris, S. E., Erzinger, J., & Dick, H. J. B. (2001). The geochemical consequences of late-stage low-grade alteration of lower ocean crust at the SW Indian Ridge: Results from ODP Hole 735B (Leg 176).

Geochimica et Cosmochimica Acta, 65(19), 3267–3287. https://doi.org/10.1016/S0016-7037(01)00677-9.

Bach, W., Peucker-Ehrenbrink, B., Hart, S. R., & Blusztajn, J. S. (2003). Geochemistry of hydrothermally altered oceanic crust: DSDP/ODP Hole 504B - Implications for seawater-crust exchange budgets and Sr- and Pb-isotopic evolution of the mantle. *Geochemistry, Geophysics, Geosystems*, 4(3), 8904. https://doi.org/10.1029/2002GC000419.

Bali, E., Keppler, H., & Audetat, A. (2012). The mobility of W and Mo in subduction zone fluids and the Mo-W-Th-U systematics of island arc magmas. *Earth and Planetary Science Letters*, 351–352, 195–207. https://doi.org/10.1016/j.epsl.2012.07.032.

Bebout, G. E., Ryan, J. G., Leeman, W. P., & Bebout, A. E. (1999). Fractionation of trace elements by subduction-zone metamorphism – Effect of convergent-margin thermal evolution. *Earth and Planetary Science Letters*, 171(1), 63–81. https://doi.org/10.1016/S0012-821X(99)00135-1.

Becker, H., Horan, M. F., Walker, R. J., Gao, S., Lorand, J. -P., & Rudnick, R. L. (2006). Highly siderophile element composition of the Earth's primitive upper mantle: Constraints from new data on peridotite massifs and xenoliths. *Geochimica et Cosmochimica Acta*, 70(17), 4528–4550. https://doi.org/10.1016/j.gca.2006.06.004.

Bennett, N. R., & Brenan, J. M. (2013). Controls on the solubility of rhenium in silicate melt: Implications for the osmium isotopic composition of Earth's mantle. *Earth and Planetary Science Letters*, 361, 320–332. https://doi.org/10.1016/j.epsl.2012.10.028.

Bézos, A., Lorand, J. -P., Humler, E., & Gros, M. (2005). Platinum-group element systematics in Mid-Oceanic Ridge basaltic glasses from the Pacific, Atlantic, and Indian Oceans. *Geochimica et Cosmochimica Acta*, 69(10), 2613–2627. https://doi.org/10.1016/j.gca.2004.10.023.

Borisov, A., & Palme, H. (1997). Experimental determination of the solubility of platinum in silicate melts. *Geochimica et Cosmochimica Acta*, 61(20), 4349–4357. https://doi.org/10.1016/S0016-7037(97)00268-8.

Bose, K., & Ganguly, J. (1995). Experimental and theoretical studies of the stabilities of talc, antigorite and phase A at high pressures with applications to subduction processes. *Earth and Planetary Science Letters*, 136(3–4), 109–121. https://doi.org/10.1016/0012-821X(95)00188-I.

Brandon, A. D., Humayun, M., Puchtel, I. S., & Zolensky, M. E. (2005). Re-Os isotopic systematics and platinum group element composition of the Tagish Lake carbonaceous chondrite. *Geochimica et Cosmochimica Acta*, 69(6), 1619–1631. https://doi.org/10.1016/j.gca.2004.10.005.

Brandon, A. D., Walker, R. J., Puchtel, I. S., Becker, H., Humayun, M., & Revillon, S. (2003). ^{186}Os-^{187}Os systematics of Gorgona Island komatiites: Implications for early growth of the inner core. *Earth and Planetary Science Letters*, 206(3–4), 411–426. https://doi.org/10.1016/S0012-821X(02)01101-9.

Brandon, A. D., Walker, R. J., Morgan, J. W., Norman, M. D., & Prichard, H. M. (1998). Coupled ^{186}Os and ^{187}Os evidence for core-mantle interaction. *Science*, 280(5369), 1570–1573. https://doi.org/10.1126/science.280.5369.1570, PubMed: 9616113.

Brenan, J. M. (2008). Re-Os fractionation by sulfide melt-silicate melt partitioning: A new spin. *Chemical Geology*, 248(3–4), 140–165. https://doi.org/10.1016/j.chemgeo.2007.09.003.

Brenan, J. M., Bennett, N. R., & Zajacz, Z. (2016). Experimental results on fractionation of the highly siderophile elements (HSE) at variable pressures and temperatures during planetary and magmatic differentiation. *Reviews in Mineralogy and Geochemistry*, 81(1), 1–87. https://doi.org/10.2138/rmg.2016.81.1.

Caracausi, A., Avice, G., Burnard, P.G., Füri, E. & Marty, B. (2016). Chondritic xenon in the Earth's mantle. *Nature 533*, 82–85. https://10.1038/nature17434.

Chou, C. -L. (1978). Fractionation of siderophile elements in the Earth's upper mantle. *Proceedings of the Lunar and Planetary Science Conference 9*[th], 219–230.

Cook, D. L., & Schönbächler, M. (2016). High-precision measurement of W isotopes in Fe-Ni alloy and the effects from the nuclear field shift. *Journal of Analytical Atomic Spectrometry*, 31(7), 1400–1405. https://doi.org/10.1039/C6JA00015K.

Dale, C. W., Pearson, D. G., Starkey, N. A., Stuart, F. M., Ellam, R. M., Larsen, L. M., et al. (2009). Osmium isotopes in Baffin Island and West Greenland picrites: Implications for the ^{187}Os/^{188}Os composition of the convecting mantle and the nature of high ^{3}He/^{4}He mantle. *Earth and Planetary Science Letters*, 278(3), 267–277. https://doi.org/10.1016/j.epsl.2008.12.014.

Dale, C. W., Kruijer, T. S., & Burton, K. W. (2017). Highly siderophile element and ^{182}W evidence for a partial late veneer in the source of 3.8 Ga rocks from Isua, Greenland. *Earth and Planetary Science Letters*, 458, 394–404. https://doi.org/10.1016/j.epsl.2016.11.001.

Delpech, G., Lorand, J. -P., Grégoire, M., Cottin, J. -Y., & O'Reilly, S. Y. (2012). In-situ geochemistry of sulfides in highly metasomatized mantle xenoliths from Kerguelen, southern Indian Ocean. *Lithos*, 154, 296–314. https://doi.org/10.1016/j.lithos.2012.07.018.

Deng, J., & Stixrude, L. (2021). Deep fractionation of Hf in a solidifying magma ocean and its implications for tungsten isotopic heterogeneities in the mantle. *Earth and Planetary Science Letters*, 562. https://doi.org/10.1016/j.epsl.2021.116873, PubMed: 116873.

DePaolo, D. J., & Wasserburg, G. J. (1976). Nd isotopic variations and petrogenetic models. *Geophysical Research Letters*, 3(5), 249–252. https://doi.org/10.1029/GL003i005p00249.

Deschamps, F., Godard, M., Guillot, S., & Hattori, K. (2013). Geochemistry of subduction zone serpentinites: A review. *Lithos*, 178, 96–127. https://doi.org/10.1016/j.lithos.2013.05.019.

Fischer-Gödde, M., Becker, H., & Wombacher, F. (2011). Rhodium, gold and other highly siderophile elements in orogenic peridotites and peridotite xenoliths. *Chemical Geology*, 280(3–4), 365–383. https://doi.org/10.1016/j.chemgeo.2010.11.024.

Fonseca, R. O. C., Laurenz, V., Mallmann, G., Luguet, A., Hoehne, N., & Jochum, K. P. (2012). New constraints on the genesis and long-term stability of Os-rich alloys in the Earth's mantle. *Geochimica et Cosmochimica Acta*, 87, 227–242. https://doi.org/10.1016/j.gca.2012.04.002.

Fonseca, R. O. C., Mallmann, G., O'Neill, H. St. C., Campbell, I. H., & Laurenz, V. (2011). Solubility of Os and Ir in sulfide melt: Implications for Re/Os fractionation during mantle melting. *Earth and Planetary Science Letters*, *311*(3–4), 339–350. https://doi.org/10.1016/j.epsl.2011.09.035.

Fukao, Y., To, A., & Obayashi, M. (2003). Whole mantle P wave tomography using P and PP-P data. *Journal of Geophysical Research: Solid Earth*, *108*(B1), ESE8–ESE1. https://doi.org/10.1029/2001JB000989.

Fukao, Y., Obayashi, M., Nakakuki, T., & the Deep Slab Project Group (2009). Stagnant slab: A review. *Annual Review of Earth and Planetary Sciences*, *37*(1), 19–46. https://doi.org10.1146/annurev.earth.36.031207.124224.

Gale, A., Dalton, C. A., Langmuir, C. H., Su, Y. J., & Schilling, J. G. (2013). The mean composition of ocean ridge basalts. *Geochemistry, Geophysics, Geosystems*, *14*(3), 489–518. https://doi.org10.1029/2012GC004334.

Garnero, E. J., & McNamara, A. K. (2008). Structure and dynamics of earth's lower mantle. *Science*, *320*(5876), 626–628. https://doi.org/10.1126/science.1148028, PubMed: 18451293.

Garth, T., & Rietbrock, A. (2014). Order of magnitude increase in subducted H_2O due to hydrated normal faults within the Wadati-Benioff zone. *Geology*, *42*(3), 207–210. https://doi.org10.1130/G34730.1.

Green, T. H., & Adam, J. (2003). Experimentally determined trace element characteristics of aqueous fluid from partially dehydrated mafic oceanic crust at 3.0 GPa, 650–700 °C. *European Journal of Mineralogy*, *15*(5), 815–830. https://doi.org10.1127/0935-1221/2003/0015-0815.

Hacker, B. R. (2008). H_2O subduction beyond arcs. *Geochemistry, Geophysics, Geosystems*, *9*(3). https://doi.org/10.1029/2007GC001707.

Hacker, B. R., Abers, G. A., & Peacock, S. M. (2003). Subduction factory 1. Theoretical mineralogy, densities, seismic wave speeds, and H_2O contents. *Journal of Geophysical Research: Solid Earth*, *108*(B1), 2029. https://doi.org/10.1029/2001JB001127.

Hanan, B. B., & Graham, D. W. (1996). Lead and helium isotope evidence from oceanic basalts for a common deep source of mantle plumes. *Science*, *272*(5264), 991–995. https://doi.org/10.1126/science.272.5264.991, PubMed: 8662585.

Handler, M. R., Bennett, V. C., & Dreibus, G. (1999). Evidence from correlated Ir/Os and Cu/S for late-stage Os mobility in peridotite xenoliths: Implications for Re-Os systematics. *Geology*, *27*, 75–78.

Hart, S. R., Hauri, E. H., Oschmann, L. A., & Whitehead, J. A. (1992). Mantle plumes and entrainment: Isotopic evidence. *Science*, *256*(5056), 517–520. https://doi.org/10.1126/science.256.5056.517, PubMed: 17787949.

Harvey, J., Warren, J. M., & Shirey, S. B. (2016). Mantle sulfides and their role in re–Os and Pb isotope geochronology. *Reviews in Mineralogy and Geochemistry*, *81*(1), 579–649. https://doi.org10.2138/rmg.2016.81.10.

Hayden, L. A., & Watson, E. B. (2007). A diffusion mechanism for core–mantle interaction. *Nature*, *450*(7170), 709–711. https://doi.org/10.1038/nature06380, PubMed: 18046408.

Helffrich, G., & Kaneshima, S. (2010). Outer-core compositional stratification from observed core wave speed profiles. *Nature*, *468*(7325), 807–810. https://doi.org/10.1038/nature09636, PubMed: 21150995.

Hermann, J., Spandler, C., Hack, A., & Korsakov, A. V. (2006). Aqueous fluids and hydrous melts in high-pressure and ultra-high pressure rocks: Implications for element transfer in subduction zones. *Lithos*, *92*(3–4), 399–417. https://doi.org10.1016/j.lithos.2006.03.055.

Hill, E., Wood, B. J., & Blundy, J. D. (2000). The effect of Ca-Tschermaks component on trace element partitioning between clinopyroxene and silicate melt. *Lithos*, *53*, 203–215.

Hofmann, A. W. (2014). 3.3. Sampling mantle heterogeneity through oceanic basalts: Isotopes and trace elements. In H. D. Holland, & K. K. Turekian (Eds.), *Treatise on Geochemistry* (2nd ed) (pp. 67–101). Oxford: Elsevier.

Huang, D., Siebert, J., & Badro, J. (2021). High pressure partitioning behavior of Mo and W and late sulfur delivery during Earth's core formation. *Geochimica et Cosmochimica Acta*, *310*, 19–31. https://doi.org10.1016/j.gca.2021.06.031.

Humayun, M., Qin, L., & Norman, M. D. (2004). Geochemical evidence for excess iron in the mantle beneath Hawaii. *Science*, *306*(5693), 91–94. https://doi.org/10.1126/science.1101050, PubMed: 15459385.

Iizuka, T., Nakai, S. i, Sahoo, Y. V., Takamasa, A., Hirata, T., & Maruyama, S. (2010). The tungsten isotopic composition of Eoarchean rocks: Implications for early silicate differentiation and core–mantle interaction on Earth. *Earth and Planetary Science Letters*, *291*(1–4), 189–200. https://doi.org10.1016/j.epsl.2010.01.012.

Ireland, T. J., Walker, R. J., & Brandon, A. D. (2011). ^{186}Os-^{187}Os systematics of Hawaiian picrites revisited: New insights into Os isotopic variations in ocean island basalts. *Geochimica et Cosmochimica Acta*, *75*(16), 4456–4475. https://doi.org10.1016/j.gca.2011.05.015.

Ishikawa, A., Maruyama, S., & Komiya, T. (2004). Layered lithospheric mantle beneath the Ontong Java Plateau: Implications from xenoliths in alnöite, Malaita, Solomon Islands. *Journal of Petrology*, *45*(10), 2011–2044. https://doi.org10.1093/petrology/egh046.

Ishikawa, A., Maruoka, T., Dale, C. W., & Pearson, D. G. (2011a). Trans-lithospheric variations in highly siderophile elements beneath the Ontong Java Plateau. *Mineralogical Magazine*, *75*, 1086.

Ishikawa, A., Pearson, D. G., & Dale, C. W. (2011b). Ancient Os isotope signatures from the Ontong Java Plateau lithosphere: Tracing lithospheric accretion history. *Earth and Planetary Science Letters*, *301*(1–2), 159–170. https://doi.org10.1016/j.epsl.2010.10.034.

Iwamori, H. (2000). Deep subduction of H_2O and deflection of volcanic chain towards backarc near triple junction due to lower temperature. *Earth and Planetary Science Letters*, *181*(1–2), 41–46. https://doi.org10.1016/S0012-821X(00)00180-1.

Iwamori, H. (2004). Phase relations of peridotites under H_2O-saturated conditions and ability of subducting plates for transportation of H_2O. *Earth and Planetary Science Letters*, *227*(1–2), 57–71. https://doi.org/10.1016/j.epsl.2004.08.013.

Iwamori, H., Albarède, F., & Nakamura, H. (2010). Global structure of mantle isotopic heterogeneity and its implications for mantle differentiation and convection. *Earth and Planetary Science Letters*, *299*(3–4), 339–351. https://doi.org10.1016/j.epsl.2010.09.014.

Jackson, C. R. M., Bennett, N. R., Du, Z., Cottrell, E., & Fei, Y. (2018). Early episodes of high-pressure core formation preserved in plume mantle. *Nature*, *553*(7689), 491–495. https://doi.org/10.1038/nature25446, PubMed: 29368705.

Jackson, M. G., Blichert-Toft, J., Halldórsson, S. A., Mundl-Petermeier, A., Bizimis, M., Kurz, M. D. et al. (2020). Ancient helium and tungsten isotopic signatures preserved in mantle domains least modified by crustal recycling. *Proceedings of the National Academy of Sciences of the United States of America*, *117*(49), 30993–31001. https://doi.org/10.1073/pnas.2009663117, PubMed: 33229590.

Jacobsen, S. B., & Wasserburg, G. J. (1979). The mean age of mantle and crustal reservoirs. *Journal of Geophysical Research: Solid Earth*, *84*(B13), 7411–7427. https://doi.org10.1029/JB084iB13p07411.

Jones, J. H., & Drake, M. J. (1982). An experimental geochemical approach to early planetary differentiation. *Proceedings of the Lunar and Planetary Science Conference*, *13*, 369–370.

Kessel, R., Schmidt, M. W., Ulmer, P., & Pettke T. (2005). Trace element signature of subduction-zone fluids, melts and supercritical liquids at 120 – 180 km depth. *Nature*, *437*, 724–727.

Kimura, J. -I., Gill, J. B., Skora, S., van Keken, P. E., & Kawabata, H. (2016). Origin of geochemical mantle components: Role of subduction filter. *Geochemistry, Geophysics, Geosystems*, *17*, 3289–3325. https://doi.org10.1002/2016GC006362.

Kimura, K., Lewis, R. S., & Anders, E. (1974). Distribution of gold and rhenium between nickel-iron and silicate melts: Implications for the abundance of siderophile elements on the Earth and Moon. *Geochimica et Cosmochimica Acta*, *38*(5), 683–701. https://doi.org10.1016/0016-7037(74)90144-6.

Kleine, T., Münker, C., Mezger, K., & Palme, H. (2002). Rapid accretion and early core formation on asteroids and the terrestrial planets from Hf–W chronometry. *Nature*, *418*(6901), 952–955. https://doi.org/10.1038/nature00982, PubMed: 12198541.

Kogiso, T., Suzuki, K., Suzuki, T., Shinotsuka, K., Uesugi, K., Takeuchi, A., et al. (2008). Detecting micrometer-scale platinum-group minerals in mantle peridotite with microbeam synchrotron radiation X-ray fluorescence analysis. *Geochemistry, Geophysics, Geosystems*, *9*(3), Q03018. https://doi.org10.1029/2007GC001888.

Kogiso, T., Tatsumi, Y., & Nakano, S. (1997). Trace element transport during dehydration processes in the subducted oceanic crust: 1. experiments and implications for the origin of ocean island basalts. *Earth and Planetary Science Letters*, *148*, 193–205.

Komabayashi, T., Hirose, K., Funakoshi, K., & Takafuji, N. (2005). Stability of phase A in antigorite (serpentine) composition determined by in situ X-ray pressure observations. *Physics of the Earth and Planetary Interiors*, *151*(3–4), 276–289. https://doi.org/10.1016/j.pepi.2005.04.002.

König, S., Münker, C., Schuth, S., & Garbe-Schönberg, D. (2008). Mobility of tungsten insubduction zones. *Earth and Planetary Science Letters*, *274*, 82–92.

König, S., Schuth, S., & Luguet, A. (2010). Boninites as windows into trace element mobility in subduction zones. *Geochimica et Cosmochimica Acta*, *74*(2), 684–704. https://doi.org10.1016/j.gca.2009.10.011.

König, S., Münker, C., Hohl, S., Paulick, H., Barth, A. R., Lagos, M., et al. (2011). The Earth's tungsten budget during mantle melting and crust formation. *Geochimica et Cosmochimica Acta*, *75*, 2119–2136.

Kruijer, T. S., & Kleine, T. (2018). No ^{182}W excess in the Ontong Java Plateau source. *Chemical Geology*, *485*, 24–31. https://doi.org10.1016/j.chemgeo.2018.03.024.

Kruijer, T. S., Kleine, T., Fischer-Gödde, M., & Sprung, P. (2015). Lunar tungsten isotopic evidence for the late veneer. *Nature*, *520*(7548), 534–537. https://doi.org/10.1038/nature14360, PubMed: 25855296.

Kruijer, T. S., Touboul, M., Fischer-Godde, M., Bermingham, K. R., Walker, R. J., & Kleine, T. (2014). Protracted core formation and rapid accretion of protoplanets. *Science*, *344*, 1150–1154. https://doi.org10.1126/science.1251766.

Le Roux, V. L., Bodinier, J. -L., Tommasi, A., Alard, O., Dautria, J. -M., Vauchez, A., et al. (2007). The Lherz spinel lherzolite: Refertilized rather than pristine mantle. *Earth and Planetary Science Letters*, *259*(3–4), 599–612. https://doi.org10.1016/j.epsl.2007.05.026.

Liu, J., Touboul, M., Ishikawa, A., Walker, R. J., & Graham Pearson, D. (2016). Widespread tungsten isotope anomalies and W mobility in crustal and mantle rocks of the Eoarchean Saglek Block, northern Labrador, Canada: Implications for early Earth processes and W recycling. *Earth and Planetary Science Letters*, *448*, 13–23. https://doi.org10.1016/j.epsl.2016.05.001.

Liu, J. G., Rudnick, R. L., Walker, R. J., Gao, S., Wu, F. Y., & Piccoli, P. M. (2010). Processes controlling highly siderophile element fractionations in xenolithic peridotites and their influence on Os isotopes. *Earth and Planetary Science Letters*, *297*(1–2), 287–297. https://doi.org10.1016/j.epsl.2010.06.030.

Lorand, J. -P., Alard, O., & Luguet, A. (2010). Platinum-group element micronuggets and refertilization process in Lherz orogenic peridotite (northeastern Pyrenees, France). *Earth and Planetary Science Letters*, *289*(1–2), 298–310. https://doi.org10.1016/j.epsl.2009.11.017.

Lorand, J. -P., & Luguet, A. (2016). Chalcophile and siderophile elements in mantle rocks: Trace elements controlled by trace minerals. *Reviews in Mineralogy and Geochemistry*, *81*(1), 441–488. https://doi.org10.2138/rmg.2016.81.08.

Luguet, A., Alard, O., Lorand, J. P., Pearson, N. J., Ryan, C., & O'Reilly, S. Y. (2001). Laser-ablation microprobe (LAM)-ICPMS unravels the highly siderophile element geochemistry of the oceanic mantle. *Earth and Planetary Science Letters*, *189*(3–4), 285–294. https://doi.org10.1016/S0012-821X(01)00357-0.

Luguet, A., Pearson, D. G., Selby, D., Meisel, T., & Brenan, J. (2008). Highly siderophile element geochemistry. *Chemical Geology*, *248*(3–4), 115–118. https://doi.org10.1016/S0009-2541(08)00059-4.

Luguet, A., & Pearson, G. (2019). Dating mantle peridotites using Re-Os isotopes: The complex message from whole rocks, base metal sulfides, and platinum group minerals. *American Mineralogist, 104*(2), 165–189. https://doi.org10.2138/am-2019-6557.

Luguet, A., & Reisberg, L. (2016). Highly siderophile element and ^{187}Os signatures in non-cratonic basalt-hosted peridotite xenoliths: Unravelling the origin and evolution of the post-archean lithospheric mantle. *Reviews in Mineralogy and Geochemistry, 81*(1), 305–367. https://doi.org10.2138/rmg.2016.81.06.

Mallmann, G., & O'Neill, H. St. C. (2007). The effect of oxygen fugacity on the partitioning of Re between crystals and silicate melt during mantle melting. *Geochimica et Cosmochimica Acta, 71*(11), 2837–2857. https://doi.org10.1016/j.gca.2007.03.028.

Mann, U., Frost, D. J., Rubie, D. C., Becker, H., & Audétat, A. (2012). Partitioning of Ru, Rh, Pd, Re, ir and Pt between liquid metal and silicate at high pressures and high temperatures—Implications for the origin of highly siderophile element concentrations in the Earth's mantle. *Geochimica et Cosmochimica Acta, 84*(0), 593–613. https://doi.org10.1016/j.gca.2012.01.026.

Manning, C. E. (2004). The chemistry of subduction-zone fluids. *Earth and Planetary Science Letters, 223*(1–2), 1–16. https://doi.org10.1016/j.epsl.2004.04.030.

Mavrogenes, J. A., & O'Neill, H. St. C. (1999). The relative effects of pressure, temperature and oxygen fugacity on the solubility of sulfide in mafic magmas. *Geochimica et Cosmochimica Acta, 63*(7–8), 1173–1180. https://doi.org10.1016/S0016-7037(98)00289-0.

McDade, P., Blundy, J. D., & Wood, B. J. (2003). Trace element partitioning on the Tinaquillo solidus at 1.5 GPa. *Physics of the Earth and Planetary Interiors, 139*, 129–147.

McDonough, W. F. & Sun, S.-s. (1995). The composition of the Earth. *Chemical Geology, 120*(3–4), 223–253. https://doi.org10.1016/0009-2541(94)00140-4.

Mei, Q. -F., Yang, J. -H., & Yang, Y. -H. (2018). An improved extraction chromatographic purification of tungsten from a silicate matrix for high precision isotopic measurements using MC-ICPMS. *Journal of Analytical Atomic Spectrometry, 33*(4), 569–577. https://doi.org10.1039/C8JA00024G.

Meibom, A., & Anderson, D. L. (2004). The statistical upper mantle assemblage. *Earth and Planetary Science Letters, 217*(1–2), 123–139. https://doi.org10.1016/S0012-821X(03)00573-9.

Meisel, T., Walker, R. J., Irving, A. J., & Lorand, J. -P. (2001). Osmium isotopic compositions of mantle xenoliths: A global perspective. *Geochimica et Cosmochimica Acta, 65*(8), 1311–1323. https://doi.org10.1016/S0016-7037(00)00566-4.

Morris, J. D., & Hart, S. R. (1983). Isotopic and incompatible element constraints on the genesis of island arc volcanics from Cold Bay and Amak Island, Aleutians, and implications for mantle structure. *Geochimica et Cosmochimica Acta, 47*(11), 2015–2030. https://doi.org10.1016/0016-7037(83)90217-X.

Moynier, F., Yin, Q. Z., Irisawa, K., Boyet, M., Jacobsen, B., & Rosing, M. T. (2010). Coupled ^{182}W-^{142}Nd constraint for early Earth differentiation. *Proceedings of the National Academy of Sciences of the United States of America, 107*(24), 10810–10814. https://doi.org/10.1073/pnas.0913605107; PubMed: 20534492.

Mukhopadhyay, S. (2012). Early differentiation and volatile accretion recorded in deep-mantle neon and xenon. *Nature, 486*, 101–104. https://doi.org10.1038/nature11141.

Mundl, A., Touboul, M., Jackson, M. G., Day, J. M. D., Kurz, M. D., Lekic, V., et al. (2017). Tungsten-182 heterogeneity in modern ocean island basalts. *Science, 356*(6333), 66–69. https://doi.org/10.1126/science.aal4179, PubMed: 28386009.

Mundl-Petermeier, A., Walker, R. J., Fischer, R. A., Lekic, V., Jackson, M. G., & Kurz, M. D. (2020). Anomalous ^{182}W in high ^{3}He/^{4}He ocean island basalts: Fingerprints of Earth's core? *Geochimica et Cosmochimica Acta, 271*, 194–211. https://doi.org10.1016/j.gca.2019.12.020.

Mundl-Petermeier, A., Walker, R. J., Jackson, M. G., Blichert-Toft, J., Kurz, M. D., & Halldórsson, S. A. (2019). Temporal evolution of primordial tungsten-182 and ^{3}He/^{4}He signatures in the Iceland mantle plume. *Chemical Geology, 525*, 245–259. https://doi.org10.1016/j.chemgeo.2019.07.026.

Mungall, J. E., & Brenan, J. M. (2014). Partitioning of platinum-group elements and Au between sulfide liquid and basalt and the origins of mantle-crust fractionation of the chalcophile elements. *Geochimica et Cosmochimica Acta, 125*, 265–289. https://doi.org10.1016/j.gca.2013.10.002.

Murakami, M., Ohishi, Y., Hirao, N., & Hirose, K. (2012). A perovskitic lower mantle inferred from high-pressure, high-temperature sound velocity data. *Nature, 485*(7396), 90–94. https://doi.org10.1038/nature11004, PubMed: 22552097.

Murthy, V. R. (1991). Early differentiation of the Earth and the problem of mantle siderophile elements: A new approach. *Science, 253*(5017), 303–306. https://doi.org/10.1126/science.253.5017.303, PubMed: 17794697.

Mutter, C. Z., & Mutter, J. C. (1993). Variations in thickness of layer 3 dominated oceanic crustal structure. *Earth and Planetary Science Letters, 117*(1–2), 295–317. https://doi.org10.1016/0012-821X(93)90134-U.

Nakanishi, N., Giuliani, A., Carlson, R., Horan, M. F., Woodhead, J., Pearson, D. G., et al. (2021). Tungsten-182 evidence for an ancient kimberlite source. *Proceedings of the National Academy of Science, 118*, e2020680118. https://doi.org10.1073/pnas.2020680118.

Nakamura, Y., & Kushiro, I. (1974). Composition of the gas phase in Mg_2SiO_4–SiO_2–H_2O at 15 kbar. *Year Book Carnegie Institute, 73*, 255–258.

O'Nions, R. K., Evensen, N. M., & Hamilton, P. J. (1979). Geochemical modeling of mantle differentiation and crustal growth. *Journal of Geophysical Research, 84*(B11), 6091–6101. https://doi.org10.1029/JB084iB11p06091.

Peacock, S. M. (2001). Are the lower planes of double seismic zones caused by serpentine dehydration in subducting oceanic mantle? *Geology, 29*(4), 299–302. https://doi.org10.1130/0091-7613(2001)029<0299:ATLPOD>2.0.CO;2.

Peacock, S. M., & Wang, K. (1999). Seismic consequences of warm versus cool subduction metamorphism: Examples from southwest and northeast Japan. *Science, 286*(5441), 937–939. https://doi.org/10.1126/science.286.5441.937, PubMed: 10542143.

Pearson, D. G., Irvine, G. J., Ionov, D. A., Boyd, F. R., & Dreibus, G. E. (2004). Re–Os isotope systematics and platinum group element fractionation during mantle melt extraction: A study of massif and xenolith peridotite suites. *Chemical Geology, 208*(1–4), 29–59. https://doi.org10.1016/j.chemgeo.2004.04.005.

Peto, M. K., Mukhopadhyay, S., & Kelley, K. A. (2013). Heterogeneities from the first 100 million years recorded in deep mantle noble gases from the Northern Lau Back-arc Basin. *Earth and Planetary Science Letters, 369*/370, 13–23. https://doi.org10.1016/j.epsl.2013.02.012.

Peucker-Ehrenbrink, B., Bach, W., Hart, S. R., Blusztajn, J. S., & Abbruzzese, T. (2003). Rhenium–osmium isotope systematics and platinum group element concentrations in oceanic crust from DSDP/ODP sites 504 and 417/418. *Geochemistry, Geophysics, Geosystems, 4*(7), 8911. https://doi.org10.1029/2002GC000414.

Plank, T., Cooper, L. B., & Manning, C. E. (2009). Emerging geothermometers for estimating slab surface temperatures. *Nature Geoscience, 2*(9), 611–615. https://doi.org10.1038/ngeo614.

Shipboard Scientific Party (2000). Leg 185 summary: Inputs to the Izu-Mariana subduction system. In T. Plank, J. N. Ludden, & C. Escutia, et al. (Eds.). *Proceedings of the Ocean Drilling Program, Initial Reports, 185*. College Station, TX (Ocean Drilling Program), 1–63.

Poli, S., & Schmidt, M. W. (1995). H_2O transport and release in subduction zones: Experimental constraints on basaltic and andesitic systems. *Journal of Geophysical Research: Solid Earth, 100*(B11), 22299–22314. https://doi.org10.1029/95JB01570.

Puchtel, I. S., Blichert-Toft, J., Touboul, M., Horan, M. F., & Walker, R. J. (2016). The coupled ^{182}W-^{142}Nd record of early terrestrial mantle differentiation. *Geochemistry, Geophysics, Geosystems, 17*(6), 2168–2193. https://doi.org10.1002/2016GC006324.

Puchtel, I. S., Blichert-Toft, J., Touboul, M., & Walker, R. J. (2018). ^{182}W and HSE constraints from 2.7 Ga komatiites on the heterogeneous nature of the Archean mantle. *Geochimica et Cosmochimica Acta, 228*, 1–26. https://doi.org10.1016/j.gca.2018.02.030.

Puchtel, I. S., Walker, R. J., Brandon, A. D., & Nisbet, E. G. (2009). Pt–Re–Os and Sm–Nd isotope and HSE and REE systematics of the 2.7 Ga Belingwe and Abitibi komatiites. *Geochimica et Cosmochimica Acta, 73*(20), 6367–6389. https://doi.org10.1016/j.gca.2009.07.022.

Rehkämper, M., Halliday, A. N., Fitton, J. G., Lee, D. -C., Wieneke, M., & Arndt, N. T. (1999). Ir. Ru, Pt, and Pd in basalts and komatiites: New constraints for the geochemical behavior of the platinum-group elements in the mantle. *Geochimica et Cosmochimica Acta, 63*(22), 3915–3934. https://doi.org10.1016/S0016-7037(99)00219-7.

Reifenröther, R., Münker, C., & Scheibner, B. (2021). Evidence for tungsten mobility during oceanic crust alteration. *Chemical Geology, 584*, 12504.

Reimink, J. R., Chacko, T., Carlson, R. W., Shirey, S. B., Liu, J., Stern, R. A., et al. (2018). Petrogenesis and tectonics of the Acasta gneiss Complex derived from integrated petrology and ^{142}Nd and ^{182}W extinct nuclide-geochemistry. *Earth and Planetary Science Letters, 494*, 12–22. https://doi.org10.1016/j.epsl.2018.04.047.

Reisberg, L. (2021). Osmium isotope constraints on formation and refertilization of the non-cratonic continental mantle lithosphere. *Chemical Geology, 574*, 120245. https://doi.org10.1016/j.chemgeo.2021.120245.

Ricolleau, A., Fei, Y., Cottrell, E., Watson, H., Deng, L., Zhang, L. et al. (2009). Density profile of Pyrolite under the lower mantle conditions. *Geophysical Research Letters, 36*(6). https://doi.org/10.1029/2008GL036759.

Ringwood, A. E. (1966). A model for the upper mantle. *Journal of Geophysical Research, 67*, 57–66. https://doi.org10.1029/JZ067i002p00857.

Ringwood, A. E. (1977). Composition of the core and implications for origin of the earth. *Geochemical Journal, 11*(3), 111–135. https://doi.org10.2343/geochemj.11.111.

Rizo, H., Andrault, D., Bennett, N. R., Humayun, M., Brandon, A., Vlastelic, I., et al. (2019). ^{182}W evidence for core-mantle interaction in the source of mantle plumes. *Geochemical Perspectives Letters, 11*, 6–11. https://doi.org10.7185/geochemlet.1917.

Rizo, H., Walker, R. J., Carlson, R. W., Horan, M. F., Mukhopadhyay, S., Manthos, V., et al. (2016a). Preservation of Earth-forming events in the tungsten isotopic composition of modern flood basalts. *Science, 352*(6287), 809–812. https://doi.org/10.1126/science.aad8563, PubMed: 27174983.

Rizo, H., Walker, R. J., Carlson, R. W., Touboul, M., Horan, M. F., Puchtel, I. S., et al. (2016b). Early Earth differentiation investigated through ^{142}Nd, ^{182}W, and highly siderophile element abundances in samples from Isua, Greenland. *Geochimica et Cosmochimica Acta, 175*, 319–336. https://doi.org10.1016/j.gca.2015.12.007.

Rudge, J. F. (2006). Mantle pseudo-isochrons revisited. *Earth and Planetary Science Letters, 249*(3–4), 494–513. https://doi.org10.1016/j.epsl.2006.06.046.

Schmidt, M. W., & Poli, S. (1998). Experimentally based water budgets for dehydrating slabs and consequences for arc magma generation. *Earth and Planetary Science Letters, 163*(1–4), 361–379. https://doi.org10.1016/S0012-821X(98)00142-3.

Schoenberg, R., Kamber, B. S., Collerson, K. D., & Eugster, O. (2002). New W-isotope evidence for rapid terrestrial accretion and very early core formation. *Geochimica et Cosmochimica Acta, 66*(17), 3151–3160. https://doi.org10.1016/S0016-7037(02)00911-0.

Shaw, D. M. (1970). Trace element fractionation during anatexis. *Geochimica et Cosmochimica Acta, 34*(2), 237–243. https://doi.org10.1016/0016-7037(70)90009-8.

Shimoda, G., & Kogiso, T. (2019). Effect of serpentinite dehydration in subducting slabs on isotopic diversity in recycled oceanic crust and its role in isotopic heterogeneity of the mantle. *Geochemistry, Geophysics, Geosystems, 20*(11), 5449–5472. https://doi.org10.1029/2019GC008336.

Snow, J. E., & Schmidt, G. (1998). Constraints on Earth accretion deduced from noble metals in the oceanic mantle. *Nature, 391*(6663), 166–169. https://doi.org10.1038/34396.

Stagno, V., Ojwang, D. O., McCammon, C. A., & Frost, D. J. (2013). The oxidation state of the mantle and the extraction of carbon from Earth's interior. *Nature*, *493*(7430), 84–88. https://doi.org/10.1038/nature11679, PubMed: 23282365.

Staudigel, H., Davies, G. R., Hart, S. R., Marchant, K. M., & Smith, B. M. (1995). Large scale isotopic Sr, Nd and O isotopic anatomy of altered oceanic crust: DSDP/ODP sites 417/418. *Earth and Planetary Science Letters*, *130*(1–4), 169–185. https://doi.org10.1016/0012-821X(94)00263-X.

Staudigel, H., Plank, T., White, B., & Schmincke, H. -U. (1996). Geochemical fluxes during seafloor alteration of the basaltic upper crust: DSDP Sites 417 and 418. In G. E. Bebout, D. W. Scholl, S. H. Kirby, & J. P. Platt (Eds.), *Subduction: Top to Bottom, Geophysical Monograph Series, 96*. Washington, DC: American Geophysical Union.

Stracke, A. (2012). Earth's heterogeneous mantle: A product of convection-driven interaction between crust and mantle. *Chemical Geology*, *330–331*, 274–299. https://doi.org10.1016/j.chemgeo.2012.08.007.

Stracke, A., Bizimis, M., & Salters, V. J. M. (2003). Recycling oceanic crust: Quantitative constraints. *Geochemistry, Geophysics, Geosystems*, *4*, 8003. doi:1029/2001GC000223

Stracke, A., Hofmann, A. W., & Hart, S. R. (2005). FOZO, HIMU, and the rest of the mantle zoo. *Geochemistry, Geophysics, Geosystems*, *6*(5). https://doi.org/10.1029/2004GC000824.

Suer, T. A., Siebert, J., Remusat, L., Day, J. M. D., Borensztajn, S., Doisneau, B., et al. (2021). Reconciling metal-silicate partitioning and late accretion in the Earth. *Nature Communications*, *12*(1), 2913. https://doi.org/10.1038/s41467-021-23137-5, PubMed: 34006864.

Sun, S.-s, & McDonough, W. F. (1989). Chemical and isotopic systematics of oceanic basalts: Implications for mantle composition and processes [Geological Society special publication]. *Geological Society, London, Special Publications*, *42*(1), 313–345. https://doi.org10.1144/GSL.SP.1989.042.01.19.

Syracuse, E. M., van Keken, P. E., & Abers, G. A. (2010). The global range of subduction zone thermal models. *Physics of the Earth and Planetary Interiors*, *183*(1–2), 73–90. https://doi.org10.1016/j.pepi.2010.02.004.

Tackley, P. J. (2000). Mantle convection and plate tectonics: Toward an integrated physical and chemical theory. *Science*, *288*(5473), 2002–2007. https://doi.org/10.1126/science.288.5473.2002, PubMed: 10856206.

Takamasa, A., Nakai, S., Sahoo, Y., Hanyu, T., & Tatsumi, Y. (2009). W isotope compositions of oceanic islands basalts from French Polynesia and their meaning for core–mantle interaction. *Chemical Geology*, *260*(1–2), 37–46. https://doi.org10.1016/j.chemgeo.2008.11.018.

Takamasa, A., Suzuki, K., Fukami, Y., Iizuka, T., Tejada, M. L. G., Fujisaki, W., et al. (2020). Improved method for highly precise and accurate $^{182}W/^{184}W$ isotope measurements by multiple collector inductively coupled plasma mass spectrometry and application for terrestrial samples. *Geochemical Journal*, *54*(3), 117–127. https://doi.org/10.2343/geochemj.2.0594.

Tanaka, S. (2002). Very low shear wave velocity at the base of the mantle under the South Pacific Superswell. *Earth and Planetary Science Letters*, *203*(3–4), 879–893. https://doi.org10.1016/S0012-821X(02)00918-4.

Tanaka, S. (2007). Possibility of a low P-wave velocity layer in the outermost core from global SmKS waveforms. *Earth and Planetary Science Letters*, *259*(3–4), 486–499. https://doi.org10.1016/j.epsl.2007.05.007.

Tappe, S., Budde, G., Stracke, A., Wilson, A., & Kleine, T. (2020). The tungsten-182 record of kimberlites above the African superplume: Exploring links to the core-mantle boundary. *Earth and Planetary Science Letters*, *547*, 116473. https://doi.org10.1016/j.epsl.2020.116473.

Taylor, S. R., & McLennan, S. M. (1995). The geochemical evolution of the continental crust. *Reviews of Geophysics*, *33*(2), 241–265. https://doi.org10.1029/95RG00262.

Tejada, M. L. G., Suzuki, K., Hanyu, T., Mahoney, J. J., Ishikawa, A., Tatsumi, Y., et al. (2013). Cryptic lower crustal signature in the source of the Ontong Java Plateau revealed by Os and Hf isotopes. *Earth and Planetary Science Letters*, *377–378*, 84–96. https://doi.org10.1016/j.epsl.2013.07.022.

Thirlwall, M. F. (1997). Pb isotopic and elemental evidence for OIB derivation from young HIMU mantle. *Chemical Geology*, *139*(1–4), 51–74. https://doi.org10.1016/S0009-2541(97)00033-8.

Thorne, M. S., Garnero, E. J., Jahnke, G., Igel, H., & McNamara, A. K. (2013). Mega ultra low velocity zone and mantle flow. *Earth and Planetary Science Letters*, *364*, 59–67. https://doi.org10.1016/j.epsl.2012.12.034.

Touboul, M., Liu, J., O'Neil, J., Puchtel, I. S., & Walker, R. J. (2014). New insights into the Hadean mantle revealed by ^{182}W and highly siderophile element abundances of supracrustal rocks from the Nuvvuagittuq Greenstone Belt, Quebec, Canada. *Chemical Geology*, *383*, 63–75. https://doi.org10.1016/j.chemgeo.2014.05.030.

Touboul, M., Puchtel, I. S., & Walker, R. J. (2012). ^{182}W evidence for long-term preservation of early mantle differentiation products. *Science*, *335*(6072), 1065–1069. https://doi.org/10.1126/science.1216351, PubMed: 22345398.

Touboul, M., Puchtel, I. S., & Walker, R. J. (2015). Tungsten isotopic evidence for disproportional late accretion to the Earth and Moon. *Nature*, *520*(7548), 530–533. https://doi.org/10.1038/nature14355, PubMed: 25855299.

Touboul, M., & Walker, R. J. (2012). High precision tungsten isotope measurement by thermal ionization mass spectrometry. *International Journal of Mass Spectrometry*, *309*, 109–117. https://doi.org10.1016/j.ijms.2011.08.033.

Ulmer, P., & Trommsdorff, V. (1995). Serpentine stability to mantle depths and subduction-related magmatism. *Science*, *268*(5212), 858–861. https://doi.org/10.1126/science.268.5212.858, PubMed: 17792181.

van Keken, P. E., Hacker, B. R., Syracuse, E. M., & Abers, G. A. (2011). Subduction factory: 4. Depth-dependent flux of H_2O from subducting slabs worldwide. *Journal of Geophysical Research*, *116*(B1). https://doi.org/10.1029/2010JB007922.

Van Orman, J. A., Keshav, S., & Fei, Y. (2008). High-pressure solid/liquid partitioning of Os, Re and Pt in the Fe-S system. *Earth and Planetary Science Letters*, *274*(1), 250–257. https://doi.org10.1016/j.epsl.2008.07.029.

Vockenhuber, C., Oberli, F., Bichler, M., Ahmad, I., Quitté, G., Meier, M., et al. (2004). New half-life measurement of ^{182}Hf: Improved chronometer for the early solar system. *Physical Review Letters*, *93*(17), 172501. https://doi.org/10.1103/PhysRevLett.93.172501.

White, R. S., McKenzie, D., & O'Nions, R. K. (1992). Oceanic crustal thickness from seismic measurements and rare earth element inversions. *Journal of Geophysical Research*, *97*(B13), 19683–19715. https://doi.org10.1029/92JB01749.

Willbold, M., Elliott, T., & Moorbath, S. (2011). The tungsten isotopic composition of the Earth's mantle before the terminal bombardment. *Nature*, *477*(7363), 195–198. https://doi.org/10.1038/nature10399, PubMed: 21901010.

Willbold, M., Mojzsis, S. J., Chen, H. -W., & Elliott, T. (2015). Tungsten isotope composition of the Acasta gneiss Complex. *Earth and Planetary Science Letters*, *419*, 168–177. https://doi.org10.1016/j.epsl.2015.02.040.

Willbold, M., & Stracke, A. (2006). Trace element composition of mantle end-members: Implications for recycling of oceanic and upper and lower continental crust. *Geochemistry, Geophysics, Geosystems*, *7*(4), Q04004. https://doi.org10.1029/2005GC001005.

Willbold, M., & Stracke, A. (2010). Formation of enriched mantle components by recycling of upper and lower continental crust. *Chemical Geology*, *276*(3–4), 188–197. https://doi.org10.1016/j.chemgeo.2010.06.005.

Yin, Q., Jacobsen, S. B., Yamashita, K., Blichert-Toft, J., Télouk, P., & Albarède, F. (2002). A short timescale for terrestrial planet formation from Hf–W chronometry of meteorites. *Nature*, *418*(6901), 949–952. https://doi.org/10.1038/nature00995, PubMed: 12198540.

Yoshino, T., Makino, Y., Suzuki, T., & Hirata, T. (2020). Grain boundary diffusion of W in lower mantle phase with implications for isotopic heterogeneity in oceanic island basalts by core-mantle interactions. *Earth and Planetary Science Letters*, *530*. https://doi.org/10.1016/j.epsl.2019.115887, PubMed: 115887.

Yuen, D., Monnereau, M., Hansen, U., Kameyama, M., & Matyska, C. (2007). *Dynamics of superplumes in the lower mantle*. In M. S. Yuen DA, S. I. Karato, & B. F. Windley (Eds.), Superplumes*: Beyond Plate Tectonics*. Springer.

Zhang, Z., & Hirschmann, M. M. (2016). Experimental constraints on mantle sulfide melting up to 8 GPa. *American Mineralogist*, *101*(1), 181–192. https://doi.org10.2138/am-2016-5308.

Zindler, A., & Hart, S. (1986). Chemical geodynamics. *Annual Review of Earth and Planetary Sciences*, *14*(1), 493–571. https://doi.org10.1146/annurev.ea.14.050186.002425.

Zindler, A., Staudigel, H., & Batiza, R. (1984). Isotope and trace element geochemistry of young Pacific seamounts: Implications for the scale of upper mantle heterogeneity. *Earth and Planetary Science Letters*, *70*(2), 175–195. https://doi.org10.1016/0012-821X(84)90004-9.

6

Numerical Examination of the Dynamics of Subducted Crustal Materials with Different Densities

Taku Tsuchiya[1], Takashi Nakagawa[2,3], and Kenji Kawai[4]

ABSTRACT

Crustal materials subducted into the deep Earth are the major source of the mantle chemical heterogeneity, and their dynamics are expected to strongly depend on the density contrasts between the subducted crustal materials and surrounding mantle materials. Here, we estimate the bulk density variations of some different lithologies, pyrolite, mid-ocean ridge basalt (MORB), tonalite-trondhjemite-granodiorite (TTG), and also more hypothetical anorthosite and KREEP (basalt enriched in potassium, rare Earth elements, and phosphorus), in the lower mantle pressure range at static temperature, and test numerically the evolution of these subducted crustal materials within the mantle convection. We identify the following four features. (1) Since the TTG crust gains buoyancy in the lower mantle, it eventually ascends through time. (2) Although the cold anorthositic crust could sink to the core-mantle boundary (CMB), it would be heated by the heat flux from the core and finally assimilated into the lower mantle materials. (3) In contrast, dense KREEP basalt could be expected to sink to the CMB and gravitationally stabilized through geological time. (4) MORB shows an intermediate behavior between anorthosite and KREEP. The results suggest that dense KREEP could be primordial chemical reservoirs if it was generated in the early Earth as well as in the Moon.

6.1. INTRODUCTION

Following pioneering works by Aki et al. (1977) and Woodhouse and Dziewonski (1984), many studies have proved that the global seismic tomography can improve our knowledge on the dynamics and evolution of Earth's deep interior. Since as a rule of thumb seismic wave perturbations can be reconciled by lateral temperature variations (Poirier, 2000), they provide information on the distribution of hot and cold regions in the Earth. Since the cold anomalies beneath the subduction zone can be related to subducted slabs, the image of the high-velocity anomalies allows us to deduce the modality of the down-welling flow in the Earth. Beneath East Asia, for example, tomographic studies show slab downgoing from the surface to ~660 km depth, the boundary between the mantle transition zone and the lower mantle, and flattening there (e.g., Fukao et al., 1992; Borgeaud et al., 2019). Meanwhile, the large high-velocity anomalies imaged immediately above the core-mantle boundary (CMB) beneath East Asia are interpreted as the graveyard of fallen slabs (e.g., Fukao et al., 2009). These suggest the "stagnant slab" at the bottom of the mantle transition

[1] Geodynamics Research Center, Ehime University, Matsuyama, Japan
[2] Department of Earth and Planetary System Science, Hiroshima University, Higashi-Hiroshima, Japan
[3] Department of Planetology, Kobe University, Kobe, Japan
[4] Department of Earth and Planetary Sciences, The University of Tokyo, Tokyo, Japan

Core-Mantle Co-Evolution: An Interdisciplinary Approach, Geophysical Monograph 276, First Edition.
Edited by Takashi Nakagawa, Taku Tsuchiya, Madhusoodhan Satish-Kumar, and George Helffrich.
© 2023 American Geophysical Union. Published 2023 by John Wiley & Sons, Inc.
DOI: 10.1002/9781119526919.ch06

zone and the subsequent "slab avalanche" to the lower mantle.

In contrast to downwelling flows, detecting upwelling flows in the deep Earth's interior is not easy due to difficulty in seismological analysis and interpretation. As the buoyancy induced by hot and light materials drives upwelling flows in the mantle, those flows might be associated with the low-velocity anomalies. It is well known that the low-velocity anomalies are more difficult to see seismologically than high-velocity anomalies because of diffraction. Especially, short-wavelength low-velocity anomalies associated with a thin hot material, such as plume tails, are difficult to image with traveltime tomography within the seismic ray approximation because the low-velocity anomalies have only small effects on traveltimes due to the wavefront healing effect (e.g., Nolet and Dahlen, 2000). In spite of the correlation of hotspot location with the large low velocity provinces (LLVPs) above the CMB (e.g., Thorne et al., 2004), the origin of the hotspot has been controversial due to the limited resolution of low-velocity anomalies in traveltime tomography (Courtillot et al., 2003; Anderson, 2005). Recently, finite frequency tomography (e.g., Montelli et al., 2004) and waveform inversion (e.g., French and Romanowicz, 2015) successfully imaged low-velocity anomalies in the lower mantle beneath some hotspots.

In addition to the technical development in global seismic tomography, recently deployed array networks such as F-net and USArray have enhanced the seismic wavefield resolution in the Earth. Analysis of receiver functions using waveforms recorded by Hi-net stations deployed in Japan (>600 stations) revealed variations in seismic velocity near the subducting slab and proposed the thermal structure and the distribution of hydrous minerals in a subduction zone (Kawakatsu & Watada, 2007; Tonegawa et al., 2008). Tomographic studies using USArray data (~2,000 observation sites) inferred the seismic velocity structure in the upper mantle beneath the North America and found a sequence of Farallon slabs and the lithospheric root of the Colorado Plateau (Obrebski et al., 2011; Hamada and Yoshizawa, 2015). A modern waveform inversion technique using waveform data recorded by full deployment of the USArray seismic stations during 2004–2015 inferred the three-dimensional (3-D) shear wave velocity structure of the lowermost 400 km of the mantle beneath Central America and the Caribbean (Kawai et al., 2014; Borgeaud et al., 2017) and beneath the northern and western Pacific (Suzuki et al., 2016, 2020, 2021) and found that subducted paleoslabs interact with the hot materials in the thermal boundary layer (TBL) above the CMB. Although almost all the high- and low-velocity anomalies can be explained by temperature variations combined with a phase transition of the most major constituent mineral phase $(Mg,Fe)SiO_3$ (Kawai & Tsuchiya, 2009; Schuberth et al., 2009), those observed in some places of the lowermost 100–200 km of the mantle seem to be too strong and sharp to be explained by temperature variations alone, suggesting the possible existence of chemical heterogeneity in the lowermost mantle (Wen, 2001; Wen et al., 2001; Borgeaud et al., 2017; Suzuki et al., 2016, 2020, 2021). These indicate that in order to interpret the seismic velocity anomalies, it is required to take into account not only thermal history of the mantle but also its chemical evolution.

The major chemical differentiation of the Earth's mantle occurs where the solidus temperatures of constituent materials approach the mantle geotherm and partial melting occurs. This is most likely in the TBL in the mantle with steep temperature gradients, that is, below the Earth's surface and above the base of mantle. At mid-ocean ridges, decompression melting of mantle peridotite causes chemical differentiation, producing basaltic rocks enriched in incompatible elements (e.g., Kushiro, 2001). Similar phenomena can be expected at high temperatures near the CMB, which might be related to the dense magma ocean or the ultralow seismic velocity zones (e.g., Labrosse et al., 2007). In the subduction zone, the partial melting yields felsic lithology such as granite. Then, subduction of cold, thus dense, slabs brings such chemically differentiated crustal materials to the deep mantle. This might be one of the important sources of the mantle chemical heterogeneity, and it is thought that through mantle convection, some of the subducted materials are recycled up to the surface and some get buried deep into the bottom of the mantle. The density contrast between the subducted crustal materials and the surrounding mantle is clearly a striking parameter to control their dynamics and hence to understand the evolution of the mantle chemical heterogeneity.

As well as the present-day oceanic and continental crusts, it is desirable to take into account primordial crustal materials (e.g., Wood et al., 1970; McKay et al., 1979). It is generally thought that on the cooling of the Moon magma ocean, crystallized lighter minerals (plagioclases) were floated toward the surface and formed highland, whose constituent rock is called anorthosite, while heavier minerals (olivine and pyroxenes) sank and formed the lunar mantle (e.g., Shearer et al., 2006). Also, KREEP (basalt enriched in potassium, rare Earth elements, and phosphorus) is thought to have formed in this process as a leftover of the formation of the lunar crust. In the case of the Moon, due to the short-lived igneous activity and no subsequent weathering, these ancient productions have been preserved to date. In an analogous fashion to this lunar geology and on the concept of the giant impact, primordial terrestrial crusts have been proposed to be anorthositic and basaltic enriched in KREEP (Kawai et al., 2009; Maruyama et al., 2013; Gréaux et al., 2018; Nishi et al., 2018). Thus, in this study, in addition to the present-day crustal compositions such as mid-ocean ridge basalt (MORB) and granitic Tonalite-trondhjemite-granodiorite (TTG), we consider more primordial compositions of KREEP basalt and anorthosite. Here, although the density of granite, the major composition of upper continental crusts, is substantially smaller than the mantle density, some studies suggest its subduction involved with subducted plates (Scholl and von Huene, 2007; Yamamoto et al., 2009; Ichikawa et al., 2013; Kawai et al., 2013). These crustal materials might have different densities at mantle pressures depending on the different lithologies, in particular, fractions of felsic and mafic components, so that the evolution in the mantle convection would also be different. Numerical mantle convection simulations to see what happens when the chemical anomalies in the deep mantle are not of the MORB composition have however not been performed so far. In previous numerical modeling on generation of chemical heterogeneity in mantle convection, basaltic material including its recycling process associated with the partial melting was investigated (Christensen and Hoffman, 1994; Xie and Tackley, 2004a, 2004b; Nakagawa and Buffett, 2005; Nakagawa et al., 2010). Those studies were designed to assess the residential time of trace element in the deep mantle (e.g., Xie and Tackley, 2004a, 2004b) as well as the interpretations on large-scale dynamics (e.g., Nakagawa and Buffett, 2005) and its relationship to the seismic anomalies in the deep mantle (e.g., Nakagawa et al., 2010). As pointed out above, the origin of chemical heterogeneity is still conjectural and not restricted to only recycled MORB material addressed in the previous studies. Although the compositions of primordial crusts are not fully understood and still more hypothetical, it is worth testing the behaviors of some representative candidates. First, we estimate the densities of these materials in the lower mantle pressure range using literature data, then their dynamics is investigated by conducting mantle convection simulation with obtained density contrasts, where the density of pyrolite, a representative mantle composition, is taken as a reference.

6.2. MODELING DENSITY

We employ published equation of state data of the SiO_2 phases (stishovite, $CaCl_2$-type, and α-PbO_2-type structure) (Karki et al., 1997), $MgSiO_3$ bridgmanite and post-perovskite, (Mg,Fe)O ferropericlase, $CaSiO_3$ perovskite, $MgAl_2O_4$ calcium ferrite-type phase (Tsuchiya, 2011; Tsuchiya and Kawai, 2013; Wang et al., 2015), $NaAlSiO_4$ NAL phase, and Al_2O_3 polymorphs (Tsuchiya et al., 2005; Kawai et al., 2009) to determine the density variations with pressure (depth). These equations of state were calculated using the *ab initio* quantum mechanical theory at the static (0 K) temperature condition, and we assume the density contrasts are insensitive to temperature. Since iron is well known to dissolve into the $MgSiO_3$, MgO, and $MgAl_2O_4$ phases, we assume that iron dissolution only affects their masses. The effects of iron on the volumes are neglected because they are expected to have only minor effects on the densities. Other minor impurities are also not considered for the same reason. The densities of all lithologies are estimated without including temperature effects because the mutual density contrasts at the same temperature, which are fundamental to control their dynamics in the fluid dynamics simulations, are expected to be less sensitive to temperature.

Using these data, we model the density of each lithology at lower mantle pressures. In this modeling, volume fractions of major constituent mineral phases are set according to previous reports (Irifune and Tsuchiya, 2007; Ricolleau et al., 2010) to 55 vol% (Mg,Fe)SiO_3 bridgmanite (post-perovskite), 40 vol% (Mg,Fe)O ferropericlase, and 5 vol% $CaSiO_3$ perovskite for pyrolite; 30 vol% $MgSiO_3$ bridgmanite (post-perovskite), 30 vol% $CaSiO_3$ perovskite, 20 vol% SiO_2, and 20 vol% (Mg,Fe)Al_2O_4 for basalt. On the other hand, the mineral proportion in TTG is approximated to be a molar fraction with SiO_2:$NaAlSiO_4$ of 9:1 (Komabayashi et al., 2009; Kawai and Tsuchiya, 2015) and the anorthosite composition is approximated to be identical to anorthite ($CaAl_2Si_2O_8$), which corresponds to $CaSiO_3$:Al_2O_3:SiO_2 of the mineral assemblage of three phases with 1:1:1 in the lower mantle condition. The constituent mineral phases for the individual rocks in the lower mantle condition are summarized in Table 6.1.

The modeled density profiles are shown in Figure 6.1, indicating that at the same temperature condition, KREEP basalt and MORB are ~4.5% and ~1.5% denser than the reference pyrolite, respectively, anorthosite has a comparable density with pyrolite, and TTG is in contrast ~4.5% less dense than pyrolite. In the present model, KREEP basalt is quite dense and the density contrast between KREEP basalt and MORB is caused simply by the difference in the bulk iron content as described in the caption of Table 6.1. Small density jumps can be seen in the modeled density profiles. Those at 100~120 GPa in KREEP basalt, MORB, and pyrolite correspond to the post-perovskite phase transition in bridgmanite (Tsuchiya et al., 2004), while jumps at ~60 GPa in anorthosite and ~120 GPa in TTG correspond to the corundum-to-Rh_2O_3(II) transition in Al_2O_3 (Tsuchiya et al., 2005) and the $CaCl_2$-to-seifertite transition in SiO_2

Table 6.1 Mineralogy for pyrolite, basalt (MORB and KREEP), TTG, and anorthosite in the lower mantle condition

Rock model	Mineral phases	Composition	Volume fraction (%)
Pyrolite	Bridgmanite (post-perovskite)	$(Mg_{0.95},Fe_{0.05})SiO_3$	55
	Ferropericlase	$(Mg_{0.82},Fe_{0.18})O$	40
	Ca-perovskite	$CaSiO_3$	5
MORB	Bridgmanite (post-perovskite)	$(Mg_{0.85},Fe_{0.15})SiO_3$	30
	Ca-perovskite	$CaSiO_3$	30
	Stishovite (seifertite)	SiO_2	20
	Calcium ferrite-type	$(Mg_{0.85},Fe_{0.15})Al_2O_4$	20
KREEP	Bridgmanite (post-perovskite)	$(Mg_{0.4},Fe_{0.6})SiO_3$	30
	Ca-perovskite	$CaSiO_3$	30
	Stishovite (seifertite)	SiO_2	20
	Calcium ferrite-type	$(Mg_{0.4},Fe_{0.6})Al_2O_4$	20
TTG	Stishovite (seifertite)	SiO_2	90
	NAL phase	$NaAlSiO_4$	10
Anorthosite	Ca-perovskite	$CaSiO_3$	100/3
	Corundum (Rh_2O_3(II)-type)	Al_2O_3	100/3
	Stishovite (seifertite)	SiO_2	100/3

Notes: Phase names in parentheses are the high-pressure forms. The iron fractions in bridgmanite and ferropericlase in pyrolite are set to 5% and 18%, in bridgmanite and calcium ferrite in MORB and KREEP basalt are set to 15% and 60%. Source for the pyrolite and MORB composition: Adapted from Irifune and Tsuchiya 2007. Source for the TTG composition: Adapted from Komabayashi et al. 2009. Source for the KREEP composition: Adapted from Mckay et al. 1979.

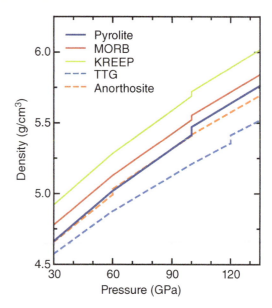

Figure 6.1 Calculated density profile in the lower mantle pressure range between 30 and 135 GPa at static temperature for pyrolite, MORB, KREEP basalt, TTG, and anorthosite. A color figure is available in the online version.

(Karki et al., 1997), respectively. Using these static (0 K) density profiles obtained from mineral physics as inputs, we next conduct mantle convection simulations to test their dynamics in the lower mantle.

6.3. GEODYNAMICS MODELING

6.3.1. Model Setup

Geodynamical modeling is a powerful tool to explore the thermo-chemical structure of the deep mantle using constraints on density from mineral physics (Fig. 6.1). Here, we assess the evolution of crustal material of different composition by running numerical simulations. We use a numerical code StagYY (Tackley, 2008) for thermo-chemical mantle convection simulations in a quarter of a 2-D spherical annulus. The mantle material is here modeled as the truncated anelastic fluid with temperature-, pressure-, and yield strength-dependent viscosity, which is decomposed for two endmember components (depleted harzburgite and enriched basaltic material). In the basaltic material, the heat source is enhanced by 10 times compared to that in the depleted harzburgite composition. In StagYY, a finite-volume multigrid solver is used for solving Stokes equation coupled with the mass conservations, and the MPDATA technique is employed for temperature equation and tracer particle approach for the chemical composition. The governing equations are given as:

$$\nabla \cdot (\rho \underline{v}) = 0,$$

$$\underline{\nabla} \cdot \underline{\underline{\sigma}} - \underline{\nabla} p = \rho g,$$

$$\underline{\underline{\sigma}} = \eta \left(\nabla \underline{v} + (\nabla \underline{v})^T - \frac{2}{3} \nabla \cdot \underline{v} \right),$$

$$\rho C_p \frac{DT}{Dt} = \nabla \cdot (k \nabla T) + \rho H - \alpha \rho g v_r T + \sigma : \dot{e},$$

$$\frac{DC}{Dt} = 0, \quad (6.1)$$

where ρ is the density of the mantle, v is the velocity and v_r is the radial component of velocity, σ is the stress, p is the pressure, g is acceleration due to gravity, η is the viscosity, C_p is the heat capacity, H is the radioactive heating, α is the thermal expansivity, $\dot{e}$ is the second invariant value of the strain rate, k is the thermal conductivity, T is the absolute temperature, C is the chemical composition of the silicate mantle, and t is the time. Note that $DA/Dt = \partial A/\partial t + \nabla \cdot (\underline{v}A)$.

The significant material property is the rheological properties of the silicate mantle mathematically given as:

$$\eta = \left(\frac{1}{\eta_m} + \frac{2\dot{e}}{\sigma_Y(p)}\right)^{-1} \quad (6.2)$$

$$\eta_m = A_0 \sum_{i,j=1}^{n=3,4} \Delta\eta_{ij}^{f\Gamma_{ij}} \exp\left(\frac{E+pV}{RT}\right) \quad (6.3)$$

$$\sigma_Y(p) = C_Y + \mu p \quad (6.4)$$

where A_0 is a constant defined at $p = 0$ and $T = 1600$ K, $\Delta\eta_{ij}$ is a viscosity change caused by the phase transition, f is the fraction of a subducted crustal material in the mantle (varying from 0 to 1), Γ_{ij} is the phase function, E and V are activation energy and volume, respectively, R is the ideal gas constant, p and T are pressure and temperature, respectively, $\sigma_Y(p)$ is the yield strength, C_Y is the cohesion of the yield strength, μ is the friction coefficient, and $\dot{e}$ is the second invariant value of the strain-rate tensor (Nakagawa and Tackley, 2014).

The other significant physical and chemical processes are to generate the deep mantle heterogeneities in thermal and chemical effects. In order to create such heterogeneities in the deep mantle, the melt-induced differentiation is allowed to create the (oceanic) crust and subduct into the deep mantle caused by the plate tectonics-like behavior in pseudo-plastic yielding (e.g., Xie and Tackley, 2004a, 2004b). On the generation of melting, as described in Xie and Tackley (2004a, 2004b), the melt fraction is determined as:

$$f = \min \begin{pmatrix} \frac{T - T_{sol}(z)}{L}, C \\ T \rightarrow T - Lf \end{pmatrix} \quad (6.5)$$

where $T_{sol}(z)$ is the solidus temperature as a function of depth, L is the latent heat, and f is the melt fraction. The solidus temperature is simply given as the following formula:

$$T_{sol}(z)$$
$$= \begin{cases} 2050 + 0.62z + 660\left(\text{erf}\left(\frac{z}{220}\right) - 1\right) & z < 660 \, km \\ 2760 + 0.45z + 1700\left(\text{erf}\left(\frac{z}{1000}\right) - 1\right) & z \geq 660 \, km \end{cases}$$
$$(6.6)$$

where z is the depth of mantle with km unit. This solidus temperature is as a fitted result of Herzberg et al. (2000) for the upper mantle and of Zerr et al. (1998) for the lower mantle.

The most important point in this study is to incorporate into the density profile shown in Figure 6.1. We use the reference density of each component of the modeled mantle along with the mantle adiabat. This is computed by the following equations taken from Tackley (1996):

$$\frac{\partial \rho_i}{\partial z} = K_s \frac{\alpha}{c_p} \left(\frac{\gamma_s}{\gamma_i}\right) \rho_i; \; \gamma_i \rho_i = const., \quad (6.7)$$

where K_s is the compressibility number defined at the surface, α is the thermal expansivity, c_p is the heat capacity, γ_s is the Grüneisen parameter at the surface, γ_i is the Grüneisen parameter, and ρ_i is the reference density of two endmembers along with the reference mantle adiabat (i = hz as harzburgite and i = bas as the enriched basaltic material). To determine the Grüneisen parameter, equation (6.7) is used, which can control the density of each endmember at the CMB. The reference density is shown in Figure 6.2. The different density profiles for each basaltic composition are caused by using the different Grüneisen parameter which affects the compressibility of the mantle silicates (Nakagawa and Tackley, 2008). As the Grüneisen parameter of the basaltic component is smaller, the density difference between harzburgite and basalt becomes greater. Using this effect, we model the density profiles of each material shown in Figure 6.1. The density of the average composition of mantle is determined by the mechanical mixing between ambient mantle and subducted crustal compositions, which is given as:

$$\rho_{ave} = \rho_{\text{ambient mantle}}(1 - C) + \rho_{\text{subducted crust}}C, \quad (6.8)$$

where C represents the fraction of the crustal composition and density of ambient mantle is composed of 75% of olivine and 25% of pyroxene. As a result, the density difference between the ambient mantle and subducted crustal materials using four types of rock indicated in Figure 6.1 ranges from −0.8% to 3.49% when the reference chemical fraction (anorthosite, TTG, MORB, and KREEP) is assumed as 20% in the modeled mantle. The density of each material and its difference from the average mantle composition at the CMB pressure are listed at Table 6.2.

Some studies suggested that the basaltic rocks must have been present at the surface of the Earth at 4.56-4.4 Ga (e.g., Smithies et al., 2009; Johnson et al., 2017), such as the equivalent of primordial continents on the Moon (e.g., Maruyama et al., 2013). Geological studies showed that the Moon is predominantly covered by a ~50 km thick anorthositic crust and KREEP basalts (e.g., Wieczorek and Phillips, 1998) crystallized from erupted lava flow. The lunar magma ocean is believed to have solidified at 4.56-4.4 Ga by first crystallizing 60–75 vol% olivine and orthopyroxenes, then Ca-plagioclase, which due to its lesser density, rapidly floated on top of residual Fe-rich basaltic melt, resulting in the formation of a thick

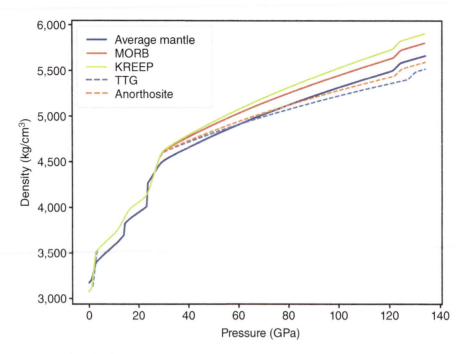

Figure 6.2 Density profiles for the geodynamics simulations. On the pyroxene composition, the density contrast labeled in this plot corresponds to the lithologies shown in Figure 6.1. The average mantle profile is generated by the mixture of subducted crustal material and ambient mantle material (equation (6.8)). (A color figure is available in the online version.)

Table 6.2 Density at the CMB and density contrast at the CMB compared to the average composition used in geodynamic modeling

Lithology	Density of pyrolite at the CMB (ρ_{ave} kg/m³)	Density at the CMB (ρ_{px} kg/m³)	Density contrast at the CMB (%)	Density contrast at the CMB scaled by the average mantle composition (%)
MORB	5,664.01	5,801.23	2.42	2.42
KREEP	5,705.32	5,904.53	3.49	4.25
TTG	5,542.70	5,497.96	−0.81	−2.93
Anorthosite	5,579.87	5,590.87	0.20	−1.29

Note: The average mantle composition is computed as $\rho_{ave} = 0.8(0.75\rho_{ol} + 0.25\rho_{px}) + 0.2\rho_{px}$.

mono-mineralic anorthosite upper crust and a dense KREEP lower mafic crust.

Meanwhile, some studies proposed that TTG rocks from the Archean (4.0 Ga to 3.2 Ga) were produced from ~25% melting of low-magnesium basalts, which were themselves derived from an earlier generation of high-Mg basaltic crust (Smithies et al., 2009; Johnson et al., 2017). Although the generation of TTG rocks is controversial (Moyen and Martin, 2012), there is a widespread belief that most TTG were generated via slab melting (Martin, 1993). Therefore, MORB is a mixture of TTG and its restite. These multiple differentiation processes generating TTG, KREEP, and anorthosite are still difficult to be implemented accurately in mantle convection simulations. We therefore simplified the differentiation processes by replacing the density of the MORB by densities of the other crustal lithologies in equation (6.8).

For the simulation setup, the following resolution and initial state of the mantle are assumed: the numerical resolution of 256 (horizontal) × 128 (vertical) is used with 4 million tracers to track the chemical compositions and melt fraction. The averaged resolution is about 20 km. In order to resolve the dynamics in boundary regions, the more grid points are used in those regions, which range 1–5 km grid interval. The initial condition is assumed as adiabatic temperature defined as 1600 K at the surface value plus two TBLs at both surface and CMB with the small perturbation with the uniform chemical composition (20% of differentiated crustal materials). The core cooling condition is included, which the temperature at the CMB is determined as a function of the heat flow across the CMB based on thermodynamics formulation of core energetics by Labrosse (2015). Other physical parameters are listed in Table 6.3.

6.3.2. Numerical Results

Figure 6.3 shows the final snapshot of compositional structure for four compositional models, which is taken at t = 4.6 billion years. For the basaltic material (Fig. 6.3a), the dense chemical pile is found above the CMB because the subducted oceanic crust may segregate in the deep mantle. However, two compositional models (TTG and anorthosite shown in Fig. 6.3b,c) do not lead to large-scale chemical anomalies above the CMB because their density differences at the CMB indicate the negative values compared to the pyrolytic material; hence, the chemically-distinct material is not likely to be settled above the CMB. For KREEP model (Fig. 6.3d), large-scale chemical piles are found with the blue region inside of piles, which will be discussed below. Basically, the density of each rock at the CMB affects the formation of dense thermo-chemical piles. When the density of crustal component is denser than the pyrolitic composition (see Fig. 6.2), the crustal component is likely to accumulate above the CMB. Whereas, the crustal component is lighter, the crustal component may not form the large-scale thermo-chemical structure.

Table 6.3 Physical parameters of geodynamics simulations

Symbol	Meaning	Value
η_0	Reference viscosity	1×10^{21} Pa s
ρ_0	Surface density	3,300 kg m^{-3}
g	Surface gravity	9.8 m s^{-2}
α_0	Surface thermal expansivity	5×10^{-5} K^{-1}
κ_0	Surface thermal diffusivity	7×10^{-7} m^2 s^{-1}
ΔT_{sa}	Temperature scale	2500 K
C_p	Heat capacity	1,250 J kg^{-1} K^{-1}
L_m	Latent heat	6.25×10^5 J kg^{-1}
H	The present-day internal heating rate	5×10^{-12} W kg^{-1}
λ	The half-life of radioactive elements	2.43×10^9 yr
C	Cohesion stress	10 MPa
μ	Friction coefficient	0.02
E	The activation energy	290 kJ mol^{-1}
V	Activation volume	2.4×10^{-6} m^3 mol^{-1}

Notes: The activation energy and dry mantle volume are from Yamazaki and Karato (2001). The latent heat caused by partial melting is taken from Xie and Tackley (2004a, 2004b).

On the reason why such a blue region in the thermo-chemical pile is found for KREEP material (see Fig. 6.3d), this region corresponds to the partially molten region shown in Figure 6.4, which is caused by the accumulation

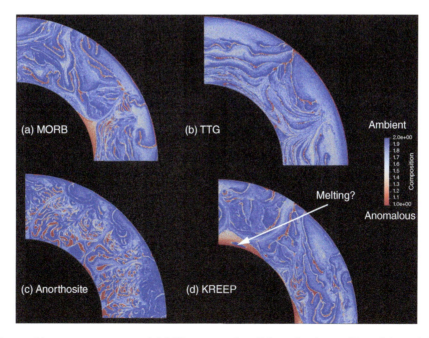

Figure 6.3 Composition structure at t = 4.6 billion years for all four density profiles of the subducted crustal material shown in Figure 6.2. A color figure is available in the online version.

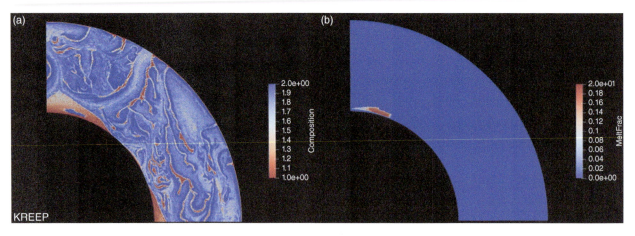

Figure 6.4 Comparison between compositional structure and melt fraction for the KREEP case. The melting region can be found inside the dense piles. A color figure is available in the online version.

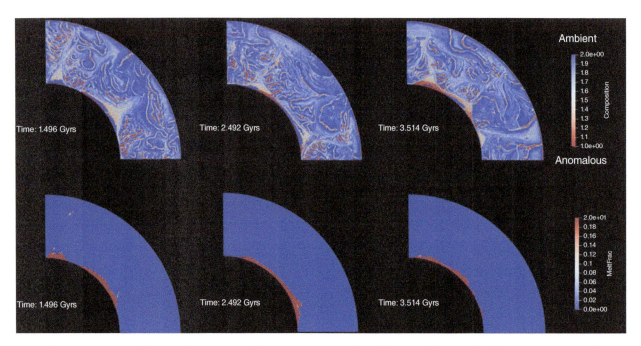

Figure 6.5 Time evolution of compositional structure (top) and melt fraction (bottom) for the KREEP case. Dark blue region inside of piles corresponds to the melting zone. A color figure is available in the online version.

of the basaltic material that enriches the heat source compared to the ambient mantle composition. The expected melt fraction in this region would be ~20%. In order to check how such a melting region is evolved over billion years of timescales, Figure 6.5 shows the time evolution of chemical structure and melt fraction as a function of time for the KREEP case, which are taken at every 1.5 billion years approximately. The global-scale melting region in the deep mantle is found until 3.5 billion years at least, which is inside of the dense piles. This global-scale melting region is transited into the patchy melting region shown in Figure 6.4 due to the cooling down of the Earth's mantle.

6.4. DISCUSSION AND CONCLUSIONS

Subducted crustal materials into the deep Earth are thought to be the major source of the mantle chemical heterogeneity (e.g., Tackley, 2012). In this chapter, we modeled the densities of several different crustal lithologies in the lower mantle pressure range at static temperature and investigated their dynamics and the effects of density contrasts on the evolution of lower mantle chemical heterogeneity. The modeled density profiles showed that two basalts (KREEP and MORB) are denser than the reference pyrolite, anorthosite is in contrast comparable, and TTG is less dense. Mantle dynamics simulations including these density contrasts demonstrated that the style of convection changes depending on the lithology of crustal materials due to their different densities. Since the TTG crust gains buoyancy in the lower mantle, it eventually ascends through time. Although the cold anorthositic crust could sink to the CMB, it would be heated by the heat flux from the core, entrained by the upwelling of convection, and finally assimilated into the lower mantle materials. In contrast, taken into account results of recent geodynamic studies, dense KREEP basalt could be expected to sink to the CMB and dense enough to survive through geological time. In this case, it is also suggested that the base mantle could be partially molten due to the radiogenic heat production. This partial melt might initiate further chemical differentiation in the deepest mantle and could be a generation source of upwelling plumes originated at the CMB (e.g., French et al., 2013; French and Romanowicz, 2015).

The chemical structure in deep mantle is still quite controversial (e.g., Tackley, 2000). Although seismological (e.g., Takeuchi, 2007; French et al., 2013; French and Romanowicz, 2015) and mineral physical studies (e.g., Cottaar et al., 2014; Wang et al., 2015) have suggested a global and vigorous mantle circulation, hidden primordial or primitive chemical reservoirs have been geochemically believed to exist somewhere in the deep mantle (e.g., Kellogg et al., 1999; Anderson, 2005). If such long-lived regions really exist in the mantle, those must keep sufficient gravitational stability throughout the Earth's history. Otherwise, after heated up from the core, they would finally get buoyancy to start floating. The present calculations indicate that the materials are denser than the surrounding mantle at the isothermal condition (the MORB and KREEP cases), some piles form above the CMB, but the volume of piles is greater in the KREEP case because the density contrast between KREEP and pyrolite is more than twice larger than that between MORB and pyrolite. Therefore, if dense crusts having the composition like KREEP had formed and subducted in the early Earth as well as the Moon, those could be the primordial chemical reservoirs. In addition, the present calculations suggest that such piles could be partially molten due to radiogenic heat and generate upwelling plumes. The plumes originated from the KREEP piles could contain some primordial signatures, which seem consistently related to some geochemical observations (e.g., Jackson et al., 2017). It has been proposed that dense materials could be produced associated with the solidification of basal magma ocean (Labrosse et al., 2007). For an alternative idea, the core is also suggested to be the primordial chemical reservoir (e.g., Rizo et al., 2019; Roth et al., 2019; Xiong et al., 2021). It has been proposed that a strong rheological gap could also create long-lived stable domains in the mantle (Ballmer et al., 2017). To understand more details on the nature of primordial reservoirs, further integrative investigations from mineral physical, petrological, seismological analyses, and geodynamic modeling are clearly required.

For example, to discuss the possible existence of the reservoirs of segregated KREEP material at the bottom of the mantle, computing seismic velocity anomalies is meaningful because the larger density of the subducted KREEP material suggests its slower seismic wave speeds. It was reported that the MORB material with ~1.5% denser than pyrolite has ~−2% slower S-wave velocity than pyrolite at the bottom of mantle (Tsuchiya, 2011; Tsuchiya and Kawai, 2013). Based on this result and the ~4.5% excess density of the KREEP material, an ~−3.5% low-velocity anomaly is roughly estimated against pyrolite, assuming insignificant differences in the shear moduli of the MORB and KREEP materials. This velocity anomaly is comparable to a −4% S-wave velocity reduction observed at the bottom of the mantle underneath the central Pacific (Lay et al., 2006). For more precise determination of the seismic velocities of the KREEP material, calculations of elastic constants of each mineral component are clearly required. This might be an important research subject in the future studies.

ACKNOWLEDGMENTS

This work was completed under the supports of KAKENHI (Grant Nos. 15H05826, 15H05832, 15H05834, 15K17744, and 16K05531). We thank C. Shiraishi for her assistance in manuscript preparation.

REFERENCES

Aki, K., Christoffersson, A., & Husebye, E. S. (1977). Determination of the three-dimensional seismic structure. *Journal of Geophysical Research*, *82*, 277–296.

Anderson, D. L. (2005). Scoring hotspots: The plume and plate paradigms. *Geological Society Special Papers*, *388*, 31–54.

Ballmer, M. D., Houser, C., Hernlund, J. W., Wentzcovitch, R. M., & Hirose, K. (2017). Persistence of strong silica-enriched domains in the Earth's lower mantle. *Nature Geoscience*, *10*, 236–240.

Borgeaud, A. F. E., Kawai, K., Konishi, K., & Geller, R. J. (2017). Imaging paleoslabs in the D″ layer beneath Central America and the Caribbean using seismic waveform inversion. *Science Advances*, *3*, e1602700.

Borgeaud, A. F. E., Kawai, K., & Geller, R. J. (2019). Three-dimensional S velocity structure of the mantle transition zone beneath Central America and the Gulf of Mexico inferred using waveform inversion. *Journal of Geophysical Research: Solid Earth*, *124*, 9664–9681.

Christensen, U. R., & Hofmann, A. W. (1994). Segregation of subducted oceanic crust in the convecting mantle. *Journal of Geophysical Research*, *99*, 19867–19884.

Cottaar, S., Heister, T., Rose, I., & Unterborn, C. (2014). Burn-Man: A lower mantle mineral physics toolkit. *Geochemistry, Geophysics, Geosystems*, *15*, 1164–1179.

Courtillot, V., Davaille, A., Besse, J., & Stutzmann, E. (2003). Three distinct types of hotspots in the Earth's mantle. *Earth and Planetary Science Letters*, *205*, 295–308.

French, S. W., Lekic, V., & Romanowicz, B. (2013). Waveform tomography reveals channeled flow at the base of the oceanic asthenosphere. *Science*, *342*, 227–230.

French, S. W., & Romanowicz, B. (2015). Broad plumes rooted at the base of the Earth's mantle beneath major hotspots. *Nature*, *525*, 95–99.

Fukao, Y., Obayashi, M., Inoue, H., & Nebai, M. (1992). Subducting slabs stagnant in the mantle transition zone. *Journal of Geophysical Research*, *97*, 4809–4822.

Fukao, Y., Obayashi, M., Nakakuki, T., & the Deep Slab Project Group, (2009). Stagnant slab: A review. *Annual Review of Earth and Planetary Sciences*, *37*, 19–46.

Gréaux, S., Nishi, M., Tateno, S., Kuwayama, Y., Hirao, N., Kawai, K., et al. (2018). High-pressure phase relation of KREEP basalts: A clue for finding the lost Hadean crust? *Physics of the Earth and Planetary Interiors*, *274*, 184–194.

Hamada, K., & Yoshizawa, K. (2015). Interstation phase speed and amplitude measurements of surface waves with nonlinear waveform fitting: Application to US Array. *Geophysical Journal International*, *202*, 1463–1482.

Herzberg, C., Raterron, P., & Zhang, J. (2000). New experimental observations on the anhydrous solidus for peridotite KLB-1. *Geochemistry, Geophysics, Geosystems*, *1*, 2000GC000089.

Ichikawa, H., Kawai, K., Yamamoto, S., & Kameyama, M. (2013). Supply rate of continental materials to the deep mantle through subduction channels. *Tectonophysics*, *592*, 46–52.

Irifune, T., & Tsuchiya, T. (2007). Phase transitions and mineralogy of the lower mantle. In D. Price (Ed.), *Treatise on Geophysics*, vol. *2*, pp. 33–62. Elsevier.

Jackson, M. G., Konter, J. G., & Becker, T. W. (2017). Primordial helium entrained by the hottest mantle plumes. *Nature*, *542*, 340–343.

Johnson, T. E., Brown, M., Gardiner, N. J., Kirkland, C. L., & Smithies, R. H. (2017). Earth's first stable continents did not form by subduction. *Nature*, *543*, 239–242.

Kawai, K., & Tsuchiya, T. (2009). Temperature profile in the lowermost mantle from seismological and mineral physics joint modeling. *Proceedings of the National Academy of Sciences, U.S.A.*, *106*, 22119–22123.

Kawai, K., & Tsuchiya, T. (2015). Elasticity of continental crust around the mantle transition zone. In A. Khan, and F. Deschamps, (ed.), *The Earth's Heterogeneous Mantle*, Chap. 8, pp. 259–274. Springer.

Kawai, K., Konishi, K. Geller, R. J., & Fuji, N. (2014). Methods for inversion of body-wave waveforms for localized three-dimensional seismic structure and an application to D″ beneath Central America. *Geophysical Journal International*, *197*, 495–524.

Kawai, K., Tsuchiya T., Tsuchiya J., & Maruyama, S. (2009). Lost primordial continents. *Gondwana Research*, *16*, 581–586.

Kawakatsu, H., & Watada, S. (2007). Seismic evidence for deep water transportation in the mantle. *Science*, *316*, 1468–1471.

Karki, B. B., Warren, M. C., Stixrude, L., Ackland, G. J., & Crain, J. (1997). Ab initio studies of high-pressure structural transformations in silica. *Physical Review B*, *55*, 3465–3471.

Kawai, K., Yamamoto, S., Tsuchiya, T., & Maruyama, S. (2013). The second continent: Existence of granitic continental materials around the bottom of the mantle transition zone. *Geoscience Frontiers*, *4*, 1–6.

Kellogg, L. H., Hager, B. H., & van der Hilst, R. D. (1999). Compositional stratification in the deep mantle. *Science*, *283*, 1881–1884.

Komabayashi, T., Maruyama, S., & Rino, S. (2009). A speculation on the structure of the D″ layer: The growth of anti-crust at the core-mantle boundary through the subduction history of the Earth. *Gondwana Research*, *15*, 342–353.

Kushiro, I. (2001). Partial melting experiments on peridotite and origin of mid-ocean ridge basalt. *Annual Review of Earth and Planetary Sciences*, *29*, 71–107.

Labrosse S. (2015). Thermal evolution of the core with a high thermal conductivity. *Physics of the Earth and Planetary Interiors*, *247*, 36–55.

Labrosse, S., Hernlund, J. W., & Coltice, N. (2007). A crystallizing dense magma ocean at the base of the Earth's mantle. *Nature*, *450*, 866–869.

Lay, T., Hernlund, J., Garnero, E. J., & Thorne, M. S. (2006). A post-perovskite lens and D″ heat flux beneath the central Pacific. *Science*, *314*, 1272–1276.

Martin, H. (1993). The mechanisms of petrogenesis of the Archaean continental crust—Comparison with modern processes. *Lithos*, *30*, 373–388.

Maruyama, S., Ikoma, M., Genda, H., Hirose, K., Yokoyama, T., & Santosh, M. (2013). The naked planet Earth: Most essential pre-requisite for the origin and evolution of life. *Geoscience Frontiers*, *4*, 141–165.

McKay, G. A., Wiesmann, H., Bansal, B. M., & Shih, C.-Y. (1979). Petrology, chemistry, and chronology of Apollo 14 KREEP basalts. *Proceedings of the Tenth Lunar and Planetary Science Conference*, 181–205.

Moyen, J. -F., & Martin, H. (2012). Forty years of TTG research. *Lithos*, *148*, 312–336.

Montelli, R., Nolet, G., Dahlen, F. A., Masters, G., Engdahl, E. R., & Hung, S.-H. (2004). Finite-frequency tomography reveals a variety of plumes in the mantle. *Science*, *303*, 338–343.

Nakagawa, T., & Buffett, B. A. (2005). Mass transport mechanism between the upper and lower mantle in numerical simulations of thermochemical mantle convection with multicomponent phase changes. *Earth and Planetary Science Letters*, *230*, 11–27.

Nakagawa, T., & Tackley, P. J. (2008). Lateral variations in CMB heat flux and deep mantle seismic velocity caused by a thermal-chemical-phase boundary layer in 3D spherical convection. *Earth and Planetary Science Letters*, *271*, 348–358.

Nakagawa, T., & Tackley, P. J. (2014). Influence of combined primordial layering and recycled MORB on the coupled thermal evolution of Earth's mantle and core. *Geochemistry, Geophysics, Geosystems*, *15*, 619–633.

Nakagawa, T., Tackley, P. J., Deschamps, F., & Connolly, J. A. D. (2010). The influence of MORB and harzburgite composition on thermo-chemical mantle convection in a 3-D spherical shell with self-consistently calculated mineral physics. *Earth and Planetary Science Letters*, *296*, 403–412.

Nishi, M., Gréaux, S., Tateno, S., Kuwayama, Y., Kawai, K., Irifune, T., et al. (2018). High-pressure phase transitions of anorthosite crust in the Earth's deep mantle. *Geoscience Frontiers*, *9*, 1859–1870.

Nolet, G., & Dahlen, F. A. (2000). Wave front healing and the evolution of seismic delay times. *Journal of Geophysical Research*, *105*, 19043–19054.

Obrebski, M., Allen, R. M., Pollitz, F., & Hung, S.-H. (2011). Lithosphere- asthenosphere interaction beneath the western United States from the joint inversion of body-wave traveltimes and surface-wave phase velocities. *Geophysical Journal International*, *185*, 1003–1021.

Poirier, J.-P. (2000). *Introduction to the Physics of the Earth's Interior*. Cambridge: Cambridge University Press.

Ricolleau, A., Perrillat, J. -P., Fiquet, G., Daniel, I., Matas, J., Addad, A., et al. (2010). Phase relations and equation of state of a natural MORB: Implications for the density profile of subducted oceanic crust in the Earth's lower mantle. *Journal of Geophysical Research*, *115*, B08202.

Rizo, H., Andrault, D., Bennett, N. R., Humayun, M., Brandon, A., Vlastelic, I., et al. (2019). ^{182}W evidence for core-mantle interaction in the source of mantle plumes. *Geochemical Perspectives Letters*, *11*, 6–11.

Roth, A. S. G., Liebske, C., Maden, C., Burton, K. W., Schönbächler, M., & Busemann, H. (2019). The primordial He budget of the Earth set by percolative core formation in planetesimals. *Geochemical Perspectives Letters*, *9*, 26–31.

Scholl, D. W., & von Huene, R. (2007). Crustal recycling at modern subduction zones applied to the past-issues of growth and preservation of continental basement crust, mantle geochemistry, and supercontinent reconstruction. *Mem Geological Society of America*, *200*, 9–32.

Schuberth, B. S. A., Bunge, H.-P., & Ritsema, J. (2009). Tomographic filtering of high-resolution mantle circulation models: Can seismic heterogeneity be explained by temperature alone? *Geochemistry, Geophysics, Geosystems*, *10*, Q05W03.

Shearer, C. K., Hess, P. C., Wieczorek, M. A., Pritchard, M. E., Parmentier, E. M., Borg, L. E., et al. (2006). Thermal and magmatic evolution of the Moon. In B. L. Jolliff, M. A. Wieczorek, C. K. Shearer, and C. R. Neal (eds.), *New Views of the Moon. Reviews in Mineralogy & Geochemistry*, vol. 60, pp. 365–518. Mineralogical Society of America and Geochemical Society.

Smithies, R. H., Champion, D. C., & Van Kranendonka, M. J. (2009). Formation of Paleoarchean continental crust through infracrustal melting of enriched basalt. *Earth and Planetary Science Letters*, *281*, 298–306.

Suzuki, Y., Kawai, K., Konishi, K., Borgeaud, A. F. E., & Geller, R. J. (2016). Waveform inversion for 3–D shear velocity structure of D" beneath the Northern Pacific: Possible evidence for a remnant slab and a 'passive plume'. *Earth, Planet. Space*, *68*, 198.

Suzuki, Y., Kawai, K., Geller, R. J., Tanaka, S., Siripunvaraporn, W., Boonchaisuk, S., et al. (2020). High-resolution 3-D S-velocity structure in the D" region at the western margin of the Pacific LLSVP: Evidence for small-scale plumes and paleoslabs. *Physics of the Earth and Planetary Interiors*, *307*, 106544.

Suzuki, Y., Kawai, K., & Geller, R. J. (2021). Imaging paleoslabs and inferring the Clapeyron slope in D" beneath the northern Pacific based on high-resolution inversion of seismic waveforms for 3-D transversely isotropic structure. *Physics of the Earth and Planetary Interiors*, *321*, 106751.

Tackley, P. J. (1996). Effects of strongly variable viscosity on three-dimensional compressible convection in planetary mantles. *Journal of Geophysical Research*, *101*, 3311–3332.

Tackley, P. J. (2000). Mantle convection and plate tectonics: Toward an integrated physical and chemical theory. *Science*, *288*, 2002–2006.

Tackley, P. J. (2008). Modelling compressible mantle convection with large viscosity contrasts in a three-dimensional spherical shell using the yin-yang grid. *Physics of the Earth and Planetary Interiors*, *171*, 7–18.

Tackley, P. J. (2012). Dynamics and evolution of the deep mantle resulting from thermal, chemical, phase and melting effects. *Earth-Science Reviews*, *110*, 1–25.

Takeuchi, N. (2007). Whole mantle SH velocity model constrained by waveform inversion based on three-dimensional Born kernels. *Geophysical Journal International*, *169*, 1153–1163.

Thorne, M. S. Garnero, E. J., & Grand, S. P. (2004). Geographic correlation between hot spots and deep mantle lateral shear-wave velocity gradients. *Physics of the Earth and Planetary Interiors*, *146*, 47–63.

Tonegawa, T., Hirahara, K., Shibutani, T., Iwamori, H., Kanamori, H., & Shiomi, K. (2008). Water flow to the mantle transition zone inferred from a receiver function image of the Pacific slab. *Earth and Planetary Science Letters*, *274*, 346–354.

Tsuchiya, J., Tsuchiya, T., & Wentzcovitch, R. M. (2005). Transition from the Rh_2O_3(II)-to-$CaIrO_3$ structure and the high-pressure-temperature phase diagram of alumina. *Physical Review B*, *72*, 020103(R).

Tsuchiya, T. (2011). Elasticity of subducted basaltic crust at the lower mantle pressures: Insights on the nature of deep mantle heterogeneity. *Physics of the Earth and Planetary Interiors*, *188*, 142–149.

Tsuchiya, T., & Kawai, K. (2013). Ab initio mineralogical model of the Earth's lower mantle. In S. Karato (ed.), *Physics and Chemistry of the Deep Earth*, pp. 213–243. Wiley.

Tsuchiya, T., Tsuchiya, J., Umemoto, K., & Wentzcovitch, R. M. (2004). Phase transition in $MgSiO_3$ perovskite in the Earth's lower mantle. *Earth and Planetary Science Letters*, *224*, 241–248.

Wang, X., Tsuchiya, T., & Hase, A. (2015). Computational support for a pyrolitic lower mantle containing ferric iron. *Nature Geoscience*, *8*, 556–559.

Wen, L. (2001). Seismic evidence for a rapidly varying compositional anomaly at the base of the Earth's mantle beneath the Indian Ocean. *Earth and Planetary Science Letters, 194,* 83–95.

Wen, L., Silver, P., James, D., & Kuehnel, R. (2001). Seismic evidence for a thermo-chemical boundary at the base of the Earth's mantle. *Earth and Planetary Science Letters, 189,* 141–153.

Wieczorek, M. A., & Phillips, R. J. (1998). Potential anomalies on a sphere: Applications to the thickness of the lunar crust. *Journal of Geophysical Research, 103,* 1715–1724.

Wood, J. A., Dickey, J. S., Marvin, U. B., & Powell, B. N. (1970). Lunar anorthosites and a geophysical model of the Moon. *Proceedings of the Apollo 11 Lunar Science Conference Geochimica et Cosmochimica Acta,* (Suppl 1), 965–988.

Woodhouse, J. H., & Dziewonski, A. M. (1984). Mapping the upper mantle: Three-dimensional modelling of Earth structure by inversion of seismic waveforms. *Journal of Geophysical Research, 87,* 5953–5986.

Xie, S., & Tackley, P. J. (2004a). Evolution of U-Pb and Sm-Nd systems in numerical models of mantle convection and plate tectonics. *Journal of Geophysical Research, 109,* D11204.

Xie, S., & Tackley, P. J. (2004b). Evolution of helium and argon isotopes in a convecting mantle. *Physics of the Earth and Planetary Interiors, 146,* 417–439.

Xiong, Z., Tsuchiya, T., & Van Orman, J. (2021). Helium and argon partitioning between liquid iron and silicate melt at high pressure. *Geophysical Research Letters, 48,* e2020GL090769.

Yamamoto, S., Senshu, H., Rino, S., Omori, S., & Maruyama, S. (2009). Granite subduction: Arc subduction, tectonic erosion and sediment subduction. *Gondwana Research, 15,* 443–453.

Yamazaki, D, & Karato, S. (2001). Some mineral physics constraints on the rheology and geothermal structure of Earth's lower mantle. *American Mineralogist, 86,* 385–301.

Zerr, A., Diegeler, A., & Boehler, R. (1998). Solidus of Earth's deep mantle. *Science, 281,* 243–246.

Part II
Core-Mantle Interaction: An Interdisciplinary Approach

7

Some Issues on Core-Mantle Chemical Interactions: The Role of Core Formation Processes

Shun-ichiro Karato

ABSTRACT

A model of core formation is reviewed and its consequences on mantle chemistry and core-mantle interaction are discussed. A growing planet forms a cold proto-core made of a mixture of various materials including primitive materials rich in highly siderophile elements (HSE) as well as volatiles. These primitive materials are squeezed out to the bottom of the magma ocean when Fe accumulated above the proto-core sinks to the center. After heated, the proto-core materials are mixed with the magma ocean. Consequently, mantle is made of a mixture of materials equilibrated with Fe at modest pressure and temperature and a small amount of primitive materials from the proto-core. This model explains the observed abundance pattern of all siderophile elements and predicts that most of HSE (+ volatiles) came from the proto-core formed early in the process of Earth formation rather than added in the later stage. The model implies that most of the core materials are in equilibrium with the mantle at the lower pressure and temperature than those of the current core-mantle boundary (CMB). Therefore, the core is undersaturated with volatile and siderophile elements at the CMB and, consequently, the core is likely a sink (not a source) of these elements even if the concentrations of these elements in the bulk of the core exceed those in the mantle. The transport of light elements from the mantle to the core provides a mechanism for a low-velocity layer on top of the core. The core-mantle disequilibrium promotes the migration of molten Fe into the mantle by the morphological instability in regions of the CMB where (Mg,Fe)O is interconnected. In these regions (presumably the ultralow-velocity regions), volatile (and siderophile) elements are carried by the molten Fe ~tens km into the mantle providing a window for the core chemistry.

7.1. INTRODUCTION

The core of Earth is made largely of metallic Fe and the mantle is made of silicates (+ oxides). These two types of materials have largely different chemical (and physical) properties and therefore the nature of chemical (and physical) interactions between these regions through Earth's history is an important subject in Earth science. One key issue in such a study is to understand the element partitioning (the solubility ratio of an element between two materials) under the broad physical and chemical conditions. Major progress has been made on element solubility (partitioning) through high-pressure, temperature experimentations (e.g., Fukai and Suzuki, 1986; Okuchi, 1997; Siebert et al., 2011; Mann et al., 2012; Ohtani, 2013). Those studies show strong influence of pressure (and temperature) on element solubility (partitioning). In particular, Fe has much higher solubility of volatile (and siderophile) elements than silicates under high pressures and temperatures, suggesting that the core may have higher concentrations of these elements than the mantle.

These results have been used to infer the conditions in which core-mantle separation (core formation) occurred (e.g., Siebert et al., 2012) or to infer the composition of

Department of Earth and Planetary Sciences, Yale University, New Haven, Connecticut, USA

the core (e.g., Ohtani, 2013; Badro et al., 2015). Similarly, given a plausible inference of higher concentrations of volatile and siderophile elements in the core than in the mantle, some authors suggested that the core is the source of these elements to the mantle (e.g., Hayden & Watson, 2007, 2008; Rizo et al., 2019).

However, the validity of the conclusions of some of these studies is unclear. One of the major issues is the degree of chemical equilibrium between the core and the mantle during core formation. For example, the direction of material transport across the core-mantle boundary (CMB) depends strongly on the nature of deviation of element concentrations in each region from the equilibrium values that depends on the processes of core formation.

Processes of core formation can be constrained by the observed abundance of siderophile (Fe-loving) elements. For example, the observed abundance of moderately siderophile elements (Ni, Co etc.) is used to place constraints on the depth of the magma ocean in which core was likely formed (e.g., Siebert et al., 2012). In contrast, observed abundance pattern of highly siderophile elements (HSE) cannot be explained by the chemical equilibrium, and it is often suggested that a small amount of materials rich in HSE and volatile elements were added after core formation (i.e., the "late veneer" model; Jones & Drake, 1986; Albarède, 2009; Walker, 2009). However, the validity of the late veneer model is questioned based on the observations on the abundance pattern of volatile and siderophile elements (Wood et al., 2010) and on the abundance ratio of various volatile elements (Hirschmann, 2016).

In this chapter, I first review a plausible core formation model (based largely on Karato & Murthy, 1997) and discuss a possible model to explain the HSE abundance pattern without invoking the late veneer. The model suggests an early acquisition of volatiles in a growing planet and a large degree of disequilibrium between the bulk of the core and the mantle that provides a basis for understanding the nature of chemical interactions between the core and the mantle through geologic history.

7.2. CORE FORMATION AND THE COMPOSITION OF THE CORE AND THE MANTLE

7.2.1. A Core Formation Model

Source materials from which Earth and other planets may have been formed can be inferred from the composition of meteorites and of asteroids. The age measurements of meteorites (mainly from W-Hf and U-Pb isotopic compositions) show that most of them were formed within the first a few Myrs after the formation of CAI (calcium-aluminum-rich inclusions in carbonaceous chondrite) (e.g., Kleine & Rudge, 2011). Similar age measurements show that Earth was formed later, ~50 Myrs after CAI (e.g., Halliday et al., 2001; Kleine & Rudge, 2011). Therefore, Earth (and other planets) is likely made of a broad range of materials including those that had undergone chemical differentiation (e.g., Fe meteorites-, ordinary chondrites-, achondrites-like materials) as well as those that show little chemical differentiation (e.g., carbonaceous chondrite-like materials).

Processes of formation of planets have also been investigated mostly based on theoretical studies (e.g., Hayashi et al., 1985; Kokubo & Ida, 1998; Lichtenberg et al., 2021; Morbidelli et al., 2012; Wetherill, 1990). These studies show that planetary formation occurs by the collisional growth of planetesimals formed by the accumulation of dusts or pebbles (e.g., Lambrechts et al., 2019). Composition of dusts and pebbles is determined largely by the distance from the Sun. However, substantial radial mixing occurs by the influence of giant planets such as Jupiter and Saturn in the early stage (a few Myrs after the formation of the solar system; Raymond and Izidoro, 2017). Consequently, we expect that planets are formed by the collision of planetesimals with a variety of composition and size. During the collisional growth of planetesimals, compositions of materials will change due to chemical reactions but the degree of chemical reactions depends strongly on the temperature and the size of various materials involved.

Two sources of heating are important. In the early stage of planetary formation, the size of planetary bodies (planetesimals) is so small that the main source of energy is the decay of radioactive elements with short half-lives such as ^{26}Al (half-life time = 0.7 Myr). These radioactive elements heat planetesimals quickly (less than a few Myrs) to form Fe core (i.e., Fe meteorites; Lichtenberg et al., 2021). However, the degree of heating depends strongly on the size of planetesimals: effective heating occurs only for relatively large planetesimals (>30 km) and smaller ones (<10 km) would not have been heated much (T < 500 K), and in these bodies primitive composition (carbonaceous chondrite-like composition) will be preserved. Even the highly heated ones will be cooled down quickly (<3 Myrs) (Lichtenberg et al., 2021). Therefore, a planet is likely formed by the accumulation of relatively cold materials with a variety of compositions.

After ~10 Myrs, the main heat source for growing planet is the gravitational energy. Consequently, in the early stage of planetary growth [after ~10 Myrs but before magma ocean formation (~50 Myrs)], the central temperature is low and temperature increases toward the surface following $T(r) = T_o + A \cdot r^2$, where T_o is temperature at the center of the proto-core, r is the distance from the center, and A is a constant (Fig. 7.1). As a result, the surface of a growing planet will be melted to form a magma ocean when its size becomes large enough. The critical size for

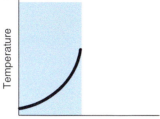
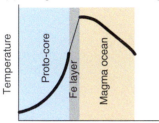
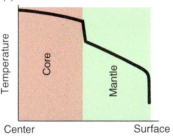

Figure 7.1 Temperature in a growing planet

(a) In the early stage of planetary growth after a few Myrs, central part has low temperature, and temperature increases toward the surface due to the increase in gravitational energy release as the planetary size increases. Some accreting planetesimals may have warm temperature due to the heating of short half-life radioactive elements. But their contribution is limited to relatively large planetesimals. Source: Adapted from Lichtenberg et al. (2021).

(b) When the size of a growing planet is large enough, the surface temperature exceeds the solidus and a magma ocean is formed. The threshold size depends on a few parameters including the nature of blanketing effect of atmosphere but it is estimated at around ~2 times the Mars size. Source: Adapted from Sasaki and Nakazawa (1986).

the formation of a magma ocean depends on the efficiency of cooling and therefore depends strongly on the nature of the blanketing atmosphere (e.g., Abe & Matsui, 1985, 1988; Sasaki & Nakazawa, 1986) but is estimated to be approximately twice the Mars size.

So in the stage where Fe-silicate separation starts in a growing planet, it should have a layered structure as shown in Figure 7.2 (step 1) where the central part (called the proto-core) is made of a mixture of proto-planetary materials (a variety of planetesimals). Temperature is low and the proto-core remains a mechanical mixture without much chemical reaction. In the outer region, there will be a magma ocean when a growing planet exceeds some size. Core formation, that is, Fe-silicate separation, is effective in the magma ocean. In the early stage of planetary formation, colliding bodies are relatively small and most of their Fe-rich cores if exist will be melted and broken into pieces with an ~1 cm size controlled by the surface tension (Karato, 1975; see also Stevenson, 1990). Fe droplets will sink through a magma ocean keeping chemical equilibrium with the surrounding silicate melts until they reach to the cold proto-core on which they will accumulate to form a layer of molten Fe. When the thickness of the layer becomes large enough, it sinks into the center via the Rayleigh-Taylor instability (step 2 in Fig. 7.2; Honda et al., 1993). The wavelength of this instability is large (~ the size of the proto-core; Honda et al., 1993; Karato & Murthy, 1997), and therefore sinking Fe and the proto-core will not react much. As an Fe layer sinks into the proto-core, proto-core materials must be squeezed out to the bottom of the magma ocean. Those proto-core materials will be entrained into the mantle and provide HSE and volatiles to the preexisting mantle.

Initially, these materials are dense because of low temperature. After some time (~20–30 Myrs) [estimated with a thickness of ~100 km and the thermal diffusivity of ~5 × 10^{-6} m^2/s (Ohta et al., 2012)], they are heated and become buoyant and will be mixed with the magma ocean. At this stage, most of colliding bodies are relatively large. When large bodies with an Fe-rich core collide, the Fe core will directly merge to the preexisting Fe core without much chemical reaction with the silicates and oxides (Benz et al., 1987; Dahl & Stevenson, 2010; Karato & Murthy, 1997). A giant impact (may be more of them) is one of these processes.

A giant impact would heat the growing Earth (and the impactor) (e.g., Canup, 2004) and would influence the Fe-silicate partitioning because temperature has an important effect on element partitioning (e.g., Murthy, 1991). Possible consequence of a giant impact to the composition of Earth is discussed by Nakajima and Stevenson (2015, 2018). However, as far as it occurred after the main part of the core was formed and the mass of the impactor is small (~10% of the Earth mass; Canup, 2004), its influence on the geochemistry of the mantle (such as the HSE abundance, the volatile content) and on the core-mantle interaction is likely minor.

Note that the mixing timescale may vary because it depends on the thickness of a layer of HSE-rich materials on top of the proto-core, and also on the vigor of convectional mixing. Maier et al. (2009) presented evidence that the amount of HSE in the Earth's mantle increased till ~3 Ga, implying that nearly complete mixing of HSE-rich

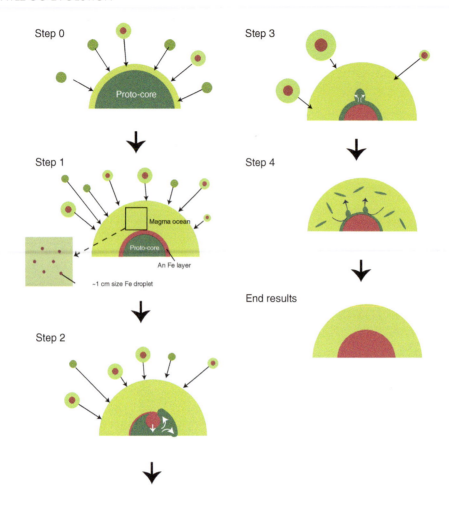

Figure 7.2 Schematic diagrams showing the processes of core formation. Source: Adapted from Karato and Murthy (1997).

Step 0: In the early stage of planetary formation, a mixture of primitive and differentiated materials will accrete to form relatively cold proto-planet.

Step 1: When a growing planet exceeds a threshold size, a magma ocean is formed on its surface, and Fe-core formation starts in the magma ocean. Fe is mostly molten and assumes a surface tension-controlled size of ~1 cm (Karato, 1975; Stevenson, 1990). These droplets sink through a magma ocean keeping chemical equilibrium until their sinking stops on top of the cold and strong proto-core.

Step 2: An Fe layer on top of the proto-core becomes unstable when enough Fe is accumulated, and a large Fe mass will sink to the center of the growing planet without chemical equilibrium. This will remove the proto-core materials to the bottom of the magma ocean.

Step 3: The proto-core materials squeezed out to the bottom of the mantle are initially cold and dense. Once the proto-core materials are heated enough (after a few tens of Myrs), they will be buoyant and mixed with the magma ocean. At this stage, a majority of Fe coming to the growing planet is the cores of relatively large planetary bodies, and not much chemical reaction will occur between primitive materials and Fe. The proto-core materials may also be mixed with the mantle through the entrainment by solid state mantle convection. Source: Adapted from Sleep (1988).

Step 4 (final result): Current planet (Earth) has a core that contains volatile and siderophile elements corresponding to the chemical equilibrium at lower pressure (P) and temperature (T) than the P-T conditions at the current core-mantle boundary. The mantle contains siderophile elements whose abundance is determined by the modest P-T corresponding to the bottom of the magma ocean in addition to the siderophile elements in the proto-core that include highly siderophile elements near chondritic abundance ratio.

materials took ~1.5 Gyrs. This long timescale of HSE addition suggests that mixing of proto-core materials with the mantle may have occurred by solid state convection through the entrainment of basal material (Sleep, 1988). It is difficult to explain the observed long timescale for HSE addition by the late veneer model because the timescale of late accretion and the amount of mass by the late accretion are both too short or too small (Morbidelli & Wood, 2015).

7.2.2. Siderophile Elements

Models such as the one shown in Figure 7.2 can be tested by the observed abundance of siderophile elements in the mantle (Fig. 7.3). The amount of siderophile elements in the mantle shows a systematic trend (e.g., Walter et al., 2000): (1) compared to the primitive materials (i.e., the CI carbonaceous chondrite), abundance of moderately siderophile elements (e.g., Ni, Co) in the Earth's mantle is depleted. The degree of depletion can be attributed to core formation in the deep magma ocean [$P \sim 50$ GPa, $T \sim 3500$ K; Siebert et al., 2012 (see Fig. 7.3a)]. (2) The degree of depletion of HSE is larger than that of modestly siderophile elements. And the abundance of HSE is ~0.3% relative to CI carbonaceous chondrite and independent of element within ~+/−50% (e.g., Day et al., 2016; Fig. 7.3). The nearly element-independent abundance is remarkable because the laboratory studies show highly element-sensitive partition coefficient (variation of a factor of ~100 among different elements; Mann et al., 2012).

It is often suggested that this remarkable observation of HSE abundance in the Earth's mantle is caused by the late addition of HSE-rich (and by inference, volatile-rich) materials: the "late veneer" model (Albarède, 2009; Jones & Drake, 1986; Walker, 2009). In addition to the rather *ad hoc* nature of this late veneer model, this model has a problem in explaining the observed siderophile and volatile element distribution. The observed pattern of abundance of these elements shows that among elements with different volatility, those that have highly siderophile nature are depleted more, suggesting that substantial

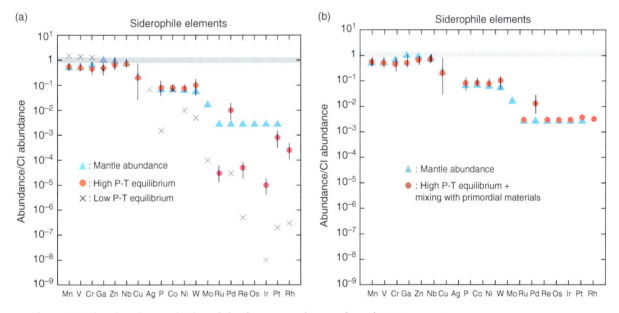

Figure 7.3 The abundance of siderophile elements in the mantle and its interpretation

(a) A comparison of the observed siderophile element abundance with equilibrium core formation model with various equilibrium pressure (P) and temperature (T).
(b) A comparison of the observed siderophile element abundance with equilibrium core formation model at high P and T + mixing with primordial materials after the overturn of the proto-core.

In the model (b), the model results depend on the amount of mixing of primitive materials. If the model were to explain the observations, the amount of primitive materials (from the proto-core) should be ~0.3% (shown in this figure).
Vertical Bars Indicate the Estimated Errors.
Sources: Walter et al. (2000), Mann et al. (2012), and Rudge et al. (2010).

amount of volatile elements was present when the core was formed (Wood et al., 2010) (for the issues on the late veneer model, see also Morbidelli & Wood, 2015).

The core formation model presented in Figure 7.2 provides an explanation of the observed pattern of HSE elements as well as the abundance pattern of siderophile and volatile elements summarized by Wood et al. (2010). In Figure 7.3b, I show the abundance of siderophile elements corresponding to a model shown in Figure 7.2. In this model, mantle materials are in chemical equilibrium with Fe at some high pressure and temperature ($P \sim 50$ GPa, $T \sim 2500$ K; Siebert et al., 2011) at steps 1 and 2. However, in step 4 where the proto-core materials are mixed with the magma ocean, HSE (and volatile elements) are added to the magma ocean. Some of HSE and volatile elements will be removed by ongoing core formation, but the core formation in this late stage occurs mostly by direct merge, and hence the removal of HSE and volatile elements will be limited. Note that since HSE are added from the proto-core before the main stage of core formation in this model, this model also explains the abundance pattern of siderophile and volatile elements as reported by Wood et al. (2010). Therefore, this is a possible model to explain the observed abundance pattern of HSE (as well as other siderophile elements).

However, this model is viable only when the amount of primitive materials in the proto-core added to the magma ocean is ~0.3%. The main question is if the fraction of primitive materials required to explain the HSE abundance is plausible or not. A discussion on this point and other implications of this model is provided in section 7.4 (Discussion).

7.3. CORE-MANTLE CHEMICAL INTERACTION

7.3.1. Is the Core a Source or a Sink of Volatile and Siderophile Elements?

The above model implies that the abundance of volatile and siderophile elements (H, C, Ni, W, etc.) of the core is determined by the chemical equilibria at relatively low pressure and temperature (core formation during the parent bodies of colliding materials, chemical equilibrium in the magma ocean) compared to the pressure and temperature at the current CMB. Consequently, it is likely that the abundance of these elements in the core is below the saturation limit corresponding to the pressure-temperature conditions at the CMB.

Keeping this in mind, let us consider the nature of chemical interaction between the core and the mantle through the geologic time long after the core formation. Although we have little evidence of the abundance of volatile elements such as H and C (or siderophile elements) in the core, it is plausible that the abundance of H and C (or siderophile elements) in the core is higher than that in the mantle because Fe has much higher solubility of these elements than silicates. Assuming that elements such as H and C (or siderophile elements) in the core have higher abundance than in the mantle, it is often suggested that the core is a source of these elements to the mantle (Hayden & Watson, 2007, 2008; see also Brandon & Walker, 2005; Rizo et al., 2019).

A concept behind this is the Fick's first law of diffusion (Fig. 7.4a):

$$J_i^{c \to m} = -D \frac{\partial C}{\partial x} \propto \left(C_i^c - C_i^m \right) \tag{7.1}$$

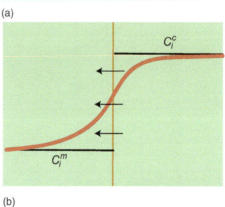

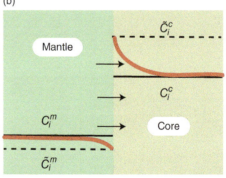

Figure 7.4 Transport of a trace element across a boundary

(a) A case of two same material
An element flows from the higher concentration region to the low concentration region.

(b) A case of two different materials (e.g., the mantle and the core)
The direction of flow of an element is not only controlled by the concentration difference (($logC_i^m - logC_i^c$)($C_i^{m,c}$: Concentration of Element i in the Mantle (m) or the Core (c))), but also by the difference in the solubility ($-\left(log\widetilde{C}_i^m - log\widetilde{C}_i^c \right)$)($\widetilde{C}_i^{m,c}$: solubility of element i in the mantle (m) or the core (c)). For the case of the core-mantle boundary, the latter dominates and since $\widetilde{C}_i^c \gg \widetilde{C}_i^m$, an element will flow from the mantle to the core even if $C_i^c > C_i^m$.

where $J_i^{c \to m}$ is flux of an element i from the core ("c") to the mantle ("m"), D is the relevant diffusion coefficient, and $C_i^{c,m}$ is concentration of element i in the core or mantle. More precisely, relation (7.1) should be modified to

$$J_i^{c \to m} \propto \left(\log C_i^c - \log C_i^m \right). \quad (7.2)$$

In both cases, these relations mean that an element flows from regions of higher concentration to regions with lower concentration, and since concentrations of various elements (H, C, etc.) in the core are likely higher than those in the mantle, these elements would flow from the core to the mantle. However, this argument cannot be used when we consider the transport of an element between two different materials.

In such a case, the driving force for element diffusion is not simply the difference in concentration $\left(\log C_i^c - \log C_i^m \right)$ but rather the gradient of the chemical potential including the excess energy associated with the dissolution of an element in a material that differs between two different materials [Fe and silicates (and oxides)]. Consequently, the driving force for diffusion must be generalized to the chemical potential difference, $\mu_i^c - \mu_i^m$ with

$$\mu_i^j = \mu_{i,o}^j + RT \log C_i^j \quad (7.3)$$

where $\mu_{i,o}^j$ is the excess chemical potential associated with dissolution of an element i in a material j [c (core) or m (mantle)], and C_i^j is the concentration of an element i in a material j. Therefore, relation (7.2) must be replaced with

$$J_i^{c \to m} \propto RT \left(\log C_i^c - \log C_i^m \right) + \left(\mu_{i,o}^c - \mu_{i,o}^m \right). \quad (7.4)$$

Here, the first term, $RT \left(\log C_i^c - \log C_i^m \right)$, corresponds to relation (7.2). In the second term, $\mu_{i,o}^j$ is related to the solubility of an element i. If we consider the chemical equilibrium of the core with the mantle at the CMB, then the chemical equilibrium with respect to the exchange of an element "i" can be written as

$$\mu_{i,o}^m + RT \log \widetilde{C}_i^m = \mu_{i,o}^c + RT \log \widetilde{C}_i^c \quad (7.5)$$

where $\widetilde{C}_i^{c,m}$ is the equilibrium concentration of element i in c (or m) in chemical equilibrium at the CMB. Hence,

$$\mu_{i,o}^c - \mu_{i,o}^m = -RT \log \left(\widetilde{C}_i^c - \widetilde{C}_i^m \right). \quad (7.6)$$

Therefore, if the actual concentration of an element i in the core and the mantle is C_i^c and C_i^m, respectively, then the chemical potential difference of this species (i) between the core and the mantle is given by

$$J_i^{c \to m} \propto \left(\log C_i^c - \log C_i^m \right) - \left(\log \widetilde{C}_i^c - \log \widetilde{C}_i^m \right). \quad (7.7)$$

This relation means that there are two sources for chemical potential difference driving diffusion an element, $\left(\log C_i^c - \log C_i^m \right)$ and $-\left(\log \widetilde{C}_i^c - \log \widetilde{C}_i^m \right)$. The first is from the chemical potential difference due to the concentration difference, and an element moves from an area of high concentration to an area of low concentration. The second is the chemical potential difference due to the difference in element solubility, and an element moves from an area of low solubility to an area of high solubility.

Because both of these terms include the concentration as $\log C$, the key is the concentration ratio. Therefore, I define two partition coefficients, $\frac{C_i^c}{C_i^m} \equiv K_{i,CF}$ and $\frac{\widetilde{C}_i^c}{\widetilde{C}_i^m} \equiv K_{i,CMB}$, where $K_{i,CF}$ is the effective partition coefficient relevant for core formation and $K_{i,CMB}$ is the partition coefficient at the CMB. For two reasons, I conclude that $K_{i,CMB} \gg K_{i,CF} > 1$ and hence $-\left(\log \widetilde{C}_i^c - \log \widetilde{C}_i^m \right)$ dominates over $\left(\log C_i^c - \log C_i^m \right)$. First, the chemical equilibrium during the core formation is between molten Fe and molten silicates, whereas the chemical equilibrium at the CMB is between molten Fe and solid minerals. Molten silicate has much higher element solubility than solid minerals (e.g., Blundy & Wood, 2003; Karato, 2016). This makes $K_{i,CF}$ small compared to $K_{i,CMB}$. Second, partition coefficient of an element between Fe (core) and silicates (mantle) increases with pressure (and temperature) (e.g., Mann et al., 2012). This makes $K_{i,CMB}$ higher than $K_{i,CF}$ (because pressure and temperature at which chemical equilibrium is attained during core formation are substantially lower than those at the CMB). Therefore, I conclude that $K_{i,CMB} \gg K_{i,CF}$ and $-\left(\log \widetilde{C}_i^c - \log \widetilde{C}_i^m \right)$ dominate over $\left(\log C_i^c - \log C_i^m \right)$.

Consequently, element will flow from the mantle to the core because the core has much higher solubility of the element than the mantle. Therefore, the core is a sink of volatile (and siderophile) elements rather than a source of volatile (and siderophile) elements even if the concentration of the relevant element is higher in the core than in the mantle. This conclusion is valid including a case where materials with different bulk composition were accumulated above/or the CMB as far as element transport occurs by diffusion (see Discussion).

The present model has an application to the models for low-velocity regions near the top of the outer core (Helffrich & Kaneshima, 2010). The low velocities observed in these regions are likely caused by the concentration of light elements such as oxygen. Gubbins & Davies (2013) proposed that this occurs due to baro-diffusion where light elements are considered to be transported to the shallow part due to gravity. This model assumes that there is no mass transport across the CMB. This is physically implausible because the core and the mantle are not in chemical equilibrium.

The present model provides an alternative explanation for the low-velocity regions near the top of the outer core. According to this model, there must be the transport of elements such as oxygen from the mantle to the core. Those elements diffuse into the core to a depth of $l \approx 2\sqrt{Dt}$ (D: diffusion coefficient of the element, t: time). For oxygen, for example, diffusion coefficient in the molten Fe near the melting point is 3×10^{-7} m^2/s (Sayadyaghoubi et al., 1995). Therefore, in ~1 (~4) Gyrs, the penetration depth will be ~200 (~400) km that agrees well with the seismological observations (Helffrich & Kaneshima, 2010). However, the role of light elements in modifying seismic wave velocity can be complicated. Most light elements increase velocity (the Birch's law) but the degree to which velocity increases is element dependent (e.g., Badro et al., 2014; Ichikawa & Tsuchiya, 2020; Ohtani, 2013; Sanloup et al., 2004; Zhang et al., 2016). Brodholt and Badro (2017) suggested a combined role of Si and O to explain the presence of a low-velocity region near the top of the core (see also Trønnes et al., 2019).

7.3.2. Meso-Scale Material Transport: The Morphological Instability

Despite above thermodynamically robust discussion showing that the core is a sink of siderophile and/or volatile elements, there are geochemical observations suggesting a leak of core materials to the mantle (e.g., Brandon & Walker, 2005; Rizo et al., 2019). In the following, I will show that these seemingly conflicting observations are two sides of a common physical process caused by the core-mantle disequilibrium.

Because the bulk of the core and the mantle is not chemical equilibrium, concentration of several elements such as oxygen in the core is undersaturated. This leads to the flow of oxygen (and other elements) from the mantle to the core leading to a region at the bottom of the mantle where concentration of these elements is depleted. This promotes the penetration of the molten Fe into the mantle because the molten Fe that penetrates into the mantle gets more oxygen and the free energy of molten Fe is reduced (Mullins & Sekerka, 1963, 1964).

Such a phenomenon was reported by Otsuka and Karato (2012) where they show that when molten Fe and solid (Mg,Fe)O are in contact under the disequilibrium initial condition, molten Fe penetrated into (Mg,Fe)O, and the molten Fe penetrated into (Mg,Fe)O is pinched off to become an isolated blob (Fig. 7.5a). Since blobs are isolated, the influence of gravity is small in comparison to Fe penetration along the grain-boundaries. Therefore, blobs migrate deep into (Mg,Fe)O presumably controlled by the dissolution and precipitation of FeO to form a layer rich in molten Fe. Otsuka and Karato (2012) showed that the growth of the Fe-rich layer in (Mg,Fe)O follows a relation $L \approx \sqrt{\widetilde{D}t}$, where L is the thickness of Fe-rich layer in (Mg,Fe)O, $\widetilde{D}$ is the effective diffusion coefficient, and t is the time. Otsuka and Karato (2012) estimated $\widetilde{D}$ from their data and found that the effective diffusion coefficient is substantially higher than the diffusion coefficient of Mg-Fe in (Mg,Fe)O. Otsuka and Karato (2012) developed a model of growth of such a layer and showed that $\widetilde{D}$ is proportional to the diffusion coefficient of Mg-Fe and also the grow rate depends on the degree of disequilibrium. Their results predict that molten Fe might penetrate to several tens km in ~1 Gyr.

Another observation is that associated with the penetration of molten Fe into (Mg,Fe)O, oxygen (FeO) diffuses into molten Fe (Fig. 7.5a). This is a demonstration of the concept that we discussed before that at the CMB: various elements diffuse from the mantle to the core if the core is undersaturated with those elements.

Recently, Yoshino (2019) extended the study by Otsuka and Karato (2012) using polycrystalline specimens of (Mg,Fe)O and bridgmanite. For bridgmanite, they observed no detectable penetration of Fe that confirms the inference made by Otsuka and Karato (2012). For (Mg,Fe)O, Yoshino (2019) observed not only the penetration of Fe blobs in (Mg,Fe)O crystals as Otsuka and Karato (2012) observed but also observed Fe penetration along grain-boundaries. Under his experimental conditions, the depth of Fe penetration along grain-boundaries is deeper than the depth of penetration of Fe blobs in (Mg,Fe)O grains. Based on this observation, Yoshino (2019) concludes that Fe penetration along grain-boundary is prerequisite for Fe penetration into the crystal by the morphological instability. Then he uses a model by Poirier and Le Mouel (1992) showing that Fe penetration along grain-boundaries is limited by gravity to be 10–100 m, and therefore he concludes that the morphological instability is ineffective.

The presence of these two modes of Fe penetration is a natural consequence of complete wetting of MgO grain-boundaries by Fe reported by Urakawa et al. (1987). However, Yoshino (2019)'s argument that Fe penetration along grain-boundaries is prerequisite to Fe blob penetration is likely incorrect, because Otsuka and Karato (2012) observed Fe blob penetration into (Mg,Fe)O single crystal. When Fe penetration occurs in the gravity field, both modes of penetration occur, but the penetration of Fe along grain-boundaries stops when gravity effect becomes important, say at 10–100 m (Poirier and Le Mouel, 1992). However, the effect of gravity is small for Fe blobs, and penetration of Fe blobs continues beyond this limit. As a consequence, at a point where Fe containing grain-boundaries ceases to exist due to gravity, Fe blobs will still be formed and migrate caused by the morphological instability. When those blobs reach grain-boundaries, continuous Fe-wetted grain-boundaries are formed again. Therefore, we expect

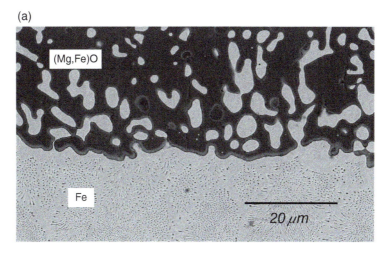

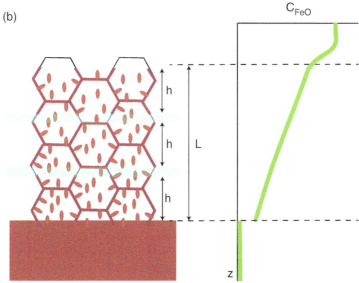

Figure 7.5 Fe penetration into (Mg,Fe)O (the lower mantle) by the morphological instability. Source: Ohtani and Ringwood (1984) / Elsevier.

(a) SEM micrographs (backscattered electron images) showing the reactions between molten Fe and $(Mg_{75},Fe_{25})O$ (P = 24 GPa, T = 2270 K, t = 20 minutes)
Fe (seen as bright regions) penetrates into (Mg,Fe)O both along grain-boundaries and into grain interior. Also, evidence of oxygen dissolution in molten Fe can be seen by slighter darker regions in the bottom half of this figure. Those regions correspond to FeO that was precipitated during cooling from the dissolved oxygen when Fe was molten; (Ohtani and Ringwood, 1984).

(b) A schematic drawing showing the likely structure of Fe (red) Penetration into a polycrystalline (Mg,Fe)O in the gravitational field (h: thickness of continuous Fe-wetted grain-boundaries (1–100 m), L: thickness of Fe penetrated layer; green bands: gaps between continuous Fe-wetted boundary regions) Together with the concentration profile of FeO (C_{FeO}) as a function of vertical axis (z).
The concentration profile is schematic but is consistent with the experimental observations (Otsuka & Karato, 2012). The thickness of a continuous Fe layer on grain-boundaries is limited to $h \sim 1$–100 m due to gravity (Poirier & Le Mouel, 1992). However, even above this level, Fe will penetrate as blobs driven by the concentration gradient. Consequently, a thick Fe layer (L ~ a few tens km) will contain many thin layers (h ~ 1–100 m) in which grain-boundaries are covered with molten Fe continuously separated by a gap, a structure similar to the one suggested for a partially molten rock by Shankland et al. (1981). Since a large fraction of grain-boundaries of (Mg,Fe)O are covered with molten Fe, low seismic wave velocity and high electrical conductivity will be expected in a region where (Mg,Fe)O is interconnected. Molten Fe will carry siderophile and volatile elements and therefore plumes coming from those regions may contain some core signature (Brandon & Walker, 2005).

that a layer of Fe penetration likely contains intermittent layers of Fe-wetted grain-boundaries together with the penetration by Fe-blobs (Fig. 7.5b).

Therefore, I conclude that Fe penetration deep into the mantle is a distinct possibility in regions where (Mg,Fe)O is interconnected. However, this mechanism is ineffective in bridgmanite or post-perovskite where diffusion of Mg (Fe) is so slow (e.g., Ammann et al., 2010). In a typical lower mantle, the concentration of (Mg,Fe)O is ~17% (e.g., Ringwood, 1991). Since the percolation threshold is ~20–25% (Stauffer & Aharony, 1992), regions where the morphological instability brings core materials deep into mantle are regions with anomalously high concentration of (Mg,Fe)O and are likely limited. Since Fe penetration (particularly along grain-boundaries) reduces seismic wave velocities dramatically, these regions may correspond to the regions where ultralow velocities are found (Lay et al., 1998). The area extent of such regions is small (a few %; Lay et al., 1998). Therefore, the net flux of volatile or siderophile elements from the core is limited. One large remaining uncertainty with this model is that the pressure effects on this process is not well constrained.

I conclude that in a majority of regions where molten Fe penetrates via the morphological instability, Fe will cover grain-boundaries. Therefore, the magnitude of reduction in seismic wave velocity and the increase in electrical conductivity will be larger than a case where molten Fe exists only in the grains as isolated blobs.

7.4. DISCUSSIONS

7.4.1. Plausibility of the Proto-Core Model for Highly Siderophile Elements (HSE)

The model presented here can explain the observed HSE abundance if the mass ratio of primitive materials and the mantle is $\xi \equiv \frac{M_{CC}}{M_{mantle}} = 3 \times 10^{-3}$, where M_{CC} is the mass of primitive materials (similar to carbonaceous chondrite) and M_{mantle} is the mass of the mantle. If we define $M_{proto-core} = \beta M_{mantle}$ and $M_{CC} = \zeta M_{proto-core}$, then $\xi = \beta \zeta$. To estimate β, we assume that proto-core is a region where no substantial chemical reactions occur during planetary formation (~10 Myrs). This would correspond to a threshold temperature of ~1000 K for the diffusion distance of ~1 m [based on diffusion coefficients of various elements (extrapolated to high pressures using the homologous temperature scaling); Van Orman et al., 2001]. Then assuming the relation $T_{surface} \propto R^2$ ($T_{surface}$: surface temperature of a growing planet, R: size of a growing planet), the mass of the proto-core is estimated to be $M_{proto-core} \approx M_{Mars} \approx 0.1 \cdot M_E \approx 0.15 \cdot M_{mantle}$ ($\beta \approx 0.15$). Then, $\zeta = 0.027$.

So the fraction of the primitive component (carbonaceous chondrite-like materials) of the proto-core needs to be ~3%. Since some fractions of the proto-core materials mixed in the magma ocean would react with Fe from the later stage bombardment (a majority of Fe in these bodies would directly merge the preexisting core and will not react with the magma ocean materials), the amount of proto-core materials needed to explain the HSE pattern will be somewhat larger than 3%. The fraction of carbonaceous chondrite among the meteorites is ~5% (Wasson, 1985). But this is likely underestimate because carbonaceous chondrites are easy to be altered compared to other meteorites and many of them are likely unrecognized as meteorites. Another estimate is from the observations on the asteroid belts from which most meteorites come (DeMeo and Carry, 2014) that gives a substantially higher fraction of carbonaceous chondrite-type material in the asteroid belt. Given the large uncertainties in the estimates of many parameters, I conclude that the required mass fraction of primitive materials (~0.3%) is not inconsistent with plausible models of solar system materials as inferred from the composition of meteorites and asteroids. Therefore, I believe that this hypothesis deserves a serious consideration as an alternative to the late veneer model for the origin of HSE (and volatiles). Note that if the primitive components of the proto-core materials are carbonaceous chondrite-type materials, then these materials will have ~10 wt% water that would provide ~2 ocean mass of water. Obviously, other materials also contain water (e.g., Jarosewich, 1990; Piani et al., 2020), so the net water content of Earth will be greater.

How about the C/H, N/H ratios? Hirschmann (2016) discussed that the C/H and N/H ratios of carbonaceous chondrite-type materials invoked in the late veneer model are too high compared to those of Earth. The present model also provides a possible explanation for this issue. The primitive materials invoked in the present model and the late veneer model are same: carbonaceous chondrite-type materials. The C/H and N/H ratios of carbonaceous chondrite are substantially larger than those of the Earth's mantle. Since late veneer is added later, it is not clear how C/H and N/H ratios can be reduced (Hirschmann, 2016).

The present model provides a plausible explanation for the observed C/H and N/H ratio. In the present model, HSE and volatile-rich materials are mostly from the proto-core, and mixed with the mantle. Either during the magma ocean freezing or through partial melting after magma ocean freezing, melt-solid separation will selectively remove C and N more than H because C and N are more incompatible than H (e.g., Shcheka et al., 2006; Speelmanns et al., 2019). In contrast, in the late veneer model, there are no obvious processes to modify these ratios from the carbonaceous chondritic values.

An alternative explanation for the "excess" HSE is the addition of core materials to the mantle (Hayden & Watson, 2007; Walker & Walker, 2005). As I demonstrated in this chapter, although the core likely contains substantially more siderophile elements (including HSE) than the mantle, the direction of diffusional transport of these elements is from the mantle to the core. In addition, even if there were some leak of HSE from the core, it is difficult to explain why HSE abundance is nearly constant if HSE is added to the mantle by diffusional transport as suggested by Hayden and Watson (2007) because diffusion coefficient is element dependent.

7.4.2. Hydrogen in the Core

In section 7.3.1, I stated that H (hydrogen) content in the core is likely higher than that in the mantle based on the experimental study showing higher H solubility in Fe than in silicates (Fukai & Suzuki, 1986), and suggested that the core is a big reservoir of water in Earth. This is supported by the first-principles calculations on H partitioning between molten Fe and molten silicates (Li et al., 2020). However, there are some publications showing the lithophile nature of H (and C) during core formation in the magma ocean (Clesi et al., 2018; Malavergne et al., 2019).

If these experimental results are applicable, H content in the core is not as high as one might expect for the partitioning between solid silicates and Fe. Then we need to reinterpret $K_{i,CF}$ as a partition coefficient between molten silicates and Fe at the pressure and temperature corresponding to the bottom of the magma ocean. This will increase $K_{i,CF}$ compared to the partition coefficient between solid silicates and Fe. However, Li et al. (2020) presented an argument that the experimental results reported by Clesi et al. (2018) and Malavergne et al. (2019) might be affected by the H (and C) loss from Fe during quenching. In any case, because $K_{i,CMB}$ is so large (due to small solubility of H and C in silicates) that the conclusion $K_{i,CMB} > K_{i,CF}$ will remain valid even if it may not be $K_{i,CMB} \gg K_{i,CF}$.

7.4.3. Other Factors Controlling the Element Transport Across the Core-Mantle Boundary

At the CMB, there is not only the concentration contrast of some elements such as H or C, but materials with different compositions may also accumulate at the CMB. For example, FeO-rich dense materials are accumulated on the mantle side of the CMB, and light materials are accumulated at the core side of the CMB. What are their roles in the material transport at the CMB?

One of which is already discussed in connection to the morphological instability that may occur in (Mg,Fe)O-rich regions at the mantle side of the CMB. Another case is the basal magma ocean (BMO) (Labrosse et al., 2007; see also Trønnes et al., 2019). Materials there likely contains high amount of volatiles. Therefore, they likely contribute transport of these elements to the core. In particular, Trønnes et al. (2019) proposed a model emphasizing the material exchange between BMO and the core. However, the influence of pressure on element partitioning was not considered in their study.

Core cooling will cause precipitation of some (light) materials (e.g., Hirose et al., 2017). Again, these materials likely contain smaller amount of volatile elements because the core is undersaturated with volatile and siderophile elements, and consequently help volatile and siderophile elements transport from the mantle to the core. So in short, accumulation of these materials enhances the tendency that volatiles are transported from the mantle to the core. Only exception is the morphological instability.

Finally, Kanda and Stevenson (2006) proposed a meso-scale Fe penetration into the CMB by the pressure gradient caused by mantle convection. The depth of penetration of Fe into the mantle by this mechanism is proportional to the viscosity of the mantle. Kanda and Stevenson (2006) assumed the viscosity of 10^{21} Pa s (average viscosity of the lower mantle; Peltier, 1998) to estimate the penetration depth of ~ 1 km. However, the viscosity in the CMB region is much less than the average viscosity of the lower mantle because of high temperature. Nakada and Karato (2012) estimated a viscosity of $\sim 10^{18}$ Pa s in this region that predicts the penetration depth of ~ 1 m.

7.5. SUMMARY AND CONCLUDING REMARKS

Our understanding of the nature of chemical interactions between the mantle and the core has been improved due to the advance in determining partition coefficients of various elements under the broad range of physical conditions. However, since the bulk of the core and the mantle is not in chemical equilibrium, understanding the processes of core mantle separation (core formation) is also important. Among others, the concept that the core is in equilibrium with silicates at lower pressure-temperature than the pressure-temperature at the CMB is well supported by experimental data and the observed abundance of modestly siderophile elements. In this chapter, I present a brief discussion of an implication of this concept for the nature of element transport across the CMB.

One new aspect of the model presented here is the role of the proto-core. The role of the proto-core was discussed by Karato and Murthy (1997), but they dismissed the important role of the proto-core for mantle geochemistry based on the argument that since the proto-core materials are cold and dense, they will not mix with the materials above (the magma ocean materials). This led to their model that only a small fraction (less than 0.5%) of the core materials is in equilibrium with the mantle materials. A recent analysis of W-Hf, U-Pb age determination suggests that the fraction of Fe in equilibrium with silicates during core formation is substantially larger (Rudge et al., 2010). The present model is a modified version of the model proposed by Karato and Murthy (1997) by including the influence of heating of proto-core materials after they are placed at the bottom of the magma ocean (steps 3–4 in Fig. 7.2). This model predicts that a substantial amount of volatile and HSE was acquired by a growing Earth in the early stage of its formation rather than in the later stage.

The present model also predicts that the core contains volatile and siderophile elements corresponding to the chemical equilibrium at lower pressure and temperature than the pressure and temperature at the current CMB. Therefore, the bulk of the core is out of chemical equilibrium with the bulk of the mantle. This promotes transport of materials between the core and the mantle. In case where transport occurs atomistic diffusion, the driving force for diffusion comes from the difference in the excess energy of an element in materials of the core and the mantle as well as the difference in concentration of elements themselves. An estimate of these two terms based on the theoretical model and the experimental data on element partitioning suggests that the excess energy term dominates and the core is a sink rather than the source of these elements.

This immediately implies that the concentrations of these elements in the core have been increasing with time, and hence the core is growing. A moving boundary such as the CMB promotes a mesoscopic material transport at the CMB via morphological instability if some conditions are met. A review of published results shows that such an instability likely occurs in regions where (Mg,Fe)O are interconnected (Otsuka & Karato, 2012). These are presumably the regions with ultralow seismic wave velocities. Those regions provide a window for core materials.

In conclusion, the present model of core formation provides a unified explanation of a variety of geophysical and geochemical observations (Fig. 7.6). They include geochemical observations on the siderophile element abundance as well as the "core signature" from some isotopes, and geophysical observations of low-velocity (and high-conductivity) regions in the D" layer as well as the low-velocity layer near the top of the outer core. The thermodynamic analysis presented here gives a guideline on the nature of materials transport (exchange) between the mantle and the core. However, various issues remain unclear including the influence of pressure on the kinetics of the morphological instability and the role of materials

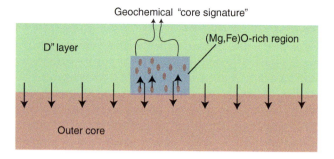

Figure 7.6 A cartoon showing the nature of mass/element transport across the core-mantle boundary
In the majority of the core-mantle boundary (CMB), volatile, siderophile elements diffuse from the mantle to the core because the core is highly undersaturated with these elements. This transport of elements results in the compositional gradient across the CMB that promotes mesoscopic transport of molten Fe into the mantle in regions where (Mg,Fe)O is interconnected. The transport of these elements from the mantle to core forms a layer near the top of the outer core that is enriched with these elements. This model explains the presence of localize ultralow-velocity regions in the mantle side of the CMB as well as a low-velocity layer near the top of the outer core although some combination of light elements may be needed to explain low seismic wave velocities. Source: Adapted from Brodholt and Badro (2017).

transport across the CMB in the chemical evolution of the whole Earth.

ACKNOWLEDGMENTS

This work is partly supported by a grant from NSF.

I thank Zhen Liu who conducted a set of experiments on the penetration of molten Fe into (Mg,Fe)O polycrystals in 2013 as a term paper project (shown in Fig. 7.5a), and Zhenting Jiang and Jennifer Girard for technical help. A discussion with Alessandro Morbidelli was helpful to improve the model for HSE acquisition and for the late veneer. Reviews by two anonymous reviewers and by an editor (Taku Tsuchiya) are also helpful in improving this chapter. Thank you all.

REFERENCES

Abe, Y., & Matsui, T. (1985). Formation of an impact-generated H_2O atmosphere and its implications for the early thermal history of the Earth. *Journal of Geophysical Research*, 90, 545–559.

Abe, Y., & Matsui, T. (1988). Evolution of an impact-generated H_2O-CO_2 atmosphere and formation of a hot proto-ocean on Earth. *Journal of Atmospheric Sciences*, 45, 3081–3101.

Albarède, F. (2009). Volatile accretion history of the terrestrial planets and dynamic implications. *Nature*, 461, 1227–1233.

Ammann, M. W., Brodhlot, J. P., Wookey, J., & Dobson D. P. (2010). First-principles constraints on diffusion in lower-mantle minerals and a weak D" layer. *Nature*, *465*, 462–465.

Badro, J., Brodhlot, J. P., Piet, H., Siebert, J., & Ryerson F. J. (2015). Core formation and core composition from coupled geochemical and geophysical constraints. *Proceedings of the National Academy of Sciences of the United States*, *112*, 12310–12314.

Badro, J., Cõté, A. S., & Brodhlot, J. P. (2014). A seismologically consistent compositional model of Earth's core. *Proceedings of the National Academy of Sciences of the United States*, *111*, 7542–7545.

Benz, W., Slattery, W. L., & Cameron, A. G. W. (1987). The origin of the moon and the single-impact hypothesis II. *Icarus*, *71*, 30–45.

Blundy, J. D., & Wood, B. J. (2003). Partitioning of trace elements between crystals and melts. *Earth and Planetary Science Letters*, *210*, 383–397.

Brandon, A., & Walker, R. J. (2005). The debate over core-mantle interaction. *Earth and Planetary Science Letters*, *232*, 211–225.

Brodholt, J., & Badro, J. (2017). Composition of the low seismic velocity E' layer at the top of Earth's core. *Geophysical Research Letters*, *44*, 8303–8310.

Canup, R. M. (2004). Simulations of a late lunar-forming impact. *Icarus*, *168*, 433–456.

Clesi, V., Bouhifd, M. A., Bolfan-Casanova, N., Manthilake, G., Schiavi, F., Raepsaet, C., et al. (2018). Low hydrogen content in the cores of terrestrial planets. *Science Advances*, *4*, e1701876.

Dahl, T. W., & Stevenson, D. J. (2010). Turbulent mixing of metal and silicate during planet accretion - And interpretation of the Hf-W chronometer. *Earth and Planetary Science Letters*, *295*, 177–186.

Day, J. M. D., Brandon, A. D., & Walker, R. J. (2016). Highly siderophile elements in Earth, Mars, the Moon and asteroids. *Reviews in Mineralogy and Geochemistry*, *81*, 161–238.

DeMeo, F. E., & Carry, B. (2014). Solar system evolution from compositional mapping of the asteroid belt. *Nature*, *505*, 629–634.

Fukai, Y., & Suzuki, T. (1986). Iron-water interaction under high pressure and its implications in the evolution of the Earth. *Journal of Geophysical Research*, *91*, 9222–9230.

Gubbins, D., & Davies, C. J. (2013). The stratified layer at the core-mantle boundary caused by barodiffusion of oxygen, sulfur and silicon. *Physics of the Earth and Planetary Interiors*, *215*, 21–28.

Halliday, A. N., Lee, D.-C., Porcelli, D., & Wiercert, U. (2001). The rates of accretion, core formation and volatile loss in the early Solar system. *Philosophical Transactions: Mathematical, Physical and Engineering Sciences*, *359*, 2111–2135.

Hayashi, C., Nakazawa, K., & Nakagawa, Y. (1985). Formation of the solar system. In D. C. Black, & M. S. Matthews (Ed.), *Protoplanets and Planets II*, pp. 1100–1153. Tuscon, AZ: University of Arizona Press.

Hayden, L. A., & Watson, E. B. (2007). A diffusion mechanism for core-mantle interaction. *Nature*, *450*, 709–712.

Hayden, L. A., & Watson, E. B. (2008). Grain boundary mobility of carbon in Earth's mantle: A possible carbon flux from the core. *Proceedings of the National Academy of Sciences of the United States*, *105*, 8537–8541.

Helffrich, G., & Kaneshima, S. (2010). Outer-core compositional stratification from observed core wave speed profile. *Nature*, *468*, 807–810.

Hirose, K., Morard, G., Sinmyo, R., Umemoto, K., Hernlund, J., Helffrich, G., & Labrosse, S. (2017). Crystallization of silicon dioxide and compositonal evolution of the Earth's core. *Nature*, *543*, 99–102.

Hirschmann, M. M. (2016). Constraints on the early delivery and fractionation of Earth's major volatiles from C/H, C/N, and C/S ratios. *American Mineralogist*, *101*, 540–553.

Honda, R., Mizutani, H., & Yamamoto, T. (1993). Numerical simulation of the Earth's core formation. *Journal of Geophysical Research*, *98*, 2075–2089.

Ichikawa, H., & Tsuchiya, T. (2020). Ab Initio thermoelasticity of liquid iron-nickel-light element alloys. *Minerals*, *10*, 50. doi:10.3390/min10010059.

Jarosewich, E. (1990). Chemical analyses of meteorites: A compilation of stony and iron meteorite analyses. *Meteoritics*, *25*, 323–337.

Jones, J. H., & Drake, M. J. (1986). Geochemical constraints on core formation in the Earth. *Nature*, *322*, 221–228.

Kanda, R. V. S., & Stevenson, D. J. (2006). Suction mechanism for iron entrainment into the lower mantle. *Geophysical Research Letters*, *33*, 10.1029/2005GL025009.

Karato, S. (1975). *Chemical equilibrium during gravitational separation : Implications for the formation processes of earth's core, paper presented at Annunal Meeting of the Seismological Society of Japan*. Tokyo: Seismological Society of Japan.

Karato, S. (2016). Physical basis of trace element partitioning: A review. *American Mineralogist*, *101*, 2577–2593.

Karato, S., & Murthy, V. R. (1997). Core formation and chemical equilibrium in the Earth I. Physical considerations. *Physics of Earth and Planetary Interiors*, *100*, 61–79.

Kleine, T., & Rudge, J. F. (2011). Chronometry of meteorites and the formation of the Earth and Moon. *Elements*, *7*, 41–46.

Kokubo, E., & Ida, S. (1998). Oligarchic growth of protoplanets. *Icarus*, *131*, 171–178.

Labrosse, S., Hernlund, J. W., & Coltice, N. (2007). A crystallizing dense magma ocean at the base of the Earth's mantle. *Nature*, *450*, 866–869.

Lambrechts, M., Morbidelli, A., Jacobson, S. A., Johansen, A., Bitsch, B., Izidoro, A., et al. (2019). Formation of planetary systems by pebble accretion and migration: How the radial pebble flux determines a terrestrial-planet or super-Earth growth mode. *Astronomy and Astrophysics*, *627*, A83.

Lay, T., Williams, Q., & Garnero, E. J. (1998). The core-mantle boundary layer and deep Earth dynamics. *Nature*, *392*, 461–468.

Li, Y., Vočadlo, L., Sun, T., & Brodholt, J. (2020). The Earth's core as a reservoir of water. *Nature Geoscience*, *13*, 453–457. doi:10.1038/s41561-020-0578-1.

Lichtenberg, T., Drążkowska, J., Schönbächler, M., Golabek, G. J., & Hands, T. O. (2021). Bifurcation of planetary building blocks dueing Solar System formation. *Science*, *371*, 6527.

Maier, W. D., Barnes, S. J., Campbell, I. H., Fiorentini, M. L., Peltonen, P., Barnes, S.-J., et al. (2009). Progressive mixing of meteoritic veneer into the early Earth's deep mantle. *Nature*, *460*, 620–623.

Malavergne, V., Bureau, H., Raepsaet, C., Gaillard, F., Poncet, M., Surblé, S., et al. (2019). Experimental constraints on the fate of H and C during planetary core-mantle differentiation. Implications for the Earth. *Icarus*, *321*, 473–485.

Mann, U., Frost, D. J., Rubie, D. C., Becker, H., & Audétat, A. (2012). Partitioning of Ru, Rh, Pd, Re, Ir and Pt between liquid metal and silicate at high pressures and high temperatures - Implications for the origin of highly siderophile element concentrations in the Earth's mantle. *Geochemica et Cosmochemica Acta*, *84*, 593–613.

Morbidelli, A., Lunine, J. I., O'Brien, D. P., Raymond, S. N., & Walsh, K. J. (2012). Building terrestrial planets. *Annual Review of Earth and Planetary Sciences*, *40*, 251–275.

Morbidelli, A., & Wood, B. J. (2015). Late accretion and late veneer. In J. Badro, & M. J. Wlaker (Eds.), *The Early Earth: Accretion and Differentiation*, pp. 71–101. Washington DC: American Gephysical Union.

Mullins, W. W., & Sekerka, R. F. (1963). Morphological stability of a particle growing by diffusion or heat flow. *Journal of Applied Physics*, *34*, 323–329.

Mullins, W. W., & Sekerka, R. F. (1964). Stability of a planar interface during solidification of a dilute binary alloy. *Journal of Applied Physics*, *35*, 444–451.

Murthy, V. R. (1991). Early differentiation of the Earth and the problem of mantle siderophile elements: A new approach. *Science*, *253*, 303–306.

Nakada, M., & Karato, S. (2012). Low viscosity of the bottom of the Earth's mantle inferred from the analysis of Chandler wobble and tidal deformation. *Physics of the Earth and Planetary Interiors*, *192/193*, 68–80.

Nakajima, M., & Stevenson, D. J. (2015). Melting and mixing states of the Earth's mantle after the Moon-forming impact. *Earth and Planetary Science Letters*, *427*, 286–295.

Nakajima, M., & Stevenson, D. J. (2018). Inefficient volatile loss from the Moon-forming disk: Reconciling the giant impact hypothesis and a wet Moon. *Earth and Planetary Science Letters*, *487*, 117–126.

Ohta, K., Yagi, T., Taketoshi, N., Hirose, K., Komabayashi, T., Baba, T., et al. (2012). Lattice thermal conductivity of $MgSiO_3$ perovskite and post-perovskite at the core-mantle boundary. *Earth and Planetary Science Letters*, *349/350*, 109–115.

Ohtani, E. (2013). Chemical and physical properties and thermal state of the core. In S. Karato (Ed.), *Physics and Chemistry of the Deep Earth*. New York: Wiley-Blackwell.

Ohtani, E., & Ringwood, A. E. (1984). Composition of the core, I. Solubility of oxygen in molten iron at high temperatures. *Earth and Planetary Science Letters*, *71*, 85–93.

Okuchi, T. (1997). Hydrogen partitioning into molten iron at high pressure: Implications for Earth's core. *Science*, *278*, 1781–1784.

Otsuka, K., & Karato, S. (2012). Deep penetration of molten iron into the mantle caused by a morphological instability. *Nature*, *492*, 243–247.

Peltier, W. R. (1998). Postglacial variation in the level of the sea: Implications for climate dynamics and solid-Earth geophysics. *Review of Geophysics*, *36*, 603–689.

Piani, L., Marrocchi, Y., Rigaudier, T., Vacher, L. G., Thomassin, D., & Marty, B. (2020). Earth's water may have been inherited from materials similar to enstatite chondrite meteorites. *Science*, *369*, 1110–1113.

Poirier, J.-P., & Le Mouel, J.-L. (1992). Does infiltration of core material into the lower mantle affect the observed geomagnetic field?. *Physics of Earth and Planetary Interiors*, *73*, 29–37.

Raymond, S. N., & Izidoro, A. (2017). Origin of water in the inner Solar System: Planetesimals scattered inward during Jupiter and Saturn's rapid gas accretion. *Icarus*, *297*, 134–148.

Ringwood, A. E. (1991). Phase transformations and their bearings on the constitution and dynamics of the mantle. *Geochemica and Cosmochemica Acta*, *55*, 2083–2110.

Rizo, H., Andrault, D., Bennett, N. R., Humayun, M., Brandon, A., Vlastelic, I., et al. (2019). ^{182}W evidence for core-mantle interaction in the source of mantle plumes. *Geochemical Perspectives Letters*, *11*, 6–11.

Rudge, J. F., Kleine, T., & Bourdon, B. (2010). Broad bounds on Earth's accretion and core formation constrained by geochemical models. *Nature Geoscience*, *3*, 439–443.

Sanloup, C., Fiquet, G., Gregoryanz, E., Morard, G., & Mezouar, M. (2004). Effect of Si on liquid Fe compressibility: Implications for sound velocity in core materials. *Geophysical Research Letters*, *31*. 10.1029/2004GL019526.

Sasaki, S., & Nakazawa, K. (1986). Metal-silicate fractionation in the growing Earth: Energy source for the terrestrial magma ocean. *Journal of Geophysical Research*, *91*, 9231–9238.

Sayadyaghoubi, Y., Sun, S., & Jahanshahi, S. (1995). Determination of the chemical diffusion of oxygen in liquid iron oxide at 1615°C. *Metallurgical and Materials Transactions*, *26B*, 795–802.

Shankland, T. J., O'Connell, R. J., & Waff, H. S. (1981). Geophysical constraints on partial melt in the upper mantle. *Review of Geophysics and Space Physics*, *19*, 394–406.

Shcheka, S. S., Wiedenbeck, M., Frost, D. J., & Keppler, H. (2006). Carbon solubility in mantle minerals. *Earth and Planetary Science Letters*, *245*, 730–742.

Siebert, J., Badro, J., Antonangeli, D., & Ryerson, F. J. (2012). Metal-silicate partitioning of Ni and Co in a deep magma ocean. *Earth and Planetary Science Letters*, *321–322*, 189–197.

Siebert, J., Corgne, A., & Ryerson, F. J. (2011). Systematics of metal-silicate partitioning for many siderophile elements applied to Earth's core formation. *Geochemica et Cosmochemica Acta*, *75*, 1451–1489.

Sleep, N. H. (1988). Gradual entrainment of a chemical layer at the base of the mantle by overlying convection. *Geophysical Journal of Royal Astronomical Society*, *95*, 437–447.

Speelmanns, I. M., Schmidt, M. W., & Liebske, C. (2019). The almost lithophile character of nitrogen during core formation. *Earth and Planetary Science Letters*, *509*, 186–197.

Stauffer, D., & Aharony, A. (1992). *Introduction to Percolation Theory*, p.181. London: Taylor and Francis.

Stevenson, D. J. (1990). Fluid dynamics of core formation. In H. E. N. a. J. H. Jones (Ed.), *Origin of the Earth*, pp. 231–249. Oxford: Oxford University Press.

Trønnes, R. G., Baron, M. A., Eigenmann, K. R., Guren, M. G., Heyn, B. H., Løken, A., et al. (2019). Core formation, mantle differentiation and core-mantle interaction within Earth and the terrestrial planets. *Tectonophysics*, *760*, 165–198.

Urakawa, S., Kato, M., & Kumazawa, M. (1987). Experimental study on the phase relations in the system Fe-Ni-O-S. In M. H. Manghnani and Y. Syono (Eds.), *High-Pressure Research in Mineral Physics*, pp. 95–111. Tokyo: Terra Scientifiic Publishing Company.

Van Orman, J. A., Grove, T. L., & Shimizu, N. (2001). Rare earth element diffusion in diopside: Influence of temperature, pressure and ionic radius, and an elastic model for diffusion in silicates. *Contributions to Mineralogy and Petrology, 141*, 687–703.

Walker, R. J. (2009). Highly siderophile elements in the Earth, Moon and Mars: Update and implications for planetary accretion and differentiation. *Chemie der Erde, 69*, 101–125.

Walker, R. J., & Walker, D. (2005). Does the core leak?. *EOS, Transactions of American Geophysical Union, 86*, 237–244.

Walter, M. J., Newsom, H. E., Ertel, W., & Holzheid, A. (2000). Siderophile elements in the Earth and Moon: Metal/silicate partitioning and implications for core formation. In R. M. Canup (Ed.), *Origin of the Earth and Moon*, pp. 265–290. The University of Arizone Press.

Wasson, J. T. (1985). *Meteorites*, p.267. New York: Freeman.

Wetherill, G. W. (1990). Formation of the Earth. *Annual Review of Earth and Planetary Sciences, 18*, 205–256.

Wood, B. J., Halliday, A., & Rehkämper, M. (2010). Volatile accretion history of the Earth. *Nature, 567*, E6–E7.

Yoshino, T. (2019). Penetration of molten iron alloy into the lower mantle phase. *Comptes Rendus Geoscience, 351*, 171–181.

Zhang, Y., Sekine, T., He, H., Yu, Y., Liu, Y., & Zhang, M. (2016). Experimental constraints on light elements in the Earth's outer core. *Scientific Reports, 6*. doi:10.1038/srep22473.

8

Heat Flow from the Earth's Core Inferred from Experimentally Determined Thermal Conductivity of the Deep Lower Mantle

Yoshiyuki Okuda and Kenji Ohta

ABSTRACT

The thermal conductivity of the minerals in the vicinity of the Earth's core-mantle boundary (CMB) affects the structure and thickness of the thermal boundary layer of the mantle, which constrains the core cooling rate and Earth's total heat budget. Herein, we review recent experimental and theoretical studies examining the thermal conductivity of the Earth's lower mantle. Existing results of our thermal conductivity measurements were combined to construct self-consistent models for the lowermost mantle conductivity. We used these models to conduct simulations of the temperature profiles at the base of the mantle and the CMB heat flux. The averaged global CMB heat flow is estimated to be approximately 10 TW, which is in line with traditional estimates. The total heat flow of approximately 10 TW with large regional flux variations promotes interaction and co-evolution of the core and mantle.

8.1. INTRODUCTION

Active heat transport at large-scale internal dynamics has continued from the Earth's formation to the present. Direct temperature gradient measurements in both land and seafloor boreholes, using information regarding the thermal conductivity of constituting crustal rocks, obtained a uniform terrestrial heat flux of 69 mW/m², corresponding to a global heat flow of 35 TW. When the influence of volcanic activities and hydrothermal circulations was added, the current total heat flow from the Earth's surface is estimated as 46 ± 3 TW (Jaupart et al., 2015). This heat flow comprises core and mantle cooling, heat from the decay of radioactive elements, and other minor influences such as tidal deformation, chemical segregation, and thermal contraction gravitational heating (Lay et al., 2008). The global radiogenic heat of ~20 TW is inferred from the chondritic bulk silicate Earth model (Arevalo et al., 2009; Jaupart et al., 2015), which approximately matches estimates by geoneutrino observation of 16–20 TW (Araki et al., 2005; The KamLAND collaboration, 2011). Accordingly, the determination of heat flow using mantle or core cooling provides a strong constraint on the rest of the components involved. The degree of mantle cooling was estimated by numerical convection modeling by reproducing the plume heat flux observed in seismic tomography images (Sleep, 1990; Davies, 1993). However, the estimated lower mantle cooling ranges from 5 to 30 TW primarily because of the uncertainty regarding applying the poorly known temperature dependence of mantle viscosity and the effect of chemical buoyancy.

The heat flux across the core-mantle boundary (CMB) represents the degree of core cooling. Heat is transported from the outer core to the lowermost mantle via conduction only because of the significant density contrast between the mantle and the core. Therefore, the CMB heat flow, that is, Q_{CMB} (W), can be given by Fourier's law as follows:

$$Q_{\mathrm{CMB}} = -\kappa \left(\frac{\partial T}{\partial z} \right) A, \quad (8.1)$$

Department of Earth and Planetary Sciences, Tokyo Institute of Technology, Tokyo, Japan

Core-Mantle Co-Evolution: An Interdisciplinary Approach, Geophysical Monograph 276, First Edition.
Edited by Takashi Nakagawa, Taku Tsuchiya, Madhusoodhan Satish-Kumar, and George Helffrich.
© 2023 American Geophysical Union. Published 2023 by John Wiley & Sons, Inc.
DOI: 10.1002/9781119526919.ch08

where A is the CMB surface area (m^2) and κ and $\partial T/\partial z$ are the thermal conductivity of the bottom of the mantle (W/m/K) and the temperature gradient in this region (K/m), respectively. Accordingly, understanding heat transportation across the CMB and structure of the bottom thermal boundary layer (TBL) is highly reliant on knowledge regarding the thermal conductivity of the lowermost mantle minerals as functions of pressure (P), temperature (T), and chemical impurities concentrations in a mineral (x). Existing studies have estimated the thermal conductivity of the D″ layer as 5–12 W/m/K, which is based on the experimental values of MgO, Al$_2$O$_3$, and SiO$_2$ at conditions that fall woefully short of the deep lower mantle (Brown, 1986; Stacey, 1992; Manga & Jeanloz, 1997; Hofmeister, 1999). As such, *in situ* high P-T measurements of the thermal conductivity of the minerals constituting the lower mantle are required.

Static experiments at the conditions in the Earth's deep lower mantle are typically performed using a diamond anvil cell (DAC). In the past few decades, novel pulsed laser optical techniques combined with a DAC have been developed for determining high P-T thermal conductivity and the diffusivity of minerals (Beck et al., 2007; Hsieh et al., 2009; Yagi et al., 2011; McWilliams et al., 2015; Lobanov et al., 2017; Hasegawa et al., 2019a; Geballe et al., 2020). Advanced computational studies have also accelerated research on the phonon components of thermal conductivity (thermal conductivity contributed by lattice vibration, i.e., lattice thermal conductivity) in lower mantle minerals. We developed the pulsed-light heating thermoreflectance technique combined with a DAC and reported the thermophysical properties of selected lower mantle minerals (Yagi et al., 2011; Imada et al., 2014; Ohta et al., 2012a, 2017; Hasegawa et al., 2019a, 2019b; Okuda et al., 2017, 2019, 2020). Despite active research in this area in recent years, a review of such mineral physics studies is lacking following Kavner and Rainey (2016) and Tsuchiya et al. (2020). Therefore, the aims of this chapter are as follows: (1) to combine our previous studies to construct P-T-x-dependent lower mantle conductivity models with brief reviews of related work and (2) to simulate the thermal structure and heat flux near the CMB based on our conductivity models.

8.2. MODELING OF LOWER MANTLE THERMAL CONDUCTIVITY WITH BRIEF REVIEWS

Bridgmanite (Bdg), Fe,Al-bearing MgSiO$_3$-rich perovskite, is thought to be the most abundant mineral in the Earth's lower mantle despite the controversial nature of lower mantle composition and mineralogy (e.g., Irifune et al., 2010; Murakami et al., 2012; Wang et al., 2015; Kurnosov et al., 2017; Mashino et al., 2020). Irrespective of any suggested lower mantle mineralogy, Bdg will be the main heat conductor in this area. Therefore, the phonon component of the thermal conductivity of Bdg has been extensively studied both experimentally and theoretically (Osako & Ito, 1991; Manthilake et al., 2011; Haigis et al., 2012; Ohta et al., 2012a; Dekura et al., 2013; Ammann et al., 2014; Tang et al., 2014; Stackhouse et al., 2015; Ghaderi et al., 2017; Hsieh et al., 2017; Zhang et al., 2017; Okuda et al., 2017, 2019). Bdg can contain multiple impurities (Al, Fe^{2+}, and Fe^{3+}) and shows the complex cite occupancy and iron spin state of these impurities (Lin et al., 2013). For this reason, the theoretical calculation of the lattice conductivity of Fe,Al-bearing Bdg remains in flux. Our high P thermoreflectance measurements using a DAC yielded data for the lattice thermal conductivity of Bdg with various Fe and Al contents for CMB pressure at 300 K (Ohta et al., 2012a; Okuda et al., 2017, 2019). On the basis of these results, the P-T-x variations in the lattice conductivity of Bdg were formulated as follows:

$$\kappa_{\text{Bdg}}(P, T) = \kappa_{\text{Bdg},x}(P, 300\ K)\left(\frac{300}{T}\right)^a, \quad (8.2)$$

$$\kappa_{\text{Bdg},x}(P, 300K) = A(P)x + B(P), \quad (8.3)$$

where x is defined as the summation of Fe and Al contents in Bdg per formula unit, $\kappa_{\text{Bdg},x}$ is the thermal conductivity of Bdg with Fe + Al = x, and a is the temperature coefficient. Here, the form of T^{-a} is according to the suggested weaker temperature dependence of silicates with chemical impurities than that of simple pure oxides which follows a T^{-1} form (Hofmeister, 1999). The coefficients in equation (8.3) are A(P) = $-(7.0 \times 10^{-4})P^2 - (6.4 \times 10^{-2})P - 2.6$ and B(P) = $(4 \times 10^{-4})P^2 + 0.13 P + 6.3$, where the pressure unit is in GPa. The a-value ranges between 0.37 and 0.43 (Okuda et al., 2017), which was estimated based on a principle that Mg-pure Bdg better conducts heat than that with impurities. The temperature response of the lattice conductivity of Bdg with 10 mol% Fe and 10 mol% Al, the plausible chemical composition in the lower mantle, at 135 GPa is modeled here and shown in Figure 8.1a with a lattice conductivity of (Mg$_{0.886}$Fe$_{0.129}$)(Al$_{0.115}$Si$_{0.906}$)O$_3$ Bdg at the same pressure (Deschamps & Hsieh, 2019; Hsieh et al., 2017). The discrepancy between them may be the result of different impurity concentrations.

Post-perovskite (PPv) phase transition is expected to occur at a few hundred kilometers above the CMB, which may influence the thermal structure at the base of the mantle and, as a result, its dynamics due to changes in various physical properties across the transition (e.g., Čížková et al., 2010; Tosi et al., 2010). We first reported experimental data regarding the lattice thermal conductivity of PPv with MgSiO$_3$ and (Mg,Fe)SiO$_3$ compositions (Ohta et al., 2012a; Okuda et al., 2020). Both studies indicated the higher conductivity of PPv compared with

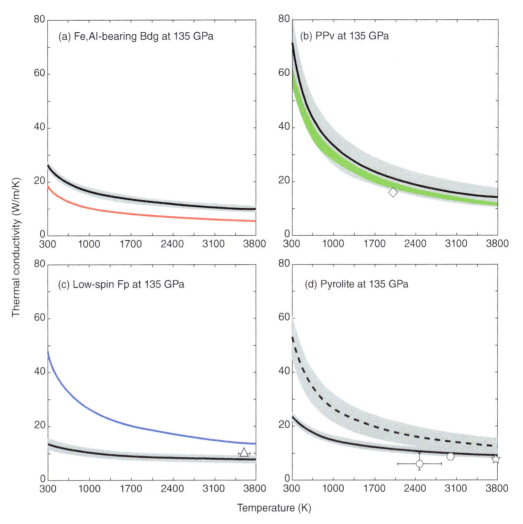

Figure 8.1 Lattice thermal conductivity of (a) Fe,Al-bearing Bdg, (b) PPv, (c) low-spin Fp with 19 mol% Fe, and (d) pyrolitic mantle as a function of temperature at 135 GPa (black lines with gray bands). Red line, conductivity model of Bdg (Hsieh et al., 2017; Deschamps & Hsieh, 2019); green band, conductivity model of $MgSiO_3$ PPv (Ammann et al., 2014); diamond, *ab initio* calculation for $MgSiO_3$ PPv (Dekura and Tsuchiya, 2019); square, classical equilibrium MD simulation for $MgSiO_3$ PPv (Haigis et al., 2012); blue line, conductivity model of low-spin Fp with 10 mol% Fe (Hsieh et al., 2018; Deschamps & Hsieh, 2019); triangle, *ab initio* calculation for low-spin Fp with 12.5 mol% Fe (Tang et al., 2014); circle, laser flash heating experiment on pyrolite (Geballe et al., 2020); pentagon, conductivity of pyrolite (Deschamps & Hsieh, 2019); star, pyrolite based on nonequilibrium molecular dynamics method (Stackhouse et al., 2015). Solid and broken lines in (d) are composite conductivities of Bdg+Fp and PPv+Fp, respectively.

Bdg at equivalent *P-T* conditions, which was anticipated based on their crystal structures and analog materials (Hofmeister, 2007; Keawprak et al., 2009; Cheng et al., 2011; Hunt et al., 2012). For the lattice conductivity of PPv, we applied the following empirical relationship (Manthilake et al., 2011):

$$\kappa = \kappa_{ref}\left(\frac{\rho}{\rho_{ref}}\right)^g \left(\frac{T_{ref}}{T}\right)^a, \quad (8.4)$$

where κ_{ref} and ρ_{ref} are the lattice conductivity and density at the reference pressure and temperature (T_{ref}) and g and a are the pressure and temperature coefficients, respectively. The chemical impurity factor was not included in this model because there is only one available lattice thermal conductivity measurement on Fe-bearing PPv (Okuda et al., 2020). Still, our measured PPv with the composition of $(Mg_{0.97}Fe_{0.03})SiO_3$ is thought to be close to that in the pyrolitic mantle (Sinmyo et al., 2011). To model the lattice conductivity of $(Mg_{0.97}Fe_{0.03})SiO_3$ PPv, the parameters of $a = 0.65$, $g = 6.0$, $\kappa_{ref} = 59.8$ W/m/K, $T_{ref} = 300$ K, and $\rho_{ref} = 5.47$ g/cm^3 were determined using our high *P-T* experiments (Okuda et al., 2020) (Fig. 8.1b). Theoretical studies examining MgSiO$_3$ PPv showed slightly lower values than ours (Ammann et al., 2014; Dekura & Tsuchiya, 2019; Haigis et al., 2012). Considering the experimental uncertainty, the effect of impurity on the lattice thermal conductivity of PPv could be very small.

In the modeling of lower mantle conductivity, (Mg,Fe)O ferropericlase (Fp) is also important. Because of its simple structure and chemical bonding, MgO periclase is regarded as a benchmark for developing a new technique with which to estimate lower mantle conductivity (Manga & Jeanloz, 1997; Katsura, 1997; Beck et al., 2007; de Koker, 2009, 2010; Stackhouse et al., 2010; Tang & Dong, 2010; Manthilake et al., 2011; Haigis et al., 2012; Dalton et al., 2013; Hofmeister, 2014a; Imada et al., 2014; Dekura & Tsuchiya, 2017; Kwon et al., 2020). Periclase is a good heat conductor, but it was found that iron substitution in periclase induces a large degree of conductivity reduction (Manthilake et al., 2011; Tang et al., 2014; Goncharov et al., 2015; Ohta et al., 2017; Hsieh et al., 2018; Hasegawa et al., 2019b). Furthermore, iron spin crossover in Fp dramatically changes the pressure response of its lattice conductivity (Hsieh et al., 2018; Ohta et al., 2017). Using equation (8.4), we estimated the lattice thermal conductivity of low-spin (Mg$_{0.81}$Fe$_{0.19}$)O (Ohta et al., 2017; Fig. 8.1c). For comparison, the reported lattice conductivities of low-spin Fp with 10 and 12.5 mol% Fe are also shown (Deschamps & Hsieh, 2019; Hsieh et al., 2018; Tang et al., 2014). A change in conductivity due to temperature and impurity concentration was observed. Interestingly, Fp with 19 mol% Fe became less conductive relative to Bdg and PPv (Fig. 8.1a–c).

A Bdg (or PPv) plus Fp composite with an 8:2 volume fraction ratio was considered herein as a pyrolitic lower mantle model using Hashin-Shtrikman averaging (Hashin & Shtrikman, 1962). The chemical compositions of these minerals are fixed throughout lower mantle conditions. Accordingly, we assumed the constant iron partitioning coefficients of Bdg/Fp and PPv/Fp to be approximately 0.5 and 0.2, respectively (Piet et al., 2016). A recent experiment and theoretical calculation have shown that the thermal conductivity of pure-CaPv is higher than that of the other lower mantle minerals, which increases the bulk thermal conductivity value by ~10% (Zhang et al., 2021). However, since the thermal conductivity of CaPv in the lower mantle may be different from that of pure-CaPv due to impurities, we did not include the effect of CaPv. Higher PPv conductivity relative to Bdg results in higher composite conductivity conditions in the pyrolitic composition. The pyrolite conductivity models obtained at 135 GPa as a function of temperature are shown in Figure 8.1d. Our Bdg and Fp mixture model is consistent with existing research within uncertainties (Stackhouse et al., 2015; Deschamps & Hsieh, 2019; Geballe et al., 2020).

As described above, we summarized the existing studies that primarily examined the phonon (lattice) contribution (see Tsuchiya et al., 2020 for a review of calculations on the phonon conductivity). The thermal conductivity of a material is the sum of the lattice, radiative, and electronic terms. The radiative thermal conductivity of lower mantle minerals is estimated based on optical absorption measurements (e.g., Hofmeister, 1999). The PPv and Fp will be opaque and show very low radiative conductivity even at a high temperature associated with the CMB (Goncharov et al., 2006, 2010; Lobanov et al., 2017, 2021). The role of radiative heat transport in the pyrolitic lower mantle is likely to be small because of the strong light absorption of temperature-enhanced Fp (Lobanov et al., 2020). However, further studies of an *in situ* high *P-T* optical absorption measurement for each lower mantle mineral are required for better understandings of radiative heat conductivity in the lower mantle (Goncharov et al., 2008; Keppler et al., 2008; Hofmeister, 2014b; Kavner & Rainey, 2016). Negligibly weak heat conduction by electrons in the lower mantle is anticipated from high *P-T* electrical conductivity measurements on Bdg, Fp, and PPv (Ohta et al., 2008a, 2010, 2017; Sinmyo et al., 2014).

Considering all conduction mechanisms, the thermal conductivity of Bdg-dominant pyrolite in CMB conditions (135 GPa and 3400–3800 K) is approximately 9 W/m/K, and PPv-dominant one exhibits a 35%–40% higher conductivity of approximately 12 W/m/K (Fig. 8.1d). The present values are in the range of traditional estimates (Brown, 1986; Stacey, 1992; Manga & Jeanloz, 1997; Hofmeister, 1999). A silica-enriched lower mantle (relative to pyrolite) has repeatedly been proposed (Murakami et al., 2012; Ballmer et al., 2017; Mashino et al., 2020). In such a case, lower mantle conductivity is enhanced because of increased Bdg or PPv fractions compared with Fp, which has lower lattice conductivity (Fig. 8.1).

8.3. TEMPERATURE PROFILES IN THE THERMAL BOUNDARY LAYER AND CORE-MANTLE BOUNDARY HEAT FLUX

8.3.1. Temperature Profiles in the Thermal Boundary Layer

Next, we calculated the temperature structures of the TBL immediately above the CMB using the modeled lower mantle conductivity ($\kappa(z(P),T)$) using the heat conduction equation:

$$\frac{\partial T(z(P),t)}{\partial t} = \frac{1}{C_P(z(P,T))\rho(z(P,T))}\frac{\partial}{\partial z} \times \left[\kappa(z(P),T)\frac{\partial T(z(P),t)}{\partial z}\right], \quad (8.5)$$

where C_P and ρ are the isobaric heat capacity and density of the lower mantle as a function of depth z, respectively, and C_P was set as a constant value of 1,300 J/kg/K. Density as a function of depth z was applied for ρ at $0.51z + 4{,}094$ kg/m^3, which was derived by linear fitting of the one-dimensional (1-D) density-depth profile in the Earth's lower mantle (Dziewonski & Anderson, 1981). We considered two lower mantle rock models: pyrolytic (Bdg(or PPv):Fp = 8:2) and perovskitic rock (Bdg or PPv individually). We used the commercial software package COMSOL Multiphysics for the finite element method calculation. Mantle geometry was set as a 2-D rectangle with a height of 1,200 km (corresponding to 1,700–2,900 km in depth) and a width of 1,000 km, but to simplify this, we assumed no variations in the input physical parameters along radial direction. Triangle meshes were applied as grids, the depth resolution for which was set as 0.75 km. We fixed the CMB temperature (T_{CMB}) at 3400 K based on the most recently provided melting temperature of a dry peridotite at CMB pressure (Kim et al., 2020). The temperature at 1,200 km above the CMB was fixed at 2353 K (T_0), that is, the adiabatic mantle temperature ($T_{adiabat}$) provided in the literature (Brown & Shankland, 1981). The location and Clapeyron slope of PPv transition, that is, 8 MPa/K, were taken from the literature (Tsuchiya et al., 2004; Ohta et al., 2008b). The initial temperature was set to $T_0 = 2353$ K for all depths (all grids), except in the case of T_{CMB} (=3400 K). At each depth, we entered the T-dependent conductivity of the pyrolite or perovskitic mantle modeled as noted above. Time integration was divided into 100 steps in $t = 10^7$–10^9 years. The "average geotherm" determined here denoted a temperature profile that deviated from $T_{adiabat}$ at a depth where PREM showed a kink, which likely occurred because of the onset of the TBL. The top of the TBL in "hot geotherm" is defined by the D″ discontinuity under the Central Pacific, which is thought to be a relatively hot domain due to the presence of upwelling plumes. The temperature gradient in the TBL was obtained by calculating the average temperature gradient from the CMB to the point where $(T - T_{adiabat})/(T_{CMB} - T_{adiabat})$ was 0.01.

The calculated average TBL geotherms of the pyrolitic and perovskitic models were almost the same (Fig. 8.2). The estimated temperature gradient in the average geotherm was 5.0 ± 0.4 K/km, which was smaller than the previous estimation based on seismic studies (10 ± 3.8 K/km) (van der Hilst et al., 2007). This may have been because they investigated the region beneath the subduction zone in Central America, which is thought to be a colder region. The calculated average geotherm derived using the Clapeyron slope for the Bdg-to-PPv transition of an 8 MPa/K well explains the D″ discontinuity beneath central Asia, the Arctic, and the Caribbean (Fig. 8.2).

The estimated hot geotherm of the pyrolite showed weaker depth dependence compared with the perovskitic model. The presence of Fp contributed to the difference between these models. The temperature gradients in the pyrolitic and perovskitic models were estimated as 0.6 ± 0.1 and 1.1 ± 0.1 K/km, respectively. However, the temperature gradient inferred from the seismic study at the northern part of the Pacific large low shear velocity province (LLSVP), which is thought to be representative of a hot geotherm, was reported as being 8.5 ± 2.5 K/km (Lay et al., 2006). It should be noted that LLSVP is believed to be chemically distinct from the surrounding mantle (Ni et al., 2002; Burke et al., 2008), and as such, the thermal conductivity there may differ from that of pyrolite. However, to date, this has been difficult to estimate because of unknown chemical composition in LLSVPs. Such a large conflicting temperature gradient was derived from the assumption of the double-crossing with a larger Clapeyron slope of 16 MPa/K used in Lay et al. (2006). In our geotherm calculation, we were unable to derive a solution that showed Bdg-PPv double-crossing even when we changed T_{CMB} in the range of 3400–3800 K with a Clapeyron slope of 8 MPa/K (Fig. 8.2). Therefore, we suggest that the back transformation of PPv to Bdg will not occur in the Earth's lowermost mantle.

8.3.2. Core-Mantle Boundary Heat Flux and its Implications

We calculated the CMB heat flux (Q_{CMB}) based on Fourier's law [equation (8.1)] using the temperature gradients obtained in the TBL and the thermal conductivity in this region (Fig. 8.3). The values of Q_{CMB} along the average geotherm were 9.8 ± 1.9 TW for the pyrolite and 10.8 ± 2.0 TW for the perovskitic mantle, indicating minimal dependence of the Q_{CMB} with the compositional model of the lower mantle. In the case of hot geotherm, Q_{CMB} was 1.2 ± 0.2 TW for the pyrolitic and 2.3 ± 0.4 TW for the perovskitic models owing to the gentle temperature gradient and depressed conductivity at a higher temperature. We also calculated Q_{CMB} with a seismically inferred temperature gradient in the cold region of 10 ± 3.8 K/km

138 CORE-MANTLE CO-EVOLUTION

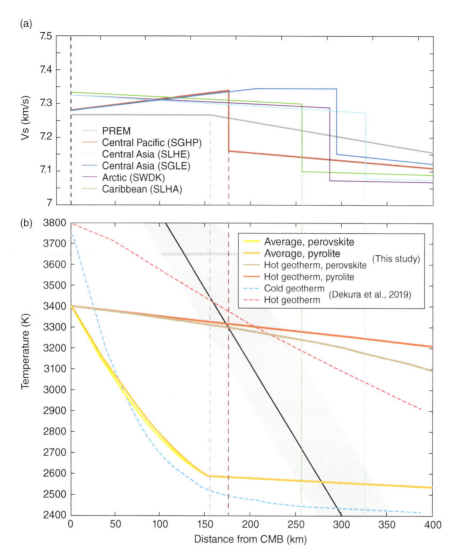

Figure 8.2 (a) Shear velocity models at the CMB in several different regions inferred from double-array stack studies [Source: Adapted from Kawai and Tsuchiya (2009)], and (b) the estimated temperature profile above the CMB. Yellow and dark yellow lines indicate the calculated average geotherm with perovskitic and pyrolitic model, respectively, and brown and orange lines indicate the calculated hot geotherm with perovskitic and pyrolitic model, respectively. Pink and blue broken lines indicate the estimated hot and cold geotherm (Dekura & Tsuchiya, 2019), respectively. Black line and gray band indicate the Clapeyron slope of the PPv transition with $MgSiO_3$ (Tsuchiya et al., 2004) and pyrolitic compositions (Ohta et al., 2008b), respectively.

(van der Hilst et al., 2007), which showed a high heat flux of 19.7 ± 8.9 TW for pyrolite and 21.6 ± 9.5 TW for perovskitic. The average heat fluxes calculated with the hot and cold geotherms were 10.4 ± 4.9 TW for pyrolite and 11.9 ± 5.3 TW for perovskitic, which coincided with the average Q_{CMB}. The perovskitic model considered here is an extreme case and lacks any influence on the part of Fp. The suggested perovskitic model indicates that the Bdg accounted for 88 vol%–92 vol% (Murakami et al., 2012; Mashino et al., 2020). Therefore, the estimated Q_{CMB} of our Bdg (or PPv) model would be in the upper bound. From the minimum and maximum Q_{CMB} values of both compositional models, we concluded that the net Q_{CMB} was 10.4 ± 2.5 TW. The estimated average Q_{CMB} was within the traditional prediction of 5–13 TW (e.g., Stacey, 1992). Moreover, the present Q_{CMB} estimate was consistent with the Q_{CMB} determined from the electronic thermal conductivity and adiabatic temperature gradient at the top of the outer core (Ohta et al., 2016; Fig. 8.3).

The terrestrial heat flow balance with well-constrained terms elucidates the unknown influence of mantle cooling. By neglecting minor heat sources, such as tidal heating (<0.4 TW), the estimated total Q_{CMB} of 10.4 ± 2.5 TW with a global radiogenic heat of 20 ± 3 TW and the

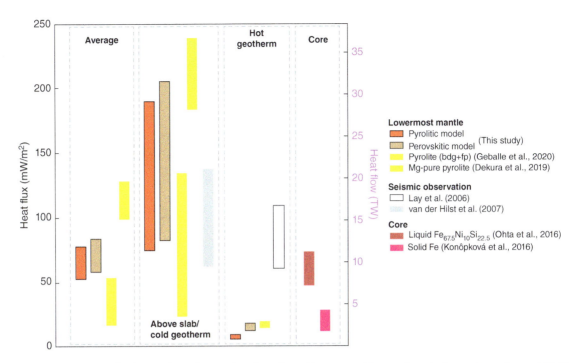

Figure 8.3 The complied CMB heat flux and surface flow. Orange and brown bars indicate the estimated CMB heat flow of pyrolitic and perovskitic composition, respectively, and yellow and green bars indicate those with the reported thermal conductivity of pyrolitic rock (Geballe et al., 2020) and MgSiO$_3$ PPv + MgO aggregate (Dekura & Tsuchiya, 2019), respectively. Gray and open bars indicate the CMB heat flow estimated from the seismically inferred temperature gradient above the CMB beneath central Pacific (Lay et al., 2006) and Central America (van der Hilst et al., 2007), respectively, and with the typically used lowermost mantle thermal conductivity of 10 W/m/K (Stacey, 1992). Red and pink bars indicate the CMB heat flow estimated from the thermal conductivity and adiabat of the top of outer core estimated by Ohta et al. (2016) and Konôpková et al. (2016), respectively.

total surface heat flow of 46 ± 3 TW yielded secular mantle cooling of ~16 TW on average but with a broad uncertainty of ~9 TW due to the uncertainties in each component (Fig. 8.4). The loss of the mantle heat resulted from the upwelling of hot plumes and the downwelling of cold slabs. Generally, estimation of the energy propagation by upwelling is difficult because of the low resolution of the deep plume heads shown in tomographic images (Montelli et al., 2004), as well as potentially small, undetectable plumes (Malamud & Turcotte, 1999). The contribution of mantle cooling by cold subducted slabs was estimated as 13–14 TW using finite frequency tomography and Stokes' flow model (Nolet et al., 2006). With mantle cooling of ~16 TW, hot plumes carry only ~2 TW from the mantle to the surface, indicating that a large portion of mantle cooling is governed by the subduction of cold slabs rather than hot plumes. This is consistent with the view that not all hot plumes are sufficiently buoyant to bring energy to the Earth's surface because of the interaction between other plumes, the negative Clapeyron slope at a 660 km depth, and heavier materials upwelling from the chemically distinct dense LLSVPs (Labrosse, 2002). Recent estimates revised the plume heat

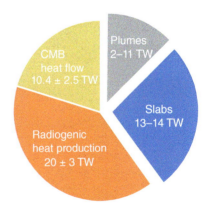

Figure 8.4 Global heat flow balance of the Earth. Tidal heating contribution is not considered here. Cooling rate via slab subduction was referred. Source: Adapted from Nolet et al. (2006).

flow up to 6 TW, which is within the uncertainty of our estimate, based on ambiguities related to each component (Hoggard et al., 2020). Therefore, we conclude that plate tectonics control cooling in the Earth's mantle (Fig. 8.4).

The estimated Q_{CMB} using hot and cold geotherms showed an extreme peak-to-peak lateral variation in the

CMB heat flux. The thermal conductivity difference in the lowermost mantle due to temperature and compositional variation was at most ~50% (Fig. 8.1), whereas the ($\partial T/\partial z$) difference between hot and cold geotherms was ~1,000%. Accordingly, the cause of lateral heat flux variation was mainly the difference in regional temperature variation results concerning mantle material circulation. The estimated Q_{CMB} variation is consistent with that derived from the temperature variation inferred from seismology, assuming that seismic wave velocity variations are merely thermal in origin and assuming a constant thermal conductivity of 0–22 TW (Stackhouse et al., 2015).

The parameter $q^* = (q_{max} - q_{min})/2q_{mean}$ indicates the strength of the lateral heat flux variation at the CMB. In our simulation, the q^* values for pyrolitic and perovskitic models were estimated as 1.1 (+0.8/−0.5) and 1.2 (+0.9/−0.6), respectively. A smaller value of ~0.2 was preferred in the past, which was inferred from seismological studies investigated beneath the central Pacific (Lay et al., 2006) and Central America (van der Hilst et al., 2007). We reiterate that the assumption of double-crossing with a large Clapeyron slope led to a relatively strong temperature gradient in hot regions corresponding to a small q^* value. Numerical dynamo simulations showed that the structure and strength of the magnetic field may strongly depend on q^* at the CMB. Gubbins et al. (2011) showed that increasing q^* from 0.15 to 0.45 generated a strongly concentrated downflow in the outer core beneath cold regions at the CMB. The researchers expected such a narrow downflow to increase the heat flux at the inner core boundary (ICB), which may have been sufficient for causing localized melting of the inner core. Our large estimated q^* value of ~1 was much higher than 0.45, and as such, it supports localized melting of the inner core (Gubbins et al., 2011). Because a solid inner core is suggested to contain less volatiles than the surrounding outer core, the melting of the inner core may form a stable dense and volatile-poor layer above the ICB (Gubbins et al., 2011), which may, in turn, explain the origin of P-wave anomalies observed immediately above this boundary (the so-called F-layer) (e.g., Song & Helmberger, 1995). Furthermore, laboratory experiment showed that such heterogeneous heat flow at the CMB resulted in a temperature difference in the core and induced anisotropic inner core crystallization (Sumita & Olson, 1999), which may explain the seismically observed west-eastern hemispherical P-wave velocity, as well as attenuation anomalies in the inner core (e.g., Tanaka & Hamaguchi, 1997).

Moreover, an increase in q^* was found to enhance the localization of outer core convection and geodynamo activity, which generated a stronger magnetic field (Takahashi et al., 2008). They observed that a broad outer core downwelling shifted from the cold region at the CMB when applying $q^* = 0.5$. By increasing q^* to ~1, the downwelling flow became strongly concentrated in the high heat flow region (i.e., a relatively cold area), which created a stronger magnetic field, indicating that a geodynamo structure is dependent on the strength of thermal heterogeneity in the CMB. As an important example, Takahashi et al. (2008) suggested that the symmetrical, thermally heterogeneous lowermost mantle along the equator could generate a stable and extremely strong magnetic field, which may be the origin of the Superchron. The large q^* of ~1 found in this study has profound implications for outer core convection, geodynamo, and the Earth's inner core structure. We expect additional simulations of geodynamo and mantle dynamics parameterizing a higher q^* value to yield better understanding of core and lower mantle dynamics.

8.4. FUTURE PERSPECTIVES OF THERMAL CONDUCTIVITY MEASUREMENTS ON LOWER MANTLE MINERALS

Selected issues in lower mantle thermal conductivity remain unclear and require further study. *In situ* high P-T lattice thermal conductivity measurements on Bdg and (Mg,Fe)O have not been reported to date, which condemns the assumption of a temperature coefficient related to conductivity (Fig. 8.1a,c,d). However, this issue is already viable as indicated in a study for PPv (Okuda et al., 2020). CaPv shows a higher lattice conductivity compared with other lower mantle major minerals (Zhang et al., 2021), and as such, it plays an important role in thermal conduction of subducted mid-ocean ridge basalt (MORB), in which its volume fraction exceeds 20% (Hirose et al., 2005). In this context, the thermophysical properties of SiO_2 and Al_2O_3 polymorphs are also important for understanding the heat conduction in MORB during subduction. Direct high P-T thermal conductivity measurements of pyrolite are limited to a phase of Bdg as a $MgSiO_3$ polymorph (Geballe et al., 2020). Therefore, for experiments under high P-T conditions, PPv-containing rocks are important, given that double-crossing does not take place at the lowermost mantle.

Opaque Fp at high P-T is reported to significantly reduce the radiative thermal conductivity of the pyrolitic lower mantle (Lobanov et al., 2020). Considering that a Bdg-predominant lower mantle better explains seismic observations than pyrolite (e.g., Mashino et al., 2020), the volume ratio of Fp can control the amount of photon heat conduction in the lower mantle, which needs to be quantified. Further *in situ* high P-T optical absorption measurement for each lower mantle mineral, especially for Fp, is beneficial for better understandings of radiative heat conductivity in the lower mantle.

Heat transported by other carriers such as electrons in the major lower mantle minerals is thought to be negligible (Ohta et al., 2008a, 2010, 2017; Sinmyo et al., 2014), but minor phases showing superionic conduction (e.g., FeOOHx; Hou et al., 2021) and metallization (e.g., FeO; Ohta et al., 2012b) could enhance this component. Thermal conductivities of such important minor phases at high P-T conditions may bring us additional insights into the thermal conductivity heterogeneity and lateral heat flux variation at the CMB.

Our high P-T thermoreflectance technique relied on the estimation of C_P, the experimental determination of which further improved the quality of the lattice's thermal conductivity data. The texture (i.e., a crystal-preferred orientation) in the lowermost mantle rock could create a variation in bulk conductivity. As a result, the thermal structure and dynamics in this region should be constrained by a combination of mineral physics studies and seismic anisotropy observations.

ACKNOWLEDGMENTS

This work was supported by JSPS KAKENHI Grant number 15H05827.

REFERENCES

Ammann, M. W., Walker, A. M., Stackhouse, S., Wookey, J., Forte, A. M., Brodholt, J. P., et al. (2014). Variation of thermal conductivity and heat flux at the Earth's core mantle boundary. *Earth and Planetary Science Letters*, *390*, 175–185. doi:10.1016/j.epsl.2014.01.009.

Araki, T., Enomoto, S., Furuno, K., Gando, Y., Ichimura, K., Ikeda, H., et al. (2005). Experimental investigation of geologically produced antineutrinos with KamLAND. *Nature*, *436*, 499–503. doi:10.1038/nature03980.

Arevalo, R., McDonough, W. F., & Luong, M. (2009). The K/U ratio of the silicate Earth: Insights into mantle composition, structure and thermal evolution. *Earth and Planetary Science Letters*, *278*, 361–369. doi:10.1016/j.epsl.2008.12.023.

Ballmer, M. D., Houser, C., Hernlund, J. W., Wentzcovitch, R. M., & Hirose, K. (2017). Persistence of strong silica-enriched domains in the Earth's lower mantle. *Nature Geoscience*, *10*, 236–240. doi:10.1038/ngeo2898.

Beck, P., Goncharov, A. F., Struzhkin, V. V., Militzer, B., Mao, H.-k., & Hemley, R. J. (2007). Measurement of thermal diffusivity at high pressure using a transient heating technique. *Applied Physics Letters*, *91*, 181914. doi:10.1063/1.2799243.

Brown, J. M. (1986). Interpretation of the D″ zone at the base of the mantle: Dependence on assumed values of thermal conductivity. *Geophysical Research Letters*, *13*, 1509–1512. doi:10.1029/GL013i013p01509.

Brown, J. M., & Shankland, T. J. (1981). Thermodynamic parameters in the Earth as determined from seismic profiles. *Geophysical Journal International*, *66*, 579–596. doi:10.1111/j.1365-246X.1981.tb04891.x.

Burke, K., Steinberger, B., Torsvik, T.H., & Smethurst, M.A. (2008). Plume Generation Zones at the margins of Large Low Shear Velocity Provinces on the core–mantle boundary. *Earth and Planetary Science Letters*, *265*, 49–60. https://doi.org/10.1016/j.epsl.2007.09.042.

Cheng, J. G., Zhou, J. S., Goodenough, J. B., Sui, Y., Ren, Y., & Suchomel, M. R. (2011). High-pressure synthesis and physical properties of perovskite and post-perovskite $Ca_{1-x}Sr_xIrO_3$. *Physical Review B*, *83*, 064401. doi:10.1103/PhysRevB.83.064401.

Čížková, H., Čadek, O., Matyska, C., & Yuen, D. A. (2010). Implications of post-perovskite transport properties for core–mantle dynamics. *Physics of the Earth and Planetary Interiors*, *180*, 235–243. doi:10.1016/j.pepi.2009.08.008.

Dalton, D. A., Hsieh, W. P., Hohensee, G. T., Cahill, D. G., & Goncharov, A. F. (2013). Effect of mass disorder on the lattice thermal conductivity of MgO periclase under pressure. *Scientific Reports*, *3*, 2400. doi:10.1038/srep02400.

Davies, G. F. (1993). Cooling the core and mantle by plume and plate flows. *Geophysical Journal International*, *115*, 132–146. doi:10.1111/j.1365-246X.1993.tb05593.x.

de Koker, N. (2009). Thermal conductivity of MgO periclase from equilibrium first principles molecular dynamics. *Physical Review Letters*, *103*, 125902. doi:10.1103/PhysRevLett.103.125902.

de Koker, N. (2010). Thermal conductivity of MgO periclase at high pressure: Implications for the D″ region. *Earth and Planetary Science Letters*, *292*, 392–398. doi:10.1016/j.epsl.2010.02.011.

Dekura, H., & Tsuchiya, T. (2017). *Ab initio* lattice thermal conductivity of MgO from a complete solution of the linearized Boltzmann transport equation. *Physical Review B*, *95*, 184303. doi:10.1103/PhysRevB.95.184303.

Dekura, H., & Tsuchiya, T. (2019). Lattice Thermal Conductivity of $MgSiO_3$ Postperovskite under the Lowermost Mantle Conditions from Ab Initio Anharmonic Lattice Dynamics. *Geophysical Research Letters*, *46*, 12919–12926, doi:10.1029/2019gl085273.

Dekura, H., Tsuchiya, T., & Tsuchiya, J. (2013). *Ab initio* lattice thermal conductivity of $MgSiO_3$ perovskite as found in Earth's lower mantle. *Physical Review Letters*, *110*, 025904. doi:10.1103/PhysRevLett.110.025904.

Deschamps, F., & Hsieh, W.-P. (2019). Lowermost mantle thermal conductivity constrained from experimental data and tomographic models. *Geophysical Journal International*, *219*, S115–S136. doi:10.1093/gji/ggz231.

Dziewonski, A.M., & Anderson, D.L. (1981). Preliminary reference Earth model. *Physics of the Earth and Planetary Interiors*, *25*, 297–356. https://doi.org/10.1016/0031-9201(81)90046-7.

Geballe, Z. M., Sime, N., Badro, J., van Keken, P. E., & Goncharov, A. F. (2020). Thermal conductivity near the bottom of the Earth's lower mantle: Measurements of pyrolite up to 120 GPa and 2500 K. *Earth and Planetary Science Letters*, *536*, 116161. doi:10.1016/j.epsl.2020.116161.

Ghaderi, N., Zhang, D. B., Zhang, H., Xian, J., Wentzcovitch, R. M., & Sun, T. (2017). Lattice Thermal Conductivity of $MgSiO_3$ Perovskite from First Principles. *Scientific Reports*, *7*, 5417. doi:10.1038/s41598-017-05523-6.

Goncharov, A. F., Haugen, B. D., Struzhkin, V. V., Beck, P., & Jacobsen, S. D. (2008). Radiative conductivity in the Earth's lower mantle. *Nature, 456*, 231–234. doi:10.1038/nature07412.

Goncharov, A. F., Lobanov, S. S., Tan, X., Hohensee, G. T., Cahill, D. G., Lin, J.-F., et al. (2015). Experimental study of thermal conductivity at high pressures: Implications for the deep Earth's interior. *Physics of the Earth and Planetary Interiors, 247*, 11–16. doi:10.1016/j.pepi.2015.02.004.

Goncharov, A. F., Struzhkin, V. V., & Jacobsen, S. D. (2006). Reduced radiative conductivity of low-spin (Mg,Fe)O in the lower mantle. *Science, 312*, 1205–1208. doi:10.1126/science.1125622.

Goncharov, A. F., Struzhkin, V. V., Montoya, J. A., Kharlamova, S., Kundargi, R., Siebert, J., et al. (2010). Effect of composition, structure, and spin state on the thermal conductivity of the Earth's lower mantle. *Physics of the Earth and Planetary Interiors, 180*, 148–153. doi:10.1016/j.pepi.2010.02.002.

Gubbins, D., Sreenivasan, B., Mound, J., & Rost, S. (2011). Melting of the Earth's inner core. *Nature, 473*, 361–363. doi:10.1038/nature10068.

Haigis, V., Salanne, M., & Jahn, S. (2012). Thermal conductivity of MgO, MgSiO$_3$ perovskite and post-perovskite in the Earth's deep mantle. *Earth and Planetary Science Letters, 355–356*, 102–108. doi:10.1016/j.epsl.2012.09.002.

Hasegawa, A., Yagi, T., & Ohta, K. (2019a). Combination of pulsed light heating thermoreflectance and laser-heated diamond anvil cell for in-situ high pressure-temperature thermal diffusivity measurements. *Review of Scientific Instruments, 90*, 074901. doi:10.1063/1.5093343.

Hasegawa, A., Ohta, K., Yagi, T., Hirose, K., Okuda, Y., & Kondo, T. (2019b). Composition and pressure dependence of lattice thermal conductivity of (Mg,Fe)O solid solutions. *Comptes Rendus Geoscience, 351*, 229–235. doi:10.1016/j.crte.2018.10.005.

Hashin, Z., & Shtrikman, S. (1962). A variational approach to the theory of the effective magnetic permeability of multiphase materials. *Journal of Applied Physics, 33*, 3125–3131. doi:10.1063/1.1728579.

Hirose, K., Takafuji, N., Sata, N., & Ohishi, Y. (2005). Phase transition and density of subducted MORB crust in the lower mantle. *Earth and Planetary Science Letters, 237*, 239–251. doi:10.1016/j.epsl.2005.06.035.

Hofmeister, A. M. (1999). Mantle values of thermal conductivity and the geotherm from phonon lifetimes. *Science, 283*, 1699–1706. doi:10.1126/science.283.5408.1699.

Hofmeister, A. M. (2007). Thermal conductivity of the Earth's deepest mantle. In D. A. Yuen, S. Murayama, S. I. Karato, & B. F. Windley (Eds.), *Superplume: Beyond Plate Tectonics* (pp. 269–292), Dordrecht, The Netherlands: Springer.

Hofmeister, A. M. (2014a). Thermal diffusivity and thermal conductivity of single-crystal MgO and Al$_2$O$_3$ and related compounds as a function of temperature. *Physics and Chemistry of Minerals, 41*, 361–371. doi:10.1007/s00269-014-0655-3.

Hofmeister, A. M. (2014b). Thermodynamic and optical thickness corrections to diffusive radiative transfer formulations with application to planetary interiors. *Geophysical Research Letters, 41*, 3074–3080. doi:10.1002/2014gl059833.

Hoggard, M. J., Parnell-Turner, R., & White, N. (2020). Hotspots and mantle plumes revisited: Towards reconciling the mantle heat transfer discrepancy. *Earth and Planetary Science Letters, 542*, 116317. doi:10.1016/j.epsl.2020.116317.

Hou, M., He, Y., Jang, B. G., Sun, S., Zhuang, Y., Deng, L., et al. (2021). Superionic iron oxide–hydroxide in Earth's deep mantle. *Nature Geoscience, 14*, 174–178. https://doi.org/10.1038/s41561-021-00696-2.

Hsieh, W.-P., Chen, B., Li, J., Keblinski, P., & Cahill, D. G. (2009). Pressure tuning of the thermal conductivity of the layered muscovite crystal. *Physical Review B, 80*, 180302. doi:10.1103/PhysRevB.80.180302.

Hsieh, W.-P., Deschamps, F., Okuchi, T., & Lin, J.-F. (2017). Reduced lattice thermal conductivity of Fe-bearing bridgmanite in Earth's deep mantle. *Journal of Geophysical Research, 122*, 4900–4917. doi:10.1002/2017jb014339.

Hsieh, W. P., Deschamps, F., Okuchi, T., & Lin, J. F. (2018). Effects of iron on the lattice thermal conductivity of Earth's deep mantle and implications for mantle dynamics. *Proceedings of the National Academy of Sciences U S A, 115*, 4099–4104. doi:10.1073/pnas.1718557115.

Hunt, S. A., Davies, D. R., Walker, A. M., McCormack, R. J., Wills, A. S., Dobson, D. P., et al. (2012). On the increase in thermal diffusivity caused by the perovskite to post-perovskite phase transition and its implications for mantle dynamics. *Earth and Planetary Science Letters, 319–320*, 96–103. doi:10.1016/j.epsl.2011.12.009.

Imada, S., Ohta, K., Yagi, T., Hirose, K., Yoshida, H., & Nagahara, H. (2014). Measurements of lattice thermal conductivity of MgO to core-mantle boundary pressures. *Geophysical Research Letters, 41*, 4542–4547. doi:10.1002/2014gl060423.

Irifune, T., Shinmei, T., McCammon, C. A., Miyajima, N., Rubie, D. C., & Frost, D. J. (2010). Iron partitioning and density changes of pyrolite in Earth's lower mantle. *Science, 327*, 193–195. doi:10.1126/science.1181443.

Jaupart, C., Labrosse, S., Lucazeau, F., & Mareschal, J. C. (2015). Temperatures, Heat, and Energy in the Mantle of the Earth. In D. Bercovici (Ed.), *Treatise on Geophysics* (pp. 223–270), Amsterdam: Elsevier. doi:10.1016/b978-0-444-53802-4.00126-3.

Katsura, T. (1997). Thermal diffusivity of periclase at high temperatures and high pressures. *Physics of the Earth and Planetary Interiors, 101*, 73–77. doi:10.1016/s0031-9201(96)03223-2.

Kavner, A., & Rainey, E. S. G. (2016). Heat Transfer in the Core and Mantle. In H. Terasaki & R. A. Fischer (Eds.), *Deep Earth: Physics and Chemistry of the Lower Mantle and Core* (pp. 31–42). Washington DC: American Geophysical Union. doi:10.1002/9781118992487.ch3.

Kawai, K., & Tsuchiya, T. (2009). Temperature profile in the lowermost mantle from seismological and mineral physics joint modeling. *Proceedings of the National Academy of Sciences U S A, 106*, 22119–22123. doi:10.1073/pnas.0905920106.

Keawprak, N., Tu, R., & Goto, T. (2009). Thermoelectricity of CaIrO$_3$ ceramics prepared by spark plasma sintering. *Journal of the Ceramic Society of Japan, 117*, 466–469. doi:10.2109/jcersj2.117.466.

Keppler, H., Dubrovinsky, L. S., Narygina, O., & Kantor, I. (2008). Optical absorption and radiative thermal conductivity of silicate perovskite to 125 gigapascals. *Science, 322*, 1529–1532. doi:10.1126/science.1164609.

Kim, T., Ko, B., Greenberg, E., Prakapenka, V., Shim, S. H., & Lee, Y. (2020). Low Melting Temperature of Anhydrous Mantle Materials at the Core–Mantle Boundary. *Geophysical Research Letters, 47*, e2020GL089345. doi:10.1029/2020gl089345.

Konôpková, Z., McWilliams, R.S., Gómez-Pérez, N., & Goncharov, A.F. (2016). Direct measurement of thermal conductivity in solid iron at planetary core conditions. *Nature, 534*, 99–101. https://doi.org/10.1038/nature18009.

Kurnosov, A., Marquardt, H., Frost, D. J., Ballaran, T. B., & Ziberna, L. (2017). Evidence for a Fe^{3+}-rich pyrolitic lower mantle from (Al,Fe)-bearing bridgmanite elasticity data. *Nature, 543*, 543–546. doi:10.1038/nature21390.

Kwon, C., Xia, Y., Zhou, F., & Han, B. (2020). Dominant effect of anharmonicity on the equation of state and thermal conductivity of MgO under extreme conditions. *Physical Review B, 102*, 184309. doi:10.1103/PhysRevB.102.184309.

Labrosse, S. (2002). Hotspots, mantle plumes and core heat loss. *Earth and Planetary Science Letters, 199*, 147–156. doi:10.1016/s0012-821x(02)00537-x.

Lay, T., Hernlund, J., & Buffett, B. A. (2008). Core-mantle boundary heat flow. *Nature Geoscience, 1*, 25–32. doi:10.1038/ngeo.2007.44.

Lay, T., Hernlund, J., Garnero, E. J., & Thorne, M. S. (2006). A post-perovskite lens and D″ heat flux beneath the central Pacific. *Science, 314*, 1272–1276. doi:10.1126/science.1133280.

Lin, J.-F., Speziale, S., Mao, Z., & Marquardt, H. (2013). Effects of the Electronic Spin Transitions of Iron in Lower Mantle Minerals: Implications for Deep Mantle Geophysics and Geochemistry. *Reviews of Geophysics, 51*, 244–275. doi:10.1002/rog.20010.

Lobanov, S. S., Holtgrewe, N., Ito, G., Badro, J., Piet, H., Nabiei, F., et al. (2020). Blocked radiative heat transport in the hot pyrolitic lower mantle. *Earth and Planetary Science Letters, 537*, 116176. doi:10.1016/j.epsl.2020.116176.

Lobanov, S. S., Holtgrewe, N., Lin, J.-F., & Goncharov, A. F. (2017). Radiative conductivity and abundance of post-perovskite in the lowermost mantle. *Earth and Planetary Science Letters, 479*, 43–49. doi:10.1016/j.epsl.2017.09.016.

Lobanov, S. S., Soubiran, F., Holtgrewe, N., Badro, J., Lin, J.-F., & Goncharov, A. F. (2021). Contrasting opacity of bridgmanite and ferropericlase in the lowermost mantle: Implications to radiative and electrical conductivity. *Earth and Planetary Science Letters, 562*, 116871. https://doi.org/10.1016/j.epsl.2021.116871.

Malamud, B. D., & Turcotte, D. L. (1999). How many plumes are there? *Earth and Planetary Science Letters, 174*, 113–124. doi:10.1016/s0012-821x(99)00257-5.

Manga, M., & Jeanloz, R. (1997). Thermal conductivity of corundum and periclase and implications for the lower mantle. *Journal of Geophysical Research, 102*, 2999–3008. doi:10.1029/96jb02696.

Manthilake, G. M., de Koker, N., Frost, D. J., & McCammon, C. A. (2011). Lattice thermal conductivity of lower mantle minerals and heat flux from Earth's core. *Proceedings of the National Academy of Sciences U S A, 108*, 17901–17904. doi:10.1073/pnas.1110594108.

Mashino, I., Murakami, M., Miyajima, N., & Petitgirard, S. (2020). Experimental evidence for silica-enriched Earth's lower mantle with ferrous iron dominant bridgmanite. *Proceedings of the National Academy of Sciences U S A, 117*, 27899–27905. doi:10.1073/pnas.1917096117.

McWilliams, R. S., Konôpková, Z., & Goncharov, A. F. (2015). A flash heating method for measuring thermal conductivity at high pressure and temperature: Application to Pt. *Physics of the Earth and Planetary Interiors, 247*, 17–26. http://dx.doi.org/10.1016/j.pepi.2015.06.002.

Montelli, R., Nolet, G., Dahlen, F. A., Masters, G., Engdahl, E. R., & Hung, S. H. (2004). Finite-frequency tomography reveals a variety of plumes in the mantle. *Science, 303*, 338–343. doi:10.1126/science.1092485.

Murakami, M., Ohishi, Y., Hirao, N. & Hirose, K. (2012). A perovskitic lower mantle inferred from high-pressure, high-temperature sound velocity data. *Nature, 485*, 90–94. doi:10.1038/nature11004.

Ni, S., Tan, E., Gurnis, M., & Helmberger, D. (2002). Sharp Sides to the African Superplume. *Science, 296*, 1850–1852. https://doi.org/10.1126/science.1070698.

Nolet, G., Karato, S., & Montelli, R. (2006). Plume fluxes from seismic tomography. *Earth and Planetary Science Letters, 248*, 685–699. doi:10.1016/j.epsl.2006.06.011.

Ohta, K., Hirose, K., Ichiki, M., Shimizu, K., Sata, N., & Ohishi, Y. (2010). Electrical conductivities of pyrolitic mantle and MORB materials up to the lowermost mantle conditions. *Earth and Planetary Science Letters, 289*, 497–502. doi:10.1016/j.epsl.2009.11.042.

Ohta, K., Onoda, S., Hirose, K., Sinmyo, R., Shimizu, K., Sata, N., et al. (2008a). The electrical conductivity of post-perovskite in Earth's D″ layer. *Science, 320*, 89–91. doi:10.1126/science.1155148.

Ohta, K., Hirose, K., Lay, T., Sata, N., & Ohishi, Y. (2008b). Phase transitions in pyrolite and MORB at lowermost mantle conditions: Implications for a MORB-rich pile above the core–mantle boundary. *Earth and Planetary Science Letters, 267*, 107–117. doi:10.1016/j.epsl.2007.11.037.

Ohta, K., Yagi, T., Hirose, K., & Ohishi, Y. (2017). Thermal conductivity of ferropericlase in the Earth's lower mantle. *Earth and Planetary Science Letters, 465*, 29–37. doi:10.1016/j.epsl.2017.02.030.

Ohta, K., Kuwayama, Y., Hirose, K., Shimizu, K., & Ohishi, Y. (2016). Experimental determination of the electrical resistivity of iron at Earth's core conditions. *Nature, 534*, 95–98. https://doi.org/10.1038/nature17957.

Ohta, K., Yagi, T., Taketoshi, N., Hirose, K., Komabayashi, T., Baba, T., et al. (2012a). Lattice thermal conductivity of $MgSiO_3$ perovskite and post-perovskite at the core–mantle boundary. *Earth and Planetary Science Letters, 349–350*, 109–115. doi:10.1016/j.epsl.2012.06.043.

Ohta, K., Cohen, R. E., Hirose, K., Haule, K., Shimizu, K., & Ohishi, Y. (2012b). Experimental and theoretical evidence for pressure-induced metallization in FeO with rocksalt-type structure. *Physical Review Letters, 108*, 026403. doi:10.1103/PhysRevLett.108.026403.

Okuda, Y., Ohta, K., Hasegawa, A., Yagi, T., Hirose, K., Kawaguchi, S. I., et al. (2020). Thermal conductivity of Fe-bearing post-perovskite in the Earth's lowermost mantle. *Earth and Planetary Science Letters, 547*, 116466. doi:10.1016/j.epsl.2020.116466.

Okuda, Y., Ohta, K., Sinmyo, R., Hirose, K., Yagi, T., & Ohishi, Y. (2019). Effect of spin transition of iron on the thermal conductivity of (Fe, Al)-bearing bridgmanite. *Earth and Planetary Science Letters, 520*, 188–198. doi:10.1016/j.epsl.2019.05.042.

Okuda, Y., Ohta, K., Yagi, T., Sinmyo, R., Wakamatsu, T., Hirose, K., et al. (2017). The effect of iron and aluminum incorporation on lattice thermal conductivity of bridgmanite at the Earth's lower mantle. *Earth and Planetary Science Letters, 474*, 25–31. doi:10.1016/j.epsl.2017.06.022.

Osako, M., & Ito, E. (1991). Thermal diffusivity of $MgSiO_3$ perovskite. *Geophysical Research Letters, 18*, 239–242. doi:10.1029/91gl00212.

Piet, H., Badro, J., Nabiei, F., Dennenwaldt, T., Shim, S. H., Cantoni, M., et al. (2016). Spin and valence dependence of iron partitioning in Earth's deep mantle. *Proceedings of the National Academy of Sciences, U S A, 113*, 11127–11130. doi:10.1073/pnas.1605290113.

Sinmyo, R., Hirose, K., Muto, S., Ohishi, Y., & Yasuhara, A. (2011). The valence state and partitioning of iron in the Earth's lowermost mantle. *Journal of Geophysical Research, 116*, B07205. doi:10.1029/2010JB008179.

Sinmyo, R., Pesce, G., Greenberg, E., McCammon, C., & Dubrovinsky, L. (2014). Lower mantle electrical conductivity based on measurements of Al, Fe-bearing perovskite under lower mantle conditions. *Earth and Planetary Science Letters, 393*, 165–172. doi:10.1016/j.epsl.2014.02.049.

Sleep, N. H. (1990). Hotspots and mantle plumes: Some phenomenology. *Journal of Geophysical Research, 95*, 6715–6736. doi:10.1029/JB095iB05p06715.

Song, X., & Helmberger, D. V. (1995). Depth dependence of anisotropy of Earth's inner core. *Journal of Geophysical Research, 100*, 9805–9816. doi:10.1029/95jb00244.

Stacey, F.D.. (1992). *Physics of the Earth*. Kenmore, Brisbane: Brookfield Press.

Stackhouse, S., Stixrude, L., & Karki, B. B. (2010). Thermal conductivity of periclase (MgO) from first principles. *Physical Review Letters, 104*, 208501. doi:10.1103/PhysRevLett.104.208501.

Stackhouse, S., Stixrude, L., & Karki, B. B. (2015). First-principles calculations of the lattice thermal conductivity of the lower mantle. *Earth and Planetary Science Letters, 427*, 11–17. doi:10.1016/j.epsl.2015.06.050.

Sumita, I., & Olson, P. (1999). A laboratory model for convection in Earth's core driven by a thermally heterogeneous mantle. *Science, 286*, 1547–1549. https://doi.org/10.1126/science.286.5444.1547.

Takahashi, F., Tsunakawa, H., Matsushima, M., Mochizuki, N., & Honkura, Y. (2008). Effects of thermally heterogeneous structure in the lowermost mantle on the geomagnetic field strength. *Earth and Planetary Science Letters, 272*, 738–746. doi:10.1016/j.epsl.2008.06.017.

Tanaka, S., & Hamaguchi, H. (1997). Degree one heterogeneity and hemispherical variation of anisotropy in the inner core from PKP (BC)- PKP (DF) times. *Journal of Geophysical Research, 102*, 2925–2938. https://doi.org/10.1029/96JB03187.

Tang, X., & Dong, J. (2010). Lattice thermal conductivity of MgO at conditions of Earth's interior. *Proceedings of the National Academy of Sciences U S A, 107*, 4539–4543. doi:10.1073/pnas.0907194107.

Tang, X., Ntam, M. C., Dong, J., Rainey, E. S. G., & Kavner, A. (2014). The thermal conductivity of Earth's lower mantle. *Geophysical Research Letters, 41*, 2746–2752. doi:10.1002/2014gl059385.

The KamLAND collaboration (2011). Partial radiogenic heat model for Earth revealed by geoneutrino measurements. *Nature Geoscience, 4*, 647–651. doi:10.1038/ngeo1205.

Tosi, N., Yuen, D. A., & Čadek, O. (2010). Dynamical consequences in the lower mantle with the post-perovskite phase change and strongly depth-dependent thermodynamic and transport properties. *Earth and Planetary Science Letters, 298*, 229–243. doi:10.1016/j.epsl.2010.08.001.

Tsuchiya, T., Tsuchiya, J., Dekura, H., & Ritterbex, S. (2020). Ab initio study on the lower mantle minerals. *Annual Review of Earth and Planetary Sciences, 48*, 99–119. https://doi.org/10.1146/annurev-earth-071719-055139.

Tsuchiya, T., Tsuchiya, J., Umemoto, K., & Wentzcovitch, R. M. (2004). Phase transition in $MgSiO_3$ perovskite in the earth's lower mantle. *Earth and Planetary Science Letters, 224*, 241–248. doi:10.1016/j.epsl.2004.05.017.

van der Hilst, R. D., de Hoop, M. V., Wang, P., Shim, S. H., Ma, P., & Tenorio, L. (2007). Seismostratigraphy and thermal structure of Earth's core-mantle boundary region. *Science, 315*, 1813–1817. doi:10.1126/science.1137867.

Wang, X., Tsuchiya, T., & Hase, A. (2015). Computational support for a pyrolitic lower mantle containing ferric iron. *Nature Geoscience, 8*, 556–559. doi:10.1038/ngeo2458.

Yagi, T., Ohta, K., Kobayashi, K., Taketoshi, N., Hirose, K., & Baba, T. (2011). Thermal diffusivity measurement in a diamond anvil cell using a light pulse thermoreflectance technique. *Measurement Science and Technology, 22*, 024011. doi:10.1088/0957-0233/22/2/024011.

Zhang, D.-B., Allen, P. B., Sun, T., & Wentzcovitch, R. M. (2017). Thermal conductivity from phonon quasiparticles with subminimal mean free path in the $MgSiO_3$ perovskite. *Physical Review B, 96*, 100302. doi:10.1103/PhysRevB.96.100302.

Zhang, Z., Zhang, D.-B., Onga, K., Hasegawa, A., Ohta, K., Hirose, K., et al. (2021). Thermal conductivity of $CaSiO_3$ perovskite at lower mantle conditions. *Physical Review B, 104*, 184101. Doi:10.1103/PhysRevB.104.184101.

9

Assessment of a Stable Region of Earth's Core Requiring Magnetic Field Generation over Four Billion Years

Takashi Nakagawa[1,2], Shin-ichi Takehiro[3], and Youhei Sasaki[4]

ABSTRACT

We discuss whether the emergence of a stable region at the top of Earth's core is consistent with the continuous generation of a magnetic field over 4 billion years using a one-dimensional (1-D) thermal and compositional evolution model. The key points of this study are to revise the recent assessment scheme of a stable region for a 1-D radial convective structure, apply for a realistic reference state of the Earth's core, and incorporate the core-mantle chemical coupling modeled as the downward chemical flux across the core-mantle boundary (CMB). First, for a given present-day thermal and compositional structure of Earth's core, a stable region can be found when the present-day CMB heat flow is lower than the isentropic heat flow, which is below 12 TW without core-mantle chemical coupling, whereas this threshold becomes as large as 14 TW when the chemical coupling operates at the CMB. Second, we perform a backward time integration of the thermal and magnetic evolution model from the present-day state to t = 4.6 Ga. With current constraints on the present-day CMB heat flow (15–17.5 TW), a stable region at the top of Earth's core would not be consistent with the continuous generation of the magnetic field. However, there are exceptional cases that may satisfy both the existence of a stable region and continuous generation of the magnetic field over 4 billion years when CMB heat flow allows lower values between 11 TW and 13.75 TW, resulting in the maximum thickness of the stable region of 50 km and 75 km with and without CMB chemical coupling, respectively. Therefore, a stable region at the top of Earth's core cannot be confirmed at the present point because of uncertainties in the present-day CMB heat flow and thermal conductivity of Earth's core.

9.1. INTRODUCTION

9.1.1. Geophysical Observations of a Stable Layer at the Top of the Earth's Outer Core

It is quite controversial to reach the condition of a stably stratified region at the top of the outer core because

[1]Department of Planetology, Kobe University, Kobe, Japan
[2]Department of Earth and Planetary System Science, Hiroshima University, Higashi-Hiroshima, Japan
[3]Research Institute for Mathematical Sciences, Kyoto University, Kyoto, Japan
[4]Department of Information Media, Hokkaido Information University, Ebetsu, Japan

such a region has not yet been resolved using seismic and geomagnetic observations (Alexandrakis & Eaton, 2010; Anderson & Isaak, 2002; Buffett, 2014; Helffrich & Kaneshima, 2010; Tanaka, 2007). Earlier seismological studies suggested that the anomalous structure near the top of the core detected by the residual differential travel time of SmKS could be interpreted as a stable region apart from the convective region (Lay & Young, 1990; Tanaka & Hamaguchi, 1993). Those studies reported seismic structures that indicated that the seismic velocity was lower than that of the reference models of Earth's core in the layer below the core-mantle boundary (CMB) with a thickness between 50 and 100 km. Their investigations also suggested that the origin of any stable stratification might be caused by the chemical effects of segregation of the light elements produced by the inner core growth, core-mantle chemical coupling, and the primordial core formation process. Recent seismological investigations have revisited the anomalous region at the outermost outer core using better datasets (Helffrich & Kaneshima, 2010; Tanaka, 2007). The outermost outer core may have approximately a 300 km thick stable region inferred from the slower seismic structure, which is also interpreted as chemical stratification. The recent update of the seismological structure was modeled using the density and bulk modulus of the liquid iron calculated using thermodynamics theory (Irving et al., 2019). This update pointed out that the seismologically slow anomalies might be found with a thickness of less than 100 km. In this update, such slow anomalies might not be interpreted as a stable region because the Bullen's parameter is nearly unity and the Brunt-Väisälä frequency is nearly zero in such a region, indicating that a chemically well-mixed and thermally isentropic structure would be expected in the entire outer core (Irving et al., 2019).

The geomagnetic investigations suggest that the uppermost outer core may have partial stratification (Gubbins, 2007). A detailed analysis of the magneto-archimedes-coriolis (MAC) waves in Earth's core also suggested that it is possible that a 140 km thick stable region explains the geomagnetic secular variations (Buffett, 2014). Chulliat et al. (2015) confirmed that a stably stratified region could extend to 140 km below the CMB by extracting the magnetic Rossby waves from the core surface flow predicted by satellite magnetic data. In that study, they tried to explain the very short-term secular variations but not for the long-term secular variations. Lesur et al. (2015) pointed out that the stable region at the top of Earth's core cannot be ruled out with toroidal motion near the core surface adding the poloidal flow.

9.1.2. Mineral Physics Interpretations

The anomalous feature near the top of Earth's outer core pointed out by geophysical observations has been attempted to be interpreted as a stable region of compositional or thermal origin through recent measurements of the high-pressure material properties. For instance, the metal-silicate partitioning experiments at the CMB condition reveal how light elements are transported from the silicate mantle to the metallic core, known as core-mantle chemical coupling (e.g., Frost et al., 2010). This effect might imply chemical stratification at the top of Earth's core using seismological analyses (e.g., Helffrich & Kaneshima, 2010). Recent measurements of the thermal conductivity of iron alloys under CMB conditions suggest that the thermal conductivity of Earth's core could be much higher than earlier expected (e.g., Gomi et al., 2013). A high thermal conductivity suggests that the stable stratification near the top of Earth's core represents the sub-isentropic shell, where the isentropic heat flow exceeds the actual heat flow across the CMB (Labrosse, 2015). More recently, high P-T experiments have also found that MgO or SiO_2 could crystallize in the molten iron alloy (Badro et al., 2016; Hirose et al., 2017). These crystallized materials should be less dense than the surrounding iron alloy and are likely to generate additional compositional convection near the top of Earth's outer core. Such convection seems to violate the emergence of a stable region at the top of Earth's core. However, this dissolution mechanism might operate in the immiscibility region of the Fe-O-Si fluid inferred from other mineral physics experiments (Arveson et al., 2019). Such findings suggest that either a stable region or double convective layers across Earth's outer core would be possible. Hence, the formation of stable stratification is still neither inevitable nor impossible from both mineral physics and geophysical observations.

9.1.3. Interpretations Using Theoretical/Numerical Models: Which Origin gives a Better Understanding of the Geophysical Observation Incorporating Mineral Physics?

Using the material properties of Earth's core obtained from high-pressure mineral experiments, the origin and formation of the stable layer below the CMB have been discussed with numerical and theoretical models. Labrosse et al. (1997) and Lister and Buffett (1998) investigated the evolution of the thermal and compositional structure of Earth's core using one-dimensional (1-D) models. They proposed the possibility of the emergence of a thermally stable layer below the CMB when the heat flow extracted by mantle convection is sufficiently smaller than the isentropic heat flow at the CMB. In contrast, Buffett and Seagle (2010) attempted to explain the slow seismic region as the chemical stratification incorporating the chemical thermodynamic model provided by Frost et al. (2010). However, the chemical stratification would cause a faster seismic velocity anomaly due to the layer's density reduction, which would not be consistent with the seismic observation mentioned in section 9.1.1. Labrosse (2015) examined the effects of the high thermal conductivity on the thermal evolution of Earth's core and showed the present-day thermal structure of Earth's core as a function of the radius, including the long-term generation of geomagnetic field. This investigation proposed that a stable layer near the top of Earth's core could emerge when the CMB heat flow is less than 11 TW, while a stable region would be expected for a few 100 km to 1,000 km thick. The issue of this investigation is to consider compositional effects only in the heat budget (e.g., thermal conductivity's compositional dependence) but not to consider the dynamical effects of convection, in particular, vigorous mixing by the compositional convection excited by the inner core growth. For this reason, the upper limit of this thickness range is not consistent with the geophysical observations. Moreover, in such an upper limit of this range, geodynamo action is likely to fail because the thickness of the convective layer would not be sufficient to regenerate the magnetic field (e.g., Nakagawa, 2015). Generally, lower CMB heat flow is preferable for the stable layer whereas it is unpreferable for magnetic field generation. Therefore, in order to discuss emergence of the stable layer, we should check continuous magnetic field generation over 4 billion years based on the paleomagnetic observations (e.g., Biggin et al., 2015) under certain possible scenarios of thermal and chemical evolution of the Earth's core.

9.1.4. What do We Investigate Here?

As discussed in section 9.1.3, theoretical and numerical assessments of the emergence of a stable layer at the top of Earth's core have not been sufficiently investigated. One of reasons for insufficiency is not adequately addressing the effects of penetration to the stable layer due to the underlying convection. To resolve this issue, Takehiro and Sasaki (2018a) proposed a new assessment scheme for the emergence and erosion of a stable layer using radial profiles of work by buoyancy forces. Their approach is summarized as follows: (1) assume that Earth's core is chemically well mixed and thermally isentropic; (2) compute the work by buoyancy fluxes using a global mass and heat balance based on Lister and Buffett (1995); (3) when the radial profile of the work by buoyancy is negative in a particular region, convection is inhibited, and a stable region is generated if penetration does not occur. Takehiro and Sasaki (2018a) concluded that a thermally stable region could exist if the heat flow across the CMB was less than the isentropic heat flow at the CMB (9.3 TW). However, their model assumption was simplified for the reference structure of Earth's core and did not include core-mantle chemical coupling.

Another reason for insufficient investigation is that it has not been examined whether the long-term generation of magnetic field may coexist a stable region at the top of Earth's core or not. There are still controversial discussions by using numerical dynamo simulations (Nakagawa, 2015; Takehiro & Sasaki, 2018b; Gastine et al., 2020). In order to approach this issue, the magnetic field generation computed from the buoyancy flux of core convection is useful to check if a stable region can be simultaneously found with the long-term magnetic field.

Therefore, we investigate a 1-D radial convective structure of Earth's core, including a realistic reference state and core-mantle chemical coupling more precisely.

We improve the thermal and chemical evolution models proposed so far (e.g., Labrosse 2015) by introducing the effects of compositional convection with the proposed scheme by Takehiro and Sasaki (2018a). In addition, we incorporate the downward chemical flux across the CMB injected by the diffusion process on the chemical evolution, resulting in core-mantle chemical coupling (Buffett & Seagle, 2010; Gubbins & Davies, 2013; Helffrich, 2014). Integrating the evolution model with certain plausible CMB heat flow as a function of time and requiring the constraint for the long-term generation of the magnetic field, it is possible to provide evolution scenarios for allowing both emergence of stable region and magnetic field generation.

9.2. MODEL AND ANALYSIS STRATEGY

The assessment scheme for the radial convective structure of Earth's core follows Takehiro and Sasaki (2018a). Figure 9.1 illustrates an assessment scheme for a stable region at the top of Earth's core in Takehiro and Sasaki (2018a). The primary concept of this assessment is to judge the sign of the kinetic energy production rate by the

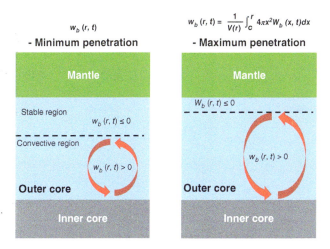

Figure 9.1 Schematic illustration of the convective structure of Earth's core. Two chemical fluxes are considered: the light element release caused by the inner core growth and the light element injection due to barodiffusion in the core-mantle chemical coupling. As quoted in the main text, the minimum thickness of the convective region (the maximum thickness of the stable region) can be determined with the change of sign of the buoyancy flux [$w_b > 0$: convective region; $w_b < 0$: stable region; a formulation of w_b is found in equation (9.27)], while the maximum thickness of the convective region (the minimum thickness of stable region) can be found with the change of sign of the integral value of buoyancy flux [$W_b > 0$: convective region; $W_b < 0$: stable region; a formulation of W_b is found in equation (9.28)]. Color version is available for the electronic version.

buoyancy force (this is equivalent to the buoyancy flux or work done by buoyancy). The entire structure of Earth's outer core is assumed to be in a chemically well-mixed and thermally isentropic state. When a region indicates that the buoyancy is positive, such a region can be determined as the convective region. Otherwise, the stable region indicates the negative value of the power by buoyancy because the negative value of the work done by buoyancy is not occurred for the convection. Hence, this region can be interpreted as a stable region. As described in the previous section, the main improvement of this study is the use of a more realistic reference state (density structure) of Earth's core. Moreover, the downward chemical flux associated with core-mantle chemical coupling is also addressed.

9.2.1. Reference Structure

We follow the radial structure of Earth's core required for computing the global energy and mass balance described by Labrosse (2015). With isentropic and compositionally uniform assumptions, the radial structure of Earth's core is given as follows:

$$\rho_c(r) = \rho_0 \left(1 - \left(\frac{r}{L_\rho}\right)^2 - A_\rho \left(\frac{r}{L_\rho}\right)^4 \right) \quad (9.1)$$

$$g(r) = \frac{4}{3}\pi G \rho_0 r \left(1 - \frac{3}{5}\left(\frac{r}{L_\rho}\right)^2 - \frac{3}{7}A_\rho\left(\frac{r}{L_\rho}\right)^4\right) \quad (9.2)$$

$$T_c(r) = T_c(c)\left(\frac{\rho_c(r)}{\rho_c(c)}\right)^\gamma \quad (9.3)$$

where $\rho_c(r)$, $g(r)$, and $T_c(r)$ are the density, gravity, and temperature, respectively. ρ_0 is the density at the center of Earth, G is the gravitational constant, c is the size of the inner core, γ is the Grüneisen parameter, and $T_c(c)$ is the temperature at the inner core boundary (ICB) given by the melting temperature of Earth's core:

$$T_c(c) = T_{m0} - K_0 \left(\frac{\partial T_m}{\partial P}\right)_X \left(\frac{c}{L_\rho}\right)^2$$
$$+ \left(\frac{\partial T_m}{\partial X}\right)_P \frac{X_0}{f_c\left(\frac{b}{L_\rho}\right)} \left(\frac{c}{L_\rho}\right)^3 \quad (9.4)$$

where T_{m0} is the melting temperature at the center of Earth's core, K_0 is the bulk modulus, $(\partial T_m/\partial P)_X$ is the pressure gradient of melting temperature, $(\partial T_m/\partial X)_P$ is the compositional gradient of melting temperature, X_0 is the reference concentration of light element of entire core and the light element is only partitioned into the liquid outer core, b is the radius of Earth's core, and $f_c(x)$ is a function defined by the integration of the density structure as follows:

$$f_c(x) = x^3 \left(1 - \frac{3}{5}x^2 - \frac{3}{7}A_\rho x^4\right) \quad (9.5)$$

L_ρ and A_ρ are the density scale height and density fitting constant, respectively.

$$L_\rho = \sqrt{\frac{3K_0}{2\pi G \rho_0^2}}; \quad A_\rho = \frac{5K_0' - 13}{10} \quad (9.6)$$

where K_0' is the radial derivative of the bulk modulus. This polynomial formulation is consistent with the density profile of Earth's outer core in the Preliminary Reference Earth Model (Dziewonski & Anderson, 1981).

The mass fraction of the light element can be supplied to both the outer and inner boundaries of the outer core. In this study, two major assumptions for the chemical evolution are used: (1) we assume that the Earth's outer core is far from chemical equilibrium (very undersaturated situation of light element); and (2) Fick's law drives the chemical diffusion from mantle to core. For the light element of Earth's core, we choose the oxygen as a major light element because the oxygen maximize the effect of the core-mantle chemical coupling compared to other major candidates of light elements of Earth's core (e.g., hydrogen, silicon, sulfur, and carbon) (Gubbins & Davies, 2013).

In order to express the change of the chemical structure associated with the core-mantle chemical coupling near CMB and with the light element release at ICB through the inner core growth, two chemical potentials, μ_{OC} (that at CMB) and μ_{IC} (chemical potential change at ICB), are needed to be introduced, which are computed from the gravitational energy release integrated from the ICB and CMB, respectively. By using the similar procedure to Labrosse (2015) and Hirose et al. (2017), the radial profiles of the chemical potentials are given as follows:

$$\mu_{OC} = \mu_{CMB} + \mu_{OC}'(r)$$
$$= \mu_{CMB} - \frac{2}{3}\pi G \rho_0 \alpha_{cO}(b^2 - r^2)\left(1 - \frac{3}{10}\frac{b^2 + r^2}{L_\rho^2}\right) \quad (9.7)$$

$$\mu_{IC} = \mu_{ICB} + \mu_{IC}'(r)$$
$$= \mu_{ICB} - \frac{2}{3}\pi G \rho_0 \alpha_{cI}(r^2 - c^2)\left(1 - \frac{3}{10}\frac{r^2 + c^2}{L_\rho^2}\right) \quad (9.8)$$

where b is the radius of Earth's core and $\alpha_{cO,cI}$ is the chemical expansion coefficients at the CMB and ICB:

$$\alpha_{cO} = \frac{\Delta_X \rho_O}{\rho_c(b)X_0} \quad (9.9)$$

$$\alpha_{cI} = \frac{\Delta_X \rho_I}{\rho_c(c)X_0} \quad (9.10)$$

where $\Delta_X \rho_O$ and $\Delta_X \rho_I$ are the density changes due to the core-mantle chemical interaction and inner core growth, respectively. The thermal conductivity of Earth's core is given by the following radial function:

$$k_c(r) = k_{c0}\left(1 - A_k\left(\frac{r}{L_\rho}\right)^2\right) \quad (9.11)$$

where A_k is the fitting constant of the thermal conductivity introduced by Gomi et al. (2013) and Labrosse (2015), and k_{c0} is the thermal conductivity at the center of Earth's core. The parameters required to compute the radial structure of Earth's core are listed in Table 9.1.

9.2.2. Global Energy and Mass Balance

Our thermal and compositional evolution model was based on the work of Labrosse (2015). The concept of model is illustrated in Figure 9.2. The difference from the original model is incorporating the effects of core-mantle chemical coupling. The light element generated by metal-silicate partitioning (e.g., Frost et al., 2010) can be injected into the outermost outer core by barodiffusion (e.g., Gubbins and Davies, 2013).

Table 9.1 Parameters for the reference structure of Earth's core

Notation	Parameter	Value	Reference
ρ_0	Density at the center	12,451 kg m^{-3}	Labrosse (2015)
L_ρ	Density scale height	8,039 km	Labrosse (2015)
A_ρ	Fourth-order polynomial fitting constant of the density	0.484	Labrosse (2015)
K_0	Bulk modulus	1.4×10^{12} Pa	Labrosse (2015)
K_0'	Pressure derivative of bulk modulus	3.567	Labrosse (2015)
b	Core radius	3,486 km	Labrosse (2015)
c	Inner core radius	1,221 km at the present	Dziewonski and Anderson (1981)
k_{c0}	Thermal conductivity at the center	163 W/m/K	Gomi et al. (2013)
A_k	Radial dependence of thermal conductivity	2.39	Gomi et al. (2013)
γ	Grüneisen parameter	1.5	Vocaldo et al. (2003)
$\left(\frac{\partial T_m}{\partial P}\right)_X$	Pressure derivative of melting temperature	9×10^{-9} K/Pa	Alfe et al. (2007)
$\left(\frac{\partial T_m}{\partial X}\right)_P$	Compositional derivative of melting temperature	-2.1×10^4 K^{-1}	Alfe et al. (1999)
X_O	Reference concentration of light element of the Earth's core	5.6%	Labrosse (2015)
α_{cO}	Compositional expansion across the CMB	1.13	Hirose et al. (2017)
α_{cI}	Compositional expansion across the ICB	0.83	Gubbins et al. (2003)
$\Delta_X\rho_O$	Density difference caused by the core-mantle chemical coupling	Computed from α_{cO}	
$\Delta_X\rho_I$	Density difference across the ICB	Computed from α_{cI}	
c_p	Heat capacity	750 J K^{-1} kg^{-1}	Labrosse (2015)
T_{m0}	Melting temperature at the center	5300 K	Within a range provided by Nomura et al. (2014) and Morard et al. (2014)
α_T	Thermal expansion	10^{-5}	Buffett and Seagle (2010)
D_c	Chemical diffusivity	3×10^{-8} m^2 s^{-1}	Maximum value in Posner et al. (2018)
μ_O	Chemical potential of oxygen	1.6×10^7 J/kg	Gubbins and Davies (2013)

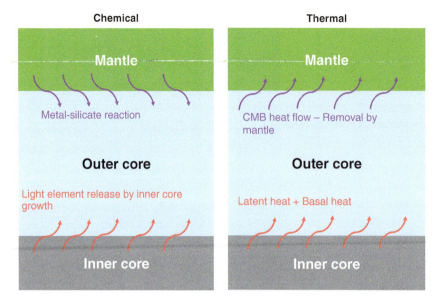

Figure 9.2 A conceptual illustration of thermal and chemical evolution of Earth's core. For chemical evolution (left), the metal-silicate partitioning reaction is assumed at the CMB and the light element release is caused by the inner solidification at the ICB. For thermal evolution (right), at the CMB, the heat is removed by the convective mantle. At the ICB, the heat is added by the latent heat in inner core solidification and basal heating. Color version is available for the electronic version.

The global mass balance in Earth's core is written as follows:

$$4\pi r^2 \rho_c(r)\frac{d\overline{X}}{dt} + \frac{d}{dr}F_c(r) = S_{ICB}\delta(r-c) + S_{CMB}\delta(b-r) \quad (9.12)$$

where F_c is the compositional flux caused by the core convection, $\delta(r)$ is the Dirac delta function, S_{ICB} is the light element release caused by the inner core growth, and S_{CMB} is the light element injection caused by barodiffusion, which are defined as follows:

$$S_{ICB} = 4\pi c^2 \rho_c(c) X_0 \frac{dc}{dt} \quad (9.13)$$

$$S_{CMB} = 4\pi b^2 \rho_c(b) D_c \frac{\alpha_{cO}}{H} g(b) \quad (9.14)$$

where D_c is the diffusivity of the light element and H is the chemical potential at the CMB pressure. Regarding S_{ICB}, X_0 is the reference concentration of the light elements in the outer core alloy and $d\overline{X}/dt$ is the average variation of light element amount in Earth's core, calculated as follows:

$$\frac{d\overline{X}}{dt} = \frac{dX_I}{dt} + \frac{dX_O}{dt} = \frac{4\pi c^2 X_0 \frac{dc}{dt} + S_{CMB}}{4\pi \int_c^b r^2 \rho_c(r) dr}. \quad (9.15)$$

The compositional convective flux is defined as follows:

$$F_c(r) = 4\pi r^2 \rho_c(r)\overline{Xu_r} \quad (9.16)$$

where $\overline{Xu_r}$ denotes the spherically averaged compositional flux. By using the global mass balance shown in equation (9.12), the compositional convective flux is calculated as follows:

$$F_c(r) = S_{ICB} - \xi(r) \quad (9.17)$$

where $\xi(r)$ is the average mass change of Earth's outer core, defined as follows:

$$\xi(r) = 4\pi \frac{d\overline{X}}{dt}\int_c^r r^2 \rho_c(r) dr. \quad (9.18)$$

This assumption on the global mass balance is valid if the effective diffusion distance of barodiffusion from the CMB is significantly shorter than the typical length scale of the compositional convection. The derivation of compositional convective flux (equation (9.17)) is provided in the Appendix.

The global energy balance of Earth's core is described as follows:

$$Q_{conv}(r) = Q_c(r) + Q_L(r) + E_G(r) + Q_S(r) \quad (9.19)$$

where Q_{conv}, Q_c, Q_L, E_G, and Q_S are the convective heat flow, secular cooling, latent heat, heat flow due to gravitational energy release, and isentropic heat flow, respectively. The heat flow across the CMB that may be determined by mantle convection is given as follows:

$$Q_{conv}(b) = Q_{CMB} = Q_c(b) + Q_L(b) + E_G(b)$$
$$= (P_c(b) + P_L(b) + P_G(b))\frac{dc}{dt}; \quad c > 0$$

$$Q_{CMB} = Q_c(b); c = 0 \quad (9.20)$$

where $P_c(b)$, $P_L(b)$, and $P_G(b)$ are prefactors expressed by the growth rate of the inner core.

Following Takehiro and Sasaki (2018a) and Labrosse (2015), each component in equation (9.19) is given as follows:

$$Q_{conv}(r) = 4\pi r^2 \rho_c(r)[\mu'(r)\overline{\xi u_r} + T_c(r)\overline{Su_r}] \quad (9.21)$$

$$Q_c(r) = -4\pi c_p \int_0^r r^2 \rho_c(r) c_p \frac{dT_c(r)}{dt} dr \quad (9.22)$$

$$Q_L(r) = 4\pi c^2 \rho_c(c) T_c(c) \Delta S \frac{dc}{dt} \quad (9.23)$$

$$E_G(r) = -4\pi \left(\int_c^r r^2 \rho_c(r) \mu'_{IC}(r) \frac{dc}{dt} dr \right) \quad (9.24)$$

$$Q_S(r) = 4\pi r^2 k_c(r) \frac{dT_c}{dr}; \quad k_c(r) = k_{c0}\left(1 - A_k\left(\frac{r}{L_\rho}\right)^2\right) \quad (9.25)$$

where $\overline{Su_r}$ is the spherically averaged entropy transport due to the radial flow, and ΔS is the entropy change caused by the freezing of the inner core. The results of the integration of equations (9.21)–(9.25) are identical to those in Labrosse (2015). The isentropic temperature profile is assumed to be the thermal structure of the inner core, as shown in the integral range in equations (9.22) and (9.24).

With the convective heat flow and compositional convective flux incorporating the chemical potential variation and isentropic temperature, the convective entropy flux can be obtained as follows:

$$F_S(r) = \frac{Q_{conv}(r) - \mu'(r)F_c(r)}{T_c(r)} \quad (9.26)$$

where $\mu'(r) = \mu'_{IC}(r) + \mu'_{OC}(r)$. $\mu'_{IC}(r)$ and $\mu'_{OC}(r)$ are defined at equations (9.7) and (9.8). By substituting the radial structure of Earth's core, all quantities composed of the global heat and mass balance can be formulated, as in Labrosse (2015).

The work by buoyancy can be calculated with two convective fluxes given as follows:

$$w_b(r) = g(r)\left(\frac{\alpha_T T_c(r)}{c_p} F_S(r) - \alpha_c F_c(r)\right) \quad (9.27)$$

where $w_b(r)$ is the radial profile of the work by convective buoyancy flux, α_T and α_c are the thermal and chemical expansivity, respectively (Lister & Buffett 1995; Takehiro & Sasaki 2018a). The convective region can be defined as: $w_b(r) > 0$, whereas $w_b(r) < 0$ is a stable region. For evaluating the maximum penetration thickness of convection, the radially integrated work by buoyancy is useful, as pointed out by Takehiro and Sasaki (2018a).

The integrated buoyancy work W_b is given as follows:

$$W_b(r) = \frac{1}{V_c(r)} \int_c^r 4\pi s^2 w_b(s) ds \quad (9.28)$$

Equation (9.28) is expressed in the averaged form and $V_c(r) = 4/3\pi(r^3 - c^3)$. With equations (9.27) and (9.28), the convective structure can be expressed as follows:

1. $w_b(r) > 0$: gives the minimum convective layer and penetration, that is, the maximum thickness of a stable region at the top of Earth's core (minimum penetration assessment).
2. $W_b(r) > 0$: gives the maximum convective layer and penetration, that is, the minimum thickness of a stable region at the top of Earth's core (maximum penetration assessment).

Therefore, the sign of these values is essential to determine whether convection occurs or not. This scheme can be visually explained by the schematic illustration given in Figure 9.1. The significant difference from the other core evolution models that allow the generation of a stable region (Labrosse et al., 1997; Lister & Buffett, 1998; Nakagawa, 2018) is not explicitly solving the thermal and chemical diffusion equations. Instead of solving diffusion equations, this scheme assesses an instantaneous snapshot of the radial profiles whether the sign of the 1-D radial structure of the buoyancy flux is negative or not.

9.2.3. Magnetic Evolution

It is important to find the continuous generation of the magnetic field over 4 billion years to identify the best fit scenario of the evolution of Earth's core (e.g., Biggin et al., 2015). To measure how the geodynamo action can generate the magnetic field, we calculate the similar scaling law of magnetic moment to that used in Aubert et al. (2009) and Driscoll and Bercovici (2014) for the magnetic evolution of Earth's core based on the analysis of geodynamo simulations in Olson and Christensen (2006), which is given as follows:

$$M = 4\pi b^3 \frac{1}{b-c} \int_c^b \left(\frac{\rho_c(r)}{2\mu_0}\right)^{\frac{1}{2}} \left(\frac{(r-c)\sigma(r)}{4\pi b^2 \rho_c(r)}\right)^{\frac{1}{3}} dr \quad (9.29)$$

where $\mu_0 = 4\pi \times 10^{-7}$ (H/m) is the magnetic permeability and $\sigma(r)$ is defined by using the buoyancy flux $w_b(r)$:

$$\sigma(r) = \begin{cases} w_b(r); & w_b(r) > 0 \\ 0; & w_b(r) \leq 0 \end{cases}. \quad (9.30)$$

The strength of the dipole moment is computed as the averaged value over the entire core radius. The convection would not be expected in $w_b(r) < 0$; thus, the magnetic field generation should be given as zero. It is, again, noted that $w_b(r)$ is an equivalent quantity to the buoyancy flux;

therefore, in equations (9.29) and (9.30), it is directly possible to apply $w_b(r)$ for the scaling relationship of the magnetic moment used in Aubert et al. (2009) and Driscoll and Bercovici (2014). Note that equation (9.29) can provide an estimate of the magnetic field strength for successful dynamos but cannot tell whether a self-exciting dynamo can exist or not only from information on the work by convective buoyancy flux ($w_b(r)$). Additionally, although this scaling relationship was created by the results of successful cases in numerical dynamo simulations and not guaranteed to express the realistic Earth's core, we expect that we can successfully evaluate the magnetic field strength with this scaling relationship in the long-term evolution computations.

9.2.4. Analysis Strategy

Here, we describe the analysis strategy for finding the best fit evolution scenario of Earth's core.

Here, the analysis steps are introduced as follows:

1. We assume that Earth's outer core is uniform and isentropic. The important parameters in this study were the thermal conductivity at the center of Earth's core, which was set as 163 W/m/K (Gomi et al., 2013), the density at the center of Earth, which was set as 12,451 kg/m^{-3} (e.g., Labrosse, 2015), and the melting temperature at the center of Earth, which was set as 5300 K; these values were consistent with recent experimental measurements (Morard et al., 2014; Nomura et al., 2014; Sinmyo et al., 2019). With these values, the present-day density, conductivity, and thermal profiles can be computed using equations (9.1), (9.25), (9.3), and (9.4), respectively (Fig. 9.3). The ICB temperature was approximately 5000 K. The isentropic heat flow at the CMB is 11.9 TW, which is significantly higher than that used in the previous investigation (9.3 TW; Takehiro and Sasaki, 2018a). Note that these reference structures are not dependent on

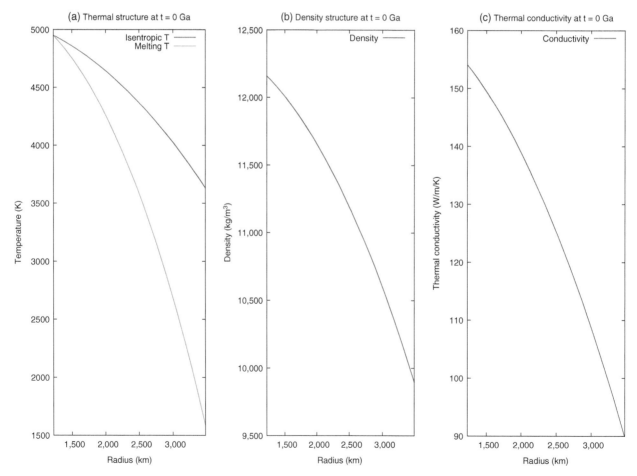

Figure 9.3 Present-day thermal (isentropic and melting temperatures), density, and thermal conductivity of Earth's core. For computing the thermal profiles, the melting temperature at the center of Earth of 5300 K and the size of the inner core of 1,221 km are applied. The density at the center of Earth is set as 12451 kg/m^{-3}. The value of thermal conductivity at the center of Earth is taken as 163 W/m/K. Color version is available for the electronic version.

the present-day CMB heat flow. Thus, they were fixed for subsequent analyses.

2. With the reference structure shown in Figure 9.3, the emergence of the stable layer is assessed for the given value of the CMB heat flow with the profiles of work by buoyancy force. $w_b(r)$ and $W_b(r)$ are calculated using equations (9.12)–(9.28) by discretizing the outer core with 200 grid points. The present-day CMB heat flow varied from 5 to 20 TW because its range has an uncertainty in the current constraints (5–15 TW; e.g., Lay et al., 2008, plus greater than 17.5 TW; Labrosse, 2015). The above assessments are performed with and without core-mantle chemical coupling, that is, the downward chemical flux $S_{CMB} = 0$ and $S_{CMB} \neq 0$ in equations (9.12) and (9.15), respectively. The chemical diffusivity for the downward chemical flux is a significant parameter set as 3×10^{-8} m^2/s (maximum values for oxygen indicated in Posner et al., 2018).

3. To find the model scenario for satisfying the continuous generation of the magnetic field over 4 billion years, we compute the evolution of Earth's core by using the backward time integration of the global energy and mass balance [equations (9.12)–(9.25)] from the present-day constraints of Earth's core by assuming the presumable prescribed CMB heat flow as a function of time. The emergence of a stable region at the top of Earth's core was evaluated using the obtained range of the present-day CMB heat flow, explaining the constraint on the magnetic evolution. The magnetic evolution is calculated as the strength of the dipole moment, given in equation (9.29). Time-stepping is performed up to 100,000 steps from the present-day state to t = 4.6 Ga, considered as 4.6×10^4 years of the time interval. Regarding the model sensitivity of the initial condition, it would not be entirely dependent on the choice of the melting temperature at the ICB, which ranges from 4500 to 6000 K (Morard et al., 2014; Nomura et al., 2014; Sinmyo et al., 2019).

Heat flow across the CMB is the most important quantity for tracking the backward time evolution of Earth's core. This quantity is generally given by the activity of mantle convection, such as in Nakagawa and Tackley (2010), because the fluctuations of CMB heat flow associated with various complicated physics in mantle convection seem important. However, for simplicity, the heat flow across the CMB is assumed as follows:

$$Q_{CMB}(t) = Q^p_{CMB} \exp(At) \qquad (9.31)$$

where t is the time measured backward from the present-day (t = 0) to 4.6 billion years ago (t = −4.6 billion years), Q^p_{CMB} is the present-day CMB heat flow, which is a significant parameter for surveying the convective structure analysis, and $A = \left(\ln\left(Q_{CMB}(4.6\,\text{Ga})/Q^p_{CMB}\right)\right)/4.6$ Ga set as $Q_{CMB}(4.6\,\text{Ga})/Q^p_{CMB} = 2$. It is noted that this factor is only given for a moderate change of the CMB heat flow as a function of time. This does not encompass the extreme change of the CMB heat flow in the early Earth in numerical mantle convection simulations (e.g., Nakagawa & Tackley, 2010).

9.3. RESULTS

9.3.1. One-Dimensional Convective Structure

Figure 9.4 shows a diagram for the assessment of a stable region in the present-day core structure for cases of $S_{CMB} = 0$. In the radial distribution of the heat flow across the outer core (Fig. 9.3a), the negative region of the heat flow is initially found when $Q^p_{CMB} < 12$ TW and almost covers the entire region of Earth's outer core when $Q^p_{CMB} = 5$ TW. Interestingly, in the diagram of buoyancy work $w_b(r)$ (Fig. 9.3b), the negative region can be found below the CMB when $Q^p_{CMB} < 11$ TW, which is recognized as a stably stratified region when the minimum penetration is assumed. However, its thickness is remarkably

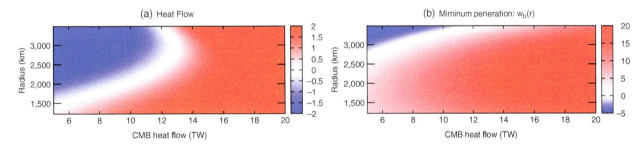

Figure 9.4 Solution diagram of the one-dimensional radial convective structure at t = 0 Ga in $S_{CMB} = 0$ as a function of Q^p_{CMB}, ranging from 5 to 20 TW. (a) Heat flow in Earth's core (the unit of the color bar is TW); (b) buoyancy flux (minimum penetration assessment; the unit of the color bar is 10^5 W/m); (c) averaged integral of the buoyancy flux (maximum penetration assessment; the unit of the color bar is 10^5 W/m); (d) thermal effect of the buoyancy flux in the minimum penetration assessment (the unit of the color bar is 10^5 W/m); (e) compositional effect of the buoyancy flux in the minimum penetration assessment (the unit of the color bar is 10^5 W/m). Color version is available for the electronic version.

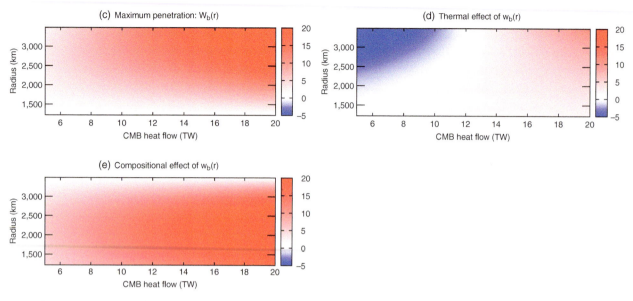

Figure 9.4 (continued)

lower than that of the negative heat flow region shown in Figure 9.3a. The maximum thickness of the stable region is approximately 700 km when Q^p_{CMB} = 5 TW. In contrast, there is no negative region in the diagram of the radially integrated buoyancy work $W_b(r)$ (Fig. 9.3c), indicating that the entire outer core is a convective region when the maximum penetration is assumed. Figures 9.3d and 9.3e show the decomposed diagrams of the thermal and chemical effects of the buoyancy work $w_b(r)$, respectively. The thermal contribution with a negative sign is compensated for the positive compositional contribution when Q^p_{CMB} < 12 TW. This finding suggests that the stable region can be found as a thermally stable region, eroded by compositional convection with the inner core growth releasing the light element. Comparing the diagrams with Figure 9.5 in Takehiro and Sasaki (2018a), the area indicating a stable region is shifted to a higher value of the present-day CMB heat flow because the isentropic heat flow at the CMB computed here (11.9 TW) is higher than that used in the previous investigation (9.3 TW; Takehiro and Sasaki, 2018a).

Figure 9.5 shows the diagram for assessing a stable region in the present-day core structure for the case of $S_{CMB} \neq 0$. The heat flow distribution in the outer core and the thermal effect of the buoyancy work shown in Figures 9.4a and 9.4d are nearly the same as those in Figures 9.3a and 9.3d. The diagram of the buoyancy work $w_b(r)$ in Figure 9.4b shows a negative region extends up to Q^p_{CMB} = 14 TW. The injection of the light element in the core-mantle chemical coupling enhances the formation of the stable region at the top of Earth's core, as thick as 1,000 km when Q^p_{CMB} = 5 TW. Thus, the core-mantle chemical coupling may enhance stable region generation at the top of Earth's core. However, there is still no negative region in the diagram of the radially averaged buoyancy work $W_b(r)$ (Fig. 9.4c), indicating that the entire outer core is a convective region even when the downward chemical flux across the CMB operates.

9.3.2. Back Trace of Core Evolution

In this section, we attempt to find a possible evolution scenario for satisfying the continuous generation of the magnetic field over 4 billion years at the thermal state of the early Earth's core using the back trace of the thermal and chemical evolution of Earth's core as a function of Q^p_{CMB}.

First, we provide examples of the back trace of the core evolution for several values of Q^p_{CMB}. Figure 9.6 shows the backward evolution profiles of CMB heat flow, CMB temperature, inner core size, and the magnetic field strength for Q^p_{CMB} =5, 10, 15, and 20 TW in S_{CMB} = 0. The CMB heat flow changes with time, as shown in equation (30) (Fig. 9.6a). The CMB temperature started at the present-day value [3635 K computed from the melting temperature at the ICB (5300 K) along with the isentropic temperature profile] and then obtained for 4200–6000 K at t = 4.6 Ga (Fig. 9.6b), depending on the present-day value of the CMB heat flow. The inner core evolution suggests that the age of the inner core ranges from 0.5 to 1.7 Ga (Fig. 9.6c), which may be consistent with the previous modeling results (e.g., Labrosse et al., 2001) and implications from the paleomagnetic data analysis (e.g., Biggin et al., 2015). Figure 9.6d indicates that a continuous magnetic field over 4 billion years and the present-day value of the dipole strength (8×10^{22} Am^2, indicated by the dashed line in Figs. 9.5d and 9.6d; e.g., Valet, 2003) requires more than 10 TW of Q^p_{CMB}. In $S_{CMB} \neq 0$ (Fig. 9.7), the main features of the evolution profiles are not significantly different from those in

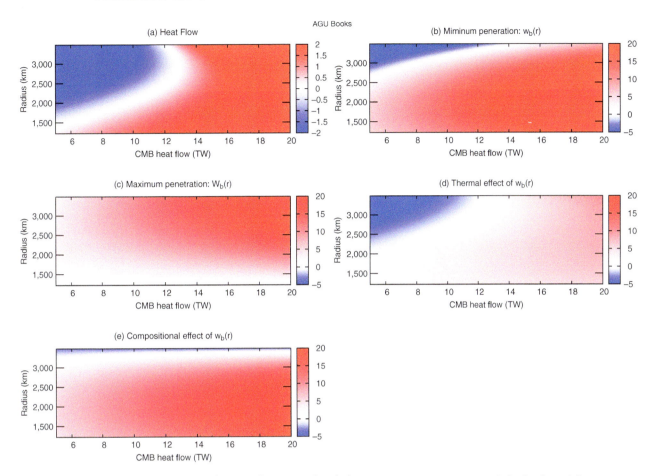

Figure 9.5 Solution diagram of the one-dimensional radial convective structure at t = 0 Ga in $S_{CMB} \neq 0$ as a function of Q^p_{CMB}, ranging from 5 to 20 TW. (a) Heat flow in Earth's core (the unit of the color bar is TW); (b) buoyancy flux (minimum penetration assessment; the unit of the color bar is 10^5 W/m); (c) averaged integral of the buoyancy flux (maximum penetration assessment; the unit of the color bar is 10^5 W/m); (d) thermal effect of the buoyancy flux in the minimum penetration assessment (the unit of the color bar is 10^5 W/m); (e) compositional effect of the buoyancy flux in the minimum penetration assessment (the unit of the color bar is 10^5 W/m). Color version is available for the electronic version.

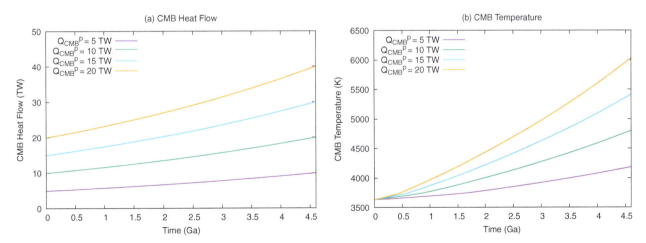

Figure 9.6 Examples of the back trace of the core evolution when $S_{CMB} = 0$. Q^p_{CMB} =5, 10, 15, and 20 TW. (a) CMB heat flow; (b) CMB temperature; (c) inner core size; (d) dipole moment strength. The dashed line indicates the present-day strength of the dipole moment, 8×10^{22} Am^2 (e.g., Valet, 2003), and the shaded region indicates the measurement uncertainty of the dipole moment strength (e.g., Biggin et al., 2015). Color version is available for the electronic version.

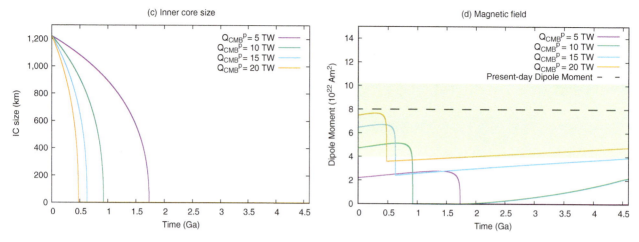

Figure 9.6 (continued)

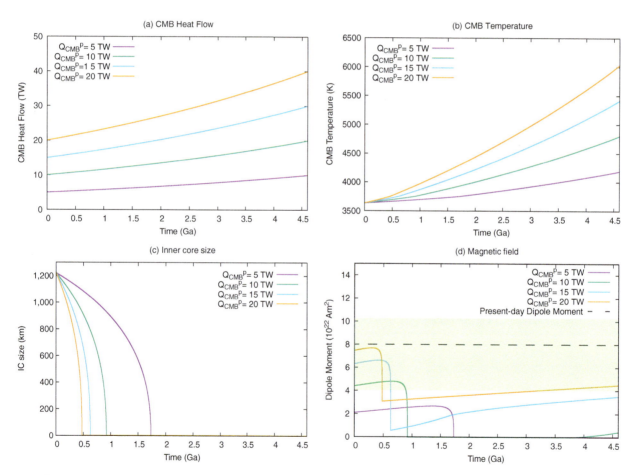

Figure 9.7 Examples of the back trace of the core evolution when $S_{CMB} \neq 0$. Q^p_{CMB} = 5, 10, 15, and 20 TW. (a) CMB heat flow; (b) CMB temperature; (c) inner core size; (d) dipole moment strength. The dashed line indicates the present-day strength of the dipole moment, 8×10^{22} Am^2 (e.g., Valet, 2003), and the shaded region indicates the measurement uncertainty of the dipole moment strength (e.g., Biggin et al., 2015). Color version is available for the electronic version.

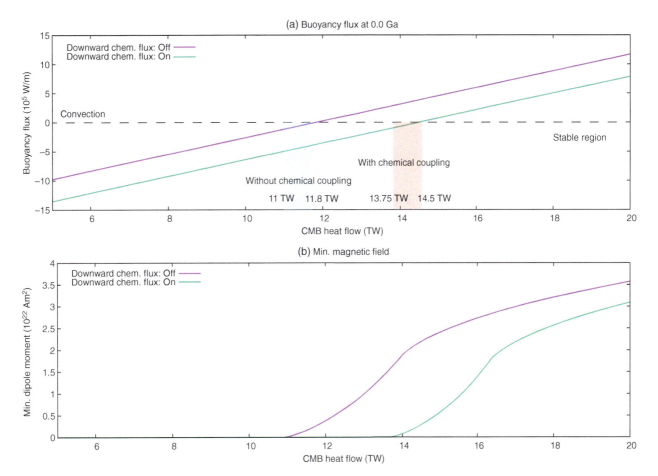

Figure 9.8 (a) The minimum value of the buoyancy flux $w_b(r)$ as a function of Q^p_{CMB}. A stable region interpreted by the minimum penetration assessment would be found until $Q^p_{CMB} = 11.8$ TW when $S_{CMB} = 0$, and $Q^p_{CMB} = 14.5$ TW when $S_{CMB} \neq 0$. (b) The minimum strength of the dipole moment in 4.6 billion years as a function of Q^p_{CMB}. The minimum value of Q^p_{CMB} is approximately 11 TW and 13.75 TW for continuous generation of the magnetic field for the cases with $S_{CMB} = 0$ (downward chem. flux: off) and $S_{CMB} \neq 0$ (downward chem. flux: on), respectively. Shaded areas in Figure 9.7a indicate the exceptional solutions allowing both a stable region at the top of Earth's core and continuous magnetic field over 4 billion years. Color version is available for the electronic version.

$S_{CMB} = 0$. The possible range of Q^p_{CMB} for the best fit value of the present-day strength of the dipole moment is greater than 10 TW.

Figure 9.8 shows the minimum value of the buoyancy flux $w_b(r)$ at present and the minimum strength of the dipole moment over 4.6 billion years as a function of Q^p_{CMB}, which can be used to determine the minimum Q^p_{CMB} for maintaining the magnetic field generation over 4 billion years and the possibility of finding a stable region. Figure 9.8a indicates the maximum value of Q^p_{CMB} for obtaining a stable region at the top of Earth's core. When $S_{CMB} = 0$, this value is approximately 11.8 TW, which corresponds to the isentropic heat flow. When $S_{CMB} \neq 0$, this value shifts to 14.5 TW. If Q^p_{CMB} is smaller than these values, a stable region at the top of Earth's core can emerge. Figure 9.7b shows the minimum magnetic field over 4.6 billion years. We judge that a continuous magnetic field fails when the geodynamo actions are shut down to generate the magnetic field. The threshold for determining the continuous generation of the magnetic field seems to be of great uncertainty because the paleointensity measurement for determining the onset timing of the inner core nucleation also has a great uncertainty as many biases depend on the sampling condition and measurement accuracy (e.g., Biggin et al., 2015). Thus, it is difficult to determine the specific value of the threshold for the minimum value of the dipole moment strength. Here, the minimum threshold of the dipole field strength is set to zero. Figure 9.8b indicates that the minimum value of Q^p_{CMB} for continuous generation of the magnetic field is greater than 11 TW when $S_{CMB} = 0$ and 13.75 TW when $S_{CMB} \neq 0$. Comparing Figures 9.7a and 9.7b, there are regions that allow finding a stable region at the top of the present Earth's core by generating a magnetic field over 4 billion years (shaded regions in Fig. 9.8a).

9.3.3. Exceptional Cases: A Stable Region with Long-Term Magnetic Field Generation

To check the situation that can allow a stable region at the top of Earth's core with finding the magnetic field generation over 4 billion years, Figure 9.9 shows the results of the back trace of the core evolution for $Q^p_{CMB} = 11$ TW with $S_{CMB} = 0$ and $Q^p_{CMB} = 13.75$ TW with $S_{CMB} \neq 0$. These are cases where the minimum Q^p_{CMB} allows both a stable region at the top of Earth's core and continuous generation of the magnetic field, resulting in the maximum thickness of the stable region. Figure 9.9a indicates that the ages of the inner core are 670 Ma and 840 Ma, respectively, consistent with earlier estimates (e.g., Labrosse, 2015). Figure 9.9b shows that the present-day strength of the dipole moment is consistent with the possible range of uncertainty provided by the paleomagnetic measurements (e.g., Biggin et al., 2015). Figures 9.10 and 9.11 show the radial profiles of $w_b(r)$ and their thermal and chemical components. When $S_{CMB} = 0$, the negative area of the buoyancy flux $w_b(r)$, which is interpreted as a stable region with the minimum penetration assessment, can be found due to the thermal effects, because Q^p_{CMB} is lower than the isentropic heat flow at the CMB (Fig. 9.10). The thickness of the stable region was eroded to 75 km by compositional convection with inner core growth. When $S_{CMB} \neq 0$, the thickness of the stable region is reduced to 50 km owing to core-mantle chemical coupling (Fig. 9.11).

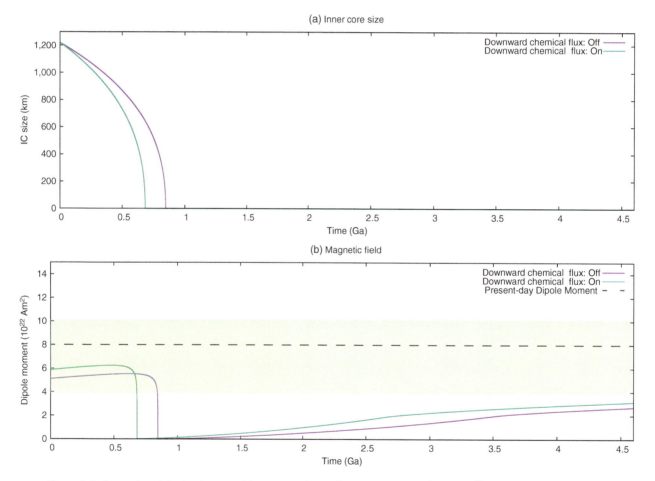

Figure 9.9 Examples of the back trace of the core evolution for two exceptional cases: $Q^p_{CMB} = 11$ TW for $S_{CMB} = 0$ and $Q^p_{CMB} = 13.75$ TW for $S_{CMB} \neq 0$. (a) Inner core size and (b) dipole moment strength. The dashed line indicates the present-day strength of the dipole moment 8×10^{22} Am^2 (e.g., Valet, 2003). The shaded region indicates the measurement uncertainty of the dipole moment strength (e.g., Biggin et al., 2015). Color version is available for the electronic version.

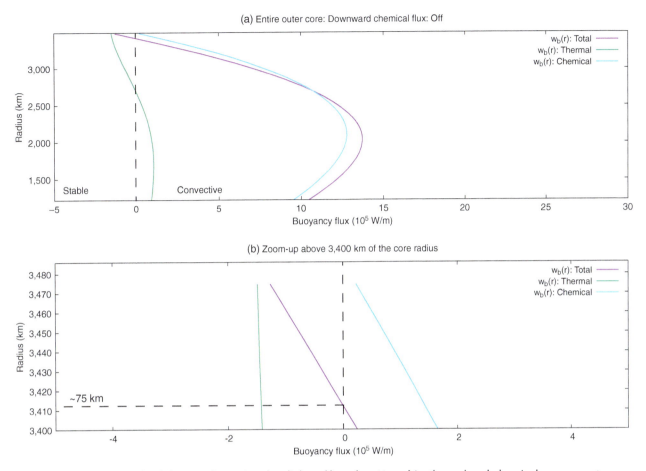

Figure 9.10 A result of the one-dimensional radial profiles of $w_b(r)$ and its thermal and chemical components for $Q^p_{CMB} = 11\,\text{TW}$ when $S_{CMB} = 0$. (a) Entire region of Earth's outer core and (b) zoom-up above $r = 3400$ km. Color version is available for the electronic version.

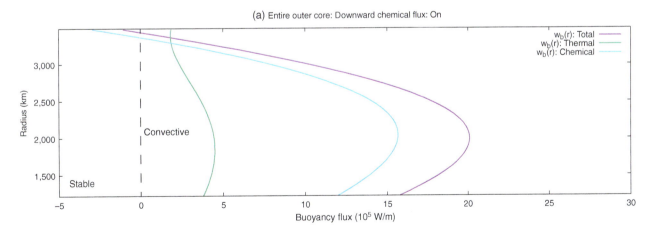

Figure 9.11 A result of the one-dimensional radial profiles of $w_b(r)$ and its thermal and chemical components for $Q^p_{CMB} = 13.75$ TW when $S_{CMB} \neq 0$. (a) Entire region of Earth's outer core and (b) zoom-up above $r = 3400$ km. Color Version is available for the electronic version.

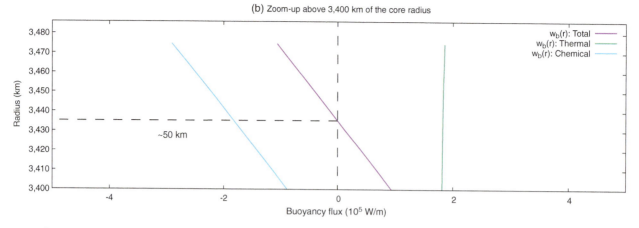

Figure 9.11 (continued)

9.4. DISCUSSION

In this chapter, we discuss the trend of finding a stable region at the top of Earth's core and continuous generation of the magnetic field as a function of Q^p_{CMB} because Q^p_{CMB} has a great uncertainty (5–15 TW; Lay et al., 2008, or more than 13 TW and favorable around 17.5 TW; Labrosse, 2015). When Q^p_{CMB} is 17.5 TW or more, the magnetic field can be generated over 4 billion years (Figs. 9.5 and 9.6), but a stable region at the top of Earth's core would not be favorable (Figs. 9.2 and 9.3). Although a stable region at the top of Earth's core seems to be incompatible for continuous generation of the magnetic field over 4 billion years, there are exceptional solutions with the emergence of a stable region and long-term generation of the magnetic field when Q^p_{CMB} is 11 TW or 14.5 TW depending on the core-mantle chemical coupling. We emphasize that a stable region at the top of Earth's core is still compatible with the continuous generation of the geomagnetic field over 4 billion years by using the value of Q^p_{CMB} in a possible range proposed by Lay et al. (2008).

There are several uncertainties that might vary findings in this study, which are, for example, the thermal structure derived from temperature at the ICB, thermal conductivity, and the effects of the choice of chemical species of light elements on chemical flux induced by the barodiffusion. First, the temperature at ICB has an uncertainty, ranging from 4800 to 5500 K (e.g., Sinmyo et al., 2019). Assuming such a range suggests that a compatible solution finding the stable region and continuous magnetic field would be slightly shifted to the lower CMB heat flow in lower ICB temperature but to the higher CMB heat flow in higher ICB temperature. When the ICB temperature is increased, the entire core temperature is also increased which enhances adiabatic temperature gradient. Then the increased isentropic heat flow reduces the kinetic energy production generated by the convection, which is favorable for stable layer formation. However, this uncertainty may only shift ±1 TW of the CMB heat flow based on the additional numerical calculations with the ICB/CMB temperature change of ±500 K which may not give a great impact to the results in this study. The second is the thermal conductivity of Earth's core, k_c, which ranges from 16 to 220 W/m/K (Konopkova et al., 2016; Ohta et al., 2016; Hsieh et al., 2020). The value of Q^p_{CMB} for the cases satisfying both emergence of stable region and continuous magnetic field generation is shifted by varying k_c. As k_c decreases, the isentropic heat flow is reduced and then the required value of Q^p_{CMB} allowing both the emergence of a stable region and continuous magnetic field generation would become lower, and vice versa. Thus, the choice of thermal conductivity may give some impact to the conclusion, which should be investigated in the next step. The third is dependence of the candidates of the light elements on the chemical flux by barodiffusion across the CMB (Gubbins and Davies, 2013). In this study, we assume the light element is oxygen only, which gives the maximum effects on the core-mantle chemical coupling (e.g., Gubbins & Davies, 2013). Other choice of light elements (e.g., Si) would not affect the present results entirely except for giving a thinner stable region.

Geophysical observations and numerical geodynamo simulations are highly controversial for the compatibility of a stable region at the top of Earth's core and long-term geomagnetic field generation (e.g., Nakagawa, 2020). Recent updates on seismological modeling across Earth's core (Irving et al., 2019) and detailed analysis in geodynamo simulations (Gastine et al., 2020) pointed out that a stable region at the top of Earth's core would not be favorable. However, there are still other possibilities for a stable region at the top of Earth's core in seismological and geomagnetic data analyses (Lesur et al., 2015; Chulliat et al., 2015; Kaneshima, 2017). Moreover, some data analyses in seismological investigations have provided a 1-D radial structure, but other seismic analyses have proposed that the location of a stable region might be associated with deep mantle heterogeneity (e.g., Tanaka & Hamaguchi, 1993; Tanaka, 2007). Other geodynamo simulations suggested that the stable region could be found at the geographical position of the deep mantle heterogeneity because the sub-isentropic condition might occur where the slower anomalies in the deep mantle would be found (e.g., Mound et al., 2019).

9.5. SUMMARY

In this study, a 1-D thermal and compositional evolution model of Earth's core with a realistic reference profile of Earth's core and the core-mantle chemical coupling was developed to find the stable region beneath the CMB based on the assessment scheme proposed by Takehiro and Sasaki (2018a). In addition, backward time integrations are performed from the plausible present structure of Earth's core, whether continuous magnetic generation over 4 billion years is possible or not. The summary of this study is as follows:

1. With the realistic reference state, a stable region can be found when the present-day CMB heat flow is lower than the isentropic heat flow, which is less than 12 TW when $S_{CMB} = 0$. With $S_{CMB} \neq 0$, that is, downward chemical flux across the CMB modeled as the core-mantle chemical interaction operates, a stable region at the top of Earth's core can emerge with a higher value of the present-day CMB heat flow, which is less than 14 TW. The downward chemical flux across the CMB works to expand the possible solutions where a stable region at the top of Earth's core can be found.

2. For explaining continuous magnetic generation over 4 billion years, Q^p_{CMB} should be greater than 11 TW at least. As indicated in Figure 9.7, a stable region at the top of Earth's core is possible with continuous magnetic field generation over 4 billion years, with thicknesses of 50 km and 75 km depending on the operation of the core-mantle chemical coupling. When the core-mantle chemical coupling is in effect, the maximum Q^p_{CMB} is expected to be 13.75 TW to satisfy both the emergence of a stable region and continuous magnetic field.

3. There are still controversial discussions about finding a stable region at the top of Earth's core in geophysical observations, geodynamo simulations, and high-pressure and temperature mineral physics experiments. In this study, Q^p_{CMB} is a key quantity for the emergence of a stable region at the top of Earth's core with the continuous generation of the magnetic field over 4 billion years. If the expected Q^p_{CMB} is greater than 13.75 TW, it is not likely to find a stable region, as pointed out by recent seismological updates (e.g., Irving et al., 2019). Nevertheless, a stable region cannot be ruled out because there is an exceptional range for Q^p_{CMB}, allowing both a stable region (up to 75 km thickness) and continuous generation of the magnetic field. Therefore, the emergence of a stable region at the top of Earth's core cannot be ruled out to explain the slow seismic anomalies there.

4. Note that several parameters of the Earth's evolution model used in this chapter are not determined. Especially, there is large uncertainty on the core thermal conductivity. We should examine influence of these uncertainties on the results presented here in near future.

Some geodynamo simulations imposing boundary heterogeneity (Mound et al., 2019) suggested lateral variations in the thickness of the stable region, which may be related to that found in earlier seismological studies (e.g., Tanaka, 2007). The lateral heterogeneity should be a key feature to converge such a controversial argument on the stable region; however, this cannot be solved in the framework of this study. More detailed investigations on geophysical observations are expected to indicate how lateral heterogeneity in the deep mantle affects heterogeneous features at the top of Earth's core.

ACKNOWLEDGMENTS

T. Nakagawa thanks Nobukazu Seama and Yoshi-Yuki Hayashi for providing the research environment at Kobe University. This work was supported by the Research Institute for Mathematical Sciences, an International Joint Usage/Research Center located in Kyoto University, and financially supported by JSPS-MEXT grant-in-aid for Scientific Research on Innovative Areas: Interaction of the Core and Mantle – Towards Integrated Deep Earth Sciences (Grant Number: 15H05834) and by JSPS KAKENHI (Grant Number: 19H01947). Authors thank to two anonymous reviewers and coeditor for improving the original manuscript significantly.

APPENDIX
DERIVATION OF EQUATION

Here, we derive equation (9.17) to assume that the characteristic length scale of the compositional convection is sufficiently longer than the typical diffusion distance of the barodiffusion at the CMB. First, the original form of the first equation in equation (9.12) is given as follows:

$$4\pi r^2 \rho_c(r) \frac{d\overline{X}}{dt} + \frac{d}{dr} F_c(r) = S_{ICB}\delta(r-c) + S_{CMB}\delta(b-r) \quad (A1)$$

where $\delta(r)$ is the Dirac's delta function. Integrating up to $r = b$, equation (A1) can be expressed as follows:

$$\xi(b) = S_{ICB} + S_{CMB} \quad (A2)$$

where $\xi(r)$ is given in equation (9.13). The global mass balance (A1) integrating up to near both the ICB ($r = c + \varepsilon$) and CMB ($r = b - \varepsilon$) can be formulated as follows:

$$\xi(c + \varepsilon) + F_c(c + \varepsilon) = S_{ICB}; \xi(b - \varepsilon) + F_c(b - \varepsilon) = S_{ICB} \quad (A3)$$

where ε is a small value; however, this should be effectively longer than the diffusion distance from the CMB. In this argument, S_{CMB} in the second equation of equation (A3) can be omitted because we assume that ε is longer than the diffusion distance. The following assumptions are made as follows:

$$\xi(c + \varepsilon) \sim 0; \xi(b - \varepsilon) \sim \xi(b) = S_{ICB} + S_{CMB} \quad (A4)$$

The boundary conditions of the compositional convective flux at both boundaries are found as follows:

$$F_c(c) = S_{ICB}; F_c(b) = -S_{CMB} \quad (A5)$$

Therefore, the global mass balance beneath the CMB can be expressed as in equation (A3):

$$\xi(r) + F_c(r) = S_{ICB} \quad (A6)$$

Equation (A6) is identical to equation (9.17) in the main text.

REFERENCES

Alexandrakis, C., & Eaton D. W. (2010). Precise seismic wave velocity stop Earth's core: No evidence for outer-core stratification. *Physics of the Earth and Planetary Interiors, 180*, 59–65.

Alfe, D., Gillan, M. J., & Price, G. D. (1999). The melting curve of iron at the pressures of the Earth's core from ab initio calculations. *Nature, 401*, 462–463.

Alfe, D., Gillan, M. J., & Price, G. D. (2007). Temperature and compositional of the Earth's core. *Contemporary Physics, 48*, 63–80.

Anderson, O. L., & Isaak, D. G. (2002). Another look at the core density deficit of Earth's outer core. *Physics of the Earth and Planetary Interiors, 131*, 19–27.

Arveson, S. M., Deng, J., Karki, B. B., & Lee, K. K. M. (2019). Evidence for Fe-Si-O liquid immiscibility at deep Earth pressures. *Proceedings of the National Academy of Sciences, 116*, 10238–10243. doi: 10.1073/pnas.1821712116.

Aubert J., Labrosse, S., & Poitou, C. (2009). Modelling in the palaeo-evolution of the geodynamo. *Geophysical Journal International, 179*, 1414–1428.

Badro, J., Siebert, J., & Nimmo, F. (2016). An early geodynamo driven by exolution of mantle components from Earth's core. *Nature, 536*, 326–328.

Biggin A. J., Piispa, E. J., Pesonen, L. J., Holme, R., Paterson, G. A., Veikkolainen, T., et al. (2015). Paleomagnetic field intensity variations suggest Mesoproterozoic inner-core nucleation. *Nature, 526*, 245–248.

Buffett, B. A., & Seagle, C. T. (2010). Stratification of the top of the core due to chemical interactions with the mantle. *Journal of Geophysical Research, 115*, B04407. doi: 10.1029/2009JB008376.

Buffett, B. A. (2014). Geomagnetic fluctuations reveal stable stratification of the Earth's core. *Nature, 356*, 329–331.

Chulliat, A., Alken, P., & Maus, S. (2015). Fast equatorial waves propagating at the top of the Earth's core. *Geophysical Research Letters, 42*, 3321–3329. doi: 10.1002/2015GL064067.

Driscoll, P., & Bercovici, D. (2014). On the thermal and magnetic histories of Earth and Venus: Influence of melting, radio activity and conductivity, *Physics of the Earth and Planetary Interiors, 236*, 36–51. Doi: 10.1016/j.pepi.2014.08.014.

Dziewonski, A. M., & Anderson, D. L. (1981). Preliminary reference Earth model. *Physics of the Earth and Planetary Interiors, 25*, 297–356.

Frost, D. J., Asahara, Y., Rubie, D. C., Miyajima, N., Dubronvinski, L. S., Holzapfel, C., et al. (2010). Partitioning of oxygen between the Earth's mantle and core. *Journal of Geophysical Research, 115*, B02202. doi: 10.1029/2009JB006302.

Gastine, T., Aubert J., & Fournier, A. (2020). Dynamo-based limit to the extent of a stable layer atop Earth's core. *Geophysical Journal International, 222*, 1433–1448. doi: 10.1093/gji/ggaa250.

Gomi, H., Ohta, K., Hirose, K., Labrosse, S., Caracas, R., Verstraete, M. J., et al. (2013). The high conductivity of iron and thermal convection in the Earth's core. *Physics of the Earth and Planetary Interiors, 247*, 2–10.

Gubbins, D. (2007). Geomagnetic constraints on stratification at the top of Earth's core. *Earth Planets Space, 59*, 661–664.

Gubbins, D., Alfe, D., Masters, G., Price, G. D., & Gillan, M. J. (2003). Can the Earth's dynamo run on heat alone? *Geophysical Journal International, 157*, 1407–1414.

Gubbins, D., & Davies, C. J. (2013). The stratified layer at the core-mantle boundary caused by baro-diffusion of oxygen, Sulphur and silicon. *Physics of the Earth and Planetary Interiors, 215*, 21–38.

Hirose, K., Morard, G., Sinmyo, R., Umemoto, K., Hernlund, J., Helffrich, G., et al. (2017). Crystallization of silicon dioxide and compositional evolution of the Earth's core. *Nature, 543*, 99–102.

Helffrich, G. (2014). Outer core compositional layering and constraints on core liquid transport properties. *Earth and Planetary Science Letters, 391*, 256–262.

Helffrich, G., & Kaneshima, S. (2010). Outer-core compositional stratification from observed core wave speed profiles. *Nature, 468*, 807–810.

Hsieh, W.-P., Goncharov, A. F., Labrosse, S., Holtgrewe, N., Lobanov, S. S., Chuvashova, I., et al. (2020). Low thermal conductivity of iron-silicon alloy at Earth's core conditions with implication for the geodynamo. *Nature Communications, 11*, 3332. doi: 10.1038/s41467-020-17106-7.

Irving, J. C. E., Cottaar, S., & Lekic, V. (2019). Seismically determined elastic parameters for Earth's outer core. *Science Advances, 4*, eaar2538. doi: 10.1126/sciadc.aar2538.

Kaneshima, S. (2017). Array analysis of SmKS waves and stratification of Earth's outermost core. *Physics of the Earth and Planetary Interiors, 223*, 2–7.

Konôpková, Z., McWilliams, R. S., Gomez-Perez, N., & Goncharov, A. F. (2016). Direct measurement of thermal conductivity in solid iron at planetary core conditions. *Nature, 534*, 99-101. doi: 10.1038/nature18009.

Labrosse, S. (2015). Thermal evolution of the core high a high thermal conductivity. *Physics of the Earth and Planetary Interiors, 247*, 36–55.

Labrosse, S., Poirier, J.-P., & Le Mouel, J.-L. (1997). On cooling of the Earth's core. *Physics of the Earth and Planetary Interiors, 99*, 1–17.

Labrosse, S., Poirier, J.-P., & Le Mouel, J.-L. (2001). The age of the inner core. *Earth and Planetary Science Letters, 190*, 111–123.

Lay T., & Young C. (1990). The stably-stratified outermost core revisited. *Geophysical Research Letters, 17*, 2001–2004.

Lay, T., Hernlund, J., & Buffett, B. A. (2008). Core-mantle boundary heat flow, *Nature Geoscience, 1*, 25–32.

Lesur, V., Whaler, K., & Wardinski, I. (2015). Are geomagnetic data consistent with stably stratified flow at the core-mantle boundary? *Geophysical Journal International, 201*, 929–946. doi: 10.1093/gji/ggv031.

Lister, J. R., & Buffett, B. A. (1995). The strength and efficiency of thermal and compositional convection in the geodynamo. *Physics of the Earth and Planetary Interiors, 91*, 17–30.

Lister, J. R., & Buffett, B. A. (1998). Stratification of the outer core at the core-mantle boundary. *Physics of the Earth and Planetary Interiors, 105*, 5–19.

Morard, G., Andrault, D., Antonangelo. D., & Bouchet, J. (2014). Properties of iron alloys under the Earth's core condutions. *Comptes Rendus Geoscience*, 130–159.

Mound, J., Davies, C., Rost, S., & Aurnou, J. (2019). Regional stratification at the top of Earth's core due to core-mantle boundary heat flux variations. *Nature Geoscience, 12*, 575–580. doi: 10.1038/s41561-019-0381-z.

Nakagawa, T. (2015). An implication for the origin of stratification below the core-mantle boundary region in numerical dynamo simulations in a rotating spherical shell. *Physics of the Earth and Planetary Interiors, 247*, 94–104. doi: 10.1016/j.pepi.2015.02.007.

Nakagawa, T. (2018). On thermo-chemical origin of stratified region at the top of Earth's core. *Physics of the Earth and Planetary Interiors, 276*, 172–181.

Nakagawa, T. (2020). A coupled core-mantle evolution: Review and future prospect. *Progress in Earth and Planetary Sciences, 7*, 57. doi: 10.1186/s40645-020-00374-8.

Nakagawa, T., & Tackley, P. J. (2010). Influence of initial CMB temperature and other parameters on the thermal evolution of Earth's core resulting from thermochemical spherical mantel convection. *Geochemistry, Geophysics, Geosystems, 11*, Q06001. doi: 10.1029/2010GC003031.

Nomura, R., Hirose, K., Uesugi, K., Ohishi, Y., Tsuchiyama, A., Miyake A., et al. (2014). Low core-mantle boundary temperature inferred from the solidus of pyrolite. *Science, 343*, 522–525.

Olson, P., & Christensen, U. R. (2006). Dipole moment scaling from convection-driven planetary dynamos. *Earth and Planetary Science Letters, 250*, 561–571.

O'Rourke, J. G., & Stevenson, D. J. (2016). Powering Earth's dynamo with magnesium precipitation from the core. *Nature, 529*, 387–389.

Ohta, K., Kuwayama, Y., Hirose, K., Shimizu, K., & Ohishi, Y. (2016). Experimental determination of the electrical resistivity of iron at Earth's core conditions. *Nature, 534*, 95–98.

Posner, E. S., Rubie, D. C., Frost, D. J., & Steinle-Neumann, G. (2018). Experimental determination of oxygen diffusion in liquid iron at high pressure. *Earth and Planetary Science Letters, 464*, 116–123. doi: 10.1016/j.epsl.2017.02.020.

Sinmyo R., Hirose, K., & Ohishi, Y. (2019). Melting curve of iron to 290 GPa determined in a resistance-heated diamond-anvil cell. *Earth and Planetary Science Letters, 510*, 45–52. doi: 10.1016/j.epsl.2019.01.006.

Takehiro, S., & Sasaki, Y. (2018a). On destruction of a thermally stable layer by compositional convection in the Earth's outer core. *Frontiers in Earth Science, 6*, 192. doi: 10.3389/feart.2018.00192.

Takehiro, S., & Sasaki, Y. (2018b). Penetration of steady fluid motions into an outer stable layer excited by thermal convection in rotating spherical shells. *Physics of the Earth and Planetary Interiors, 276*, 258–264.

Tanaka, S. (2007). Possibility of a low P-wave velocity layer in the outermost core form global SmKS waveforms. *Earth and Planetary Science Letters, 359*, 486–499.

Tanaka, S., & Hamaguchi H. (1993). Velocities and chemical stratification in the outermost outer core. *Journal of geomagnetism and geoelectricity, 45*, 1287–1301, 1993.

Valet, J. P. (2003). Time variation of geomagnetic intensity, *Reviews of Geophysics, 41*, doi: 10.1029/2001RG000141.

Vocaldo, L., Alfe, D., Gillan, M. J., & Price, G. D. (2003). The properties of iron under core conditions from first principle calculations. *Physics of the Earth and Planetary Interiors, 140*, 101–125.

10

Inner Core Anisotropy from Antipodal PKIKP Traveltimes

Hrvoje Tkalčić[1], Thuany P. Costa de Lima[1], Thanh-Son Phạm[1], and Satoru Tanaka[2]

ABSTRACT

The Earth's inner core (IC) and the innermost inner core (IMIC) are not merely subjects of academic curiosity, but they likely hold crucial information about the Earth's evolution and its geomagnetic field. In particular, the IMIC might contain a frozen anisotropic fabric different from the outer inner core (OIC). Seismological studies have probed IMIC to investigate its anisotropy – the directional dependence in seismic velocities – and reached different conclusions. Differing inferences are due to the scarcity of the seismological data sensitive to the deepest Earth's volumes and because outer layers of the Earth obscure the access to the IC by affecting the traveltime predictions and measurements. To contribute to the current efforts, we collect a global dataset of 1,150 antipodal PKIKP traveltimes for source-receiver epicentral distances between 165° and 180° sensitive to IMIC. The hand-picking of PKIKP absolute traveltimes at multiple band-pass filters is assisted by collecting P-wave records at the receivers with a similar azimuthal range. Along with relying on state-of-the-art seismological catalogs, tomographic mantle models, and OIC anisotropy models, this serves the purpose of eliminating the influences of outer layers on PKIKP traveltimes. Our new dataset is augmented by the Thai Seismic Array (TSAR) waveform data, antipodal to many South American events, and the Brazilian Seismographic Network (RSBR), antipodal to the events from Southeast Asia. The volumetric sampling with the bottoming depths between 75 and 560 km from the Earth's center provides new constraints on IMIC anisotropy and bulk IC. Both IC bulk anisotropy and the IC containing IMIC models, with the fast axis aligned with the Earth's rotation axis (ERA), the slow direction at ~60° from the ERA, and an anisotropic strength in the range 3.2–5.0 ± 0.2% explain our new PKIKP dataset.

10.1. INTRODUCTION

Anisotropy in the inner core (IC) was invoked in the late twentieth century to explain seismological observations of traveltimes of PKIKP waves and normal modes (Poupinet et al., 1983; Morelli et al., 1986; Woodhouse et al., 1986). Anisotropy occurs widely in nature, being observed in rock samples and confirmed to exist at depths through seismological observations of the well-mapped Earth's lithosphere (e.g., Crampin, 1984; Babuska & Cara, 1991). Therefore, the finding of an anisotropic IC was not surprising, although it was unexpected by some to come before the first observation of anisotropy in the lowermost mantle (Vinnik et al., 1989). Up to date, there are multiple physical mechanisms proposed for the generation of anisotropy in the Earth's deep interior (e.g., Wang & Wen, 2007; Long & Silver, 2009; Romanowicz & Wenk, 2017). In the context of the IC, these mechanisms rely on the preferred orientation of iron-nickel

[1] Research School of Earth Sciences, The Australian National University, Canberra, Australia
[2] Research Institute for Marine Geodynamics, Japan Agency for Marine-Earth Science and Technology, Yokosuka, Japan

Core-Mantle Co-Evolution: An Interdisciplinary Approach, Geophysical Monograph 276, First Edition.
Edited by Takashi Nakagawa, Taku Tsuchiya, Madhusoodhan Satish-Kumar, and George Helffrich.
© 2023 American Geophysical Union. Published 2023 by John Wiley & Sons, Inc.
DOI: 10.1002/9781119526919.ch10

alloys, the core's main constituent. However, given the IC's small volume (less than 1% of Earth's volume), its imperfect coverage by seismic waves, and the most challenging temperature-pressure conditions to replicate in high-pressure physics laboratories, there is not yet consensus among the many competing physical mechanisms of anisotropy formation in the IC (e.g., Jeanloz & Wenk, 1988; Yoshida et al., 1996; Karato, 1999) and the nature of anisotropy in the IC and its mineral physics aspects are still unclear.

There are no perfect models, and they usually cannot explain all data. It is essential to recognize that IC cylindrical anisotropy, a form of transverse isotropy, is not required by the data but works rather well in explaining them. Cylindrical anisotropy, therefore, remains the most accepted model to represent the IC seismic observations, as seen in the most recent reviews on the topic: (Deuss, 2014; Tkalčić, 2015; Romanowicz & Wenk, 2017). A model with the fast axis aligned with the Earth's rotation axis (ERA) and the slow direction parallel to the equatorial plane can explain the most but not all advanced and delayed traveltimes reported for the upper 400 km of the IC. Namely, a subset of the time arrivals of compressional waves traveling near the ERA from events in the SSI and recorded in stations in Alaska show a significant scatter in the data that would poorly fit the quadratic function that the transverse isotropy dictates. The distribution of these residuals resembles an L-shaped, when displayed as a function of the angle ξ between the PKIKP ray paths and ERA, and is referred to under the same name (e.g., Romanowicz et al., 2003). This problem leads to the proposal that either the mantle structure is unaccounted for or the IC is a conglomerate of anisotropic domains (Tkalčić, 2010; Mattesini et al., 2018).

To investigate the Earth's IC, seismologists use different tools and data, primarily based on differential traveltimes, PKPab-PKIKP or PKPbc-PKIKP (e.g., Creager, 1992; Song & Helmberger, 1993) absolute PKIKP traveltime residuals (e.g., Su & Dziewonski, 1995; Ishii & Dziewonski, 2002) or a combination of PKIKP traveltimes and normal modes (Ishii et al., 2003). The advantage of absolute PKIKP traveltimes is that they offer constraints on the innermost inner core (IMIC) seismic properties because of the depth sensitivity of PKIKP waves recorded at near-antipodal epicenter distances. However, errors associated with earthquake parameters such as source mislocation and precision of the event origin time can contribute to the uncertainties of the time measurements. Alternatively, these uncertainties can be minimized when studying PKP differential traveltimes and assuming the PKPab and PKPbc seismic waves have similar paths in the crust and mantle to the PKIKP waves. Hence, their difference in timing provides information regarding the IC. The disadvantage is that, apart from IC-sensitive waves, other seismic phases are involved when analyzing PKP differential times. For example, PKPab waves are known to be strongly contaminated by heterogeneity in the lowermost mantle (e.g., Bréger et al., 2000; Tkalčić et al., 2002), imposing a trade-off in the interpretation of PKPab-PKIKP. Also, PKPbc-PKIKP differential traveltime data are only sensitive to the outer inner core (OIC) (e.g., Shearer & Toy, 1991; Garcia & Souriau, 2000). Therefore, absolute PKIKP traveltime data are, arguably, the most reliable data type to probe the IMIC provided that they are corrected for heterogeneities in the mantle structure. Ishii et al. (2003) demonstrated that the absolute PKIKP data and normal modes are more compatible with one another than with the differential PKP traveltime data.

Another complexity is the hemispherical signature of the differential traveltime residuals for the upper IC. This observation is present in both PKIKP-PKiKP and PKPbc-PKIKP differential traveltime residuals (e.g., Tanaka & Hamaguchi, 1997; Niu & Wen, 2001; Garcia, 2002; Waszek et al., 2011). It should be mentioned that this feature is seen as the most robust feature associated with the IC in the IC seismological community, much more so than the IC differential rotation, for which the views are highly different and often disputed (Song & Richards, 1996; Souriau, 1998; Laske & Masters, 1999; Tkalčić et al., 2013). Other geodynamical models were put forward to explain the hemispherical signature, and there is no current consensus to explain its origin nor its radial extent (e.g., Garcia & Souriau, 2000; Tanaka, 2012). While the hemispherical signature near the ICB is associated with the isotropic speed and attenuation (e.g., Shearer, 1994; Song & Helmberger, 1998), some researchers found that a hemispherical pattern of anisotropy extends deeper (e.g., Waszek et al., 2011; Lythgoe et al., 2014; Frost et al., 2021). A recent study utilizing PKP differential traveltime data recorded by TSAR suggests that the shape of the hemispherical boundary is possibly an ellipse shortened in the north-south direction rather than a circle (Ohtaki et al., 2021).

Further seismological investigations have suggested increasingly more complex anisotropy models in the IC. In terms of radial dependence, it was found that the strength of anisotropy increases toward the center (e.g., Creager, 1992; Vinnik et al., 1994; McSweeney et al., 1997), and that the IC's top 100 km is isotropic, although a normal mode study argued for the existence of radial anisotropy in the same layer (Lythgoe & Deuss, 2015). Calvet and Margerin (2018) showed that the shape and orientation of grains could explain the hemispherical pattern in the uppermost IC. In recent years, the IMIC was shown to have a different form of anisotropy than the rest of the IC (Ishii & Dziewonski, 2002, 2003). This observation was supported by multiple subsequent studies of normal modes (Beghein & Trampert, 2003) and body waves (e.g., Cao & Romanowicz, 2005; Cormier & Stroujkova, 2005; Stephenson et al., 2021). However, the IMIC parameters such as its radius, the strength of anisotropy, and especially the position of its slow direction are somewhat inconsistent. All body-wave studies of the IMIC find the slow anisotropic direction in the oblique plane. Interestingly, an early normal mode study (Beghein & Trampert, 2003) indicated the slow anisotropic direction aligned with the ERA. A study of waveform correlations (Wang et al., 2015) found that the IMIC has a fast axis near the equator, although their methods and interpretation of the correlation features were recently challenged by Tkalčić et al. (2020). In summary, the IMIC models may have different implications for the evolutionary history of the IC; thus, further studies are necessary to resolve anisotropy in the IC, particularly in the IMIC.

The mineral interpretation of anisotropy in the IC involves the crystal organization of iron-nickel alloy at high pressure (e.g., Bergman, 1997; Karato, 1999). Iron's most favored crystallographic phases are the hexagonal-closed-packed (hcp) and body-centered-cubic (bcc). However, the debate for which iron phase best matches the seismological observations is ongoing (e.g., Stixrude & Cohen, 1995; Jackson et al., 2000; Belonoshko et al., 2008; Romanowicz et al., 2016; Mattesini et al., 2018; Ritterbex & Tsuchiya, 2020).

With significant advances in methodologies, expanding the global installation of broadband instruments is crucial to increasing the seismological dataset and improving our understanding of the IC and its evolution. In this study, we improve the seismic sampling coverage of the IMIC by including a newly collected, high-quality, antipodal PKIKP dataset from networks in Brazil and Thailand. First, we conduct a comprehensive analysis for considering mantle heterogeneity and OIC anisotropy models to avoid contributions from the Earth's outer layers in the dataset. We rigorously examine the IC volumetric sampling by PKIKP waves and find the justification for setting up a cylindrical anisotropy model. We then investigate anisotropy in the bulk IC, and the models with the IC divided into the OIC and IMIC and the data uncertainty, utilizing a Bayesian hierarchical inversion framework. We examine the fitting models in the light of data uncertainty and discuss the critical roles of temporary and permanent arrays and networks in the context of studying IC anisotropy.

10.2. DATA AND THEIR VOLUMETRIC COVERAGE

10.2.1. Waveform Data

Figures 10.1a and 10.1b illustrate the global distribution of stations and events used in this study, along with the projections of the source-receiver great-circle path sections sampling the IC. Most of the continuous, vertical-component broadband waveforms are collected through the IRIS DMC. We first requested waveform data for the events already a part of the PKPab-PKIKP differential traveltime dataset presented in the electronic supplements of Tkalčić (2017) due to the high quality of the data used to measure differential traveltimes. The waveforms from that dataset recorded in the epicentral distance range 165°–180° were excised from the ISC-catalog origin times to the one-minute mark after the predicted PKIKP traveltimes. Apart from importing PKIKP waveforms, we also imported the P-wave waveforms recorded in the epicentral distance range 60°–90° along the same azimuthal corridor as the stations reporting PKIKP waves (±10° from the great circle path). The rationale behind importing the P waves was to enhance the quality of the PKIKP absolute traveltime picks by comparing the waveforms.

We then repeated the same acquisition procedure for the prominent events recorded by the temporary seismic array deployed in Thailand – TSAR (Fig. 10.1c). A temporary broadband seismic network in Thailand – TSAR of 40 stations, was installed with a typical interstation distance of 100 km, except in the western part of Thailand (Tanaka et al., 2019). Initially, 34 CMG-3T sets (Güralp, Inc.) and six sets of STS2 sensors (Streckeisen AG) were installed. The array operated from January 2016 to January 2019, but nearly 50% of the initially deployed stations were damaged by flooding and lightning. Three-component ground velocities with a sampling rate of 0.025 s (40 Hz) were acquired with a 24-bit resolution by all data loggers. The data center at the Earthquake Research Institute, The University of Tokyo, provides continuous and event-edited seismograms with state-of-health information.

The acquisition procedure described above was also applied for the Brazilian Seismographic Network – RSBR (Fig. 10.1d). The permanent stations of the Brazilian Seismographic Network – RSBR (ON, BR, BL, and NB) – are operated by multiple research institutions (USP, ON, UFRN, and UNB), and the waveforms are available for download in SAC format from the database of the University of Sao Paulo Seismological Center, via ArcLink protocol. The rationale behind enhancing our global PKIKP dataset by the TSAR- and RSBR-recorded data

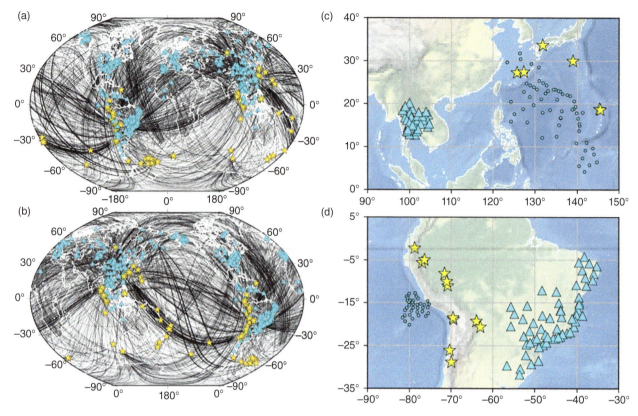

Figure 10.1 (a) Map of Seismic events (yellow stars) and stations (blue triangles) used in this study to collect antipodal PKIKP traveltime dataset. The ray paths sampling the IC are projected as dark lines to the Earth's surface. (b) Same as (a) but centered on 180° longitude. (c) Seismic array deployed in Thailand – TSAR. Cyan triangles denote stations. Yellow stars are the events recorded by the Brazilian Seismographic Network (RSBR). The antipodes to the RSBR stations are shown as small cyan circles. (d) Seismic network RSBR (cyan triangles) consisting of subnetworks: ON, BR, BL, and NB. Yellow stars are the events recorded by TSAR. The antipodes to the TSAR stations are shown as small cyan circles. The original version of this figure in color is available in supplements; the B-W version is included here by the editorial rules.

is to bring in new waveforms in quasi-equatorial planes to probe the controversial character of the IMIC anisotropy published in recent studies (see section 10.1, Introduction), and also to constrain the dataset by two networks on their antipodal sides.

Finally, we examined more recent $M_w \geq 5.7$ events to fill the gaps in spatial coverage. In total, we analyzed the waveform data of more than 150 candidate events in the time interval 1995–2019, but only about 30% of the events were deemed to have sufficiently high quality to be processed further by measuring the absolute PKIKP traveltimes. Figures 10.1c and 10.1d show the TSAR array and the RSBR network configurations, their antipodal points, and events we utilized to measure PKIKP traveltimes in this study. We used a total of 11 events from 2016 to 2019 located in South America, antipodal to TSAR, and a total of 6 events from 2011 to 2018 located in Southeast Asia, antipodal to RSBR stations.

10.2.2. Absolute PKIKP Traveltime Measurements

As mentioned above, we achieve antipodal sampling of the IC in a reverse sense with TSAR and RSBR locations. These two networks are ideally positioned relative to some of the world's largest seismicity to sample the innermost part of the IC. Figure 10.2 illustrates the waveform data quality for the two selected events: a deep (H = 606 km) southern Bolivia, Mw = 6.5, 21 February 2017, earthquake, recorded by TSAR (also featured in Tanaka et al., 2019), and a moderately deep (H = 119 km) Ryukyu Islands, Mw = 6.5, 2 March 2014, earthquake, recorded by RSBR. The PKIKP waveforms

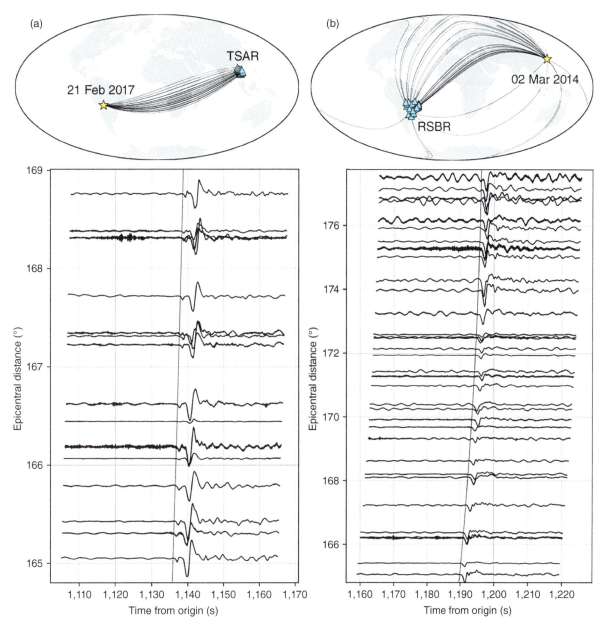

Figure 10.2 Examples of PKIKP unfiltered waveforms for TSAR and RSBR networks versus the theoretically predicted traveltime curves. (a) The location of the Mw = 6.5, 21 February 2017, H = 606 km event and TSAR network with corresponding great circle paths. In the bottom panel, PKIKP waveforms recorded by the TSAR network are plotted as a function of epicentral distances. Theoretical prediction calculated for this event is plotted as a gray curve using the model ak135. Source: Adapted from Kennett et al. (1995). (b) Similar to (a) but for the Mw = 6.5, 2 March 2014, H = 118.9 km event recorded by the RSBR network. The original version of this figure in color is available in supplements; the B-W version is included here by the editorial rules.

presented in these two examples are not of particularly exceptional quality. PKIKP onsets were the easiest to pick among all the data; they are merely chosen here as representative data.

Several control mechanisms are utilized to ensure the high quality of our dataset:

1. We do not use differential PKPab-PKIKP data because we want to avoid using PKPab traveltimes

affected by the lowermost mantle structure. The trade-off of this choice is that we lose the benefits the differential traveltime data provide, that is, eliminating the uncertainty in the hypocentral time and location and eliminating structural uncertainty near the source and receiver. Our choice is justified by the following rationale: we use some of the existing mantle models to correct for the times PKIKP waves spend in the mantle, but these mantle models are still much better for the upper mantle than for the lowermost mantle. There are multiple reasons why this is the case, and we elaborate on this in section 10.3.1 when discussing our choices of models.

2. The catalog we use is the ISC (a dataset of teleseismic events and seismic arrivals collected by multiple international institutions; Bondár and Storchak, 2011). The analysis of the record sections for more than 150 events and the subsequent measurements revealed that the catalog is of high quality, most of the P waves arriving within ±1 second from the theoretical traveltime predicted by the ak135 reference Earth model (Kennett et al., 1995). More details about hypocenter errors and catalog biases were given in Stephenson et al. (2021), where we used more than 30,000 PKIKP traveltimes. The records shown as a function of epicentral distance in Figures 10.2a and 10.2b reveal a close match between the observed onsets and theoretically predicted traveltimes for PKIKP waves.

3. We avoid too large and too small earthquakes. Too large earthquakes often have complicated moment releases, which introduce a range of uncertainty in picking the onsets of PKIKP waves. An emergent (gradual) energy release or a more complex energy release with multiple subevents may introduce uncertainties in onset times. Therefore, the drawbacks of using these events outweigh the benefits, especially if smaller events from the same area can be utilized. On the other hand, too small events have sufficiently attenuated PKIKP waves, and, consequently, they introduce uncertainty in picking. Overall, most events we utilize have $7.5 \geq M_w \geq 5.7$, and, in most cases, once the event is selected, the quality of recordings is such that all receivers that recorded PKIKP waves are used.

4. Each PKIKP waveform is acquired along with P-waves recorded on the available stations in the same azimuthal corridor from the source to the receiver, ensuring excellent control over the source-time-function character, the radiation pattern, and similarity between P and PKIKP waves. The selected event mechanisms are either reverse or normal, and the radiation pattern is highly similar for the wavefield steeply radiated from the focal sphere as P or PKIKP waves. Before measuring the onset of a PKIKP wave, we compare the PKIKP with the P waveforms, and the shape of the source-time function informs the PKIKP onset picks.

5. All waveforms are band pass filtered using multiple Butterworth filter corner frequencies. Some events are filtered with 5–10 different band-pass, low-pass, and high-pass filters before measuring the PKIKP onsets. The most dominant frequency for PKIKP waves is between 0.5 and 2.0 Hz, but the waveforms are sometimes better expressed toward lower or higher frequency ends. We estimate from these multiple selections of filters and the comparison with P waveforms that the uncertainty in picking the PKIKP onsets in most cases is <0.5 second, but a more conservative estimate of measurements uncertainty is about 1.0 second.

The theoretical traveltime predictions are corrected for the Earth's ellipticity (Kennett and Gudmundsson, 1996), and the traveltime residuals are computed by subtracting the theoretical traveltimes from the observed ones. We plot PKIKP traveltime residuals (PKIKP residuals hereafter) as a function of several control parameters to check the quality of our dataset. Those parameters are the epicentral distance, the azimuth, the back-azimuth, the longitude, and latitude of the PKIKP bottoming points, and the longitude and latitude of the piercing points at major boundaries. Finally, we compute the angle between the PKIKP ray in the IC with the ERA, aka the angle ξ. Due to this quality check procedure, only a couple of outliers were removed from our hand-picked dataset.

Figure 10.3 shows the PKIKP residuals, and the events used in this study as a function of the angle ξ, color-coded so that possible outliers could be readily appreciated. Different colors also give a unique perspective on the chronological order of the selected events and the scatter in the PKIKP traveltime measurements among various events. The scatter is a combined contribution of the measurement uncertainty (discussed above) and the theoretical errors, the most obvious stemming from the origin time and location catalog and the 1-D model of the Earth assumption. In comparison with the existing differential traveltime datasets (e.g., Tkalčić, 2017), our absolute PKIKP traveltime catalog seems to be of exceptional quality.

In Figure 10.3, we also compare the PKIKP residuals expressed in seconds (Fig. 10.3b) with the PKIKP residuals normalized by the total time PKIKP spent in the IC expressed in percentages (Fig. 10.3c). It is helpful to normalize the PKIKP residuals in this way to eliminate the time dependence on the depth of IC sampling, that is, the length of ray paths in the IC or the total time PKIKP waves spent in the IC. When the normalized PKIKP residuals are presented as a function of the angle ξ, the interpretation of anisotropy becomes more straightforward, with the direction of sampling being the only dependence we have left. The very first look at the normalized PKIKP residuals as a function of the angle ξ (Fig. 10.3c) reveals relatively significant variations

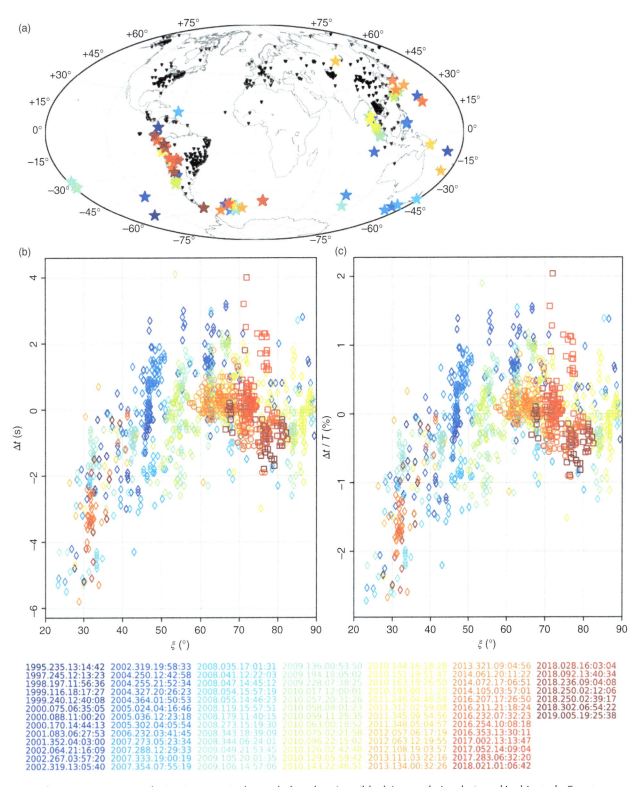

Figure 10.3 (a) Map of seismic events (color-coded) and stations (black inverted triangles) used in this study. Events are color-coded in chronological order relative to their origin times. (b) Colored diamonds show absolute PKIKP traveltime residuals with respect to the reference model ak135 and corrected for Earth's ellipticity. Source: Adapted from Kennett et al. (1995). Plotted as a function of the angle between the PKIKP ray path in the IC and the Earth's rotation axis, ξ. (c) Same as (b), but for the PKIKP traveltime residuals normalized by the traveltime PKIKP spends in the inner core (expressed in percentages). Events' color code is shown at the bottom of the figure. Squares indicate the TSAR data. Circles indicate the RSBR data. The original version of this figure in color is available in supplements; the B-W version is included here by the editorial rules.

in PKIKP residuals, with a pronounced fast axis aligned with the ERA and a slow direction at about 60°. The TSAR data (marked by squares) contribute to defining the shape of the PKIKP residual pattern, with PKIKP waves being notably slower at the oblique than at the angles perpendicular to the ERA.

10.2.3. Global Coverage of the Inner Core

Figure 10.1 shows the global coverage of the IC relative to the location of events and stations utilized in this study. A directional sampling of the IC relative to the ERA is evident from the distribution of the angle ξ. However, to gain a more comprehensive understanding of the IC sampling, we make a set of additional figures showing the relationship of the PKIKP residuals with the PKIKP bottoming points, the PKIKP piercing points at the inner core boundary (ICB), and the direction of sampling itself.

In Figure 10.4, we compare the PKIKP residuals with the PKIKP-PKPab differential residuals in terms of the distribution of their bottoming points. Note that this type of viewing of the residuals is not ideal for the antipodal data. Because of their antipodal nature, they sample the Earth near its center, so the projection of the bottoming points on a map is useful only in conjunction with a sphere of a relatively small radius around the Earth's center. Figures 10.4a and 10.4b, therefore, show the antipodal PKIKP residuals sampling very close to the

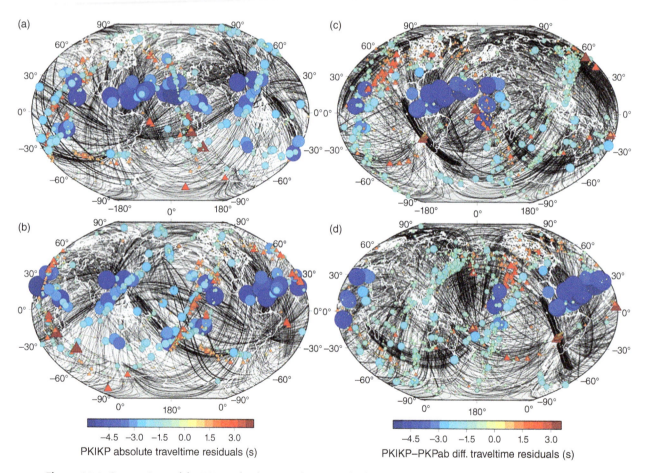

Figure 10.4 Comparison of the PKIKP absolute traveltime residuals from this study (1,150 measurements) with the PKIKP-PKPab differential traveltime residuals from Tkalčić (2017) (1,983 measurements). The traveltime residuals are color-coded and shown in varying sizes. (a) The absolute PKIKP traveltime residuals are plotted at the bottoming points of PKIKP rays in the inner core. Circles are negative (fast PKIKP propagation through the inner core), and triangles are positive (slow PKIKP in the inner core). (b) Same as (a) but centered on the 180° longitude. (c) The differential PKIKP-PKPab traveltime residuals are plotted at the bottoming points of PKIKP rays in the inner core. Circles are negative (fast PKIKP propagation through the inner core), and triangles are positive (slow PKIKP in the inner core). (d) Same as (c) but centered on the 180° longitude. The original version of this figure in color is available in supplements; the B-W version is included here by the editorial rules. (e) Frequency histograms of the angle ξ and the PKIKP bottoming point radii.

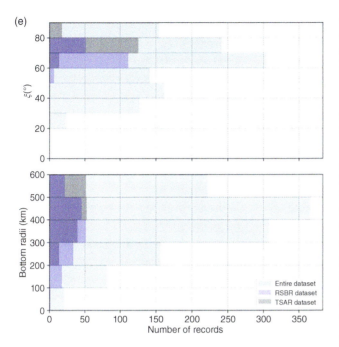

Figure 10.4 (continued)

Earth's center, color-coded so that blue and red represent fast and slow, respectively. Figures 10.4c and 10.4d show the PKIKP-PKPab dataset (Tkalčić, 2017), sampling the bulk of the IC (including nonantipodal data) for a comparison. Negative PKIKP residuals and PKIKP-PKPab residuals are proxies for fast IC, assuming that PKIKP traveltimes deviate from the ak135 predictions only in the IC and thus determine both absolute and differential traveltime residuals. Figure 10.4e shows the frequency histograms of the angle ξ and the PKIKP bottoming point radii in the IC.

Arguably, a more useful representation of the IC sampling by PKIKP waves is by means of PKIKP ray direction and the ICB piercing point location, including the depth of the IC penetration. We attempt to achieve this by plotting the residuals in polar coordinates projected onto the Southern Hemisphere of the IC, considering both the ray direction and the location of IC entry points (Fig. 10.5). We also separate the dataset relative to the sampling depth into two subsets: the rays sampling the innermost 300 km of the IC and all other remaining rays.

Note that plotting the ray direction only (Fig. 10.5a,c) provides incomplete information on the piercing latitude at the ICB and the polarity of the direction. The ray vector dip (angle ξ) resembles (although it is not identical to) the IC piercing latitude because the rays in our dataset are antipodal (this resemblance, however, does not hold for nonantipodal PKIKP waves sampling shallower in the IC). In terms of the polarity ambiguity, imagine the South American events recorded by the TSAR array or the events in the vicinity of the TSAR array recorded in South America. In both cases, the ray vectors will be projected as the points in the north-western quadrant relatively close to the map periphery as the angle ξ is 70°–80°.

Similarly, plotting the PKIKP ray location (Fig. 10.5d,f) is informative with regard to the ICB area sampled by rays but is void of the accurate ray direction. Namely, multiple rays, each with a different angle ξ can enter (or exit) the IC at the same location of the ICB. Possible coherency of the PKIKP residuals with respect to the ray location at the ICB would thus be more likely a result of heterogeneity than anisotropy. If the ICB entry (source) latitude is lower than the exit (receiver) latitude, we project the entry point to the Southern Hemisphere, and if the opposite is true, we project the exit point. This ensures that if the source and receiver exchange places, we still plot the same point on the Southern Hemisphere projection, although it means that the polarity ambiguity still exists in this type of plot. Note that there are source-receiver geometries with both points in the Northern or the Southern Hemisphere, and we apply the same convention to plot the ray location at the ICB. There is another ambiguity in this type of plot near the equator for the parallel rays entering or exiting at the same northern and southern latitude. This ambiguity, however, is minimal for the antipodal rays and exists only near the equator.

10.3. RESULTS

10.3.1. Mantle Heterogeneity Corrections

As mentioned earlier, heterogeneous 3-D structure in the Earth's mantle is a significant contributor to the differences (residuals) between the observed traveltimes of PKIKP waves and their theoretical predictions based on the chosen reference 1-D model. On the one hand, avoiding PKPab data for the IC analysis is desirable because PKPab waves exhibit long paths through the highly heterogeneous lowermost mantle (e.g., Bréger et al., 2000; Tkalčić et al., 2002). On the other hand, PKIKP waves propagate through the heterogeneous upper mantle but penetrate steeply through the lowermost mantle so that the lowermost mantle heterogeneity effect is minimal. Although the ultralow-velocity zones (ULVZs) are not accounted for in the mantle tomography models, we estimated from the recent study of ULVZs (Pachhai et al., 2022) that, in the most extreme scenario, the unaccounted traveltime correction would be 0.25 s, corresponding to a 10 km thick reduction in P-wave velocity of 25%. This is smaller than our picking error. Thus, the rationale for preferring absolute PKIKP traveltimes over differential PKPab-PKIKP traveltimes (despite the mentioned benefits of the differential traveltime measurements) is that

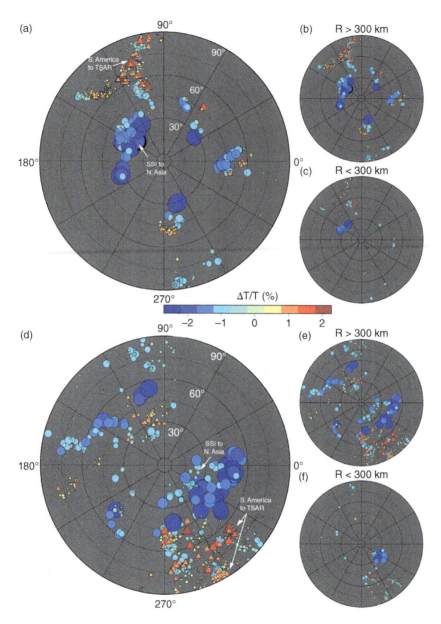

Figure 10.5 Normalized PKIKP traveltime residuals plotted in polar coordinates with respect to the direction and location of inner core sampling by PKIKP rays. (a) The Southern Hemisphere projection of the vector tangential to the PKIKP ray in the bottoming point. The longitude along the circle represents the vector azimuth, and the radius represents its dip (the angle ξ). The data due to two specific source-receiver geometries (the South Sandwich Islands events to the receivers in North Asia and the South American events to the TSAR array receivers) are shown by white arrows and text. The subsets of rays bottoming with (b) the radii <300 km and (c) >300 km. (d) The Southern Hemisphere projection of the location of the PKIKP rays at the inner core boundary. If the source has a lower latitude than the receiver, the entry point is shown, and if the source has a higher latitude than the receiver, the exit point is shown. The entry and exit points longitudes are shown along the circle, and the entry and exit points latitudes are shown along the radius. The same two source-receiver geometries as in (a) are highlighted with white arrows and text. The subsets of rays bottoming with (e) the radii <300 km and (f) >300 km. The original version of this figure in color is available in supplements; the B-W version is included here by the editorial rules.

Table 10.1 Summary of global tomographic mantle models and traveltime datasets, sensitivity, and methods used to construct them: DETOX-P3 (Hosseini et al., 2020), LLNL-G3Dv3 (Simmons et al., 2012), MIT-P08 (Li et al., 2008), and ANU-LM300Pv2 (Muir & Tkalčić, 2020)

Model	Seismic phase	#Records ($\approx$)	Epicentral distance	Frequency range	Method used to construct the model
DETOX-P3	P PP Pdiff	10.7 M	30–160°	33 mHz–1 Hz	Finite-frequency Fréchet kernels parametrized on an adaptive tetrahedral grid
LLNL-G3Dv3	P Pn	2.8 M	0–97°		Hierarchical tessellation framework that explicitly contains aspherical Earth structure
MIT-P08	P Pn Pg pP and pwP PKPab-PKIKP PKPab-PKPbc PKPdiff-PKIKP PP-P	15.1 M		40 mHz–1 Hz	Progressive Multilevel Tessellation Inversion (PMTI)
ANU-LM300Pv2	PcP-P PKPab-PKPbc	1,065	50–70° 145–155°	0.5–2 Hz	Hierarchical Hamiltonian Monte Carlo inversion (lowermost 300 km of the mantle only)

the upper mantle is better constrained than the lowermost mantle structure in the global tomographic mantle models.

The upper mantle is better constrained than the lowermost mantle in global tomographic mantle models because the seismic data used for constructing them have more sensitivity in the upper mantle (Table 10.1). Regardless, it is demanded in modern seismological practices to correct PKIKP residuals for the 3-D mantle structure, and choosing which mantle model to use can be challenging. We test several state-of-the-art global P-wave velocity mantle models (Hosseini et al., 2020; Li et al., 2008; Simmons et al., 2012), including our most recent model for the lowermost 300 km of the mantle (Muir & Tkalčić, 2020). The first two models in Table 10.1 were not constructed using PKP waves, although other body waves (e.g., Pdiff) with limited sensitivity to the lowermost mantle were used. Although Pdiff waves are sensitive to long-scale structures, they do not constrain short-scale heterogeneity that PKIKP waves at 1 Hz are sensitive to. The second two models contain PKP waves, although a subtle difference is that the ANU-LM300v2 does not contain PKIKP waves to avoid the IC effect on their traveltimes.

We first trace the PKIKP ray path, theoretically calculated using Earth reference model ak135 (Kennett et al., 1995), through selected mantle models and compute cumulative traveltime residuals due to the mantle 3-D heterogeneities. Then, we determine the correlation coefficient (CC) of the calculated mantle residuals with the observed PKIKP residuals (Fig. 10.6). In general, CC values closer to 1.0 (positive or negative) indicate a strong correlation between our observations and mantle structures, while CC values closer to 0 indicate no correlation. Figure 10.6 (upper row) reveals that the MIT-P08 model has a higher CC than the other models, which is not entirely surprising given that the PKIKP traveltimes contributed to its generation. The other three models show no correlation with the PKIKP residuals.

Additionally, we overlay normalized PKIKP residuals with the lowermost mantle models for LLNL-G3Dv3

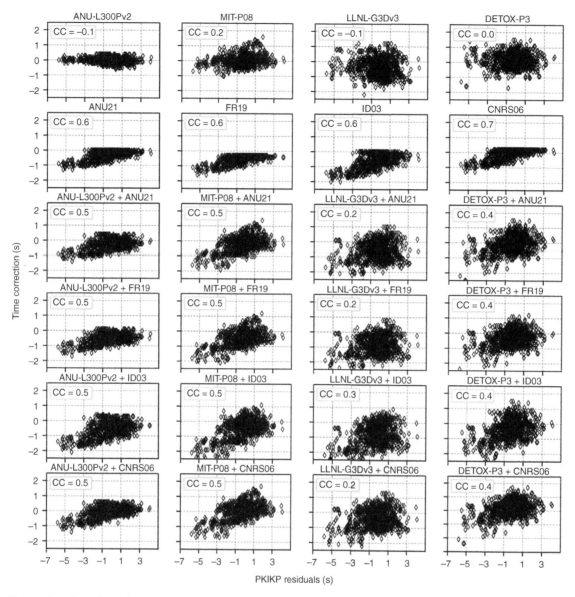

Figure 10.6 Correlation between PKIKP traveltime residuals and time corrections from global tomographic models of the whole and lowermost mantle compressional velocity perturbations and the outer inner core (OIC) anisotropy models. The traveltime residuals of PKIKP include the Earth's ellipticity corrections. The time corrections for mantle heterogeneity models ANU-LM300v2 of Muir and Tkalčić (2020), MIT-P08 of Li et al. (2008), LLNL-G3Dv3 of Simmons et al. (2012), and DETOX-P3 of Hosseini et al. (2020) are displayed in the top-row panels from left to the right, respectively. The time correlations for the OIC anisotropy models ANU21 of Stephenson et al. (2021), FR19 of Frost and Romanowicz (2019), ID03 of Ishii and Dziewónski (2003), and CNRS06 of Calvet et al. (2006) are shown in the second-row panels from left to right, respectively. Panels in the third to bottom rows display time correlations for the 16 combinations of mantle velocity heterogeneity and OIC anisotropy models. Labels on the top left side of each panel indicate the correlation coefficient (ranging from −1 to 1) defined as $CC = \left(\frac{\sum (obs - \overline{obs})(pred - \overline{pred})}{\sqrt{\sum (obs - \overline{obs})^2 \sum (pred - \overline{pred})^2}} \right)$, where $\overline{obs}$ and $\overline{pred}$ are the means of the Observations and predictions, respectively.

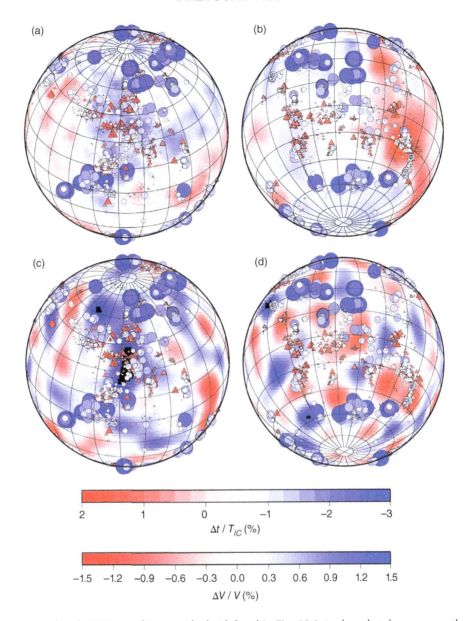

Figure 10.7 Normalized PKIKP traveltime residuals (defined in Fig. 10.2c), plotted at the core-mantle boundary entry and exit points of PKIKP rays and superimposed on the selected lowermost mantle tomographic models. Triangles and circles indicate slow and fast propagation of PKIKP waves, respectively. The traveltime residuals are color-coded and shown in varying sizes. (a) Model LLNL-G3Dv3 [Source: Adapted from Simmons et al. (2012)] seen from the north-eastern perspective. (b) Same as (a) but from the south-western perspective. (c) Model ANU-LM300Pv2 [Source: Adapted from Muir and Tkalčić (2020)] seen from the north-eastern perspective. (d) Same as (c) but from the south-western perspective. The original version of this figure in color is available in supplements; the B-W version is included here by the editorial rules.

and LM300Pv2 models (Fig. 10.7). The first model was constructed from P and Pn traveltimes, and the second model focuses explicitly on the lowermost mantle using PKP traveltimes. This exercise confirms a negligible correlation between the PKIKP residuals and the lowermost mantle structure that these tomographic models suggest. For example, let us consider the lowermost mantle under East Asia, the volume consistently characterized as fast among the majority (if not all) of the existing P-wave models of the lowermost mantle. It can be seen in Figure 10.7 that most of the positive PKIKP residuals (slow PKIKP waves) penetrate the core-mantle boundary under East Asia. Furthermore, some of the most negative PKIKP residuals (the fastest PKIKP waves) sample the lowermost mantle in the volumes that are not characterized as the fastest.

10.3.2. Outer Inner Core (OIC) Anisotropy Corrections

Next, we compute traveltime corrections for the PKIKP waves in the OIC based on selected anisotropy models available for OIC. The same analogy we used for the mantle could be applied to the IC. The main reason why anisotropy in OIC is better constrained than in the IMIC is that the seismic data used for modeling anisotropy are more abundant for the OIC. Some seismic phases, such as PKPbc and PKiKP, exist only in the epicentral distance range with shallow-sampling PKIKP waves. The differential traveltimes involving shallow-sampling PKIKP mitigate uncertainties further; thus, they are more suitable to characterize the outer path of the IC. Therefore, it is demanded in modern seismological practices to correct PKIKP residuals for the OIC anisotropy, and choosing which model to use can be challenging. Specifically, we use a mixture of pioneering and more recent OIC models for our corrections (Ishii & Dziewoński, 2003; Calvet et al., 2006; Frost & Romanowicz, 2019; Brett & Deuss, 2020; Stephenson et al., 2021). All models are shown in Table 10.2 and come with the basic information on the data used in their construction.

As before, we determine the CC between PKIKP traveltime corrections due to OIC anisotropy with the

Table 10.2 Parameters of inner core cylindrical anisotropy models proposed in different studies using body-waves: HU86 (Morelli et al., 1986), UU20 (Brett & Deuss, 2020), ID03 (Ishii & Dziewoński, 2003), CNRS06 (Calvet et al., 2006), UCB19 (Frost & Romanowicz, 2019), and ANU21 (Stephenson et al., 2021).

Bulk IC

Model	Parameters of anisotropy (%) Bulk IC			Types of data	Epicenter distances	#Measurements ($\approx$)
	ε	σ	γ			
HU86	3.20	−6.40		ISC PKIKP traveltimes modeled with spherical harmonic expansion	120°–135° 155°–180°	7,556
UU20	2.66	−3.62	−0.11	PKPab-PKIKP differential traveltimes with the increased presence of quasi-polar paths	165°–180°	?

IC containing IMIC

Model	Parameters of anisotropy (%)						Types of data	Epicenter distances	#Measurements ($\approx$)
	OIC			IMIC					
	ε_0	σ_0	γ_0	ε_1	σ_1	γ_1			
ID03; $R = 300$ km	1.80	−0.67		3.70	−19.70		ISC PKIKP absolute traveltime residuals (PREM)	173°–180°	3,000
CNRS06; $R = 430$ km	1.17	−1.75	−0.10	−0.17	−6.39	0.88	Hand-picked PKIKP absolute traveltime residuals (ak135)	150°–180°	300,000
UCB19; $R = 750$ km	1.20	−1.76	0.13	5.70	−14.40	0.53	Hand-picked PKIKP absolute and PKPab-PKIKP differential traveltime residuals (ak135)	141.8°–179.9°	3,806
ANU21; $R = 650$ km	1.30	−1.11		2.19	−7.72		ISC PKIKP absolute traveltime residuals (ak135)	150°–180°	183,257

Note: The values described in the table correspond to the coefficients for the bulk IC, defined as $\frac{\Delta t_{IC}}{T_{IC}} = \gamma + \varepsilon \cos^2 \xi + \sigma \sin^2 \xi \cos^2 \xi$, and for the IC containing IMIC, defined as $\frac{\Delta t_{IC}}{T_{IC}} = \gamma_0 + \gamma_1 + (\varepsilon_0 + \varepsilon_1)\cos^2 \xi + (\sigma_0 + \sigma_1) \sin^2 \xi \cos^2 \xi$, where ε_0, σ_0, and γ_0 are the parameters for the OIC and ε_1, σ_1, and γ_1 for the IMIC. The radius R to the inferred IMIC transition is indicated in the first column of each row.

PKIKP residuals (Fig. 10.6; second row). As the second row of Figure 10.6 shows, CC values are all 0.6–0.7, indicating a correlation between our observations and the proposed OIC anisotropy models. We anticipated these CC values to be much higher than the CC values from the mantle models, given that they were derived from the PKIKP absolute or PKP differential traveltimes involving PKIKP. However, since our data are antipodal, the cumulative traveltimes of PKIKP waves in the IMIC are relatively significant. If anisotropy in the IMIC is distinctively different from anisotropy in the OIC, it will drive the CC values in the opposite direction.

10.3.3. Interpreting PKIKP Residuals with Anisotropy

As can be seen from the PKIKP residuals plotted as a function of the ray direction in Figures 10.5a–c, there is a relatively large coherence regarding the ray direction. The most negative residuals corresponding to fast PKIKP paths coincide with polar sampling paths. We achieved an excellent azimuthal coverage for the polar paths, and there is no visible azimuthal variation. This means that azimuthal anisotropy remains the simple possible explanation for the observed traveltime variations in the IC. The azimuthal coverage of the most equatorial paths is not so good, and this emphasizes the importance of deploying more permanent stations and temporary arrays such as TSAR around the world. In summary, the oblique and equatorial data seem to exhibit more azimuthal variation, with positive and mildly negative residuals mixed. This agrees well with the pattern of PKIKP residuals also observed in Figure 10.3b.

As expected, there is slightly less coherency observed for the PKIKP residuals when plotted as a function of the ICB piercing point location (Figs. 10.5d–f). Although the data noise might be responsible for a slight departure from the azimuthal pattern, another explanation for some areas with mixed negative and positive residuals could be the presence of short-scale heterogeneities near the ICB (e.g., Yee et al., 2014). Overall, however, the observed pattern of PKIKP residuals is still indicative of anisotropy as the overarching cause affecting the PKIKP traveltimes. With this in mind, we approach modeling the PKIKP residuals using an IC cylindrical anisotropy model.

10.3.4. Bayesian Approach to the PKIKP Residuals Modeling

Instead of fitting the data using least squares, we approach this problem in a probabilistic (Bayesian) framework that also accounts for noise in the data. The noise magnitude is unknown and sampled alongside three anisotropic model coefficients in a Markov Chain Monte Carlo. We set up a parameter estimation to fit a cylindrical anisotropic model to the observed PKIKP residuals. We experiment with two cases: (1) cylindrical anisotropy in the bulk IC characterized by the fast axis aligned with the ERA, and (2) the IC divided into cylindrically anisotropic OIC and IMIC, with fast axes aligned with the ERA. Some previously obtained values for these anisotropic parameters are given in Table 10.2. For more details on the Bayesian inference and Monte Carlo methods, see, for instance, Malinverno and Briggs (2004), Aster et al. (2013), and Sambridge et al. (2013). More specific details on the hierarchical aspect of the inversion are given in recent geophysical applications, from tomography to the seismic source inversion problems (e.g., Bodin et al., 2012; Mustać & Tkalčić, 2016; Muir & Tkalčić, 2020).

The data, $d = \{d_1, d_2, \ldots, d_N\}$, are the residuals of absolute PKIKP measurements with respect to the ak135 Earth reference model (Kennett et al., 1995), corrected for Earth's ellipticity (Kennett & Gudmundsson, 1996) and $N = 1,150$ is the number of data points. The corresponding angles with respect to the ERA are given by $\xi = \{\xi_1, \xi_2, \ldots, \xi_N\}$. The forward problem predicts PKIKP residuals as a function of angle ξ corresponding to an observation given a cylindrical anisotropic model, consisting of $\varepsilon, \sigma, \gamma$ coefficients:

$$\frac{\Delta v}{v} = \gamma + \varepsilon \cos^2 \xi + \sigma \sin^2 \xi \cos^2 \xi. \quad (10.1)$$

The relative traveltime residual is negative of the velocity residual:

$$\frac{\Delta t}{T} = -\frac{\Delta v}{v}. \quad (10.2)$$

In the case of bulk IC models, the prediction is defined as follows:

$$p_i = -T_i^{IC}(\gamma + \varepsilon \cos^2 \xi_i + \sigma \sin^2 \xi_i \cos^2 \xi_i) + \Delta T_i^M. \quad (10.3a)$$

T_i^{IC}, which is the time that each ray path spends in the entire inner core, is precalculated using the station-receiver configuration of the corresponding observation in ak135. ΔT_i^M is the amount of mantle correction precalculated by tracing through 3-D P-wave mantle models (see Table 10.1 for details of the mantle models).

In the case of IC models divided in IMIC and OIC, the prediction contains two components and is defined as follows:

$$p_i = -T_i^{IMIC}(\gamma_1 + \varepsilon_1 \cos^2 \xi_i + \sigma_1 \sin^2 \xi_i \cos^2 \xi_i) \\ - T_i^{OIC}(\gamma_0 + \varepsilon_0 \cos^2 \xi_i + \sigma_0 \sin^2 \xi_i \cos^2 \xi_i) + \Delta T_i^M. \quad (10.3b)$$

where, T_i^{IMIC} and T_i^{OIC} are theoretically calculated times the ray spending in the 600 km radius of IMIC and the remaining OIC (n.b. the IMIC radius approximately a halfway through the IC is based on more recent studies with a rigorous treatment of uncertainty, e.g., Stephenson et al., 2021). Here, we invert for the coefficients, $\gamma_1, \varepsilon_1, \sigma_1$, of the IMIC as unknowns while testing the OIC coefficients, $\gamma_0, \varepsilon_0, \sigma_0$, from the five exiting models (Ishii & Dziewoński, 2003; Calvet et al., 2006; Frost & Romanowicz, 2019; Brett & Deuss, 2020; Stephenson et al., 2021). Similar to the case of bulk IC, ΔT_i^M is the mantle correction.

The likelihood function showing how likely the predictions explain the observed data is defined as follows:

$$L = -\frac{1}{\sqrt{(2\pi\delta)^N}} \times \exp\left[-\frac{1}{2\delta^2}\sum_{i=1}^N (p_i - d_i)^2\right]. \quad (10.4)$$

A hyperparameter δ represents the uncertainty in the observed data, which includes but is not limited to the picking error, earthquake origin uncertainty, and theoretical errors associated with the ray theory approximation.

We start with the first model: the bulk IC anisotropy. In Figure 10.8a, we plot the PKIKP residuals before and after the traveltime corrections for the mantle model MIT-P08 (Li et al., 2008), as a function of the angle ξ. We use a cylindrical anisotropy model for the bulk IC [equation (10.1)] with the symmetry axis aligned with the ERA and a hierarchical Bayesian method to fit the PKIKP residuals. Apart from the three model coefficients, ε, σ, and γ, the noise magnitude, δ, is obtained as a hyperparameter in the inversion. The modeled data means are plotted by the black circles. The estimated CC between the data (PKIKP residuals) and the predictions corresponding to the post-burning thinned chain models is 0.629 ± 0.004. The resulting anisotropic coefficients and the data noise are shown in the bottom left of Figure 10.8c.

In Figure 10.8b, we plot the normalized residuals (also corrected for mantle heterogeneity) to remove the dependency of the PKIKP ray lengths in the IC. We compare the ensemble of anisotropic models obtained from the post-burning thinned Markov chain (black lines) with two previous models (Morelli et al., 1986; Brett & Deuss, 2020) [see Table 10.2 for the anisotropic coefficients of the models; coefficients of the model by Brett and Deuss (2020) were obtained via personal communication with H. Brett]. In this study, we refer to the bulk IC model constrained by data sampling the innermost 690 km of the IC, that is, in their Figure 13c, which are comparable to our dataset. Although we obtain a bit stronger anisotropy (4.2%; calculated as the difference between the minimum and maximum values of $\frac{\Delta t_{IC}}{T_{IC}}$ in Fig. 10.8b), overall, our anisotropy models are similar to the model of Morelli et al. (1986). Our models are slightly less similar to the model of Brett and Deuss (2020), particularly in the coefficient σ, our values being more negative. See the section 10.4 (Discussion) for more detail on why this might be the case.

Next, we proceed with the second model setup: the IC containing the IMIC, whose modeling results are shown in Figure 10.9 (see details in section 10.3.2). The main difference between this and the previous model is that now the OIC anisotropic parameters are fixed to the values from previously published models, and our search extends for the IMIC anisotropic parameters (i.e., $\gamma_1, \varepsilon_1, \sigma_1$ as shown in Fig. 10.9c). Given that our PKIKP residuals are antipodal, with a good volumetric sampling of the IMIC, we expect robust results for this model. In Figure 10.9a, we plot the PKIKP residuals before and after the traveltime corrections for the mantle model MIT-P08 (Li et al., 2008), as a function of the angle ξ. We also plot the residuals corrected for the OIC anisotropy model of Stephenson et al. (2021). We use a cylindrical anisotropy model for the IMIC and OIC [equation 10.3b] with the symmetry axes aligned with the ERA and a hierarchical Bayesian method to fit the PKIKP residuals [equation (10.4)]. Apart from the three model coefficients, ε_1, σ_1, and γ_1, the noise magnitude, δ, is obtained as a hyperparameter in the inversion. The data noise obtained through the Bayesian inference is around 1.6 s, which could be understood as the cumulative uncertainty not only in picking practice but also in cataloged event location and time. The modeled residual means are plotted by the black circles. The obtained CC is 0.609 ± 0.004, whereas the obtained values for the anisotropic coefficients and data noise are shown in the bottom left.

Figure 10.9b is equivalent to Figure 10.8b, but instead of the bulk IC models, we now show the resulting models in this study (gray lines) consisting of the IC divided in the IMIC with the radius R = 600 km and the rest of it, the OIC. Again, the dependency of the residuals on the PKIKP ray lengths in the IC is removed by considering the normalized residuals. Our models in the thinned post-burning Markov chains are shown by gray opaque lines, and the four other models (see Table 10.2) are shown in colors [Ishii and Dziewoński, 2003 (red); Calvet et al., 2006 (purple); Frost and Romanowicz, 2019 (blue); Stephenson et al., 2021 (green)]. Overall, we obtain a stronger anisotropy (4.8%; calculated as the difference between the minimum and maximum values of $\frac{\Delta t_{IC}}{T_{IC}}$ in Fig. 10.9b) than other models, except for the model of Frost and Romanowicz (2019), which is similar to our model. Our value for σ_1 is -15.9 ± 1.2, which is similar to the values obtained by Frost and Romanowicz (2019) and Ishii and Dziewoński (2003) (see Table 10.2), and

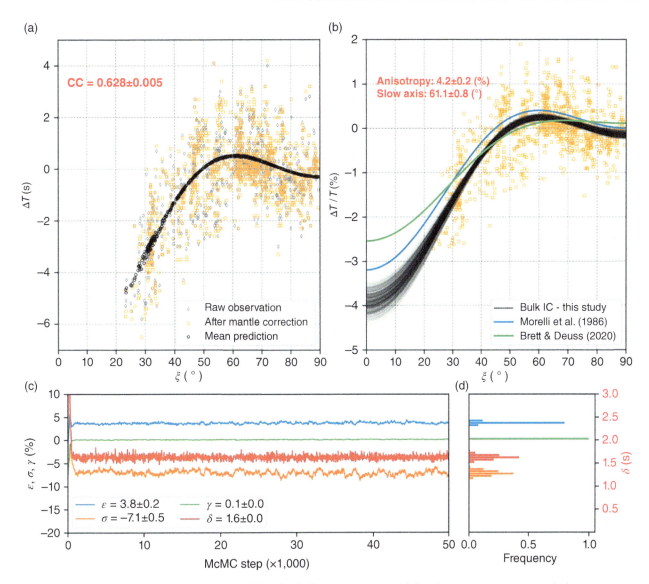

Figure 10.8 Cylindrical anisotropy model for the bulk inner core (IC), defined in equation 10.3a, with the symmetry axis along the Earth's rotation axis (ERA). The fit is obtained using a hierarchical Bayesian method with the coefficients ε, σ, and γ and noise magnitude, δ as free parameters. (a) Comparison of the observed and modeled absolute PKIKP traveltime residuals as a function of the angle between the PKIKP ray in the IC and ERA, ξ. The gray diamonds show the observed traveltime residuals corrected for Earth's ellipticity. The orange squares show the residuals after mantle heterogeneity correction using the MIT-P08 model. Source: Adapted from Li et al. (2008). The modeled residuals corresponding to the post-burning coefficients' means [see panel (c)] are shown by the black circles. The correlation coefficient (CC) between the observed and predicted residuals (see Fig. 10.6 for the definition of CC) is shown in the upper left corner. (b) Normalized PKIKP residuals, obtained by dividing the absolute residuals corrected for mantle heterogeneity from panel (a) by the IC traveltime, expressed in percentages (blue inverted triangles). The opaque black lines show bulk IC cylindrical anisotropic models in the thinned post-burning Markov Chains [see panel (c)]. Selected examples of published bulk IC models (Morelli et al., 1986; Brett & Deuss, 2020) are shown by the blue and green lines, respectively, for comparison. (c) Evolution of anisotropic model parameters ε, σ, and γ and the hyperparameter noise-amplitude, δ, in the sampling Markov Chain. The legend shows the estimated mean coefficients and one standard deviation. (d) Posterior probabilistic distribution of the four model parameters from the thinned (every twentieth sample) post-burning Markov Chain after 20,000 McMC steps. The original version of this figure in color is available in supplements; the B-W version is included here by the editorial rules.

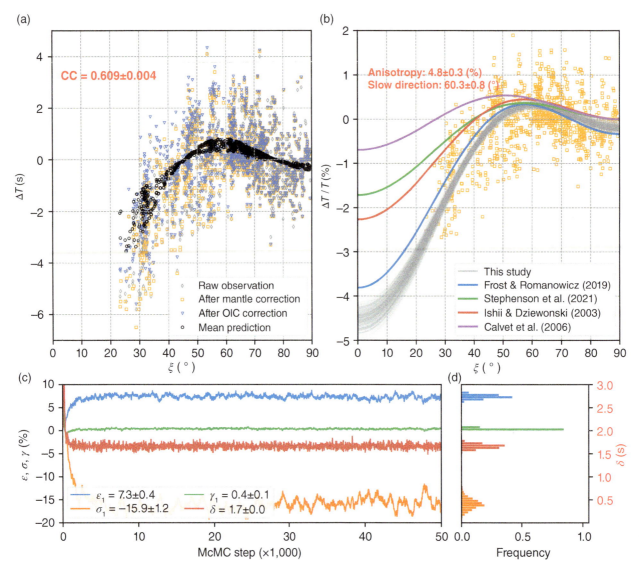

Figure 10.9 Cylindrical anisotropic model for the inner core (IC) containing the innermost inner core (IMIC), defined in equation (10.3b), for the symmetry axes aligned with the Earth's rotation axis (ERA). The fit is obtained using a hierarchical Bayesian method where ε_0, σ_0, and γ_0 are the fixed parameters for OIC determined from Brett and Deuss (2020) (see Table 10.3), and ε_1, σ_1, and γ_1 are free parameters for IMIC. (a) Comparison of the observed and predicted absolute PKIKP residuals as a function of the angle between the PKIKP ray in the IC and ERA, ξ. The gray diamonds show the observed traveltime residuals corrected for Earth's ellipticity. The orange squares show the residuals after further correction for mantle heterogeneity using the model MIT-P08 (Li et al., 2008). The residuals after further correction for the OIC using the model by Stephenson et al. (2021) are shown by the blue inverted triangles. The modeled residuals corresponding to the mean coefficients of post-burning anisotropic models [see panel (c)] are shown by the black circles. The correlation coefficient (CC) between the observed and predicted residuals is shown in the upper left corner. (b) Normalized PKIKP residuals, obtained by dividing the residuals corrected for mantle heterogeneity from panel (a) by the IC traveltime, expressed in percentages (orange squares). Opaque gray lines show the anisotropic models of the IC containing the IMIC in the thinned post-burning Markov Chains [see panel (c)]. Selected published IC models containing the IMIC from Ishii and Dziewoński (2003) (red); Calvet et al. (2006) (purple); Frost and Romanowicz (2019) (blue); Stephenson et al. (2021) (green) are shown for reference. (c) Evolution of anisotropic model parameters ε_1, σ_1, and γ_1 and the hyperparameter for noise amplitude, δ, in the sampling Markov Chain. Posterior probabilistic distribution of the four model parameters from the thinned (every twentieth sample) post-burning Markov Chain after 20,000 McMC steps. The original version of this figure in color is available in supplements; the B-W version is included here by the editorial rules.

explains the high peak-to-peak range of the curves. See the section 10.4 (Discussion) for more detail on this result.

10.4. DISCUSSION

IC anisotropy has been one of the most significant discoveries in studies of the Earth's deep interior of the twentieth century. Yet, it has remained controversial in terms of its magnitude and general form in body-wave studies until the present day, perhaps for the most obvious reason that the IC is buried deep beneath all other layers that cumulatively affect our surface observations. It is how well we can handle the upper Earth layers, for instance, in modeling seismic body-wave traveltimes, that determines the robustness of our IC models. Another reason is that different authors work with different datasets that are either measured by themselves or taken from one of the existing catalogs. This study is not different or better in that sense – it introduces a new dataset that must be scrutinized through future work. It is a carefully collected dataset using multiple constraints, but it is not unified yet with similar data. Another weakness of our study is the explicit assumption of cylindrical IC anisotropy models (i.e., bulk IC and the IC divided into OIC and IMIC). However, we make an effort to take a comprehensive approach of examining the influences of the Earth's upper layers (i.e., the mantle and the OIC) and consider the data noise in the process of fitting our new dataset, as well as the resulting uncertainty.

Given that seismological results often drive geodynamical and mineral physics interpretations, we feel that it is appropriate to focus this chapter's discussion on the modeling itself: in particular, what drives the anisotropy strength and determines the shape of the fitting curve, for example, the position of slow versus fast axes of anisotropy. What are the most robust features of the cylindrical anisotropy models, regardless of the traveltime corrections applied for the upper layers? We first discuss P-wave velocity mantle models and their influence on the result. Subsequently, we discuss the bulk IC models versus the models with the IC divided into the OIC and IMIC parts.

Table 10.3 is the summary of the two classes of models considered in this chapter under different assumptions: (1) cylindrical anisotropy in the bulk IC and (2) cylindrical anisotropy in the IC divided into the OIC and IMIC. The table contains our assumptions for the mantle (various P-wave velocity models) and the OIC (P-wave cylindrical anisotropy models). It contains the obtained anisotropic coefficients, data noise, cross-correlation, as well as the anisotropy strength and the position of the slow direction of anisotropy (the fast axis in all models is aligned with the ERA). More details of the mantle and OIC models are given in Tables 10.1 and 10.2.

We discussed the procedure for mantle corrections in section 10.3.1. We see no positive correlation between the PKIKP residuals and mantle structures (the first row of Fig. 10.6). The MIT-P08 model structure is a bit more correlated than other models, and this might be due to the fact that the MIT-P08 model was constructed using the PKP traveltime data. However, when we inspect the lowermost mantle layer of the same model (Fig. 10.7), we find that the structures inferred through tomography are not particularly compatible with our PKIKP residuals. However, when the MIT-P08 model is combined with any of the OIC models, it performs better than other mantle models, with the exception of the ANU-LM300v2 model that specifically targets the lowermost mantle (Fig. 10.6). We take a step further here and derive the IC bulk anisotropy models after correcting the residuals for all considered mantle models. Moreover, we correct the PKIKP residuals for all combinations of mantle models and OIC models under consideration and derive the OIC+IMIC anisotropy models. The second half of Table 10.3 has 20 rows corresponding to these combinations.

Table 10.3 reveals that the mantle corrections corresponding to different models influence the inversion results. In particular, the ANU-LM300v2 and MIT-P08 models yield higher CCs and stronger anisotropic strength (the anisotropic parameter ε is more positive) for both classes of IC anisotropy models, especially the latter. For the second class of IC anisotropy models, this holds regardless of which OIC anisotropy model is used for correction. Furthermore, the slow direction of anisotropy is more pronounced (the anisotropic parameter σ is more negative) for the inversions in which the ANU-LM300v2 and MIT-P08 models were used to correct the PKIKP residuals for mantle structure. For example, for the bulk IC anisotropy model, σ is significantly reduced to the value of -5.1 when the LLNL-G3Dv3 model is used to correct the PKIKP residuals. This reduces the peak-to-peak range of the fitting curve to the point that the slow direction is not more obviously positioned at the oblique than at the equatorial angles any longer. From here, we conclude that the consideration of tomographic models has a moderate influence on the IC anisotropy results and can drive the interpretation. Therefore, a comprehensive analysis of the mantle tomographic models is fully justified.

As far as the bulk IC anisotropy models go (Fig. 10.8 and top rows of Table 10.3), it is essential to remember that our residuals are antipodal, and more robust results should be expected for the bulk IC model when our dataset is combined with nonantipodal PKIKP residuals. While this is beyond the scope of this study, we can still examine what controls the strength of anisotropy and the position of its slow direction apart from the choice

Table 10.3 Resulting parameters of IC cylindrical anisotropy for the first (bulk IC) and second model (IC Divided into OIC and IMIC of R = 600 km). The first two columns show various combinations of mantle heterogeneity and OIC anisotropy models (see Tables 10.1 and 10.2 for the references) used for correcting PKIKP residuals. We also show the results with and without the TSAR data for the bulk IC model. The resulting anisotropic coefficients for the bulk IC (the upper part of the table), where the model is defined in Equation (10.3a), are ε, σ, and γ. The resulting anisotropic coefficients for the model of IC containing IMIC (the lower part of the table), defined in Equation (10.3b), are ε_0, σ_0, and γ_0 for the OIC and ε_1, σ_1, and γ_1 for the IMIC. The parameter δ is data noise expressed in seconds, inferred in the Bayesian inversion as the hyperparameter. Also shown are the correlation coefficient (defined as in Fig. 10.6), the strength of anisotropy (defined in the main text), and the position of the slow direction of anisotropy.

Bulk IC

Mantle model corrected	OIC model corrected	TSAR data included	Parameters of anisotropy (%) Bulk IC			δ (s)	CC	Slow direct. position (°)	Anisotropy strength (%)
			ε	σ	γ				
DETOX-P3	n.a.	yes	3.4	−6.0	0.0	1.7	0.586	62.4 ± 1.0	3.7 ± 0.2
		no	3.4	−6.1	0.0	1.8	0.604	62.3 ± 1.4	3.7 ± 0.2
LLNL-G3Dv3	n.a.	yes	3.0	−5.2	0.3	1.8	0.557	62.4 ± 1.3	3.2 ± 0.2
		no	2.9	−5.1	0.3	1.8	0.579	62.4 ± 1.5	3.2 ± 0.2
MIT-P08	n.a.	yes	3.8	−7.1	0.1	1.6	0.628	61.1 ± 0.8	4.2 ± 0.2
		no	3.8	−7.1	0.2	1.7	0.645	61.1 ± 0.8	4.2 ± 0.2
ANU-LM300Pv2	n.a.	yes	3.5	−6.6	0.2	1.5	0.678	61.3 ± 0.8	3.9 ± 0.2
		no	3.5	−6.5	0.2	1.6	0.693	61.2 ± 0.8	3.8 ± 0.2

IC containing IMIC of R = 600 km

Mantle model corrected	OIC model corrected	TSAR data included	Parameters of anisotropy (%) OIC			IMIC			δ (s)	CC	Slow direct. position (°)	Anisotropy strength (%)
			ε_0	σ_0	γ_0	ε_1	σ_1	γ_1				
DETOX-P3	ID03	yes	1.8	−0.67		5.4	−12.9	0.0	1.7	0.569	61.3 ± 1.2	4.0 ± 0.3
LLNL-G3D	ID03	yes	1.8	−0.67		4.0	−10.5	0.6	1.8	0.539	61.0 ± 1.5	3.2 ± 0.3
MIT-P08	ID03	yes	1.8	−0.67		6.6	−16.5	0.4	1.7	0.612	59.6 ± 0.8	4.7 ± 0.3
ANU-LM300	ID03	yes	1.8	−0.67		6.1	−15.6	0.4	1.5	0.670	59.6 ± 0.7	4.5 ± 0.2
DETOX-P3	CNRS06	yes	1.17	−1.75	−0.1	6.6	−12.1	0.0	1.8	0.564	62.3 ± 1.4	4.2 ± 0.3
LLNL-G3D	CNRS06	yes	1.17	−1.75	−0.1	5.3	−9.9	0.7	1.8	0.529	62.0 ± 1.5	3.5 ± 0.3
MIT-P08	CNRS06	yes	1.17	−1.75	−0.1	7.7	−15.6	0.4	1.7	0.608	60.5 ± 0.8	5.0 ± 0.2
ANU-LM300	CNRS06	yes	1.17	−1.75	−0.1	6.9	−13.8	0.4	1.6	0.661	60.8 ± 0.9	4.5 ± 0.3
DETOX-P3	UCB19	yes	1.20	−1.76	0.13	6.5	−11.9	−0.2	1.8	0.564	62.3 ± 1.1	4.2 ± 0.3
LLNL-G3D	UCB19	yes	1.20	−1.76	0.13	5.1	−9.4	0.5	1.8	0.530	62.5 ± 1.7	3.4 ± 0.3
MIT-P08	UCB19	yes	1.20	−1.76	0.13	7.5	−14.8	0.2	1.7	0.607	60.9 ± 0.9	4.8 ± 0.3
ANU-LM300	UCB19	yes	1.20	−1.76	0.13	6.9	−13.7	0.2	1.6	0.663	60.8 ± 0.8	4.5 ± 0.2
DETOX-P3	ANU21	yes	1.30	−1.11		6.5	−13.0	0.0	1.7	0.567	61.7 ± 1.2	4.2 ± 0.3
LLNL-G3D	ANU21	yes	1.30	−1.11		6.4	−13.0	0.0	1.7	0.567	61.7 ± 1.2	4.2 ± 0.3
MIT-P08	ANU21	yes	1.30	−1.11		7.3	−15.9	0.4	1.7	0.609	60.3 ± 0.8	4.8 ± 0.3
ANU-LM300	ANU21	yes	1.30	−1.11		6.7	−14.7	0.4	1.6	0.666	60.4 ± 0.8	4.5 ± 0.3

of the mantle model discussed above. The first-order observation from Figure 10.8 and many similar figures in the literature is that the more negative the PKIKP residuals get at the polar angles ξ, the larger the anisotropy strength. The second-order observation is that a more prominent curve maximum at oblique angles (more negative coefficient σ) also influences the coefficient ε, and, in turn, the anisotropic strength.

The cylindrical anisotropy model curve maximum is highly relevant for physical interpretations because it signifies the orientation of the slow direction of anisotropy. Mathematical reasons for the increased curve eccentricity are clear. This can be due to the quasi-polar path residuals flattening the curve downward from the polar angles side. Table 10.2 indeed reveals that the parameter σ is almost twice smaller in the model UU20 of Brett and Deuss (2020) than in the model HU86 of Morelli et al. (1986). On the other hand, the quasi-equatorial path residuals can "pull" the curve down on its equatorial angles side. For example, the curve slope introduced by the TSAR residuals (see the square symbols in Fig. 10.3c) is negative and quite significant, ~2% over the angle ξ range of ~15°. However, comparing the bulk IC models with and without the TSAR data in Table 10.3 reveals that this effect on the coefficient σ is mild. In our case for the bulk IC anisotropy, we obtain the slow direction at $61.1 \pm 0.8°$. The inferred orientation of the slow direction changes very slightly in our inversions, depending on the mantle model used for the correction. The obtained range of slow orientation is $61.1–62.4 \pm 0.8°$.

It is interesting to note that the CC does not improve when IMIC is introduced in the IC anisotropy model. The CC values in Table 10.3 do not vary significantly compared to the IC bulk anisotropy model. As discussed earlier, the highest CC values are obtained when the MIT-P08 and ANU-LM300v2 models are used for mantle corrections, and this also coincides with higher strength of anisotropy and more oblique angles for the slow direction of anisotropy. The data noise (the parameter δ) is consistently between 1.5 and 1.8 seconds, slightly more than we would expect from the measurement errors. As discussed earlier, the higher values of δ could be due to various assumptions about the physics of our problem and other imperfections, such as the uncertainty introduced by the event catalog, the reference model, the unknown Earth structure, including both heterogeneity and anisotropy.

Finally, under the assumptions made in this study, and considering all discussed factors, we can make two well-founded conclusions about the orientation of the fast and slow axes of cylindrical anisotropy: (1) the fast axis of anisotropy is aligned with the ERA and (2) the slow direction of anisotropy is in the range $59.6–62.5 \pm 0.8°$ for the bulk IC anisotropy models and $61.1–62.4 \pm 0.8°$ for the models that include IMIC and OIC as two distinctive anisotropic volumes. Some studies based on cross-correlation data, albeit with inaccurate assumptions about the sensitivity of the observed correlation features, led to the inference that the fast axis of IC anisotropy is in the equatorial plane. Clearly, our PKIKP residuals are in sharp contrast with those results. To illustrate the difference between a model obtained in a study of Wang et al. (2015) and Wang and Song (2018) and the model obtained in this study, we plot the anisotropic models projected to maps and superimpose the PKIKP residuals onto them (Fig. 10.10). The most remarkable feature in this coda-correlation derived model is the strong variation of anisotropic strength as a function of latitudes. For example, it predicts that rays sampling in Central America is significantly faster than ray sampling under the Brazilian east coast. Thanks to the inclusion of data from RSBR stations, our antipodal PKIKP residual dataset negates the presence of the alleged contrast.

10.5. CONCLUDING REMARKS

We collect a new dataset of 1,150 absolute PKIKP traveltime residuals. The most interesting features of our dataset include the highly selective set of criteria in picking PKIKP onsets and the measurements of the antipodal TSAR array and RSBR waveforms. The PKIKP residuals are then scrutinized in conjunction with their volumetric sampling of the inner core. From the azimuthal distribution of PKIKP residuals, we find a justification in the choice of cylindrical anisotropy as the simplest physical model. We perform a comprehensive analysis of the influences of mantle heterogeneity and the OIC anisotropy models on the PKIKP residuals.

Furthermore, we perform probabilistic hierarchical inversions for anisotropic parameters and data noise, treating the noise as a free parameter in the inversion. We then experiment with two classes of IC anisotropy: bulk IC models and the models in which IC is divided into the IMIC and the OIC. We find the anisotropic strength in the range $3.2–4.2 \pm 0.2\%$ for the bulk IC and $3.2–5.0 \pm 0.2\%$ for the anisotropic IC divided into IMIC and OIC, depending on the models used to correct traveltimes in the Earth's upper layers including the isotropic uppermost IC.

Last but not least, we find strong evidence for the slow direction of anisotropy ~60°, and its fast axis aligned with the ERA, not in the equatorial direction as some published models suggested. The two classes of IC anisotropy models (bulk IC models vs. the IMIC + OIC models) fit the data utilized in this study equally well. However, the stronger anisotropy we infer for the antipodal dataset than previous studies focusing on the anisotropy in the upper IC suggests either a higher degree

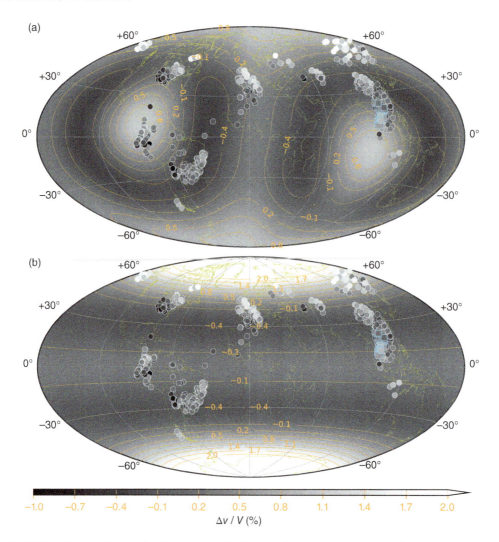

Figure 10.10 The observed normalized velocity residuals (negative of relative time residuals) plotted as a function of ray directions and superimposed over the two significantly different 2-D anisotropic models: (a) Wang et al. (2015), which has the fast axis of the IMIC pointing toward Central America, and (b) the model derived in this study, which has the fast axis of anisotropy aligned with the Earth's rotation axis. Geographical coordinates of data points in (a) and (b) are coordinates of the PKIKP ray direction vectors, pointing from the ICB entry to the ICB exit points. Data points corresponding to the TSAR network are highlighted in squares with cyan edges. The same color scheme saturation and bounds are used for both models to emphasize the anisotropic velocity-perturbation contrasts. The original version of this figure in color is available in supplements; the B-W version is included here by the editorial rules.

of crystal alignment or a distinct iron phase-stabilized in the IMIC. The temporary seismic arrays such as TSAR in Thailand and national arrays such as RSBR in Brazil play an essential role in advancing the Earth's deep interior studies.

ACKNOWLEDGMENTS

We are grateful to the Editor George Helffrich for his helpful suggestions and two anonymous reviewers whose comments improved the quality of the original manuscript. IRIS Data Services are funded through the Seismological Facilities for the Advancement of Geoscience (SAGE) Award of the National Science Foundation under Cooperative Support Agreement EAR-1851048. H. Tkalčić is grateful to the Australian Research Council for supporting seismological research on deep Earth under previous and current grants and AuScope, Inc. for continuing support. H. Tkalčić would also like to acknowledge graduate students, postdoctoral fellows, RSES faculty, and the ANU Media team for their interest in research on the IC subject. TSAR was supported by MEXT/JSPS KAKENHI 15H05832 and ERI JURP 2016-F2-07, 2017-F2-07, and 2018-F2-07.

AVAILABILITY STATEMENT

Most of the global waveform data used in this chapter are available through the Incorporated Research Institutions for Seismology (https://www.iris.edu/hq/) through the Data Management Center. Absolute PKIKP measurements are available upon request to the corresponding author.

REFERENCES

Aster, R. C., Borchers, B., & Thurber, C. H. (2013). Chapter Eleven – Bayesian Methods. *Parameter Estimation and Inverse Problems (Second Edition)*, Academic Press, 253–280. ISBN 9780123850485. https://doi.org/10.1016/B978-0-12-385048-5.00011-2

Babuska, V., & Cara, M. (1991). Seismic anisotropy in the Earth. *Modern Approaches in Geophysics Book Series*, 10, 1–219. https://doi.org/10.1007/978-94-011-3600-6

Beghein, C., & Trampert, J. (2003). Robust normal mode constraints on inner-core anisotropy from model space search. *Science*, 299(5606), 552–555. https://doi.org/10.1126/science.1078159

Belonoshko, A. B., Skorodumova, N. V., Rosengren, A., & Johansson, B. (2008). Elastic anisotropy of earth's inner core. *Science*, 319(5864), 797–800. https://doi.org/10.1126/science.1150302

Bergman, M. I. (1997). Measurements of elastic anisotropy due to solidification texturing and the implications for the Earth's inner core. *Nature*, 389(6646), 60–63. https://doi.org/10.1038/38786

Bodin, T., Sambridge, M., Rawlinson, N., & Arroucau, P. (2012). Transdimensional tomography with unknown data noise. *Geophysical Journal International*, 189(3), 1536–1556. https://doi.org/10.1111/j.1365-246X.2012.05414.x

Bondár, I., & Storchak, D. (2011). Improved location procedures at the International Seismological Centre: Improved location procedures at the ISC. *Geophysical Journal International*, 186(3), 1220–1244. https://doi.org/10.1111/j.1365-246X.2011.05107.x

Bréger, L., Tkalčić, H., & Romanowicz, B. (2000). The effect of D″ on PKP(AB–DF) travel time residuals and possible implications for inner core structure. *Earth and Planetary Science Letters*, 175(1), 133–143. https://doi.org/10.1016/S0012-821X(99)00286-1

Brett, H., & Deuss, A. (2020). Inner core anisotropy measured using new ultra-polar PKIKP paths. *Geophysical Journal International*, 223(2), 1230–1246. https://doi.org/10.1093/gji/ggaa348

Calvet, M., Chevrot, S., & Souriau, A. (2006). P-wave propagation in transversely isotropic media: II. Application to inner core anisotropy: Effects of data averaging, parametrization and a priori information. *Physics of the Earth and Planetary Interiors*, 156(1), 21–40. https://doi.org/10.1016/j.pepi.2006.01.008

Calvet, M., & Margarin, L. (2018). Shaped preferred orientation of iron grains compatible with Earth's uppermost inner core hemisphericity. *Earth and Planetary Science Letters*, 481, 395–403. https://doi.org/10.1016/j.epsl.2017.10.038

Cao, A., & Romanowicz, B. (2007). Test of the innermost inner core models using broadband PKIKP travel time residuals. *Geophysical Research Letters*, 34(8), L08303. https://doi.org/10.1029/2007GL029384

Cormier, V. F., & Stroujkova, A. (2005). Waveform search for the innermost inner core. *Earth and Planetary Science Letters*, 236(1–2), 96–105. https://doi.org/10.1016/j.epsl.2005.05.016

Crampin, S. (1984). Anisotropy in exploration seismics. *First Break*, 2(3), 19–21. https://doi.org/10.3997/1365-2397.1984006

Creager, K. C. (1992). Anisotropy of the inner core from differential travel times of the phases PKP and PKIKP. *Nature*, 356(6367), 309–314. https://doi.org/10.1038/356309a0

Deuss, A. (2014). Heterogeneity and anisotropy of Earth's inner core. *Annual Review of Earth and Planetary Sciences*, 42, 103–126. https://doi.org/10.1146/annurev-earth-060313-054658

Frost, D. A., & Romanowicz, B. (2019). On the orientation of the fast and slow directions of anisotropy in the deep inner core. *Physics of the Earth and Planetary Interiors*, 286, 101–110. https://doi.org/10.1016/j.pepi.2018.11.006

Frost, D. A., Lasbleis, M., Chandler, B., & Romanowicz, B. (2021). Dynamic history of the inner core constrained by seismic anisotropy. *Nature Geoscience*, 14(7), 531–535. https://doi.org/10.1038/s41561-021-00761-w

Garcia, R. (2002). Constraints on upper inner-core structure from waveform inversion of core phases. *Geophysical Journal International*, 150(3), 651–664. https://doi.org/10.1046/j.1365-246X.2002.01717.x

Garcia, R., & Souriau, A. (2000). Inner core anisotropy and heterogeneity level. *Geophysical Research Letters*, 27(19), 3121–3124. https://doi.org/10.1029/2000GL008520

Hosseini, K., Sigloch, K., Tsekhmistrenko, M., Zaheri, A., Nissen-Meyer, T., & Igel, H. (2020). Global mantle structure from multifrequency tomography using P, PP and P-diffracted waves. *Geophysical Journal International*, 220(1), 96–141. https://doi.org/10.1093/gji/ggz394

Ishii, M., & Dziewoński, A. M. (2002). The innermost inner core of the Earth: Evidence for a change in anisotropic behavior at the radius of about 300 km. *Proceedings of the National Academy of Sciences of the United States of America*, 99(22), 14026–14030. https://doi.org/10.1073/pnas.172508499

Ishii, M., & Dziewoński, A. M. (2003). Distinct seismic anisotropy at the centre of the Earth. *Physics of the Earth and Planetary Interiors*, 140(1–3), 203–217. https://doi.org/10.1016/j.pepi.2003.07.015

Jackson, I., Fitz Gerald, J. D., & Kokkonen, H. (2000). High-temperature viscoelastic relaxation in iron and its implications for the shear modulus and attenuation of the Earth's inner core. *Journal of Geophysical Research: Solid Earth*, 105(B10), 23605–23634. https://doi.org/10.1029/2000jb900131

Jeanloz, R., & Wenk, H. -R. (1988). Convection and anisotropy of the inner core. *Geophysical Research Letters*, 15(1), 72–75. https://doi.org/10.1029/GL015i001p00072

Karato, S. I. (1999). Seismic anisotropy of the Earth's inner core resulting from flow induced by Maxwell stresses. *Nature*, *402*(6764), 871–873. https://doi.org/10.1038/47235

Kennett, B. L. N., & Gudmundsson, O. (1996). Ellipticity corrections for seismic phases. *Geophysical Journal International*, *127*(1), 40–48. https://doi.org/10.1111/j.1365-246X.1996.tb01533.x

Kennett, B. L. N., Engdahl, E. R., & Buland, R. (1995). Constraints on seismic velocities in the Earth from travel-times. *Geophysical Journal International*, *122*(1), 108–124. https://doi.org/10.1111/j.1365-246X.1995.tb03540.x

Laske, G., & Masters, G. (1999). Limits on differential rotation of the inner core from an analysis of the Earth's free oscillations. *Nature*, *402*(6757), 66–69. https://doi.org/10.1038/47011

Li, C., van der Hilst, R. D., Engdahl, E. R., & Burdick, S. (2008). A new global model for P wave speed variations in Earth's mantle. *Geochemistry, Geophysics, Geosystems*, *9*(5), Q05018. https://doi.org/10.1029/2007GC001806

Long, M. D., & Silver, P. G. (2009). Shear wave splitting and mantle anisotropy: Measurements, interpretations, and new directions. *Surveys in Geophysics*. *30*, 407–461. https://doi.org/10.1007/s10712-009-9075-1

Lythgoe, K. H., Deuss, A., Rudge, J. F., & Neufeld, J. A. (2014). Earth's inner core: Innermost inner core or hemispherical variations? *Earth and Planetary Science Letters*, *385*, 181–189. https://doi.org/10.1016/j.epsl.2013.10.049

Lythgoe, K. H., & Deuss, A. (2015). The existence of radial anisotropy in Earth's upper inner core revealed from seismic normal mode observations. *Geophysical Research Letters*, *42*(12), 4841–4848. https://doi.org/10.1002/2015GL064326

Malinverno, A., & Briggs, V. A. (2004). Expanded uncertainty quantification in inverse problems: Hierarchical Bayes and empirical Bayes. *Geophysics*, *69*(4), 1005–1016. https://doi.org/10.1190/1.1778243

Mattesini, M., Belonoshko, A. B., & Tkalčić, H. (2018). Polymorphic Nature of Iron and Degree of Lattice Preferred Orientation Beneath the Earth's Inner Core Boundary. *Geochemistry, Geophysics, Geosystems*, *19*(1), 292–304. https://doi.org/10.1002/2017GC007285

McSweeney, T. J., Creager, K. C., & Merrill, R. T. (1997). Depth extent of inner-core seismic anisotropy and implications for geomagnetism. *Physics of the Earth and Planetary Interiors*, *101*(1–2), 131–156. https://doi.org/10.1016/S0031-9201(96)03216-5

Morelli, A., Dziewoński, A. M., & Woodhouse, J. H. (1986). Anisotropy of the inner core inferred from PKIKP travel times. *Geophysical Research Letters*, *13*(13), 1545–1548. https://doi.org/10.1029/GL013i013p01545

Muir, J. B., & Tkalčić, H. (2020). Probabilistic lowermost mantle P-wave tomography from hierarchical Hamiltonian Monte Carlo and model parametrization cross-validation. *Geophysical Journal International*, *223*(3), 1630–1643. https://doi.org/10.1093/gji/ggaa397

Mustać, M., & Tkalčić, H. (2016). Point source moment tensor inversion through a Bayesian hierarchical model. *Geophysical Journal International*, *204*(1), 311–323. doi:10.1093/gji/ggv458

Niu, F., & Wen, L. (2001). Hemispherical variations in seismic velocity at the top of the Earth's inner core. *Nature*, *410*(6832), 1081–1084. https://doi.org/10.1038/35074073

Ohtaki, T., Tanaka, S., Kaneshima, S., Siripunvaraporn, W., Boonchaisuk, S., Noisagool, S., et al. (2021). Seismic velocity structure of the upper inner core in the north polar region. *Physics of Earth and Planetary Interiors*, *311*, 106636. https://doi.org/10.1016/j.pepi.2020.106636

Pachhai, S., Li, M., Rost, S., Dettmer, J., & Tkalčić, H. (2022). Internal structure of ultralow-velocity zones consistent with origin from a basal magma ocean. *Nature Geoscience*, *15*, 79–84. https://doi.org/10.1038/s41561-021-00871-5.

Poupinet, G., Pillet, R., & Souriau, A. (1983). Possible heterogeneity of the Earth's core deduced from PKIKP travel times. *Nature*, *305*(5931), 204–206. https://doi.org/10.1038/305204a0

Ritterbex, S., & Tsuchiya, T. (2020). Viscosity of hcp iron at Earth's inner core conditions from density functional theory. *Scientific Reports*, *10*(1), 6311. https://doi.org/10.1038/s41598-020-63166-6

Romanowicz, B., & Wenk, H. R. (2017). Anisotropy in the deep Earth. *Physics of the Earth and Planetary Interiors*. *269*, 58–90. https://doi.org/10.1016/j.pepi.2017.05.005

Romanowicz, B., Cao, A., Godwal, B., Wenk, R., Ventosa, S., & Jeanloz, R. (2016). Seismic anisotropy in the Earth's innermost inner core: Testing structural models against mineral physics predictions. *Geophysical Research Letters*, *43*(1), 93–100. https://doi.org/10.1002/2015GL066734

Romanowicz, B., Tkalčić, H., & Bréger, L. (2003). On the origin of complexity in PKP travel time data. Earth's Core: Dynamics, Structure, Rotation (eds V. Dehant, K.C. Creager, S.-I. Karato and S. Zatman). *AGU Geodynamics Series*, *31*, 31–44. https://doi-org.virtual.anu.edu.au/10.1029/GD031p0031

Sambridge, M., Bodin, T., Gallagher, K., & Tkalčić, H. (2013). Transdimensional inference in the geosciences. *Philosophical Transactions of the Royal Society A*, *371*(1984), 20110547. https://doi.org/10.1098/rsta.2011.0547

Shearer, P. M. (1994). Constraints on inner core anisotropy from PKP(DF) travel times. *Journal of Geophysical Research*, *99*(B10), 19,647–19,659. https://doi.org/10.1029/94jb01470

Shearer, P. M., & Toy, K. M. (1991). PKP(BC) versus PKP(DF) differential travel times and aspherical structure in the Earth's inner core. *Journal of Geophysical Research*, *96*(B2), 2233–2247. https://doi.org/10.1029/90JB02370

Simmons, N. A., Myers, S. C., Johannesson, G., & Matzel, E. (2012). LLNL-G3Dv3: Global P wave tomography model for improved regional and teleseismic travel time prediction. *Journal of Geophysical Research: Solid Earth*, *117*(B10), B10302. https://doi.org/10.1029/2012JB009525

Song, X., & Helmberger, D. V. (1993). Anisotropy of Earth's inner core. *Geophysical Research Letters*, *20*(23), 2591–2594. https://doi.org/10.1029/93GL02812

Song, X., & Helmberger, D. V. (1998). Seismic evidence for an inner core transition zone. *Science*, *282*(5390), 924–927. https://doi.org/10.1126/science.282.5390.924

Song, X., & Richards, P. G. (1996). Seismological evidence for differential rotation of the Earth's inner core. *Nature*, *382*(6588), 221–224. https://doi.org/10.1038/382221a0

Souriau, A. (1998). Is the rotation real? *Science*, *281*, 5373. https://doi.org/10.1126/science.281.5373.55

Stephenson, J., Tkalčić, H., & Sambridge, M. (2021). Evidence for the Innermost Inner Core: Robust Parameter Search for Radially Varying Anisotropy Using the Neighborhood Algorithm. *Journal of Geophysical Research: Solid Earth, 126*(1), e2020JB020545. https://doi.org/10.1029/2020JB020545

Stixrude, L., & Cohen, R. E. (1995). High-pressure elasticity of iron and anisotropy of Earth's inner core. *Science, 267*(5206), 1972–1975. https://doi.org/10.1126/science.267.5206.1972

Su, W.-J., & Dziewoński, A. M. (1995). Inner core anisotropy in three dimensions. *Journal of Geophysical Research, 100*(B6), 9831–9852. https://doi.org/10.1029/95JB00746

Tanaka, S., Siripunvaraporn, W., Boonchaisuk, S., Noisagool, S., Kim, T., Kawai, K., et al. (2019). Thai Seismic Array (TSAR) Project. *Bulletin of Earthquake Research Institute, University of Tokyo, 94*, 1–11.

Tanaka, S. (2012). Depth extent of hemispherical inner core from PKP(DF) and PKP(Cdiff) for equatorial paths. *Physics of the Earth and Planetary Interiors, 210–211*, 50–62. https://doi.org/10.1016/j.pepi.2012.08.001

Tanaka, S., & Hamaguchi, H. (1997). Degree one heterogeneity and hemispherical variation of anisotropy in the inner core from PKP (BC)- PKP (DF) times. *Journal of Geophysical Research: Solid Earth, 102*(B2), 2925–2938. https://doi.org/10.1029/96jb03187

Tkalčić. H., Phạm, T-S., & Wang, S. (2020). The Earth's coda correlation wavefield: Rise of the new paradigm and recent advances. *Earth-Science Reviews, 208*, 103285. https://doi.org/10.1016/j.earscirev.2020.103285

Tkalčić, H. (2017). *The Earth's Inner Core: Revealed by Observational Seismology*. Cambridge, UK: Cambridge University Press. https://doi.org/10.1017/9781139583954

Tkalčić, H. (2015). Complex inner core of the Earth: The last frontier of global seismology. *Reviews of Geophysics, 53*/1, 59–94. doi:10.1002/2014RG000469

Tkalčić, H., Young, M., Bodin, T., Ngo, S., & Sambridge, M. (2013). The shuffling rotation of the Earth's inner core revealed by earthquake doublets. *Nature Geoscience, 6*(6), 497–502. https://doi.org/10.1038/ngeo1813

Tkalčić, H. (2010). Large variations in travel times of mantle-sensitive seismic waves from the South Sandwich Islands: Is the Earth's inner core a conglomerate of anisotropic domains? *Geophysical Research Letters, 37*, L14312. doi:10.1029/2010GL043841

Tkalčić, H., Romanowicz, B., & Houy, N. (2002). Constraints on D" structure using PKP(AB-DF), PKP(BC-DF) and PcP-P travel-time data from broadband records. *Geophysical Journal International, 149*(3), 599–616. https://doi.org/10.1046/j.1365-246X.2002.01603.x

Vinnik, L., Romanowicz, B., & Bréger, L. (1994). Anisotropy in the center of the inner core. *Geophysical Research Letters, 21*(16), 1671–1674. https://doi.org/10.1029/94GL01600

Vinnik, L., Farra, V., & Romanowicz, B. (1989). Observational evidence for diffracted SV in the shadow zone of the Earth's core. *Geophysical Research Letters, 16*, 519–522. https://doi.org/10.1029/Gl016I006P00519

Wang, T., & Song, X. (2018). Support for equatorial anisotropy of Earth's inner-inner core from seismic interferometry at low latitudes. *Physics of the Earth and Planetary Interiors, 276*, 247–257. https://doi.org/10.1016/j.pepi.2017.03.004

Wang, T., Song, X., & Xia, H. H. (2015). Equatorial anisotropy in the inner part of Earth's inner core from autocorrelation of earthquake coda. *Nature Geoscience, 8*(3), 224–227. https://doi.org/10.1038/ngeo2354

Wang, Y., & Wen, L. (2007). Complex seismic anisotropy at the border of a very low velocity province at the base of the Earth's mantle. *Journal of Geophysical Research: Solid Earth, 112*(9), B09305. https://doi.org/10.1029/2006JB004719

Waszek, L., Irving, J., & Deuss, A. (2011). Reconciling the hemispherical structure of Earth's inner core with its super-rotation. *Nature Geoscience, 4*(4), 264–267. https://doi.org/10.1038/ngeo1083

Woodhouse, J. H., Giardini, D., & Li, X. -D. (1986). Evidence for inner core anisotropy from free oscillations. *Geophysical Research Letters, 13*(13), 1549–1552. https://doi.org/10.1029/GL013i013p01549

Yee, T.-G., Rhie, J., & Tkalčić, H. (2014). Regionally heterogeneous uppermost inner core observed with Hi-net array. *Journal of Geophysical Research: Solid Earth, 119*(10), JB011341. https://doi.org/10.1002/2014JB011341

Yoshida, S., Sumita, I., & Kumazawa, M. (1996). Growth model of the inner core coupled with the outer core dynamics and the resulting elastic anisotropy. *Journal of Geophysical Research B: Solid Earth, 101*(12), 28085–28103. https://doi.org/10.1029/96jb02700

11

Recent Progress in High-Pressure Experiments on the Composition of the Core

Ryosuke Sinmyo[1], Yoichi Nakajima[2], and Yasuhiro Kuwayama[3]

ABSTRACT

The chemical composition of the Earth's core is fundamentally important information for understanding the structure and the dynamics of the bulk Earth. Since it is not straightforward to directly study the Earth's core materials, the chemistry of the core is mainly estimated based on a comparison between observations and the results of high-pressure and high-temperature experiments simulating core conditions in the laboratory. Here, we have summarized recent progress in the high-pressure experimental studies on the chemical composition of the core. The advanced diamond anvil cell experiments are now capable of generating higher pressures and temperatures corresponding to core conditions with high stability. The density and sound velocity of iron and iron alloys have been determined in both liquid and solid states by multiple measurements using the synchrotron X-ray. The melting phase relationships of iron alloys have been widely examined using modern analytical techniques in conjunction with nano-scale processing. We now understand the chemistry of the core better thanks to the new knowledge of the density, sound wave velocity, and phase relationships of iron and iron alloys at high pressure.

11.1. INTRODUCTION

11.1.1. Structure of the Earth's Core

It is believed that the Earth's core is mainly composed of iron, nickel, and lesser amounts of light elements (e.g., Birch, 1952). Since it is not straightforward to directly study samples of the core materials, seismic observations are the most fundamental information for understanding the structure and chemistry of the core (e.g., Dziewonski & Anderson, 1981). According to seismic observations, the core extends from 2,900 km depth to the center of the Earth at 6,400 km depth. These values correspond to 136–364 GPa in pressure. The inner core boundary (ICB) is observed at 5,100 km depth (Lehmann, 1936), where the pressure is ~330 GPa. While it is believed that the inner core is in a solid state, the surrounding outer core is in a liquid state since the shear wave does not travel through the outer core (e.g., Dziewonski & Anderson, 1981). The geomagnetic field is believed to be generated by the convection of the liquid outer core. The uppermost part of the liquid outer core is in contact with the bottom of the rocky mantle at the core-mantle boundary (CMB) of 2,900 km depth. This boundary is the greatest discontinuity in density of all boundaries in the Earth. The low compressional wave (V_p) anomaly is observed at the topmost outer core with a thickness of ~300 km, which may be due to a compositional stratification (Helffrich & Kaneshima, 2010). The V_p decreases to values a maximum of 0.3% lower than the preliminary reference Earth model (PREM) (Helffrich & Kaneshima,

---

[1] Department of Physics, Meiji University, Kawasaki, Japan
[2] Department of Physics, Kumamoto University, Kumamoto, Japan
[3] Department of Earth and Planetary Sciences, The University of Tokyo, Tokyo, Japan

---

Core-Mantle Co-Evolution: An Interdisciplinary Approach, Geophysical Monograph 276, First Edition.
Edited by Takashi Nakagawa, Taku Tsuchiya, Madhusoodhan Satish-Kumar, and George Helffrich.
© 2023 American Geophysical Union. Published 2023 by John Wiley & Sons, Inc.
DOI: 10.1002/9781119526919.ch11

2010). Additionally, an ~150 km thick layer (F-layer) is observed at the bottom of the outer core with an anomalous gradient of V_p from PREM up to 0.1% (Gubbins et al., 2008). Seismic observation suggests that the inner core shows an asymmetric structure between the Western and Eastern Hemispheres (Tanaka & Hamaguchi, 1997; Monnereau et al., 2010). The solid core is likely melting at the eastern side of the ICB, and the liquid core is frozen at the western side of the ICB (Monnereau et al., 2010). This suggests that the inner core and the outer core are in chemical equilibrium at ICB via continuous freezing and melting, while possible chemical stratification at the F-layer inhibits the chemical diffusion inside the outer core (Gubbins et al., 2008). On the other hand, the chemical reaction may be limited between core and mantle at the CMB because of the slow chemical diffusion in the minerals (Holzapfel et al., 2005), although it may be enhanced by grain boundary diffusion (Hayden & Watson, 2007; Yoshino et al., 2020) and morphological instability (Otsuka & Karato, 2012). It is believed that the core contains considerable amounts of light elements, because the density of the core is significantly lower than that of pure iron at core conditions (Birch, 1952; Poirier, 1994; Hirose et al., 2013; Alfè, 2015). According to the results of high-pressure experiments (see section 11.3), the inner core and outer core are about 1–4% and 6–9% less dense than pure iron, respectively (Dewaele et al., 2006; Kuwayama et al., 2020). This shows that the outer core must be more enriched in light elements than the inner core.

11.1.2. Light Elements in the Core

After the initial suggestion by Birch (1952), the light elements in the core have been discussed for a long time based on the sound wave velocity and density of the iron and iron-light elements alloys obtained by high-pressure experiments (Mao et al., 2008; Huang et al., 2011; Morard et al., 2013; Jing et al., 2014; Nakajima et al., 2015; Prescher et al., 2015; Kawaguchi et al., 2017; Kuwayama et al., 2020; Nishida et al., 2020) and first-principle calculations (Alfe et al., 2007; Badro et al., 2014; Ichikawa & Tsuchiya, 2020; Umemoto & Hirose, 2020). The sound wave velocities and the density of solid iron alloys have been determined by X-ray diffraction measurements and inelastic X-ray scattering measurements using the synchrotron (Fiquet et al., 2001; Mao et al., 2001; Dewaele et al., 2006; Sata et al., 2010; Antonangeli et al., 2012; Mao et al., 2012; Sakamaki et al., 2016). On the other hand, the measurements in the liquid iron have long been technically challenging, partly because of the difficulty in retaining the liquid state during measurements. Recent studies have successfully determined the density and sound wave velocity of liquid iron at high pressure corresponding to the core (Morard et al., 2013; Nakajima et al., 2015; Morard et al., 2017a; Kuwayama et al., 2020; Nakajima et al., 2020). As a result, for example, it has been suggested that carbon cannot be a dominant light element in the core based on the seismic wave velocity measurements in the iron alloy liquid (Nakajima et al., 2015). The sound wave velocities of iron alloys have also been determined by in situ ultrasonic measurements and X-ray imaging techniques (section 11.3).

The difference in the density deficit between the inner core and the outer core is key information to delimiting the chemistry of the core based on the melting phase relationships. Since the outer core must be more enriched in light elements than the inner core, the solidus and liquidus curve should be negative in the temperature-composition phase diagram in the Fe-FeX system. For instance, the melting temperature likely does not significantly decrease with increasing silicon content in the Fe-FeSi system at inner core pressure (Komabayashi, 2020). If this is the case, silicon cannot be the sole light element in the core, because the solid cannot be denser than liquid in Fe-FeSi binary system in ICB conditions. Accordingly, the melting phase relationships have been extensively studied by a number of experiments using in situ synchrotron X-ray diffraction measurements and chemical analysis using electron microscopes as discussed below (section 11.3).

11.1.3. Chemical Evolution of the Core

The chemistry of the core has evolved through the Earth's history (Wood et al., 2006; Rubie et al., 2011; Hirose et al., 2017). At the early stage of the Earth, it is believed that the liquid iron was in chemical equilibrium with silicate before core formation (Wood et al., 2006), while the core formation process may be multistage event according to the metal/silicate element partitioning experiments (Rubie et al., 2011). Siderophile (iron-loving) elements might be taken away from the bulk silicate Earth to the core in the core formation process. It has been suggested that the magma ocean and the liquid metal were in equilibrium at around 50 GPa based on the partitioning coefficient and the observed abundance of the siderophile elements in the mantle and primordial chondrite (Li & Agee, 1996; Siebert et al., 2012; Siebert et al., 2013; Fischer et al., 2015). In this case, the core would have contained 2 wt% Si and 5 wt% O at that time according to the partitioning coefficient of Si and O between metal and silicate, while the concentration of Si and O depends on the redox state of accreting materials (Siebert et al., 2013). Subsequently, the oxide phase, such as MgO or SiO_2, should have been released from the young core as it cooled down (Badro et al., 2016; O'Rourke & Stevenson, 2016; Hirose et al., 2017), because the maximum solubility of the light elements in the core significantly decreases with decreasing temperature. Such exsolution of the oxide phase in the core may have helped to initiate the early dynamo (Tarduno et al., 2015). The exsolved SiO_2 phase may be accumulated at the mid-lower mantle where the seismic scatterers were observed (Helffrich et al., 2018a). This hypothesis will be further tested by the isotopic signature (Helffrich et al., 2018b).

11.1.4. Trace Elements and Isotopic Chemistry of the Core

The isotopes and the trace elements play important roles in delimiting the chemistry and the evolution of the core based on geochemical observations. The fractionation coefficients of the stable isotopes, such as ^{30}Si, ^{29}Si, ^{13}C, and ^{57}Fe, have been studied to reconcile their abundance in the chondritic material, silicate mantle, and the core (Georg et al., 2007; Poitrasson et al., 2009; Shahar et al., 2009; Satish-Kumar et al., 2011; Shahar et al., 2011; Shahar et al., 2016). The fractionations coefficients have been determined by chemical analysis in the recovered sample of high-pressure experiments using mass spectroscopy. Recently, the fractionations of isotopes of iron have been determined by the nuclear resonant inelastic X-ray spectroscopy measurements in pure iron, iron alloys, and lower mantle minerals (Polyakov, 2009; Shahar et al., 2016). Results suggested that the ^{57}Fe is less abundant in H- or C-rich iron alloy than expected from the geochemical data, and thus H or C could not be the dominant light elements in the core (Shahar et al., 2016). The iron isotopic fractionation has also been studied between basaltic glass and iron alloys (Liu et al., 2017). It is suggested that the core formation process may not affect the variation in the iron isotopic composition (Liu et al., 2017) in contrast to previous study (Shahar et al., 2016). The discrepancy is likely due to the different silicate phases used in the studies (Shahar et al., 2016; Liu et al., 2017). The radioactive isotopes are also used for understanding the dynamics of the core. The primordial high $^3He/^4He$ ratio is observed in the modern ocean island basalt (OIB), together with the negative anomaly in the short-lived ^{182}Hf-^{182}W system (Mundl et al., 2017). While the reason for this correlation is still under debate, this may be attributed to the core-mantle interaction. It has been suggested that the core may be a reservoir of the primordial helium based on the partitioning coefficients of helium between silicate and metal at high pressure (Matsuda et al., 1993; Bouhifd et al., 2013). In ^{182}Hf-^{182}W system, the parent ^{182}Hf is a lithophile element and thus was hardly incorporated into the core during the core-mantle equilibrium process before the core formation, and thus the core material should contain a lower amount of ^{182}W compared to the bulk silicate Earth. The core formation process has been studied based on the abundance of the siderophile elements in the mantle and primordial chondrite, such as Ni, Co, V, and Cr as mentioned above (Li & Agee, 1996; Siebert et al., 2013). The concentration of the radioactive elements in the core is important for understanding the thermal history of the core (Labrosse, 2015). Nevertheless, it is believed that the core contains a very limited amount of potassium (Hirao et al., 2006; Xiong et al., 2018), uranium, and thorium (Chidester et al., 2017; Faure et al., 2020) based on the partitioning coefficient between silicate and metal at high pressure.

11.2. HIGH-PRESSURE EXPERIMENTS

11.2.1. High-Pressure Apparatuses

The interior of the Earth is under extremely high pressure and temperature, and thus high-pressure experiments have been widely conducted in order to understand the structure and dynamics of the Earth. The high-pressure experiments are divided into two types, static compression and dynamic compression. In general, the static compressions have been developed to meet the requirement of retaining the heat for a long time in order to achieve equilibrium, while the pressure conditions that can be achieved are relatively lower than dynamic compression. In contrast, dynamic compressions are capable of generating high pressure up to several TPa, while it is difficult to keep high-pressure and high-temperature conditions for a longer time to establish chemical equilibrium (see section 11.2.4).

The diamond anvil cell (DAC) and large volume press are widely used apparatuses for static compression studies of the Earth's core (Fig. 11.1). While it was technically challenging to measure the properties of the materials under core pressure, recent studies have successfully determined the crystal structure, elasticity, transport properties, and phase relationships using advanced in situ synchrotron X-ray measurements and electron microscopic analysis (see section 11.3). Dynamic compression experiments have been conducted with two-stage light gas guns (Brown & McQueen, 1986) or a high power laser (Sakaiya et al., 2014; Hwang et al., 2020; White et al., 2020) to study the core material. Recent studies have conducted experiments at a relatively lower temperature than the Hugoniot curve (section 11.2.4). In addition to these experiments, theoretical calculation has also been developed to predict the physical and chemical properties of the iron and the alloy at high-pressure and high-temperature conditions (Alfè, 2009; de Koker et al., 2012; Belonoshko et al., 2017; Ichikawa & Tsuchiya, 2020).

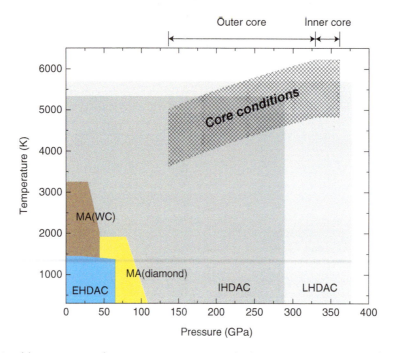

Figure 11.1 Achievable pressure and temperature range using high-pressure apparatuses. The multi-anvil (MA) apparatus with the tungsten-carbide (WC) anvils (Ishii et al., 2016), MA with the sintered diamond anvils (Yamazaki et al., 2014), laser-heated DAC (LH-DAC) (Tateno et al., 2010), internal resistance heated DAC (IH-DAC) (Sinmyo et al., 2019), and external resistance heated DAC (EH-DAC) (Kawaguchi et al., 2017; Immoor et al., 2018). The Earth's core conditions are shown in the hatched area.

11.2.2. Diamond Anvil Cell (DAC)

The DAC can generate high pressures, even as high as the center of the Earth (Tateno et al., 2010; Sakai et al., 2015; Dubrovinskaia et al., 2016). While achievable pressure is very high in the DAC experiments, it is a trade-off with the size of the sample (Fig. 11.2). Typically, samples with several hundred µm^3 (~100 µm^2 area and 1 µm depth) volumes are prepared for generating core pressure of >135 GPa. A milestone pressure and temperature condition of 377 GPa and 5700 K have been achieved by using diamond anvils with beveled culet (= the tip of the diamond anvils) (Tateno et al., 2010). The highest static pressure of 1 TPa was recently generated by double-stage semispherical diamond anvils, while the size of the sample is limited to several µm diameters (Dubrovinskaia et al., 2016). A more recent study has succeeded in compressing a relatively large sample with ~5 µm diameters up to 600 GPa using the single-stage diamond anvil with toroidal-type culets (Dewaele et al., 2018). While such ultrahigh-pressure conditions have been only generated at ambient temperature, heating experiments may be done at TPa range in the near future to explore the internal structure of the giant planets. Recently, various types of DACs have been used for novel experiments and measurements. For example, they have been adapted for single-crystal X-ray diffraction and nuclear resonant inelastic X-ray scattering (NIS) measurements in the core materials (Mao et al., 2001; Boehler, 2006). The cells are designed for the use of conical diamond anvils to collect diffractions from a wider angle for the single-crystal X-ray diffraction measurements (Boehler, 2006; Prescher et al., 2015). For the NIS measurements, the cells with large side windows are used to place the detectors as close as possible to the sample (Mao et al., 2001; Mao et al., 2004; Prescher et al., 2015). Nano-polycrystalline diamonds (NPD) are expected to be used for better anvils than the single-crystal diamond. The hardness of the NPD is independent of crystallographic orientation and is possibly higher than a single-crystal diamond at high temperatures (Irifune et al., 2003; Irifune et al., 2019). NPD anvils are now used for glitch-free X-ray absorption spectroscopy (Ishimatsu et al., 2012) and rotational deformation experiments (Nomura et al., 2017).

The lasers and resistive heaters are used in the DAC experiments to generate high temperature (Fig. 11.3). Achievable temperature conditions are summarized in Figure 11.1. The infrared continuous-mode lasers are widely used for the laser-heated DAC (LH-DAC; Fig. 11.4). The temperatures are determined by the spectroradiometric method during heating (Fig. 11.4). The radiation light from the sample is collected through

RECENT PROGRESS IN HIGH-PRESSURE EXPERIMENTS ON THE COMPOSITION OF THE CORE   195

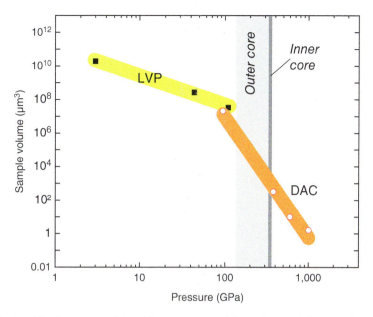

Figure 11.2 The relationships between achievable pressure and the volume of the sample. Large volume press (LVP); Piston-cylinder apparatus (Hirose & Kushiro, 1993), multi-anvil apparatus with WC anvils (Ishii et al., 2016), and multi-anvil sintered diamond anvils (Yamazaki et al., 2014). Diamond anvil cell (DAC); DAC with single-stage beveled anvil (Tateno et al., 2010); large-volume DAC (Boehler et al., 2013); double-stage semispherical anvil (Dubrovinskaia et al., 2016); single-stage toroidal DAC (Dewaele et al., 2018).

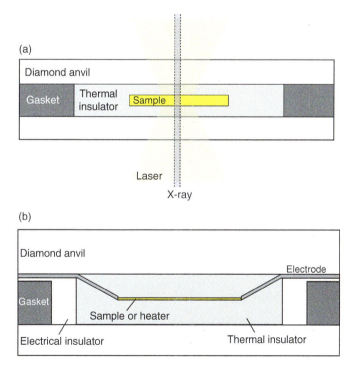

Figure 11.3 Schematic cross-sections of the DACs for high-pressure and high-temperature experiments. (a) Laser-heated DAC, (b) internal resistance heated DAC, and (c) external resistance heated DAC.

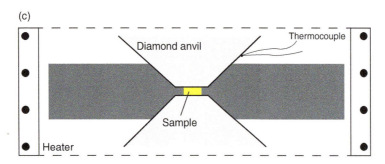

Figure 11.3 (*continued*)

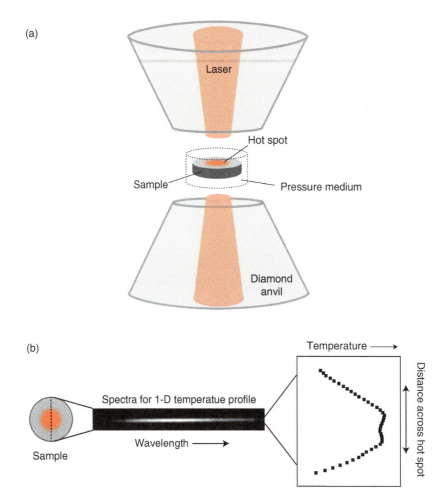

Figure 11.4 (a) Schematic illustrations of the laser-heated DAC and (b) the determination of the 1-D temperature profile across the hot spot by the spectroradiometric method.

the pinhole (Mezouar et al., 2017) or slit (Hirao et al., 2020) placed in front of the spectrometer for temperature determination. One-dimensional (1-D) temperature profiles can be obtained by the measurements through the slit (Fig. 11.4). Two-dimensional (2-D) temperature mapping can be obtained by using optics with a multispectral imaging system (Campbell, 2008; Du et al., 2013a). It has been noted that the achromatic effect may cause the uncertainty in the temperature determination during laser heating (Mezouar et al., 2017). Reflective lenses are used to avoid achromatic aberration (Mezouar et al., 2017).

The pulsed laser has recently been used to reduce the damage to the diamond anvils and the chemical contamination of the sample (Yang et al., 2012; Du et al., 2013a; Aprilis et al., 2017; Aprilis et al., 2019). The pulsed laser is synchronized with the temperature measurements and the detectors for X-ray (Kupenko et al., 2015; Aprilis et al., 2017) to selectively collect the signal at high temperatures. On the other hand, in the pulsed laser heating, the laser duration (~10 microseconds) is much shorter than the estimated chemical equilibration time (~0.1 seconds) in the core materials at the core conditions with an assumption of the sample width of 10 μm and estimated diffusivity of 10^{-9} m^2s^{-1} (Helffrich, 2014; Hirose et al., 2017). The pulsed laser heating would be more suitable for studying physical parameters at high temperatures than studying the chemical equilibrium state. For instance, the structure of the metallic glass can be studied at high pressure and high temperature as an analog of the liquid metal using a pulsed laser. The pulsed laser may be combined with double-stage and toroidal DACs for ultrahigh-pressure experiments in the future (Sakai et al., 2015; Dubrovinskaia et al., 2016; Dewaele et al., 2018).

External resistive heating is capable of generating a more homogeneous high-temperature field compared to laser heating (Dubrovinskaia & Dubrovinsky, 2003; Du et al., 2013b; Kawaguchi et al., 2017; Immoor et al., 2018). The diamond anvils are surrounded by the heater, which is made of metal wire, graphite, or semiconductor to heat up the whole sample (Fig. 11.3). The heating experiments are conducted under air in experiments below 800 K (Lai et al., 2020b), and reducing or inert gas atmosphere below ~1273 K. For higher temperatures, the experiments have been conducted inside a vacuum chamber for better thermal insulation and to avoid oxidation of the heater (Kawaguchi et al., 2017; Immoor et al., 2018). Using a platinum heater, the sound wave velocity of the liquid iron alloy was determined up to 20 GPa and 1450 K by inelastic X-ray scattering measurements (Kawaguchi et al., 2017). Thermocouples are usually attached to the surface of the diamond anvil to determine the temperature in the external resistive heating. In general, a relatively larger temperature gradient would be caused in a smaller heater for higher temperature generation. The thermocouple temperature should be double checked by the spectroradiometric method or melting of standard materials at ambient pressure (Du et al., 2013b). Recently, the internally resistive heated DAC has been developed for the heating of iron and iron alloys (Komabayashi et al., 2009; Komabayashi et al., 2012; Komabayashi et al., 2019; Sinmyo et al., 2019; Geballe et al., 2020) (Fig. 11.3). The internally resistive heating is stable for a long time, and the temperature distributions are homogeneous compared to laser heating (Sinmyo et al., 2019). Compared to the pioneering studies (Liu & Bassett, 1975; Dubrovinsky et al., 1998), today much higher pressure and temperature can be generated using an internally resistive heated DAC thanks to nano-scale processing technology such as the focused ion beam (FIB) system. The heater is inside the sample chamber and is connected to the power source by electrodes (Fig. 11.3). Temperature is determined by the spectroradiometric method in a similar manner to the LH-DAC. While the surface is the hottest part in the laser heating, the center is the hottest in the sample of the internal resistance heated DAC (IH-DAC) as discussed below (section 11.3). Any conductive materials, including iron alloys, can be used as a heater and a sample in the IH-DAC, and thus the internally heated diamond cell has great potential for the study of the Earth's core materials at core conditions. Indeed, the phase relationships of iron and iron alloys are precisely determined by a combination with X-ray diffraction measurements at high pressure and temperature (Komabayashi et al., 2009; Komabayashi et al., 2012; Komabayashi et al., 2019; Sinmyo et al., 2019). While the heating is very stable in the internal resistive heated DAC, the decomposition and melting reaction in iron alloys may cause instability in the heating (Sinmyo et al., 2019). Consequently, in the future, the heater should be separated from the sample as in the large volume press, although this is still technically challenging. Boron-doped diamond is a promising material as the heater in IH-DAC (Ozawa et al., 2018). In the internal resistive heating, insulating hard materials, such as amorphous boron, cubic boron nitride, and diamond, are used as an electrical insulator and a gasket material. Such materials were originally used for electrical conductivity measurements and X-ray absorption and diffraction measurements (Zou et al., 2001; Merkel & Yagi, 2005; Funamori & Sato, 2008; Petitgirard et al., 2015).

11.2.3. Large Volume Press

The large volume press is literally a high-pressure apparatus to compress a relatively large volume sample. Kawai-type multi-anvil and Paris-Edinburgh presses are commonly used to study the Earth's core. In general, a large volume press can generate a homogeneous high-pressure and high-temperature field compared to the DAC and shock compression. The temperature distributions are within ±50 K in the multi-anvil sample with ~1.5 mm diameter (Canil, 1994). This is much smaller than ~100 K/μm in the LH-DAC (Dewaele et al., 1998; Rainey et al., 2013). Temperature is usually monitored by a thermocouple introduced into the sample chamber of the large volume press. The heating power can be precisely controlled by monitoring the thermocouple temperature. Recently, advanced multi-anvil apparatuses can generate much higher pressures and temperatures (Yamazaki et al., 2014; Ishii et al., 2016; Xie et al., 2020). Sintered diamond anvils (Yamazaki et al., 2014) and NPD anvils (Irifune et al., 2019) have recently been used for higher pressure generation toward core pressure by the large volume press. Conventional tungsten-carbide

(WC) anvils have been improved to conduct experiments routinely at around 50 GPa (Ishii et al., 2016). The guide block systems are precisely optimized, and the WC anvils are slightly tapered toward truncation (Ishii et al., 2016). Novel boron-doped diamond heaters have been used to generate more than 3000 K (Yamada et al., 2008; Shatskiy et al., 2009; Xie et al., 2020). While the achievable pressure and temperature are still relatively lower than DAC, a large volume press has a great advantage in the stability and homogeneity of the pressure and temperature inside the samples. Since the liquid state can be maintained for a relatively long time with a stable temperature field, the elasticity and the density of the liquid iron alloys have been measured with ultrasonic sound velocity measurements using the multi-anvil press (Jing et al., 2014; Nishida et al., 2020) and with X-ray diffuse scattering using the Paris-Edinburgh press (Morard et al., 2018b; Shibazaki & Kono, 2018).

11.2.4. Shock Compression

Shock compression experiments are able to generate much higher pressure than static compression experiments. The elasticity and the phase transition boundary of the materials have been determined at extremely high pressure up to several TPa (Asimow, 2015; Duffy et al., 2015; Duffy & Smith, 2019). Density and bulk sound velocities of core materials are estimated by means of shock compression experiments (Huang et al., 2011; Sakaiya et al., 2014). The melting temperature of iron has been determined by the observed discontinuity in the Hugoniot curve at around ~250 GPa (Brown & McQueen, 1986; Yoo et al., 1993; Nguyen & Holmes, 2004; Li et al., 2020; Turneaure et al., 2020). In conventional shock experiments, the pressure and temperature increase along the Hugoniot curve (e.g., Brown & McQueen, 1986). Consequently, the melting temperature of iron has been repeatedly determined at a specific pressure of around ~250 GPa. However, recent studies have conducted experiments at a relatively lower temperature than the Hugoniot curve by ramp compression (Wang et al., 2013), precompression of the sample (Millot et al., 2018), and adjusting the energy and timing of the drive-laser beams (Millot et al., 2019). Moreover, novel measurements have recently been applied to the shock experiments for a better understanding of the elasticity and the structure of the iron alloys. The X-ray diffraction measurements have been developed for the direct determination of the crystal structure of the iron and the Fe-Si alloys (Denoeud et al., 2016; Wicks et al., 2018). Since the shock compression finishes in a very brief time, it is not straightforward to estimate the equilibrium state of the materials at high-pressure conditions. Indeed, a recent X-ray scattering study showed that over-heated solid iron was observed during laser-driven shock experiments at a higher temperature than the expected melting temperature (White et al., 2020). The kinetics of the phase transition will be further studied by ultrafast X-ray diffraction measurements using an X-ray free-electron laser (Hwang et al., 2020). The kinetics of rearrangement of atoms should be studied further under shock compression conditions (Gleason et al., 2015; Hwang et al., 2020; Morard et al., 2020; Turneaure et al., 2020; White et al., 2020).

11.3. EXPERIMENTAL RESULTS AND IMPLICATIONS FOR THE CORE

11.3.1. Measurements Using Synchrotron X-ray

Brilliant synchrotron X-ray is a powerful tool for the in situ measurements of core materials at high-pressure conditions, since the samples are smaller at higher pressure (Fig. 11.2). Moreover, the constituent materials of the high-pressure cell, such as anvils, pressure mediums, and the heater, can be a large source of the background noise. The brilliance of the synchrotron X-ray is many orders of magnitude higher than the laboratory-grade X-ray tubes. For example, the flux is 4×10^9 photons/s at a focused beam size of 1 μm^2 and 30 keV in the X-ray diffraction beamline of SPring-8 (Hirao et al., 2020).

X-ray diffraction measurements have been widely used to determine crystal structure, phase relationships, and the elasticity of the materials at high pressure. The beamlines have been constructed for the combinational studies of DAC and X-ray diffraction measurements at the synchrotron facilities, such as SPring-8, APS, ESRF, and PETRA III (Mezouar et al., 2005; Petitgirard et al., 2014; Liermann et al., 2015; Mezouar et al., 2017; Shen & Mao, 2017; Hirao et al., 2020). X-ray diffraction measurements at core pressure are now widely conducted using DAC experiments at the synchrotron facility. The crystal structure of the iron has been determined to be hexagonal close packing at 377 GPa and 5700 K, corresponding to the conditions of the center of the inner core (Tateno et al., 2010). The crystal structures have also been determined in iron-light elements alloys (Fischer et al., 2012; Prescher et al., 2015; Tateno et al., 2019). A new phase with a chemical composition of Fe_2S was found at 306 GPa in the Fe-S system (Tateno et al., 2019). The crystal structure of Fe_7C_3 was determined up to 205 GPa by means of single-crystal X-ray diffraction measurements (Prescher et al., 2015). The compressibility of the iron and its alloys is determined based on the volumes of the unit cell determined by X-ray diffraction measurements (Dewaele et al., 2006; Sata et al., 2010; Fischer et al., 2012; Miozzi et al., 2020). In contrast to solid iron, it is technically difficult to determine the structure of liquid iron using X-ray diffraction at pressure conditions of the core, because the samples need to be kept melted for the longer measurements needed to collect weaker diffraction signals from the liquid. A multichannel collector is used for X-ray diffraction measurements in amorphous materials to minimize scattering from the background materials (Weck et al., 2013). Recently, the structure of the liquid iron has been investigated using X-ray diffuse scattering signals obtained at 116 GPa and 4350 K using LH-DAC (Kuwayama et al., 2020) (see section 11.3.4).

The sound wave velocities of the solid iron and iron alloys are determined by inelastic X-ray spectroscopy (IXS) at high pressure (Antonangeli et al., 2012; Mao et al., 2012; Sakamaki et al., 2016). According to the results of IXS measurements in solid-state iron alloys at 163 GPa and 3000 K, the density deficit can be better explained by alloying hydrogen in the inner core (Sakamaki et al., 2016). The sound wave velocities of the liquid have been determined in Fe-C alloys by IXS (Nakajima et al., 2015) (see section 11.3.5). The sound velocities of the liquid Fe-S have been determined using ultrasonic measurements and synchrotron X-ray imaging with a multi-anvil apparatus up to ~20 GPa (Jing et al., 2014; Nishida et al., 2020). The X-ray imaging technique has been applied to determine the density and viscosity of the amorphous silicate materials (Petitgirard et al., 2015; Xie et al., 2020). Such measurements might also be applicable for liquid iron alloys at high pressure and temperature, while the measurements are so far limited to room temperature or relatively low-pressure conditions (Terasaki et al., 2012; Zeng et al., 2014). X-ray absorption spectroscopy (XAS) has been used for detecting the onset of melting during heating (Aquilanti et al., 2015; Morard et al., 2018a; Boccato et al., 2020; Torchio et al., 2020). The sound wave velocities of pure iron and iron alloys have been measured by NIS measurements (Mao et al., 2001; Mao et al., 2004; Chen et al., 2014; Prescher et al., 2015; Chen et al., 2018; Lai et al., 2020a). The Debye sound velocity of iron has been determined by parabolic fitting of phonon DOS spectra under the assumption of a harmonic solid with Debye-like low-frequency dynamics (Hu et al., 2003). Isotopic fractionation coefficients of iron alloys have been determined by NIS measurements (Shahar et al., 2016). Results showed that H or C cannot be the dominant light elements in the core (Shahar et al., 2016).

11.3.2. Melting Temperature

The melting temperatures of iron alloys are important information to constrain the chemistry and thermal state of the core. The temperature of the ICB should be at the melting temperature of the iron alloys, since the liquid outer core coexists with solid iron. Therefore, many studies have reported the melting temperature of the iron and iron alloys. In pioneering LH-DAC experiments, the melting temperature of iron was determined by visual observation of the movement of the sample during heating up to 200 GPa (Boehler, 1993). The microscopic optical measurements have been developed to observe plateaus or kinks in the relationships between the laser power and the temperature of the sample using radiation spectra (Lord et al., 2009; Lord et al., 2010; Anzellini et al., 2013; Fischer et al., 2013). The kink in the power-temperature curve was thought to be caused by the latent heat during melting, while it was noted that such latent heat might be detectable only by very fast measurements during melting (Geballe & Jeanloz, 2012). The optical system was further developed to obtain a 2-D mapping of the temperature during melting experiments (Campbell, 2008; Du et al., 2013a). The streak camera was recently used for the ultrahigh-frequency measurements of the radiation (Konopkova et al., 2016). The melting temperature of iron has been revisited by the X-ray diffraction measurements in the iron using a synchrotron facility (Shen et al., 1998). The disappearance and appearance of the diffraction lines were attributed to the melting and crystallizing of the iron. A fast X-ray diffraction measurement technique has been developed to observe the melting of iron (Anzellini et al., 2013). While the melting temperature of pure iron has been repeatedly studied as above, the results are not consistent with each other (Fig. 11.5). This discrepancy may be partly due to the thermal structures of the sample in the LH-DAC experiments (Rainey et al., 2013). Morard et al. (2018a) reported that the discrepancy in the melting temperature of iron is partly due to the contamination of carbon from the diamond anvils. The internal resistively heated DAC was recently used to determine the melting temperature of iron at high pressure (Sinmyo et al., 2019). As described above, the stability and the homogeneity of the heating is much better in resistance heating than in laser heating. The melting temperatures were determined based on the (1) disappearance of the X-ray diffraction, (2) melting texture in the recovered sample, and (3) relationships between temperature and the applied power. The melting temperature of pure iron was 5500 K at the ICB, and the temperatures of the ICB and CMB are estimated to be 5120 K and 3760 K, respectively, with the depression accounted for by the light elements in the core (Sinmyo et al., 2019).

Melting temperatures obtained by resistive heating were considerably lower than those of previous laser heating experiments (Anzellini et al., 2013) (Fig. 11.5). Temperature distributions of the samples have been

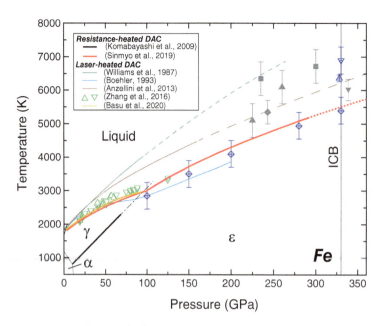

Figure 11.5 Melting temperature of iron at high pressure. Red curve, melting curve determined by internal resistance heated DAC (Sinmyo et al., 2019); black line, fcc-hcp boundary by resistance heating (Komabayashi et al., 2009). Previous studies using laser-heated DAC are shown by green line (Williams et al., 1987), blue line (Boehler, 1993), brown line (Anzellini et al., 2013), green triangles (Zhang et al., 2016), and yellow line (Basu et al., 2020). Theoretical predictions: blue diamonds (Laio et al., 2000), triangle (Alfè, 2009), and inverse triangle (Sola & Alfè, 2009). Shock data: filled diamond (Brown & McQueen, 1986), filled squares (Yoo et al., 1993), filled normal triangles (Nguyen & Holmes, 2004), and filled inverse triangles (Li et al., 2020). α, bcc; γ, fcc; ε, hcp iron.

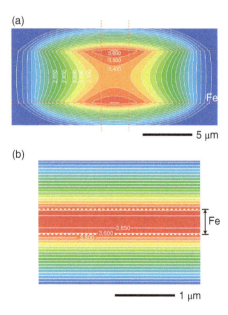

Figure 11.6 Temperature distribution around the iron sample in the (a) laser-heated DAC and (b) internal resistance heated DAC simulated by finite element modeling. Source: Modified after Sinmyo et al. (2019).

estimated by finite-element modeling to explain a part of this discrepancy (Sinmyo et al., 2019; Fig. 11.6). The results showed that the surface of the iron should be the hottest in the LH-DAC, while it is the coldest in the IH-DAC. The inside of the sample was likely molten to a large extent when a diffuse X-ray signal was observed during the laser heating (Anzellini et al., 2013), which would lead to an overestimation of melting temperature as the temperature was measured at the surface; if this is the case, the melting temperature may have been overestimated by LH-DAC and X-ray measurements. On the other hand, the melting temperature could be slightly underestimated by the IH-DAC experiments (Sinmyo et al., 2019). Notably, axial temperature variations can be smaller in a LH-DAC sample when the sample is thinner. In this case, however, it will be difficult to observe a diffuse X-ray signal due to the limited amount of the sample. Instead, the electrical resistance measurements may be more sensitive to melting in the thinner sample at higher pressure (Ohta et al., 2016; Basu et al., 2020). It is reported that the electrical resistivity of iron increased upon melting (Ohta et al., 2016; Basu et al., 2020).

11.3.3. Chemical Analysis and the Melting Phase Relationships

The melting phase relationships have been widely investigated to understand the chemistry and the dynamics of the core. The liquid outer core must be more

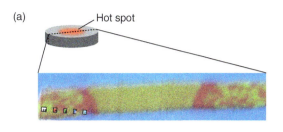

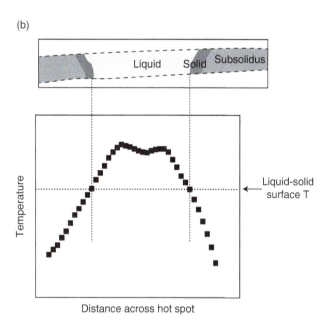

Figure 11.7 Experimental procedure to determine the temperature at the liquid-solid surface in the sample recovered from laser-heated DAC experiments. (a) Typical chemical mapping around the hot spot after melting experiments. (b) Reconstruction of the temperature profile by comparison with an observed melting texture.

enriched in the light elements than the solid inner core. Accordingly, the solidus and liquidus should be negative in the temperature-composition phase diagram in the Fe-FeX system. Moreover, the eutectic composition should be iron-poor compared to the expected bulk concentration of the core. Otherwise, the inner core would be the second iron-rich endmember phase in the Fe-FeX system, such as the B2 phase in Fe-FeSi system and the Fe_2S phase in the Fe-FeS system, which is far less dense than observations have shown (Sata et al., 2010; Fischer et al., 2012; Tateno et al., 2019). Melting experiments in Fe-FeX systems have been conducted in recent decades mostly by using multi-anvil and DAC and the results had to be extrapolated to core pressure in the multi-anvil experiments (e.g., Kuwayama & Hirose, 2004; Stewart et al., 2007; Fei & Brosh, 2014). In contrast, melting experiments are now routinely conducted at core pressure by LH-DACs. Although the samples are very small in the DAC experiments, modern electron microscope systems are capable of analyzing the chemical composition of the liquid and solid phases. Because the samples are as small as several $10*10*1$ μm^3 at core pressure (Fig. 11.2), it was very difficult to make thin sections by conventional polishing techniques from the recovered sample. Instead, the FIB system has recently been used to make thin films from the sample recovered from DAC experiments (Irifune et al., 2005; Miyahara et al., 2008) (Fig. 11.7). The thin films with very smooth surfaces can be easily taken from the region of interest by the FIB system. While the surface was often damaged by the Ga^+ ion beam, the damaged layer can be removed by further thinning with a gentle bombardment of ions using an Ar^+ ion milling apparatus (Miyajima et al., 2010). The thin films are then analyzed by scanning electron microscopy (SEM), transmission electron microscopy (TEM), and electron probe micro-analyzer (EPMA). Today a field-emission type electron-gun is widely used in the electron microscopes for better spatial and energy resolution. The energy-dispersive X-ray spectroscopy (EDS) systems are commonly attached in the SEM and TEM for chemical analysis. The electron microprobe is an apparatus for wavelength dispersive X-ray spectroscopy (WDS) using an electron beam. Although the WDS has a better ability to analyze light elements, a recent EDS system with an ultrathin window or windowless detector can detect light elements heavier than boron, such as oxygen and carbon (e.g., Mashino et al., 2019). The concentration of the oxygen in the iron has also been determined by electron energy loss spectroscopy in the TEM (Frost et al., 2010; Miyajima et al., 2010). The mass spectroscopy has been used to analyze trace elements and the isotopes in the large volume press sample (Poitrasson et al., 2009; Shahar et al., 2009; Satish-Kumar et al., 2011). It is recently tried to analyze the DAC samples by mass spectroscopy (e.g., Badro et al., 2007b; Bouhifd et al., 2013; Fischer et al., 2020).

Thanks to the precise thinning process by FIB, the temperature at the liquid-solid surface can be rigidly determined by a comparison between melting texture and the temperature profile during heating (e.g., Hirose et al., 2017) (Fig. 11.7). Chemically distinct portions are observed around the laser heating spot by the X-ray mapping using SEM and EDS systems (Fig. 11.7a). The temperature profiles can be precisely aligned to the boundary between solid and liquid in the X-ray mapping (Fig. 11.7b). The chemical composition may not show a large contrast between solid and liquid depending on the phase diagram. In this case, ion imaging by FIB potentially provides additional information on the melting texture, since the contrast in the ion imaging is sensitive to the crystal orientation (Canovic et al., 2008). Indeed, the crystallographic orientations of the solidus phases show clear similarities in the sample (Mashino et al., 2019).

The melting phase relationships have been widely investigated by LH-DAC in the binary systems of Fe-Si (Morard et al., 2011; Fischer et al., 2013; Ozawa et al., 2016), Fe-S (Morard et al., 2008; Mori et al., 2017), Fe-O (Seagle et al., 2008; Morard et al., 2017a; Oka et al., 2019), Fe-C (Morard et al., 2017a; Mashino et al., 2019), Fe-N (Breton et al., 2019; Kusakabe et al., 2019), and Fe-H (Kato et al., 2020). More recently, Boccato et al. (2020) reported melting phase relationships in Fe-C, Fe-O, Fe-S, and Fe-Si systems. The liquidus phases were determined by in situ X-ray diffraction measurements, and the chemical compositions of each phase have been determined by EDS and WDS systems. In addition, the temperature at the liquid-solid surface can be determined by a comparison of texture and temperature profiles as mentioned above (Fig. 11.7). The eutectic point can be constrained by the temperature and the chemistry of the solid and the liquid (e.g., Mori et al., 2017). Considering an ideal solution, the molar ratio of Fe in liquid at T, $X_{Fe\ liq}$, is given by the following equation:

$$X_{Fe\ liq} = exp\left\{\frac{\overline{G}^0_{Fe\ sol} - \overline{G}^0_{Fe\ liq}}{RT}\right\}$$

$$= exp\left\{-\int_{T_{m\ Fe}}^{T} \frac{\overline{H}^0_{Fe\ sol} - \overline{H}^0_{Fe\ liq}}{RT^2}\right\}$$

where $\overline{G}^0_{Fe\ i}$ and $\overline{H}^0_{Fe\ i}$ are the standard Gibbs free energy and the enthalpy of phase i, respectively, $T_{m\ Fe}$ is the melting temperature of the pure iron, and R is the gas constant. Assuming that the enthalpy difference in fusion, $\overline{H}^0_{Fe\ liq} - \overline{H}^0_{Fe\ sol}$, is independent of temperature, the equation above is written as follows:

$$X_{Fe\ liq} = exp\left\{\left(\frac{\overline{H}^0_{Fe\ liq} - \overline{H}^0_{Fe\ sol}}{RT_{m\ Fe}}\right) \times \left(1 - \frac{T_{m\ Fe}}{T}\right)\right\}$$

$$= exp\left\{A \times \left(1 - \frac{T_{m\ Fe}}{T}\right)\right\}$$

Here, parameter A is calculated from $T_{m\ Fe}$, eutectic temperature and eutectic composition at a given pressure. Figure 11.8 summarizes the eutectic point in the Fe-FeX systems at 330 GPa as determined by previous studies (Mori et al., 2017; Mashino et al., 2019; Oka et al., 2019; Sinmyo et al., 2019; Komabayashi, 2020). The eutectic temperatures are lower than the melting temperature of the pure iron in Fe-S, Fe-O, and Fe-C binary systems (Fig. 11.8). In contrast, it has been suggested that the eutectic composition of the Fe-Si system is very close to pure iron, and the eutectic temperature is also close to pure iron (Ozawa et al., 2016; Komabayashi, 2020). This suggests that the silicon cannot solely be a light element in the core. More recently, the melting phase relationships have been studied in ternary Fe-Si-S (Tateno et al., 2018), Fe-S-O (Yokoo et al., 2019), Fe-C-H (Hirose et al., 2019), and Fe-Si-O (Hirose et al., 2017) systems at core conditions. An integrated thermodynamic model will be constructed for multicomponent systems in the future (Hirose et al., 2017).

11.3.4. Density of Liquid Iron

The laboratory determination of the density (ρ) of liquid iron alloys at high pressure and high temperature is of great importance for estimating the core composition and behavior, because it is one of the primary observables of

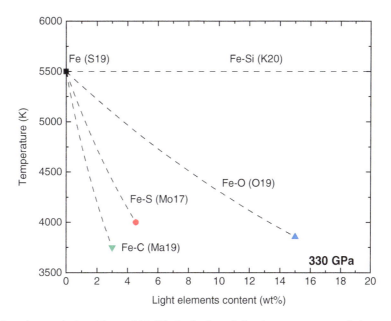

Figure 11.8 Melting phase relationships at 330 GPa in the iron-light elements systems. Estimated eutectic melting temperatures: circle, Fe-S (Mori et al., 2017); triangle, Fe-O (Oka et al., 2019); inverse triangle, Fe-C (Mashino et al., 2019); diamond, Fe-Si (Komabayashi, 2020). Square is the melting temperature of Fe (Sinmyo et al., 2019).

Earth's liquid outer core (Hirose et al., 2013). To date, the in situ X-ray absorption method, the sink/float method, and the XRD method have been applied for density measurements of liquid metals at high pressures (Sanloup, 2016; Terasaki, 2016, 2018). For example, the density of liquid Fe-S, Fe-Si, and Fe-C alloys was determined by the X-ray absorption method in a large volume press up to 6.2 GPa and 1780 K (Sanloup et al., 2000), 5 GPa and 1725 K (Sanloup et al., 2004), and 9.5 GPa and 1973 K (Terasaki et al., 2010), 9 GPa and 2123 K (Zhu et al., 2021), respectively. The density of Fe-S and FeSi alloys was also determined by the sink/float method in the large volume press up to 17.5 GPa (Balog et al., 2003) and 12 GPa (Yu & Secco, 2008), respectively. However, the experiment on liquid pure iron is more difficult than on iron alloys, because the melting temperature of pure iron is much higher. Thus, density measurements at high pressure have previously been limited to iron alloys. In addition, the achievable pressure-temperature range of the large volume press is much lower than that in the Earth's core.

The XRD method using a LH-DAC can potentially be applied to higher pressures; the density of liquid Fe-Ni-S, Fe-Ni-Si, and Fe-C alloys has been reported up to 94 GPa and 3200 K by in situ XRD techniques using a LH-DAC (Morard et al., 2013; Morard et al., 2017b). More recently, the density of liquid pure iron has been determined up to 116 GPa and 4350 K, close to the conditions at the top of Earth's core (Kuwayama et al., 2020). In order to generate a high enough temperature to melt iron and keep

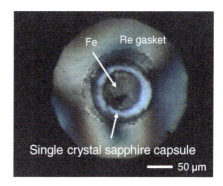

Figure 11.9 A microscope image of a sample chamber viewed through a diamond anvil and a thermal insulation layer of single-crystal sapphire.

its molten state during XRD measurements, single-crystal sapphire plates were placed between diamond anvils and the sample as thermal insulation layers. A single-crystal sapphire sleeve Al_2O_3 was also placed between the sample and the side wall of the gasket hole for better thermal insulation (Fig. 11.9). In some additional experiments, KCl powder was used as a thermal insulator, instead of Al_2O_3. Both Al_2O_3 and KCl were found to be a good leak-proof container for liquid iron, which kept the molten sample in the heating spot during collection of a diffuse scattering signal in XRD data.

One of the major difficulties with liquid analyses to extract accurate density is the proper subtraction of the background contributions. In the high-pressure experiments, the measured signal consists of not only the coherent scattering signal from the sample but also the background signals including the incoherent (Compton) scattering from the sample, the coherent and incoherent scattering from the pressure medium and the diamond anvil, and signals from any other surrounding materials. One way to subtract background is to take a diffraction signal of an empty cell and utilize it as the background (Eggert et al., 2002). However, the empty cell background data do not contain the contribution from the pressure medium. Even if the shape of the signal from the pressure medium is known, its thickness changes with increasing pressure and temperature. Temperature-diffuse scattering, which varies with temperature, should also be taken into account. In addition, the gasket hole shape, size, and thickness for the empty cell should be identical to those used for the liquid spectra, because the metallic gasket serves as an aperture for the X-ray scattering signal from the first diamond anvil and its shape is largely related to the aperture effect. The diffuse signal from the diamond also changes with pressure due to its deformation. These uncertainties of the background signal cause a critical error in determination of a density. The better way to avoid these uncertainties is a "solid-sample background method" (Eggert et al., 2002), in which a solid-sample background signal is measured just below the melting point for 1–2 seconds, subsequently the sample temperature is increased to 100–400 K above the melting temperature, and then a signal for liquid for another 1–2 seconds is collected (Kuwayama et al., 2020). The sequence is concluded within ~10 seconds. In this method, both the coherent and incoherent signals from the surrounding materials including the pressure medium and diamond anvils in the solid-sample background spectrum are same as that for liquid, because the sizes, volumes, and shapes of the surrounding materials are the same. The aperture effect of the gasket hole is also the same. The difference in the temperature effect is limited because the temperature around the sample when the background spectrum is collected is different from that for liquid spectrum by only several hundred K. The only difference between the solid-sample background spectrum and the liquid-sample spectrum is that the coherent scattering in a crystalline solid consists of a sharp Bragg peaks while that in a liquid consists of a broad diffuse scattering signal. Furthermore, only few single-crystal Bragg spots were observed around the melting temperature, due to the grain growth and the fast recrystallization (Anzellini et al., 2013): these spots are easily removed by masking. Therefore, we can easily extract only the coherent signal of liquid sample by subtracting the solid-sample background spectrum from the liquid-sample spectrum (Fig. 11.10),

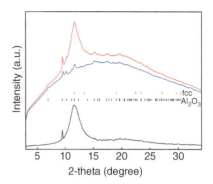

Figure 11.10 An X-ray diffraction spectrum for liquid iron collected at 2600 K at 21.5 GPa (red), together with the corresponding solid-sample background Spectrum collected at 2520 K (blue). The solid-sample background spectrum was taken at just below the melting temperature of iron where the all Bragg peaks from solid fcc (face-centered cubic) iron disappeared and the diffuse scattering signal from liquid iron started to appear. The peak positions of fcc iron (red) and Al_2O_3 insulators (black) are indicated. The bottom spectrum (black) is the difference between the liquid and the solid, corresponding to the diffuse scattering signal from liquid iron. An intense peak at around 2-theta = 9.5° is the diffraction peak from the single-crystal Al_2O_3 thermal insulator. Since this peak is very intense and exhibits blown-out highlights, it cannot be removed by subtracting background signals. Instead, it was removed using a cubic spline for data processing (dashed line). The high frequency noises hardly affect the final results on the calculated $g(r)$ in the small r region.

which is a critical advantage for obtaining an accurate density.

Another difficulty arises from the analytical method used to extract density from a diffuse XRD signal. After subtraction of the background spectrum, the X-ray diffuse scattering spectrum from liquid is normalized into atomic units and then converted to structure factor $S(Q)$, whereby $Q = 4\pi \sin\theta/\lambda$ is the momentum transfer. The $S(Q)$ of a monoatomic liquid is related to the distribution function $F(r)$ and radial distribution function $g(r)$ by the following equations:

$$F(r) = 4\pi r \{\rho(r) - \rho\},$$

$$Q\{S(Q) - 1\} = \int_0^\infty F(r) \sin(Qr) \, dr,$$

$$F(r) = \frac{2}{\pi} \int_0^\infty Q\{S(Q) - 1\} \sin(Qr) \, dQ,$$

$$g(r) = \frac{\rho(r)}{\rho} = 1 + \frac{F(r)}{4\pi r \rho},$$

where $\rho(r)$ is atomic density at a radial distance r (atoms per unit volume) and ρ is average atomic density. Thus, the density of the liquid can, in principle, be determined from the slope of $F(r)$ for r smaller than the interatomic spacing, where $F(r) = -4\pi\rho r$ and $g(r) = 0$. However, the transformation from Q to r requires integration over $Q \to \infty$. Because of the limited opening window of the high-pressure apparatus, the experimental limits on the Q range result in oscillations in $F(r)$ and $g(r)$ that lead to large uncertainty in the determination of density if not corrected. Previously, an iterative analytical procedure originally developed in Eggert et al. (2002) has been applied for liquid density determinations at high pressure (Morard et al., 2013; Sanloup et al., 2013). However, the iterative procedure alters the original $S(Q)$ during analysis. It may be a possible source of uncertainty in the density. Furthermore, it is not straightforward to evaluate how the $S(Q)$ and density values were altered by the iterative method quantitatively. The iterative procedure might show true values, only when the method has altered the $S(Q)$ to a small extent sufficiently (Morard et al., 2013; Sanloup et al., 2013). On the contrary, our method does not alter the $S(Q)$ during analysis (see following sections). It should also be noted that the background subtraction sometimes causes a critical error in the determination of a density. While some previous studies employed the spectra from the empty cell as a background (Sato et al., 2010; Ikuta et al., 2016), we used the spectra from the solid sample, which may be a better reference to the liquid sample than the brank cell.

In order to overcome the above problem, Kuwayama et al. (2020) developed an alternative analytical method in which the observed $S(Q)$ is extended beyond Q_{max} (the maximum Q in experimental data) so that the corresponding $g(r)$ is physically reasonable; $g(r) = 0$ for $r < r_{min}$ region (r_{min}, the distance between the nearest neighboring atoms), instead of iterative procedures. They found that $S(Q)$ can be extended by the following equation:

$$S_{extend}(Q) = \begin{cases} S(Q) & (Q \leq Q_{max}) \\ 1 - \dfrac{1}{Q} \int_0^{r_{min}} \left\{ 4\pi r\rho + \dfrac{2}{\pi} \int_0^{Q_{max}} \times Q\{S(Q)-1\}\sin(Qr)dQ \right\}\sin(Qr)dr & (Q > Q_{max}) \end{cases}$$

which can be calculated only from the experimental $S(Q)$ up to Q_{max} and a simple requirement that no atom exists inside the interatomic distance. The extension of $S(Q)$ successfully reduces the artificial oscillations in the calculated $F(r)$ and $g(r)$ (Figs. 11.11 and 11.12), enabling a precise determination of density. In contrast to the iterative procedure, the new method does not modify measured $S(Q)$, thus not causing any loss of information from raw data. The new analytical method was verified by the XRD data previously collected for a Ce-based metallic glass ($Ce_{70}Al_{10}Ni_{10}Cu_{10}$) and a synthetic XRD pattern,

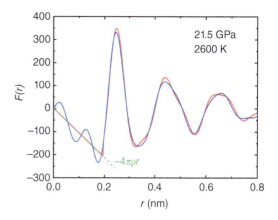

Figure 11.11 Distribution function $F(r)$ for liquid iron at 21.5 GPa and 2600 K calculated from extended $S(Q)$ (red), demonstrating that the new method successfully reduces the oscillations at $r < r_{min}$ and gives a precise liquid density from the slope. $F(r)$ calculated without extension of $S(Q)$ with assuming $s = 1$ is shown by the blue line for comparison. Source: Kuwayama et al. (2020)/American Physical Society.

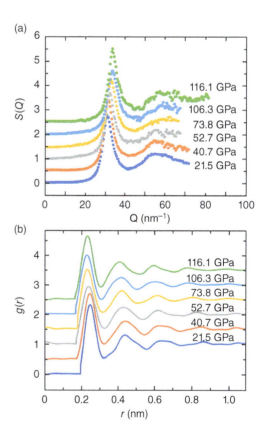

Figure 11.12 Measured structure factor, $S(Q)$, of liquid iron up to 116.1 GPa (a) and corresponding radial distribution functions, $g(r)$, determined by the new analytical method. Vertical scales are offset for clarity for both $S(Q)$ and $g(r)$ plots. Source: Kuwayama et al. (2020)/American Physical Society.

showing that this method is universally applicable for density determinations of amorphous materials.

In addition to density of liquid iron up to 116 GPa, the longitudinal sound velocity was also measured to 45 GPa and 2700 K based on inelastic X-ray scattering measurements (section 11.3.5). Combining these results with previous shock-wave data, it was possible to determine a thermal equation of state for liquid iron across the Earth's entire outer core conditions based on the Mie-Grüneisen EoS (Fig. 11.13). The adiabatic density and compressional wave velocity profiles calculated from the thermal equation of state for liquid iron with an assumed temperature at the ICB of 5400 K show that the Earth's liquid outer core exhibits low ρ by 0.99–0.81 g/cm^3 (7.5%–7.6%) and high V_P by 0.43–0.29 km/s (4.3%–3.7%). Such ρ deficit is about 1% smaller than the previous estimates of 8.4%–8.6% (Anderson & Ahrens, 1994). In contrast, it also shows that the observed K_S of the outer core is almost identical to that of liquid iron. The experimentally based equation of state of liquid iron would provide an important constraint for interpreting the composition of the outer core.

Note that the equation of state of liquid iron proposed by Anderson and Ahrens (1994) is one of the pioneering works. However, the shock experiments that they used in their equation of state (EoS) did not measure temperature. Therefore, their results, in particular for the thermal terms, largely depend on theoretical studies from the 1990s or earlier regarding the model of internal energy of liquid iron. Indeed, they overestimated the temperature by ~900 K at 278 GPa. Recent theoretical studies (Ichikawa et al., 2014; Wagle & Steinle-Neumann, 2019) also suggested that the temperatures in Anderson and Ahrens (1994) were overestimated.

11.3.5. Sound Velocity of Iron Alloys

The sound velocity of Fe alloys at high pressure and temperature is of great interest to constrain the chemical composition of the Earth's core because it can be directly compared with seismological observations. Shock compression experiments enable us to measure density and bulk sound velocity of solid and liquid Fe alloys along the Hugoniot curve to extreme pressure-temperature conditions (Brown & McQueen, 1986; Huang et al., 2011; Sakaiya et al., 2014). However, such a dynamic compression cannot control the pressure-temperature conditions and shock waves propagate through target samples within a few hundred nanoseconds such that it is unclear if chemical equilibrium has been achieved in a multicomponent sample (Brown & McQueen, 1986). The initial chemical bonding and arrangement of atoms of the shock-compression sample strongly affect the achievement of the multicomponent equilibrium (Gleason et al., 2015; Hwang et al., 2020; Morard et al., 2020; Turneaure et al., 2020; White et al., 2020).

Static compression experiments with a multi-anvil apparatus and DAC have also been performed and developed in recent decades to investigate the sound velocity of Fe alloys at high pressure and temperature. The conventional pulse-echo ultrasonic technique has been applied for both solid and liquid Fe alloys in a multi-anvil apparatus (Jing et al., 2014; Shibazaki & Kono, 2018; Terasaki et al., 2019; Nishida et al., 2020). However, this method requires the mm-size sample to resolve the overlapping acoustic pulse echoes with several tens of MHz frequency, so that the pressure range of the velocity measurements is limited to 20 GPa. Alternatively, the sound velocity measurements of solid Fe alloys at much higher pressures have been performed by using DAC combined with several techniques such as gigahertz interferometry, laser pump-probe methods, nuclear resonant inelastic X-ray scattering (NIS), and inelastic X-ray scattering (IXS). In particular, the IXS technique with DAC is now applied for both solid and liquid Fe alloys. Here, we focus on recent progress in IXS measurements on Fe alloys at high pressure and temperature.

The sound velocity V_P and V_S of samples can be determined from the phonon dispersion relations of longitudinal acoustic (LA) and transverse acoustic (TA) modes, respectively, obtained from IXS measurements. The incident photons with an energy $E_i = \hbar\omega_i$ and wave vector $\mathbf{k}_i$ are scattered by the sample and then transferred into the scattered photons with $E_s = \hbar\omega_s$ and $\mathbf{k}_s$. The energy transfer $E = E_i - E_s$ corresponds to phonon excitation energies of the sample at the momentum transfers $\mathbf{Q} = \hbar\mathbf{k}_i - \hbar\mathbf{k}_s$. In general, the phonon excitation energies of samples are 1–100 meV, which is much smaller than the incident photon energies of 15–26 keV used for IXS measurements. Therefore, the high energy meV resolution, typically 1.5–5 meV, is required to observe phonon excitation energies. For polycrystalline and liquid samples, the $\mathbf{Q}$ is defined by the scattering angles as $Q = 2k_i\sin(\theta/2)$. The propagation speed of an acoustic phonon is given by the slope of the acoustic dispersion relation (group velocity). At low Q, that is, long wavelength, the dispersion relation is almost linear and then the group velocity corresponds to the sound speed of the bulk material. In the case of polycrystalline solid Fe samples, which are typically used, the slope at the $Q = 0$ limit is often determined by a sine fit to the dispersion (e.g., Fiquet et al. 2001) because the dispersion relation at lower Q-range <4 nm^{-1} is hard to measure due to the large quasi-elastic peak or acoustic phonons originating from a diamond. Moreover, the TA mode is almost missing when using polycrystal samples due to the loss of the directional information of crystals, so that V_S can hardly be determined with polycrystalline

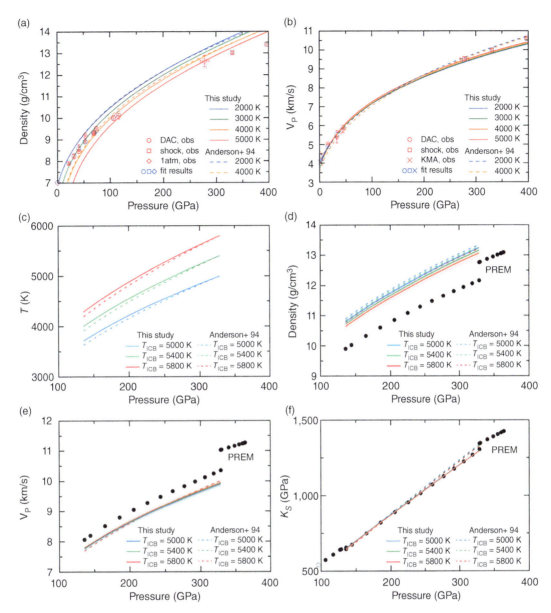

Figure 11.13 Density (ρ), p-wave velocity (V_P), and adiabatic bulk modulus (K_S) of liquid iron (Kuwayama et al., 2020/American Physical Society). (a), (b) Isothermal P-ρ and P-V_P relations calculated from our EoS for 2000 K (blue), 3000 K (green), 4000 K (yellow), and 5000 K (red). Dashed lines, 2000 K (red) and 4000 K (yellow), are from shock-compression study (Anderson & Ahrens, 1994). Red symbols represent experimental data (circles, this study; squares, shock experiments (Anderson & Ahrens, 1994); crosses, multi-anvil experiments (Nishida et al., 2016); diamonds, 1 bar data at 1811 K (Assael et al., 2006)). Consistency between the red and blue (fit results) symbols indicates that our EoS well reproduces all experimental data points. (c) Calculated isentropic temperature profiles with T_{ICB} = 5800 K (red), 5400 K (green), and 5000 K (blue). Dashed lines are those proposed by a previous study with a different Grüneisen parameter (Anderson & Ahrens, 1994). (d), (e), (f) Comparison of seismic observations (black circles, PREM; Dziewonski & Anderson, 1981) with the ρ, V_P, and K_S of liquid Fe under core pressures along the isentropic temperature profiles in (c). Uncertainties in the present estimates of ρ and V_P are ~1% (see the uncertainty band around each solid curve). Dashed lines represent those proposed on the basis of earlier shock-wave data (Anderson & Ahrens, 1994). Source: Kuwayama et al., 2020 / American Physical Society.

samples. For liquid samples, the sine fit to higher Q-range cannot be applied because the anomalous dispersion can appear at higher $Q \gg 3\,\text{nm}^{-1}$ (Scopigno et al., 2005). Therefore, a linear fit to the dispersion at $Q < \sim 3.5\,\text{nm}^{-1}$ is used to determine the sound velocity of liquid samples (Komabayashi et al., 2015; Nakajima et al., 2015).

The pioneering work of sound velocity determinations in geoscience using IXS by Fiquet et al. (2001) measured V_P of hcp-Fe in DAC to 110 GPa at 300 K, at ESRF. Subsequently, the IXS technique has been applied for Fe alloys with possible light elements (Fiquet et al., 2001; Antonangeli et al., 2004; Badro et al., 2007a). It took more than half a day to collect data at each pressure condition due to the tiny IXS signals from the high-pressure samples in DAC, which was extended to a couple of days with increasing pressure and decreasing sample sizes. This is because the intensity of inelastic signals is much less by approximately 4 orders of magnitude relative to the elastic scattering. Moreover, the beam flux can be reduced by 3–4 orders of magnitude relative to those supplied at the conventional XRD beamtime line to achieve high energy resolutions of the order of 3–5 meV, which is much smaller than 1–2 eV used in XRD measurements.

The high-pressure IXS measurements on solid Fe alloys have also been performed at APS (Mao et al., 2012) and SPring-8 (Ohtani et al., 2013). Mao et al. (2012) reported the P-wave velocity of hcp-Fe and $Fe_{85}Si_{15}$ up to 105 GPa and 700 K using external resistance heated DAC (EH-DAC). Ohtani et al. (2013) also measured the P-wave velocity of hcp-Fe under high pressure and temperature to 1000 K at SPring-8. Very recently, the pressure and temperature ranges of the IXS measurements on solid hcp-Fe were expanded to 163 GPa and 3000 K by using the LH-DAC technique at SPring-8 (Sakamaki et al., 2016). The results of those data on hcp-Fe at 300 K by several groups and at different beamlines exhibit relatively good consistency within the uncertainties of each measurement. The V_P data of hcp-Fe obtained by the IXS method is also in agreement with those from the picosecond ultrasonic method (Decremps et al., 2014), which are direct measurements of surface waves and acoustic echoes of the samples. On the other hand, the V_P data of Fe by the IXS method are often inconsistent with those from the NIS method. Because the NIS method provides the Debye velocity (V_D), not directly the V_P or V_S, the inconsistency can be due to the uncertainties of EoSs used in NIS methods for yielding V_P and V_S from the V_D data set. Those measurements suggested that the V_P of hcp-Fe can be larger than that of the solid inner core when considering a simple temperature effect (Sakamaki et al., 2016). The IXS technique has been also applied for Fe alloys with possible lighter elements under high pressures to ~140 GPa, e.g., dhcp-FeH_x, hcp-Fe-Si, FeO, FeS, FeS_2, Fe_3S, and Fe_3C (Badro et al., 2007a; Shibazaki et al., 2011, Kamada et al., 2014; Liu et al., 2016; Antonangeli et al., 2018; Takahashi et al., 2019). However, those studies reveal that the above-named light elements increase the V_P of Fe and cannot explain the slower V_P of the inner core. Most of those previous studies were carried out at 300 K such that the temperature effects on the V_P of Fe alloys are still unclear. Theoretical calculations show that abnormal reductions in V_P of hcp-Fe occur at high temperatures very close to its melting point (Martorell et al., 2013). The temperature of the inner core corresponds to the liquidus temperature of the liquid core, so that such a reduction may occur under the inner core pressure-temperature conditions. However, theoretical calculations reported that the premelting anomaly could be suppressed by the addition of light elements such as Si and C to hcp-Fe (Martorell et al., 2016; Li et al., 2018). In the future, more investigation of the temperature effect on the velocity of Fe alloys under high pressure and temperature near the melting temperature is needed to test the theoretical calculations. Moreover, the V_S of Fe alloys is rarely reported by IXS measurements because of the use of polycrystalline samples. The V_S of hcp-Fe at relevant high-pressure conditions has been suggested from NIS measurements (Mao et al., 2001) to be much faster than that of the inner core. Recently, NIS measurements on Fe_7C_3 and Fe_3C revealed that C can reduce the V_S of hcp-Fe to match with seismological observations (Chen et al., 2014; Prescher et al., 2015; Chen et al., 2018). These results are still unclear because the V_P and V_S from NIS measurements highly depend on the EoSs used, while the effect on V_S can be insignificant (Gao et al., 2011). Liu et al. (2016) determined the V_S of hcp-$Fe_{87}Ni_9Si_5$ alloy from both V_P from IXS and V_D from NIS measurements for the same starting composition at very similar pressure conditions to 133 GPa. This unique work with multiple complementary techniques shows that Ni can largely reduce the V_S of hcp-Fe but the possible temperature effect still needs to satisfy the seismological observations. So far, the V_S data of Fe alloys are essentially not sufficient to compare with inner core seismological observations. It is very challenging, but the direct measurements of both V_P and V_S for hcp-Fe and its alloys under core pressures by using IXS with a single-crystalline sample will give more valuable constraints on inner core compositions.

In contrast to the work on solid Fe alloys, the IXS measurements of liquids are still challenging, so that less work has been reported. This is because of the difficulty in handling liquid samples due to their high chemical reactivity and mobility. The first report on the sound velocity of liquid metals based on IXS measurements under high pressure and temperature was given by Alatas et al. (2008), who determined the longitudinal sound velocity of liquid In up to 4 GPa and 633 K using the EH-DAC technique at APS. Later, Komabayashi et al. (2015) also used the EH-DAC technique at SPring-8 and investigated the sound velocity of liquid In at higher pressure-temperature conditions of 6.7 GPa and 710 K. These results on liquid In by IXS measurements are in good agreement with the ultrasonic measurements at 1 bar. The EH-DAC technique was also applied for liquid Fe-C and Fe-Ni-S alloys (Nakajima et al., 2015; Kawaguchi et al., 2017). However, the heating method limits the temperature range to 1500 K (Kawaguchi et al., 2017), which only enables researchers to melt these Fe alloys up to 20 GPa even for low eutectic temperature alloys.

To expand the pressure range of IXS measurements on Fe alloys, Nakajima et al. (2015) have used the LH-DAC system at BL35XU of SPring-8 (Baron et al., 2000). Later, the system was also installed at BL43LXU of SPring-8 (Baron, 2010; Baron, 2016). In the system, the DAC is located in a vacuum chamber to minimize the X-ray scattering from the air. Contrary to the EH-DAC experiments, a thermal insulator is needed to heat samples by laser heating. They used single-crystal Al_2O_3 discs as the thermal insulator, which has a relatively high melting temperature, a fast sound speed, and a low chemical affinity with Fe alloys under high pressures. Unlike polycrystalline materials, single-crystal materials do not cause a quasi-elastic signal to the entire radial direction, which interferes with the IXS signals from the samples. Moreover, the single-crystal insulator was found to minimize the leak of liquid Fe alloys compared with polycrystalline ones, which enables researchers to keep the liquid for 1 hour or more at high pressure and temperature. With the LH-DAC system combined with the IXS technique, the sound velocity of liquid Fe alloys was enabled to be measured up to 70 GPa and 3200 K (Nakajima et al., 2015; Kawaguchi et al., 2017; Kinoshita et al., 2020; Kuwayama et al., 2020; Nakajima et al., 2020).

Figure 11.14 summarizes the V_P of liquid Fe alloys by LH-DAC with the IXS technique at high pressures to 70 GPa: Fe (Kuwayama et al., 2020), $Fe_{84}C_{16}$ (Nakajima et al., 2015), $(FeNi)_{75}S_{25}$ (Kawaguchi et al., 2017), $Fe_{84}Si_{16}$ (Nakajima et al., 2020), and $Fe_{75}P_{25}$ (Kinoshita et al., 2020). The V_P of liquid $(FeNi)_{75}S_{25}$ by IXS is in good agreement with liquid $Fe_{80}S_{20}$ obtained by ultrasonic measurements at 10–20 GPa with a multi-anvil press (Terasaki et al., 2019; Nishida et al., 2020), which can support the validity of the previous IXS measurements on liquid Fe alloys. The data of liquid $(FeNi)_{75}S_{25}$ by IXS are absent below 10 GPa and extrapolated to 1 bar. The difference from the ultrasonic data for $Fe_{80}S_{20}$ can be due to a possible magnetic transition around 1 GPa (Kawaguchi et al., 2017; Nishida et al., 2020). The V_P of liquid $Fe_{84}Si_{16}$ based on IXS is also in agreement with the ultrasonic data for liquid Fe-Si alloys. The V_P of liquid

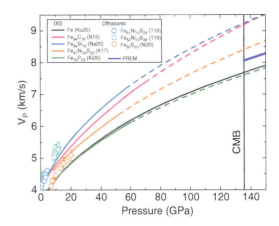

Figure 11.14 Sound wave velocity of liquid Fe alloys based on IXS measurements under high pressures. Solid and broken curves represent experimental pressure ranges and extrapolated ones, respectively, for Fe (Kuwayama et al., 2020), $Fe_{84}C_{16}$ (Nakajima et al., 2015), $Fe_{84}Si_{16}$ (Nakajima et al., 2020), $Fe_{47}Ni_{28}S_{25}$ (Kawaguchi et al., 2017), and $Fe_{75}P_{25}$ (Kinoshita et al., 2020). The open symbols are results of ultrasonic measurements for liquid $Fe_{60}Ni_{10}Si_{29}$, $Fe_{52}Ni_{10}Si_{38}$ (Terasaki et al., 2019), and $Fe_{80}S_{20}$ (Nishida et al., 2020). PREM represents seismological observations for the outer core (Dziewonski & Anderson, 1981).

$Fe_{84}Si_{16}$ and $Fe_{84}C_{16}$ is much faster than that of liquid Fe at high pressures, while S and P have a relatively small effect on the V_P of liquid Fe. The tendency among liquid Fe-light elements and the similarity of liquid Fe-Si and Fe-C could not be simply explained by the atomic sizes and mass of each light element in liquid Fe as proposed by recent theoretical calculations at the core conditions (Brodholt & Badro, 2017).

Figure 11.15 exhibits the effects of each light element on the V_P of liquid Fe at 136 GPa and 4000 K relevant to the conditions at the top of the outer core based on the IXS measurements (Nakajima et al., 2015; Kawaguchi et al., 2017; Kinoshita et al., 2020; Kuwayama et al., 2020; Nakajima et al., 2020). While P has an almost negligible effect, other light elements increase the V_P of liquid Fe. Compared with seismological observations in PREM, the amount of Si and C are limited in the present outer core, while a relatively large amount of S and P can be allowed. The pressure-temperature conditions of the present IXS measurements on liquid Fe alloys are still half of the outer core conditions. Further expansion of the pressure-temperature conditions for the velocity measurements is required to tightly constrain the chemical composition of the outer core.

11.3.6. Candidate of the Core Light Elements

As discussed above, we can better understand the core's chemical composition by the melting phase relationship,

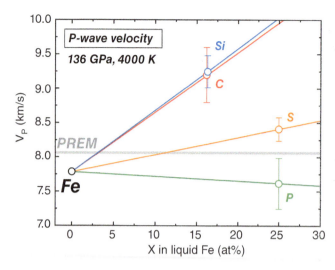

Figure 11.15 Effects of light elements on V_P of liquid Fe at the CMB conditions. The V_P values of Fe (Kuwayama et al., 2020), $Fe_{84}C_{16}$ (Nakajima et al., 2015), $Fe_{84}Si_{16}$ (Nakajima et al., 2020), $Fe_{47}Ni_{28}S_{25}$ (Kawaguchi et al., 2017), and $Fe_{75}P_{25}$ (Kinoshita et al., 2020) at 136 GPa and 4000 K are plotted as a function of the concentration of each light element in liquid Fe. PREM represents seismological observations for the outer core (Dziewonski & Anderson, 1981).

velocity, and density. We have summarized the results of the melting phase relationship (Fig. 11.8) and the velocity of the liquid Fe alloys (Fig. 11.15) above. In addition, we compared the effect of light elements on the density of the liquid based on the previous studies (Fig. 11.16). The slopes for Fe-Si, Fe-C, and Fe-S are taken based on previous studies using X-ray diffraction measurements (Morard et al., 2013; Morard et al., 2017b; Kuwayama et al., 2020). The densities of liquid $Fe_{74}Si_{26}$, $Fe_{81}S_{19}$, and $Fe_{88}C_{12}$ were calculated at 136 GPa and the possible CMB temperatures of 3760 ± 290 K (Sinmyo et al., 2019), based on the reported isothermal EoSs on $Fe_{70}Ni_4Si_{26}$, $Fe_{77}Ni_4S_{19}$ (Morard et al., 2013), and $Fe_{88}C_{12}$ (Morard et al., 2017b). The temperature and Ni effect on the reported EoSs were corrected with the thermal expansion coefficient of liquid Fe at 136 GPa (Kuwayama et al., 2020) and the mass change due to the replacement of Ni by Fe.

We summarized these three comparisons (Figs 11.8, 11.15, and 11.16) in Table 11.1. By looking at the density and V_P, Si or C cannot sorely reconcile both parameters (Table 11.1). For instance, while the density of the outer core is consistent with 4.5–5.3 wt% of Si, the amount is too much for V_P. On the contrary to Si and C, both density and V_P are simultaneously consistent with 6.2–7.3 wt% of S (Figs. 11.15 and 11.16). However, the concentration of the elements in the core should not exceed the eutectic composition; otherwise, the light elements are more enriched in the inner core than the outer core. According to the melting phase relationships, the maximum concentrations are limited to 4.6 wt% for S (Fig. 11.8). Accordingly, the elements S, Si, and C are not the sole light element of the core, although the experimental data are still limited. It should also be noted that the effects of O and H on the density and V_P are not well understood so far (Table 11.1). It is necessary to study further the density, sound wave velocity, and phase

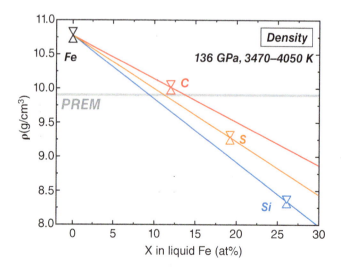

Figure 11.16 Effects of light elements on the density of liquid Fe at the CMB conditions. Triangle and inverse-triangle symbols are the densities of liquid Fe-light element alloys at 3470 K and 4050 K at 136 GPa, based on XRD measurements of Fe (Kuwayama et al., 2020), $Fe_{70}Ni_4Si_{26}$ (Morard et al., 2013), $Fe_{77}Ni_4S_{19}$ (Morard et al., 2013), and $Fe_{88}C_{12}$ (Morard et al., 2017b). The solid lines are the possible effects of each light element on the density of liquid Fe. Those are compared with seismological observations of the outer core (PREM: Dziewonski & Anderson, 1981).

Table 11.1 Candidates of the light elements in the core

	Depression of melting temperature at ICB	Eutectic composition at ICB	Required amount to match density at CMB	Required amount to match V_P at CMB
S	330 K/wt%[a]	4.6 wt%[a]	6.2–7.3 wt%[e]	6.8 wt%[g]
Si	0 K/ wt%[b]	-[b]	4.5–5.3 wt%[e]	1.6 wt%[h]
O	110 K/ wt%[c]	15 wt%[c]	?	?
C	580 K/ wt%[d]	3.0 wt%[d]	3.1–3.6 wt%[f]	0.7 wt%[i]
H	?	?	?	?
P	?	?	?	no solution[j]

[a] Mori et al., 2017
[b] Komabayashi et al., 2020
[c] Oka et al., 2019
[d] Mashino et al., 2019
[e] Morard et al., 2013
[f] Morard et al., 2017
[g] Kawaguchi et al., 2017
[h] Nakajima et al. 2020
[i] Nakajima et al., 2015
[j] Kinoshita et al., 2020.

relationships of iron and iron alloys in Fe-O, Fe-H, and multicomponent systems by experiments and theoretical calculations.

ACKNOWLEDGMENTS

This work was supported by JSPS KAKENHI Grant Numbers JP16H01115, 16H06285, 17K14418, JP18H04366, JP19H01989, 19K04040, and 21H04968.

REFERENCES

Alatas, A., Sinn, H., Zhao, J., Said, A. H., Leu, B. M., Sturhahn, W., et al. (2008). Experimental aspects of inelastic X-ray scattering studies on liquids under extreme conditions (P–T). *High Pressure Research*, *28*(3), 175–183.

Alfè, D. (2009). Temperature of the inner-core boundary of the Earth: Melting of iron at high pressure from first-principles coexistence simulations. *Physical Review B*, *79*(6), 060101(R).

Alfè, D. (2015). 2.15 – The Ab Initio Treatment of High-Pressure and High-Temperature Mineral Properties and Behavior. In G. Schubert (Ed.), *Treatise on Geophysics*, 2nd Edn, pp. 369–392. Oxford: Elsevier.

Alfe, D., Gillan, M. J., & Price, G. D. (2007). Temperature and composition of the Earth's core. *Contemporary Physics*, *48*(2), 63–80.

Anderson, W. W., & Ahrens, T. J. (1994). An equation of state for liquid iron and implications for the Earth's core. *Journal of Geophysical Research: Solid Earth*, *99*(B3), 4273–4284.

Antonangeli, D., Komabayashi, T., Occelli, F., Borissenko, E., Walters, A. C., Fiquet, G., et al. (2012). Simultaneous sound velocity and density measurements of hcp iron up to 93 GPa and 1,100 K: An experimental test of the Birch's law at high temperature. *Earth and Planetary Science Letters*, *331*, 210–214.

Antonangeli, D., Morard, G., Paolasini, L., Garbarino, G., Murphy, C. A., Edmund, E., et al. (2018). Sound velocities and density measurements of solid hcp-Fe and hcp-Fe-Si (9 wt.%) alloy at high pressure: Constraints on the Si abundance in the Earth's inner core. *Earth and Planetary Science Letters*, *482*, 446–453.

Antonangeli, D., Occelli, F., Requardt, H., Badro, J., Fiquet, G., & Krisch, M. (2004). Elastic anisotropy in textured hcp-iron to 112 GPa from sound wave propagation measurements. *Earth and Planetary Science Letters*, *225*(1–2), 243–251.

Anzellini, S., Dewaele, A., Mezouar, M., Loubeyre, P., & Morard, G. (2013). Melting of iron at Earth's inner core boundary based on fast X-ray diffraction. *Science*, *340*(6131), 464–466.

Aprilis, G., Kantor, I., Kupenko, I., Cerantola, V., Pakhomova, A., Torchio, I., et al. (2019). Comparative study of the influence of pulsed and continuous wave laser heating on the mobilization of carbon and its chemical reaction with iron in a diamond anvil cell. *Journal of Applied Physics*, *125*(9).

Aprilis, G., Strohm, C., Kupenko, I., Linhardt, S., Laskin, A., Vasiukov, D. M., et al. (2017). Portable double-sided pulsed laser heating system for time-resolved geoscience and materials science applications. *Review of Scientific Instruments*, *88*(8), 084501.

Aquilanti, G., Trapananti, A., Karandikar, A., Kantor, I., Marini, C., Mathon, O., et al. (2015). Melting of iron determined by X-ray absorption spectroscopy to 100 GPa. *Proceedings of the National Academy of Sciences U S A*, *112*(39), 12042–12045.

Asimow, P. (2015). Dynamic compression. In G. Schubert (Ed.), *Treatise on Geophysics*, 2nd Edn, pp. 393–416. Oxford: Elsevier.

Assael, M. J., Kakosimos, K., Banish, R. M., Brillo, J., Egry, I., Brooks, R., et al. (2006). Reference data for the density and viscosity of liquid aluminum and liquid iron. *Journal of Physical and Chemical Reference Data*, *35*(1), 285–300.

Badro, J., Cote, A. S., & Brodholt, J. P. (2014). A seismologically consistent compositional model of Earth's core. *Proceedings of the National Academy of Sciences U S A*, *111*(21), 7542–7545.

Badro, J., Fiquet, G., Guyot, F., Gregoryanz, E., Occelli, F., Antonangeli, D., et al. (2007a). Effect of light elements on the sound velocities in solid iron: Implications for the composition of Earth's core. *Earth and Planetary Science Letters*, *254*(1–2), 233–238.

Badro, J., Ryerson, F. J., Weber, P. K., Ricolleau, A., Fallon, S. J., & Hutcheon, I. D. (2007b). Chemical imaging with NanoSIMS: A window into deep-Earth geochemistry. *Earth and Planetary Science Letters*, *262*(3–4), 543–551.

Badro, J., Siebert, J., & Nimmo, F. (2016). An early geodynamo driven by exsolution of mantle components from Earth's core. *Nature*, *536*(7616), 326–328.

Balog, P. S., Secco, R. A., Rubie, D. C., & Frost, D. J. (2003). Equation of state of liquid Fe-10 wt % S: Implications for the metallic cores of planetary bodies. *Journal of Geophysical Research: Solid Earth*, *108*(B2), 2124.

Baron, A., Tanaka, Y., Goto, S., Takeshita, K., Matsushita, T., & Ishikawa, T. (2000). An X-ray scattering beamline for studying dynamics. *Journal of Physics and Chemistry of Solids*, *61*(3), 461–465.

Baron, A. Q. R. (2010). Status of the RIKEN Quantum NanoDynamics Beamline (BL43LXU): The Next Generation for Inelastic X-Ray Scattering. *SPring-8 Information Newsletter*, *15*, 14–19.

Baron, A. Q. R. (2016). High-Resolution Inelastic X-Ray Scattering I: Context, Spectrometers, Samples, and Superconductors. In E. J. Jaeschke, S. Khan, J. R. Schneider, & J. B. Hastings (Eds.), *Synchrotron Light Sources and Free-Electron Lasers: Accelerator Physics, Instrumentation and Science Applications*, pp. 1643–1719. Cham: Springer International Publishing.

Basu, A., Field, M. R., McCulloch, D. G., & Boehler, R. (2020). New measurement of melting and thermal conductivity of iron close to outer core conditions. *Geoscience Frontiers*, *11*(2), 565–568.

Belonoshko, A. B., Lukinov, T., Fu, J., Zhao, J., Davis, S., & Simak, S. I. (2017). Stabilization of body-centred cubic iron under inner-core conditions. *Nature Geoscience*, *10*(4), 312–316.

Birch, F. (1952). Elasticity and constitution of the Earth's interior. *Journal of Geophysical Research*, *57*(2), 227–286.

Boccato, S., Torchio, R., Anzellini, S., Boulard, E., Guyot, F., Irifune, T., et al. (2020). Melting properties by X-ray absorption spectroscopy: Common signatures in binary Fe-C, Fe-O, Fe-S and Fe-Si systems. *Scientific Reports*, *10*(1), 11663.

Boehler, R. (1993). Temperatures in the Earth's core from melting-point measurements of iron at high static pressures. *Nature*, *363*(6429), 534–536.

Boehler, R. (2006). New diamond cell for single-crystal x-ray diffraction. *Review of Scientific Instruments*, *77*(11), 115103.

Boehler, R., Guthrie, M., Molaison, J. J., dos Santos, A. M., Sinogeikin, S., Machida, S.-i., et al. (2013). Large-volume diamond cells for neutron diffraction above 90 GPa. *High Pressure Research*, *33*(3), 546–554.

Bouhifd, M. A., Jephcoat, A. P., Heber, V. S., & Kelley, S. P. (2013). Helium in Earth's early core. *Nature Geoscience*, *6*(11), 982–986.

Breton, H., Komabayashi, T., Thompson, S., Potts, N., McGuire, C., Suehiro, S., et al. (2019). Static compression of Fe$_4$N to 77 GPa and its implications for nitrogen storage in the deep Earth. *American Mineralogist*, *104*(12), 1781–1787.

Brodholt, J., & Badro, J. (2017). Composition of the low seismic velocity E′ layer at the top of Earth's core. *Geophysical Research Letters*, *44*(16), 8303–8310.

Brown, J. M., & McQueen, R. G. (1986). Phase transitions, grüneisen parameter, and elasticity for shocked iron between 77 GPa and 400 GPa. *Journal of Geophysical Research*, *91*(B7), 7485–7494.

Campbell, A. J. (2008). Measurement of temperature distributions across laser heated samples by multispectral imaging radiometry. *Review of Scientific Instruments*, *79*(1), 015108.

Canil, D. (1994). Stability of clinopyroxene at pressure-temperature conditions of the transition region. *Physics of the Earth and Planetary Interiors*, *86*(1–3), 25–34.

Canovic, S., Jonsson, T., & Halvarsson, M. (2008). Grain contrast imaging in FIB and SEM. *Journal of Physics: Conference Series*, *126*(1), 012054.

Chen, B., Lai, X., Li, J., Liu, J., Zhao, J., Bi, W., et al. (2018). Experimental constraints on the sound velocities of cementite Fe3C to core pressures. *Earth and Planetary Science Letters*, *494*, 164–171.

Chen, B., Li, Z., Zhang, D., Liu, J., Hu, M. Y., Zhao, J., et al. (2014). Hidden carbon in Earth's inner core revealed by shear softening in dense Fe$_7$C$_3$. *Proceedings of the National Academy of Sciences*, *111*(50), 17755–17758.

Chidester, B. A., Rahman, Z., Righter, K., & Campbell, A. J. (2017). Metal–silicate partitioning of U: Implications for the heat budget of the core and evidence for reduced U in the mantle. *Geochimica et Cosmochimica Acta*, *199*, 1–12.

de Koker, N., Steinle-Neumann, G., & Vlcek, V. (2012). Electrical resistivity and thermal conductivity of liquid Fe alloys at high P and T, and heat flux in Earth's core. *Proceedings of the National Academy of Sciences of the United States of America*, *109*(11), 4070–4073.

Decremps, F., Antonangeli, D., Gauthier, M., Ayrinhac, S., Morand, M., Marchand, G. L., et al. (2014). Sound velocity of iron up to 152 GPa by picosecond acoustics in diamond anvil cell. *Geophysical Research Letters*, *41*(5), 1459–1464.

Denoeud, A., Ozaki, N., Benuzzi-Mounaix, A., Uranishi, H., Kondo, Y., Kodama, R., et al. (2016). Dynamic X-ray diffraction observation of shocked solid iron up to 170 GPa. *Proceedings of the National Academy of Sciences U S A*, *113*(28), 7745–7749.

Dewaele, A., Fiquet, G., & Gillet, P. (1998). Temperature and pressure distribution in the laser-heated diamond-anvil cell. *Review of Scientific Instruments*, *69*(6), 2421–2426.

Dewaele, A., Loubeyre, P., Occelli, F., Marie, O., & Mezouar, M. (2018). Toroidal diamond anvil cell for detailed measurements under extreme static pressures. *Nature Communications*, *9*(1), 2913.

Dewaele, A., Loubeyre, P., Occelli, F., Mezouar, M., Dorogokupets, P. I., & Torrent, M. (2006). Quasihydrostatic equation of state of iron above 2 Mbar. *Physical Review Letters*, *97*(21), 215504.

Du, Z. X., Amulele, G., Benedetti, L. R., & Lee, K. K. M. (2013a). Mapping temperatures and temperature gradients during flash heating in a diamond-anvil cell. *Review of Scientific Instruments*, *84*(7), 075111.

Du, Z. X., Miyagi, L., Amulele, G., & Lee, K. K. M. (2013b). Efficient graphite ring heater suitable for diamond-anvil cells to 1,300 K. *Review of Scientific Instruments*, *84*(2), 024502.

Dubrovinskaia, N., & Dubrovinsky, L. (2003). Whole-cell heater for the diamond anvil cell. *Review of Scientific Instruments*, *74*(7), 3433–3437.

Dubrovinskaia, N., Dubrovinsky, L., Solopova, N. A., Abakumov, A., Turner, S., Hanfland, M., et al. (2016). Terapascal static pressure generation with ultrahigh yield strength nanodiamond. *Science Advances*, *2*(7), 1–12.

Dubrovinsky, L. S., Saxena, S. K., & Lazor, P. (1998). High-pressure and high-temperature in situ X-ray diffraction study of iron and corundum to 68 GPa using an internally heated diamond anvil cell. *Physics and Chemistry of Minerals*, *25*(6), 434–441.

Duffy, T., Madhusudhan, N., & Lee, K. (2015). 2.07 Mineralogy of super-earth planets. In G. Schubert (Ed.), *Treatise on Geophysics*, 2nd Edn, pp. 149–178. Oxford: Elsevier.

Duffy, T. S., & Smith, R. F. (2019). Ultra-high pressure dynamic compression of geological materials. *Frontiers in Earth Science*, *7*, 23.

Dziewonski, A. M., & Anderson, D. L. (1981). Preliminary reference earth model. *Physics of the Earth and Planetary Interiors*, *25*(4), 297–356.

Eggert, J. H., Weck, G., Loubeyre, P., & Mezouar, M. (2002). Quantitative structure factor and density measurements of high-pressure fluids in diamond anvil cells by x-ray diffraction: Argon and water. *Physical Review B*, *65*(17), 1–12.

Faure, P., Bouhifd, M. A., Boyet, M., Manthilake, G., Clesi, V., & Devidal, J.-L. (2020). Uranium and thorium partitioning in the bulk silicate Earth and the oxygen content of Earth's core. *Geochimica et Cosmochimica Acta*, *275*, 83–98.

Fei, Y., & Brosh, E. (2014). Experimental study and thermodynamic calculations of phase relations in the Fe–C system at high pressure. *Earth and Planetary Science Letters*, *408*, 155–162.

Fiquet, G., Badro, J., Guyot, F., Requardt, H., & Krisch, M. (2001). Sound velocities in iron to 110 gigapascals. *Science*, *291*(5503), 468–471.

Fischer, R. A., Campbell, A. J., Caracas, R., Reaman, D. M., Dera, P., & Prakapenka, V. B. (2012). Equation of state and phase diagram of Fe–16Si alloy as a candidate component of Earth's core. *Earth and Planetary Science Letters*, *357*, 268–276.

Fischer, R. A., Campbell, A. J., Reaman, D. M., Miller, N. A., Heinz, D. L., Dera, P., et al. (2013). Phase relations in the Fe–FeSi system at high pressures and temperatures. *Earth and Planetary Science Letters*, *373*, 54–64.

Fischer, R. A., Cottrell, E., Hauri, E., Lee, K. K. M., & Le Voyer, M. (2020). The carbon content of Earth and its core. *Proceedings of the National Academy of Sciences of the United States of America*, *117*(16), 8743–8749.

Fischer, R. A., Nakajima, Y., Campbell, A. J., Frost, D. J., Harries, D., Langenhorst, F., et al. (2015). High pressure metal-silicate partitioning of Ni, Co, V, Cr, Si, and O. *Geochimica et Cosmochimica Acta*, *167*, 177–194.

Frost, D. J., Tsuno, K., Rubie, D. C., & Nakajima, Y. (2010). Si and O in the Earth's core and their effects on the metal-silicate partitioning of other siderophile elements. *Geochimica et Cosmochimica Acta*, *74*(12), A308–A308.

Funamori, N., & Sato, T. (2008). A cubic boron nitride gasket for diamond-anvil experiments. *Review of Scientific Instruments*, *79*(5), 053903.

Gao, L., Chen, B., Zhao, J., Alp, E. E., Sturhahn, W., & Li, J. (2011). Effect of temperature on sound velocities of compressed Fe_3C, a candidate component of the Earth's inner core. *Earth and Planetary Science Letters*, *309*(3), 213–220.

Geballe, Z. M., Holtgrewe, N., Karandikar, A., Greenberg, E., Prakapenka, V. B., & Goncharov, A. F. (2020). A latent heat method to detect melting and freezing of metals at megabar pressures. *Physical Review Materials*, *5*, 033803.

Geballe, Z. M., & Jeanloz, R. (2012). Origin of temperature plateaus in laser-heated diamond anvil cell experiments. *Journal of Applied Physics*, *111*(12), 123518.

Georg, R. B., Halliday, A. N., Schauble, E. A., & Reynolds, B. C. (2007). Silicon in the Earth's core. *Nature*, *447*(7148), 1102–1106.

Gleason, A. E., Bolme, C. A., Lee, H. J., Nagler, B., Galtier, E., Milathianaki, D., et al. (2015). Ultrafast visualization of crystallization and grain growth in shock-compressed SiO_2. *Nature Communications*, *6*(1), 8191.

Gubbins, D., Masters, G., & Nimmo, F. (2008). A thermochemical boundary layer at the base of Earth's outer core and independent estimate of core heat flux. *Geophysical Journal International*, *174*(3), 1007–1018.

Hayden, L. A., & Watson, E. B. (2007). A diffusion mechanism for core–mantle interaction. *Nature*, *450*(7170), 709–711.

Helffrich, G. (2014). Outer core compositional layering and constraints on core liquid transport properties. *Earth and Planetary Science Letters*, *391*, 256–262.

Helffrich, G., Ballmer, M. D., & Hirose, K. (2018a). Core-exsolved SiO_2 dispersal in the earth's mantle. *Journal of Geophysical Research-Solid Earth*, *123*(1), 176–188.

Helffrich, G., & Kaneshima, S. (2010). Outer-core compositional stratification from observed core wave speed profiles. *Nature*, *468*(7325), U807–U896.

Helffrich, G., Shahar, A., & Hirose, K. (2018b). Isotopic signature of core-derived SiO_2. *American Mineralogist*, *103*(7), 1161–1164.

Hirao, N., Kawaguchi, S., Hirose, K., Shimizu, K., Ohtani, E., & Ohishi, Y. (2020). New developments in high-pressure X-ray diffraction beamline for diamond anvil cell at SPring-8. *Matter and Radiation at Extremes*, *5*(1), 018403.

Hirao, N., Ohtani, E., Kondo, T., Endo, N., Kuba, T., Suzuki, T., et al. (2006). Partitioning of potassium between iron and silicate at the core-mantle boundary. *Geophysical Research Letters*, *33*(8), 1–4.

Hirose, K., & Kushiro, I. (1993). Partial melting of dry peridotites at high pressures: Determination of compositions of melts segregated from peridotite using aggregates of diamond. *Earth and Planetary Science Letters*, *114*(4), 477–489.

Hirose, K., Labrosse, S., & Hernlund, J. (2013). Composition and state of the core. *Annual Review of Earth and Planetary Sciences*, *41*, 657–691.

Hirose, K., Morard, G., Sinmyo, R., Umemoto, K., Hernlund, J., Helffrich, G., et al. (2017). Crystallization of silicon dioxide and compositional evolution of the Earth's core. *Nature*, *543*(7643), 99–102.

Hirose, K., Tagawa, S., Kuwayama, Y., Sinmyo, R., Morard, G., Ohishi, Y., et al. (2019). Hydrogen Limits Carbon in Liquid Iron. *Geophysical Research Letters*, *46*(10), 5190–5197.

Holzapfel, C., Rubie, D. C., Frost, D. J., & Langenhorst, F. (2005). Fe-Mg interdiffusion in $(Mg,Fe)SiO_3$ perovskite and lower mantle reequilibration. *Science*, *309*(5741), 1707–1710.

Hu, M. Y., Sturhahn, W., Toellner, T. S., Mannheim, P. D., Brown, D. E., Zhao, J., et al. (2003). Measuring velocity of sound with nuclear resonant inelastic X-ray scattering. *Physical Review B*, *67*(9), 094304.

Huang, H. J., Fei, Y. W., Cai, L. C., Jing, F. Q., Hu, X. J., Xie, H. S., et al. (2011). Evidence for an oxygen-depleted liquid outer core of the Earth. *Nature*, *479*(7374), 513–U236.

Hwang, H., Galtier, E., Cynn, H., Eom, I., Chun, S. H., Bang, Y., et al. (2020). Subnanosecond phase transition dynamics in laser-shocked iron. *Science Advances*, *6*(23), eaaz5132.

Ichikawa, H., & Tsuchiya, T. (2020). Ab initio thermoelasticity of liquid iron-nickel-light element alloys. *Minerals*, *10*(1), 59.

Ichikawa, H., Tsuchiya, T., & Tange, Y. (2014). The P-V-T equation of state and thermodynamic properties of liquid iron. *Journal of Geophysical Research: Solid Earth*, *119*(1), 240–252.

Ikuta, D., Kono, Y., & Shen, G. (2016). Structural analysis of liquid aluminum at high pressure and high temperature using the hard sphere model. *Journal of Applied Physics*, *120*(13), 1–10.

Immoor, J., Marquardt, H., Miyagi, L., Lin, F., Speziale, S., Merkel, S., et al. (2018). Evidence for {100}<011> slip in ferropericlase in Earth's lower mantle from high-pressure/high-temperature experiments. *Earth and Planetary Science Letters*, *489*, 251–257.

Irifune, T., Isshiki, M., & Sakamoto, S. (2005). Transmission electron microscope observation of the high-pressure form of magnesite retrieved from laser heated diamond anvil cell. *Earth and Planetary Science Letters*, *239*(1–2), 98–105.

Irifune, T., Kunimoto, T., Shinmei, T., & Tange, Y. (2019). High pressure generation in Kawai-type multianvil apparatus using nano-polycrystalline diamond anvils. *Comptes Rendus Geoscience*, *351*(2–3), 260–268.

Irifune, T., Kurio, A., Sakamoto, S., Inoue, T., & Sumiya, H. (2003). Ultrahard polycrystalline diamond from graphite (vol 421, pg 599, 2003). *Nature*, *421*(6925), 806–806.

Ishii, T., Shi, L., Huang, R., Tsujino, N., Druzhbin, D., Myhil, R., et al. (2016). Generation of pressures over 40 GPa using Kawai-type multi-anvil press with tungsten carbide anvils. *Review of Scientific Instruments*, *87*(2), 024501.

Ishimatsu, N., Matsumoto, K., Maruyama, H., Kawamura, N., Mizumaki, M., Sumiya, H., et al. (2012). Glitch-free X-ray absorption spectrum under high pressure obtained using nano-polycrystalline diamond anvils. *Journal of Synchrotron Radiation*, *19*, 768–772.

Jing, Z., Wang, Y., Kono, Y., Yu, T., Sakamaki, T., Park, C., et al. (2014). Sound velocity of Fe–S liquids at high pressure: Implications for the Moon's molten outer core. *Earth and Planetary Science Letters*, *396*, 78–87.

Kamada, S., Ohtani, E., Fukui, H., Sakai, T., Terasaki, H., Takahashi, S., et al. (2014). The sound velocity measurements of Fe_3S. *American Mineralogist*, *99*(1), 98–101.

Kato, C., Umemoto, K., Ohta, K., Tagawa, S., Hirose, K., & Ohishi, Y. (2020). Stability of fcc phase FeH to 137 GPa. *American Mineralogist*, *105*(6), 917–921.

Kawaguchi, S. I., Nakajima, Y., Hirose, K., Komabayashi, T., Ozawa, H., Tateno, S., et al. (2017). Sound velocity of liquid Fe-Ni-S at high pressure. *Journal of Geophysical Research-Solid Earth*, *122*(5), 3624–3634.

Kinoshita, D., Nakajima, Y., Kuwayama, Y., Hirose, K., Iwamoto, A., Ishikawa, D., et al. (2020). Sound Velocity of Liquid Fe–P at High Pressure. *Physica Status Solidi B*, *257*(11), 2000171.

Komabayashi, T. (2020). Thermodynamics of the system Fe–Si–O under high pressure and temperature and its implications for Earth's core. *Physics and Chemistry of Minerals*, *47*(7), 32.

Komabayashi, T., Fei, Y., Meng, Y., & Prakapenka, V. (2009). In-situ X-ray diffraction measurements of the gamma-epsilon transition boundary of iron in an internally-heated diamond anvil cell. *Earth and Planetary Science Letters*, *282*(1–4), 252–257.

Komabayashi, T., Hirose, K., & Ohishi, Y. (2012). In situ X-ray diffraction measurements of the fcc–hcp phase transition boundary of an Fe–Ni alloy in an internally heated diamond anvil cell. *Physics and Chemistry of Minerals*, *39*(4), 329–338.

Komabayashi, T., Kato, J., Hirose, K., Tsutsui, S., Imada, S., Nakajima, Y., et al. (2015). Temperature dependence of the velocity-density relation for liquid metals under high pressure: Implications for the Earth's outer core. *American Mineralogist*, *100*(11–12), 2602–2609.

Komabayashi, T., Pesce, G., Sinmyo, R., Kawazoe, T., Breton, H., Shimoyama, Y., et al. (2019). Phase relations in the system Fe-Ni-Si to 200 GPa and 3,900 K and implications for Earth's core. *Earth and Planetary Science Letters*, *512*, 83–88.

Konopkova, Z., McWilliams, R. S., Gomez-Perez, N., & Goncharov, A. F. (2016). Direct measurement of thermal conductivity in solid iron at planetary core conditions. *Nature*, *534*(7605), 99–101.

Kupenko, I., Strohm, C., McCammon, C., Cerantola, V., Glazyrin, K., Petitgirard, S., et al. (2015). Time differentiated nuclear resonance spectroscopy coupled with pulsed laser heating in diamond anvil cells. *Review of Scientific Instruments*, *86*(11), 114501.

Kusakabe, M., Hirose, K., Sinmyo, R., Kuwayama, Y., Ohishi, Y., & Helffrich, G. (2019). Melting curve and equation of state of $-Fe_7N_3$: Nitrogen in the core? *Journal of Geophysical Research-Solid Earth*, *124*(4), 3448–3457.

Kuwayama, Y., & Hirose, K. (2004). Phase relations in the system Fe-FeSi at 21 GPa. *American Mineralogist*, *89*(2–3), 273–276.

Kuwayama, Y., Morard, G., Nakajima, Y., Hirose, K., Baron, A. Q. R., Kawaguchi, S. I., et al. (2020). Equation of state of liquid iron under extreme conditions. *Physical Review Letters*, *124*(16), 165701.

Labrosse, S. (2015). Thermal evolution of the core with a high thermal conductivity. *Physics of the Earth and Planetary Interiors*, *247*, 36–55.

Lai, X., Zhu, F., Liu, Y., Bi, W., Zhao, J., Alp, E. E., et al. (2020a). Elastic and magnetic properties of Fe_3P up to core pressures: Phosphorus in the Earth's core. *Earth and Planetary Science Letters*, *531*, 115974.

Lai, X., Zhu, F., Zhang, J. S., Zhang, D., Tkachev, S., Prakapenka, V. B., et al. (2020b). An Externally-Heated Diamond Anvil Cell for Synthesis and Single-Crystal Elasticity Determination of Ice-VII at High Pressure-Temperature Conditions. *Journal of Visualized Experiments*, *160*, e61389.

Laio, A., Bernard, S., Chiarotti, G. L., Scandolo, S., & Tosatti, E. (2000). Physics of iron at Earth's core conditions. *Science*, *287*(5455), 1027–1030.

Lehmann, I. (1936). P'. Publications Du Bureau Central Séismologique International. *Série A*, *14*, 87–115.

Li, J., & Agee, C. B. (1996). Geochemistry of mantle-core differentiation at high pressure. *Nature*, *381*(6584), 686–689.

Li, J., Wu, Q., Li, J., Xue, T., Tan, Y., Zhou, X., et al. (2020). Shock melting curve of iron: A consensus on the temperature at the earth's inner core boundary. *Geophysical Research Letters*, *47*(15), 1–11.

Li, Y., Vočadlo, L., & Brodholt, J. P. (2018). The elastic properties of hcp-Fe alloys under the conditions of the Earth's inner core. *Earth and Planetary Science Letters*, *493*, 118–127.

Liermann, H.-P., Konôpková, Z., Morgenroth, W., Glazyrin, K., Bednarčik, J., McBride, E., et al. (2015). The extreme conditions beamline P02. 2 and the extreme conditions science infrastructure at PETRA III. *Journal of Synchrotron Radiation*, *22*(4), 908–924.

Liu, J., Dauphas, N., Roskosz, M., Hu, M. Y., Yang, H., Bi, W., et al. (2017). Iron isotopic fractionation between silicate mantle and metallic core at high pressure. *Nature Communications*, *8*(1), 14377.

Liu, J., Lin, J. F., Alatas, A., Hu, M. Y., Zhao, J., & Dubrovinsky, L. (2016). Seismic parameters of hcp-Fe alloyed with Ni and Si in the Earth's inner core. *Journal of Geophysical Research: Solid Earth*, *121*(2), 610–623.

Liu, L. G., & Bassett, W. A. (1975). Melting of iron up to 200 kbar. *Journal of Geophysical Research*, *80*(26), 3777–3782.

Lord, O. T., Walter, M. J., Dasgupta, R., Walker, D., & Clark, S. M. (2009). Melting in the Fe-C system to 70 GPa. *Earth and Planetary Science Letters*, *284*(1–2), 157–167.

Lord, O. T., Walter, M. J., Dobson, D. P., Armstrong, L., Clark, S. M., & Kleppe, A. (2010). The FeSi phase diagram to 150 GPa. *Journal of Geophysical Research-Solid Earth*, *115*, 1–9.

Mao, H., Xu, J., Struzhkin, V., Shu, J., Hemley, R., Sturhahn, W., et al. (2001). Phonon density of states of iron up to 153 gigapascals. *Science*, *292*(5518), 914–916.

Mao, W. L., Struzhkin, V. V., Baron, A. Q. R., Tsutsui, S., Tommaseo, C. E., Wenk, H. R., et al. (2008). Experimental determination of the elasticity of iron at high pressure. *Journal of Geophysical Research-Solid Earth*, *113*(B9), 1–14.

Mao, W. L., Sturhahn, W., Heinz, D. L., Mao, H. K., Shu, J. F., & Hemley, R. J. (2004). Nuclear resonant x-ray scattering of iron hydride at high pressure. *Geophysical Research Letters*, *31*(15), 1–4.

Mao, Z., Lin, J. F., Liu, J., Alatas, A., Gao, L. L., Zhao, J. Y., et al. (2012). Sound velocities of Fe and Fe-Si alloy in the Earth's core. *Proceedings of the National Academy of Sciences of the United States of America*, *109*(26), 10239–10244.

Martorell, B., Vočadlo, L., Brodholt, J., & Wood, I. G. (2013). Strong premelting effect in the elastic properties of hcp-Fe under inner-core conditions. *Science*, *342*(6157), 466–468.

Martorell, B., Wood, I. G., Brodholt, J., & Vočadlo, L. (2016). The elastic properties of hcp-Fe1− xSix at Earth's inner-core conditions. *Earth and Planetary Science Letters*, *451*, 89–96.

Mashino, I., Miozzi, F., Hirose, K., Morard, G., & Sinmyo, R. (2019). Melting experiments on the Fe–C binary system up to 255 GPa: Constraints on the carbon content in the Earth's core. *Earth and Planetary Science Letters*, *515*, 135–144.

Matsuda, J., Sudo, M., Ozima, M., Ito, K., Ohtaka, O., & Ito, E. (1993). Noble gas partitioning between metal and silicate under high pressures. *Science*, *259*(5096), 788–790.

Merkel, S., & Yagi, T. (2005). X-ray transparent gasket for diamond anvil cell high pressure experiments. *Review of Scientific Instruments*, *76*(4), 046109.

Mezouar, M., Crichton, W., Bauchau, S., Thurel, F., Witsch, H., Torrecillas, F., et al. (2005). Development of a new state-of-the-art beamline optimized for monochromatic single-crystal and powder X-ray diffraction under extreme conditions at the ESRF. *Journal of Synchrotron Radiation*, *12*(5), 659–664.

Mezouar, M., Giampaoli, R., Garbarino, G., Kantor, I., Dewaele, A., Weck, G., et al. (2017). Methodology for in situ synchrotron X-ray studies in the laser-heated diamond anvil cell. *High Pressure Research*, *37*(2), 170–180.

Millot, M., Coppari, F., Rygg, J. R., Correa Barrios, A., Hamel, S., Swift, D. C., et al. (2019). Nanosecond X-ray diffraction of shock-compressed superionic water ice. *Nature*, *569*(7755), 251–255.

Millot, M., Hamel, S., Rygg, J. R., Celliers, P. M., Collins, G. W., Coppari, F., et al. (2018). Experimental evidence for superionic water ice using shock compression. *Nature Physics*, *14*(3), 297–302.

Miozzi, F., Matas, J., Guignot, N., Badro, J., Siebert, J., & Fiquet, G. (2020). A new reference for the thermal equation of state of iron. *Minerals*, *10*(2), 100.

Miyahara, M., Sakai, T., Ohtani, E., Kobayashi, Y., Kamada, S., Kondo, T., et al. (2008). Application of FIB system to ultra-high-pressure Earth science. *Journal of Mineralogical and Petrological Sciences*, *103*(2), 88–93.

Miyajima, N., Holzapfel, C., Asahara, Y., Dubrovinsky, L., Frost, D. J., Rubie, D. C., et al. (2010). Combining FIB milling and conventional Argon ion milling techniques to prepare high-quality site-specific TEM samples for quantitative EELS analysis of oxygen in molten iron. *Journal of Microscopy*, *238*(3), 200–209.

Monnereau, M., Calvet, M., Margerin, L., & Souriau, A. (2010). Lopsided growth of Earth's inner core. *Science*, *328*(5981), 1014–1017.

Morard, G., Andrault, D., Antonangeli, D., Nakajima, Y., Auzende, A., Boulard, E., et al. (2017a). Fe–FeO and Fe–Fe$_3$C melting relations at Earth's core–mantle boundary conditions: Implications for a volatile-rich or oxygen-rich core. *Earth and Planetary Science Letters*, *473*, 94–103.

Morard, G., Andrault, D., Guignot, N., Sanloup, C., Mezouar, M., Petitgirard, S., et al. (2008). In situ determination of Fe-Fe$_3$S phase diagram and liquid structural properties up to 65 GPa. *Earth and Planetary Science Letters*, *272*(3–4), 620–626.

Morard, G., Andrault, D., Guignot, N., Siebert, J., Garbarino, G., & Antonangeli, D. (2011). Melting of Fe-Ni-Si and Fe-Ni-S alloys at megabar pressures: Implications for the core-mantle boundary temperature. *Physics and Chemistry of Minerals*, *38*(10), 767–776.

Morard, G., Boccato, S., Rosa, A. D., Anzellini, S., Miozzi, F., Henry, L., et al. (2018a). Solving controversies on the iron phase diagram under high pressure. *Geophysical Research Letters*, *45*(20), 11074–11082.

Morard, G., Bouchet, J., Rivoldini, A., Antonangeli, D., Roberge, M., Boulard, E., et al. (2018b). Liquid properties in the Fe-FeS system under moderate pressure: Tool box to model small planetary cores. *American Mineralogist: Journal of Earth and Planetary Materials*, *103*(11), 1770–1779.

Morard, G., Hernandez, J.-A., Guarguaglini, M., Bolis, R., Benuzzi-Mounaix, A., Vinci, T., et al. (2020). In situ X-ray diffraction of silicate liquids and glasses under dynamic and static compression to megabar pressures. *Proceedings of the National Academy of Sciences*, *117*(22), 11981.

Morard, G., Nakajima, Y., Andrault, D., Antonangeli, D., Auzende, A. L., Boulard, E., et al. (2017b). Structure and density of Fe-C liquid alloys under high pressure. *Journal of Geophysical Research: Solid Earth*, *122*(10), 7813–7823.

Morard, G., Siebert, J., Andrault, D., Guignot, N., Garbarino, G., Guyot, F., et al. (2013). The Earth's core composition from high pressure density measurements of liquid iron alloys. *Earth and Planetary Science Letters*, *373*, 169–178.

Mori, Y., Ozawa, H., Hirose, K., Sinmyo, R., Tateno, S., Morard, G., et al. (2017). Melting experiments on Fe-Fe_3S system to 254 GPa. *Earth and Planetary Science Letters*, *464*, 135–141.

Mundl, A., Touboul, M., Jackson, M. G., Day, J. M., Kurz, M. D., Lekic, V., et al. (2017). Tungsten-182 heterogeneity in modern ocean island basalts. *Science*, *356*(6333), 66–69.

Nakajima, Y., Imada, S., Hirose, K., Komabayashi, T., Ozawa, H., Tateno, S., et al. (2015). Carbon-depleted outer core revealed by sound velocity measurements of liquid iron-carbon alloy. *Nature Communications*, *6*, 8942.

Nakajima, Y., Kawaguchi, S. I., Hirose, K., Tateno, S., Kuwayama, Y., Sinmyo, R., et al. (2020). Silicon-depleted present-day earth's outer core revealed by sound velocity measurements of liquid Fe-Si alloy. *Journal of Geophysical Research: Solid Earth*, *125*(6), e2020JB019399.

Nguyen, J. H., & Holmes, N. C. (2004). Melting of iron at the physical conditions of the Earth's core. *Nature*, *427*(6972), 339–342.

Nishida, K., Shibazaki, Y., Terasaki, H., Higo, Y., Suzuki, A., Funamori, N., et al. (2020). Effect of sulfur on sound velocity of liquid iron under Martian core conditions. *Nature Communications*, *11*(1), 1954.

Nishida, K., Suzuki, A., Terasaki, H., Shibazaki, Y., Higo, Y., Kuwabara, S., et al. (2016). Towards a consensus on the pressure and composition dependence of sound velocity in the liquid Fe–S system. *Physics of the Earth and Planetary Interiors*, *257*, 230–239.

Nomura, R., Azuma, S., Uesugi, K., Nakashima, Y., Irifune, T., Shinmei, T., et al. (2017). High-pressure rotational deformation apparatus to 135 GPa. *Review of Scientific Instruments*, *88*(4), 044501.

O'Rourke, J. G., & Stevenson, D. J. (2016). Powering Earth's dynamo with magnesium precipitation from the core. *Nature*, *529*(7586), 387–389.

Ohta, K., Kuwayama, Y., Hirose, K., Shimizu, K., & Ohishi, Y. (2016). Experimental determination of the electrical resistivity of iron at Earth's core conditions. *Nature*, *534*(7605), 95–98.

Ohtani, E., Shibazaki, Y., Sakai, T., Mibe, K., Fukui, H., Kamada, S., et al. (2013). Sound velocity of hexagonal close-packed iron up to core pressures. *Geophysical Research Letters*, *40*(19), 5089–5094.

Oka, K., Hirose, K., Tagawa, S., Kidokoro, Y., Nakajima, Y., Kuwayama, Y., et al. (2019). Melting in the Fe-FeO system to 204 GPa: Implications for oxygen in Earth's core. *American Mineralogist*, *104*(11), 1603–1607.

Otsuka, K., & Karato, S. (2012). Deep penetration of molten iron into the mantle caused by a morphological instability. *Nature*, *492*(7428), 243–246.

Ozawa, H., Hirose, K., Yonemitsu, K., & Ohishi, Y. (2016). High-pressure melting experiments on Fe-Si alloys and implications for silicon as a light element in the core. *Earth and Planetary Science Letters*, *456*, 47–54.

Ozawa, H., Tateno, S., Xie, L., Nakajima, Y., Sakamoto, N., Kawaguchi, S. I., et al. (2018). Boron-doped diamond as a new heating element for internal-resistive heated diamond-anvil cell. *High Pressure Research*, *38*(2), 120–135.

Petitgirard, S., Malfait, W. J., Sinmyo, R., Kupenko, I., Hennet, L., Harries, D., et al. (2015). Fate of $MgSiO_3$ melts at core-mantle boundary conditions. *Proceedings of the National Academy of Sciences of the United States of America*, *112*(46), 14186–14190.

Petitgirard, S., Salamat, A., Beck, P., Weck, G., & Bouvier, P. (2014). Strategies for in situ laser heating in the diamond anvil cell at an X-ray diffraction beamline. *Journal of Synchrotron Radiation*, *21*(1), 89–96.

Poirier, J. P. (1994). Light-elements in the earths outer core – a critical-review. *Physics of the Earth and Planetary Interiors*, *85*(3–4), 319–337.

Poitrasson, F., Roskosz, M., & Corgne, A. (2009). No iron isotope fractionation between molten alloys and silicate melt to 2,000 C and 7.7 GPa: Experimental evidence and implications for planetary differentiation and accretion. *Earth and Planetary Science Letters*, *278*(3–4), 376–385.

Polyakov, V. B. (2009). Equilibrium iron isotope fractionation at core-mantle boundary conditions. *Science*, *323*(5916), 912–914.

Prescher, C., Dubrovinsky, L., Bykova, E., Kupenko, I., Glazyrin, K., Kantor, A., et al. (2015). High Poisson's ratio of Earth's inner core explained by carbon alloying. *Nature Geoscience*, *8*(3), 220–223.

Rainey, E., Hernlund, J., & Kavner, A. (2013). Temperature distributions in the laser-heated diamond anvil cell from 3-D numerical modeling. *Journal of Applied Physics*, *114*(20), 204905.

Rubie, D. C., Frost, D. J., Mann, U., Asahara, Y., Nimmo, F., Tsuno, K., et al. (2011). Heterogeneous accretion, composition and core–mantle differentiation of the Earth. *Earth and Planetary Science Letters*, *301*(1–2), 31–42.

Sakai, T., Yagi, T., Ohfuji, H., Irifune, T., Ohishi, Y., Hirao, N., et al. (2015). High-pressure generation using double stage micro-paired diamond anvils shaped by focused ion beam. *Review of Scientific Instruments*, *86*(3), 033905.

Sakaiya, T., Takahashi, H., Kondo, T., Kadono, T., Hironaka, Y., Irifune, T., et al. (2014). Sound velocity and density measurements of liquid iron up to 800 GPa: A universal relation between Birch's law coefficients for solid and liquid metals. *Earth and Planetary Science Letters*, *392*, 80–85.

Sakamaki, T., Ohtani, E., Fukui, H., Kamada, S., Takahashi, S., Sakairi, T., et al. (2016). Constraints on Earth's inner core composition inferred from measurements of the sound velocity of hcp-iron in extreme conditions. *Science Advances*, 2(2), e1500802.

Sanloup, C. (2016). Density of magmas at depth. *Chemical Geology*, 429, 51–59.

Sanloup, C., Drewitt, J. W. E., Konôpková, Z., Dalladay-Simpson, P., Morton, D. M., Rai, N., et al. (2013). Structural change in molten basalt at deep mantle conditions. *Nature*, 503(7474), 104–107.

Sanloup, C., Fiquet, G., Gregoryanz, E., Morard, G., & Mezouar, M. (2004). Effect of Si on liquid Fe compressibility: Implications for sound velocity in core materials. *Geophysical Research Letters*, 31(7), 1–4.

Sanloup, C., Guyot, F., Gillet, P., Fiquet, G., Mezouar, M., & Martinez, I. (2000). Density measurements of liquid Fe-S alloys at high-pressure. *Geophysical Research Letters*, 27(6), 811–814.

Sata, N., Hirose, K., Shen, G., Nakajima, Y., Ohishi, Y., & Hirao, N. (2010). Compression of FeSi, Fe_3C, $Fe_{0.95}O$, and FeS under the core pressures and implication for light element in the Earth's core. *Journal of Geophysical Research-Solid Earth*, 115, 1–13.

Satish-Kumar, M., So, H., Yoshino, T., Kato, M., & Hiroi, Y. (2011). Experimental determination of carbon isotope fractionation between iron carbide melt and carbon: ^{12}C-enriched carbon in the Earth's core? *Earth and Planetary Science Letters*, 310(3–4), 340–348.

Sato, T., Funamori, N., & Kikegawa, T. (2010). High-pressure in situ structure measurement of low-Z noncrystalline materials with a diamond-anvil cell by an x-ray diffraction method. *Review of Scientific Instruments*, 81(4), 1–8.

Scopigno, T., Ruocco, G., & Sette, F. (2005). Microscopic dynamics in liquid metals: The experimental point of view. *Reviews of Modern Physics*, 77(3), 881.

Seagle, C., Heinz, D., Campbell, A., Prakapenka, V., & Wanless, S. (2008). Melting and thermal expansion in the Fe–FeO system at high pressure. *Earth and Planetary Science Letters*, 265(3–4), 655–665.

Shahar, A., Hillgren, V. J., Young, E. D., Fei, Y. W., Macris, C. A., & Deng, L. W. (2011). High-temperature Si isotope fractionation between iron metal and silicate. *Geochimica et Cosmochimica Acta*, 75(23), 7688–7697.

Shahar, A., Schauble, E. A., Caracas, R., Gleason, A. E., Reagan, M. M., Xiao, Y., et al. (2016). Pressure-dependent isotopic composition of iron alloys. *Science*, 352(6285), 580–582.

Shahar, A., Ziegler, K., Young, E. D., Ricolleau, A., Schauble, E. A., & Fei, Y. W. (2009). Experimentally determined Si isotope fractionation between silicate and Fe metal and implications for Earth's core formation. *Earth and Planetary Science Letters*, 288(1–2), 228–234.

Shatskiy, A., Yamazaki, D., Morard, G., Cooray, T., Matsuzaki, T., Higo, Y., et al. (2009). Boron-doped diamond heater and its application to large-volume, high-pressure, and high-temperature experiments. *Review of Scientific Instruments*, 80(2), 023907.

Shen, G., Mao, H. k., Hemley, R. J., Duffy, T. S., & Rivers, M. L. (1998). Melting and crystal structure of iron at high pressures and temperatures. *Geophysical Research Letters*, 25(3), 373–376.

Shen, G. Y., & Mao, H. K. (2017). High-pressure studies with x-rays using diamond anvil cells. *Reports on Progress in Physics*, 80(1), 1–53.

Shibazaki, Y., & Kono, Y. (2018). Effect of silicon, carbon, and sulfur on structure of liquid iron and implications for structure-property relations in liquid iron-light element alloys. *Journal of Geophysical Research: Solid Earth*, 123(6), 4697–4706.

Shibazaki, Y., Ohtani, E., Terasaki, H., Tateyama, R., Sakamaki, T., Tsuchiya, T., et al. (2011). Effect of hydrogen on the melting temperature of FeS at high pressure: Implications for the core of Ganymede. *Earth and Planetary Science Letters*, 301(1–2), 153–158.

Siebert, J., Badro, J., Antonangeli, D., & Ryerson, F. J. (2012). Metal-silicate partitioning of Ni and Co in a deep magma ocean. *Earth and Planetary Science Letters*, 321, 189–197.

Siebert, J., Badro, J., Antonangeli, D., & Ryerson, F. J. (2013). Terrestrial accretion under oxidizing conditions. *Science*, 339(6124), 1194–1197.

Sinmyo, R., Hirose, K., & Ohishi, Y. (2019). Melting curve of iron to 290 GPa determined in a resistance-heated diamond-anvil cell. *Earth and Planetary Science Letters*, 510, 45–52.

Sola, E., & Alfè, D. (2009). Melting of iron under Earth's core conditions from diffusion monte carlo free energy calculations. *Physical Review Letters*, 103(7), 078501.

Stewart, A. J., Schmidt, M. W., van Westrenen, W., & Liebske, C. (2007). Mars: A new core-crystallization regime. *Science*, 316(5829), 1323–1325.

Takahashi, S., Ohtani, E., Sakamaki, T., Kamada, S., Fukui, H., Tsutsui, S., et al. (2019). Sound velocity of Fe3C at high pressure and high temperature determined by inelastic X-ray scattering. *Comptes Rendus Geoscience*, 351(2–3), 190–196.

Tanaka, S., & Hamaguchi, H. (1997). Degree one heterogeneity and hemispherical variation of anisotropy in the inner core from PKP(BC)–PKP(DF) times. *Journal of Geophysical Research: Solid Earth*, 102(B2), 2925–2938.

Tarduno, J. A., Cottrell, R. D., Davis, W. J., Nimmo, F., & Bono, R. K. (2015). A Hadean to Paleoarchean geodynamo recorded by single zircon crystals. *Science*, 349(6247), 521–524.

Tateno, S., Hirose, K., Ohishi, Y., & Tatsumi, Y. (2010). The structure of iron in earth's inner core. *Science*, 330(6002), 359–361.

Tateno, S., Hirose, K., Sinmyo, R., Morard, G., Hirao, N., & Ohishi, Y. (2018). Melting experiments on Fe–Si–S alloys to core pressures: Silicon in the core? *American Mineralogist: Journal of Earth and Planetary Materials*, 103(5), 742–748.

Tateno, S., Ozawa, H., Hirose, K., Suzuki, T., I-Kawaguchi, S., & Hirao, N. (2019). Fe_2S: The most Fe-rich iron sulfide at the earth's inner core pressures. *Geophysical Research Letters*, 46(21), 11944–11949.

Terasaki, H. (2016). Physical Properties of the Outer Core. In H. Terasaki & R. A. Fischer (Eds.), *Deep Earth: Physics and Chemistry of the Lower Mantle and Core*, pp. 129–140. Washington: American Geophysical Union.

Terasaki, H. (2018). Density and elasticity measurements for liquid materials. In Y. Kono & C. Sanloup (Eds.), *Magmas Under*

Pressure Advances in High-Pressure Experiments on Structure and Properties of Melts, pp. 237–260. Amsterdam: Elsevier.

Terasaki, H., Nishida, K., Shibazaki, Y., Sakamaki, T., Suzuki, A., Ohtani, E., et al. (2010). Density measurement of Fe_3C liquid using X-ray absorption image up to 10 GPa and effect of light elements on compressibility of liquid iron. *Journal of Geophysical Research*, *115*(B6), B06207–B06207.

Terasaki, H., Rivoldini, A., Shimoyama, Y., Nishida, K., Urakawa, S., Maki, M., et al. (2019). Pressure and composition effects on sound velocity and density of core-forming liquids: Implication to core compositions of terrestrial planets. *Journal of Geophysical Research: Planets*, *124*(8), 2272–2293.

Terasaki, H., Urakawa, S., Rubie, D. C., Funakoshi, K., Sakamaki, T., Shibazaki, Y., et al. (2012). Interfacial tension of Fe-Si liquid at high pressure: Implications for liquid Fe-alloy droplet size in magma oceans. *Physics of the Earth and Planetary Interiors*, *202*, 1–6.

Torchio, R., Boccato, S., Miozzi, F., Rosa, A. D., Ishimatsu, N., Kantor, I., et al. (2020). Melting curve and phase relations of Fe-Ni alloys: Implications for the earth's core composition. *Geophysical Research Letters*, *47*(14), 1–7.

Turneaure, S. J., Sharma, S. M., & Gupta, Y. (2020). Crystal Structure and Melting of Fe Shock Compressed to 273 GPa: In Situ X-Ray Diffraction. *Physical Review Letters*, *125*(21), 215702.

Umemoto, K., & Hirose, K. (2020). Chemical compositions of the outer core examined by first principles calculations. *Earth and Planetary Science Letters*, *531*, 116009.

Wagle, F., & Steinle-Neumann, G. (2019). Liquid iron equation of state to the terapascal regime from ab initio simulations. *Journal of Geophysical Research: Solid Earth*, *124*(4), 3350–3364.

Wang, J., Smith, R. F., Eggert, J. H., Braun, D. G., Boehly, T. R., Reed Patterson, J., et al. (2013). Ramp compression of iron to 273 GPa. *Journal of Applied Physics*, *114*(2), 023513.

Weck, G., Garbarino, G., Ninet, S., Spaulding, D., Datchi, F., Loubeyre, P., et al. (2013). Use of a multichannel collimator for structural investigation of low-Z dense liquids in a diamond anvil cell: Validation on fluid H_2 up to 5 GPa. *Review of Scientific Instruments*, *84*(6), 063901.

White, S., Kettle, B., Vorberger, J., Lewis, C. L. S., Glenzer, S. H., Gamboa, E., et al. (2020). Time-dependent effects in melting and phase change for laser-shocked iron. *Physical Review Research*, *2*(3), 033366.

Wicks, J. K., Smith, R. F., Fratanduono, D. E., Coppari, F., Kraus, R. G., Newman, M. G., et al. (2018). Crystal structure and equation of state of Fe-Si alloys at super-Earth core conditions. *Science Advances*, *4*(4), eaao5864.

Williams, Q., Jeanloz, R., Bass, J., Svendsen, B., & Ahrens, T. J. (1987). The melting curve of iron to 250 gigapascals: A constraint on the temperature at Earth's center. *Science*, *236*(4798), 181–182.

Wood, B. J., Walter, M. J., & Wade, J. (2006). Accretion of the Earth and segregation of its core. *Nature*, *441*(7095), 825–833.

Xie, L. J., Yoneda, A., Liu, Z. D., Nishida, K., & Katsura, T. (2020). Boron-doped diamond synthesized by chemical vapor deposition as a heating element in a multi-anvil apparatus. *High Pressure Research*, *40*(3), 369–378.

Xiong, Z., Tsuchiya, T., & Taniuchi, T. (2018). Ab initio prediction of potassium partitioning into Earth's core. *Journal of Geophysical Research: Solid Earth*, *123*(8), 6451–6458.

Yamada, A., Irifune, T., Sumiya, H., Higo, Y., Inoue, T., & Funakoshi, K.-I. (2008). Exploratory study of the new B-doped diamond heater at high pressure and temperature and its application to in situ XRD experiments on hydrous Mg-silicate melt. *High Pressure Research*, *28*(3), 255–264.

Yamazaki, D., Ito, E., Yoshino, T., Tsujino, N., Yoneda, A., Guo, X., et al. (2014). Over 1 Mbar generation in the Kawai-type multianvil apparatus and its application to compression of $(Mg_{0.92}Fe_{0.08})SiO_3$ perovskite and stishovite. *Physics of the Earth and Planetary Interiors*, *228*, 262–267.

Yang, L., Karandikar, A., & Boehler, R. (2012). Flash heating in the diamond cell: Melting curve of rhenium. *Review of Scientific Instruments*, *83*(6), 063905.

Yokoo, S., Hirose, K., Sinmyo, R., & Tagawa, S. (2019). Melting experiments on liquidus phase relations in the Fe-S-O ternary system under core pressures. *Geophysical Research Letters*, *46*(10), 5137–5145.

Yoo, C. S., Holmes, N. C., Ross, M., Webb, D. J., & Pike, C. (1993). Shock temperatures and melting of iron at Earth core conditions. *Physical Review Letters*, *70*(25), 3931–3934.

Yoshino, T., Makino, Y., Suzuki, T., & Hirata, T. (2020). Grain boundary diffusion of W in lower mantle phase with implications for isotopic heterogeneity in oceanic island basalts by core-mantle interactions. *Earth and Planetary Science Letters*, *530*, 115887.

Yu, X., & Secco, R. A. (2008). Equation of state of liquid Fe–17 wt% Si to 12 GPa. *High Pressure Research*, *28*(1), 19–28.

Zeng, Q. S., Kono, Y., Lin, Y., Zeng, Z. D., Wang, J. Y., Sinogeikin, S. V., et al. (2014). Universal fractional noncubic power law for density of metallic glasses. *Physical Review Letters*, *112*(18), 185502.

Zhang, D., Jackson, J. M., Zhao, J., Sturhahn, W., Alp, E. E., Hu, M. Y., et al. (2016). Temperature of Earth's core constrained from melting of Fe and $Fe_{0.9}Ni_{0.1}$ at high pressures. *Earth and Planetary Science Letters*, *447*, 72–83.

Zhu, F., Lai, X., Wang, J., Amulele, G., Kono, Y., Shen, G., et al. (2021). Density of Fe-Ni-C Liquids at High Pressures and Implications for Liquid Cores of Earth and the Moon. *Journal of Geophysical Research: Solid Earth*, *126*(3), e2020JB021089.

Zou, G., Ma, Y., Mao, H.-k., Hemley, R. J., & Gramsch, S. A. (2001). A diamond gasket for the laser-heated diamond anvil cell. *Review of Scientific Instruments*, *72*(2), 1298.

12

Dynamics in Earth's Core Arising from Thermo-Chemical Interactions with the Mantle

Christopher J. Davies and Sam Greenwood

ABSTRACT

Thermo-chemical interactions at the core-mantle boundary (CMB) play an integral role in determining the dynamics and evolution of Earth's deep interior. This review considers the processes in the core that arise from heat and mass transfer at the CMB, with particular focus on thermo-chemical stratification and the precipitation of oxides. A fundamental parameter is the thermal conductivity of the core, which we estimate as $k^c = 70$–110 W m^{-1} K^{-1} at CMB conditions based on consistent extrapolation from a number of recent studies. These high conductivity values imply the existence of an early basal magma ocean (BMO) overlying a hot core and rapid cooling potentially leading to a loss of power to the dynamo before the inner core formed around 0.5–1 Gyrs ago, the so-called "new core paradox." Coupling core thermal evolution modeling and calculations of chemical equilibrium between liquid iron and silicate melts suggests that FeO dissolved from the BMO into the core after its formation, creating a stably stratified chemical layer below the CMB, while precipitation of MgO and SiO$_2$ was delayed until the last 2–3 Gyrs and was therefore not available to power the early dynamo; however, once initiated, precipitation supplied ample power for field generation. We also present a possible solution to the new core paradox without requiring precipitation or radiogenic heating using $k^c = 70$ W m^{-1} K^{-1}. The model matches the present inner core size and heat flow and temperature at the top of the convecting mantle. It predicts a present-day CMB heat flow of 8.5 TW, a chemically stable layer 100 km thick, and a BMO lifetime of 2 Gyrs.

12.1. INTRODUCTION

The core-mantle boundary (CMB) accommodates the most significant internal transition in the structure and dynamics of the Earth system. The Preliminary Reference Earth Model (PREM; Dziewonski & Anderson, 1981) shows that the horizontally averaged density ρ and compressional wave speed V_p change by ~40% across the CMB. In terms of physical properties, the lower mantle is a poor thermal and electrical conductor and has a viscosity that is 10^{15}–10^{20} times larger than that of the core, which allows it to sustain variations in temperature T of thousands of Kelvin and support large-scale dynamic structures such as the Large Low Velocity Provinces (LLVPs) that sit on the CMB (e.g., Garnero et al., 2016). The core, by contrast, is an excellent thermal and electrical conductor, while the low viscosity, similar to that of water (Pozzo et al., 2013), implies that convection in the core is turbulent. This stark contrast between structural and dynamical properties leads to thermo-chemical interactions at the CMB that provide power for generating the geomagnetic field and are important for determining the long-term evolution of the core and mantle systems.

In this chapter, we review recent progress in understanding core mantle interactions with a focus on the thermodynamics and fluid dynamics of the upper core; a

School of Earth and Environment, University of Leeds, Leeds, UK

Core-Mantle Co-Evolution: An Interdisciplinary Approach, Geophysical Monograph 276, First Edition.
Edited by Takashi Nakagawa, Taku Tsuchiya, Madhusoodhan Satish-Kumar, and George Helffrich.
© 2023 American Geophysical Union. Published 2023 by John Wiley & Sons, Inc.
DOI: 10.1002/9781119526919.ch12

complementary perspective from the mantle side can be found in Nakagawa (2020). Many excellent reviews of the CMB region already exist, and so we focus on the main developments since the authoritative Treatise on Geophysics reviews by Nimmo (2015a,b), Buffett (2015), Hernlund and McNamara (2015), and Jaupart et al. (2015). Relevant background on geodynamo simulations has also been recently reviewed by Wicht and Sanchez (2019). For brevity, we further focus on thermal and chemical interactions. Core-mantle interactions also influence the rotational dynamics of the Earth, a topic that was reviewed by Tilgner (2015) and more recently by Dumberry (2018), and the shape of the CMB, which has recently been discussed in connection with the anomalously low (Koelemeijer et al., 2017) or high (Lau et al., 2017) density of LLVPs. Here, we will assume that the CMB is spherical and that the core and mantle are corotating.

The dynamo process that maintains the geomagnetic field is ultimately driven by heat extracted across the CMB. Syntheses of paleointensity data show that the field has been continuously generated for at least the last 3.5 Gyrs (Tarduno et al., 2010; Biggin et al., 2015; Tauxe & Yamazaki, 2015; Bono et al., 2019), while recordings dating back to 4.2 Ga (Tarduno et al., 2015) are currently debated (Tang et al., 2019; Tarduno et al., 2020). Heat loss at the CMB drives vigorous convection that maintains the bulk core close to an adiabatic temperature and uniform composition (e.g., Braginsky & Roberts, 1995; Nimmo, 2015a). Compared to the mean CMB temperature of ~4000 K (Lay et al., 2009; Davies et al., 2015), the thermal anomalies associated with core convection are $O(10^{-3})$ K (Stevenson, 1987; Bloxham & Jackson, 1990), while the convective chemical anomalies are many orders of magnitude smaller than the mean light element mass fraction of ~10 wt%. Consequently, even small thermo-chemical anomalies resulting from interactions at the CMB can significantly affect core dynamics.

The dynamics that result from thermo-chemical core-mantle coupling are dictated by the fluxes of heat and mass at the CMB. The total CMB heat flow Q^c is controlled by mantle dynamics – it is poorly constrained even for the present day, with current estimates suggesting the range $Q^c = 7$–17 TW (Lay et al., 2009; Nimmo, 2015a), which amounts to ~15% – 50% of Earth's total heat budget (Jaupart et al., 2015). Back in time Q^c must be inferred from parameterized models or simulations of mantle dynamics (Jaupart et al., 2015; Nakagawa, 2020). The key quantity for core dynamics is the superadiabatic heat flow $Q^c - Q_a^c$. The adiabatic heat flow on the core side of the CMB (denoted by superscript c) is

$$Q_a^c = -4\pi r_c^2 k^c \left. \frac{\partial T_a^c}{\partial r} \right|_{r=r_c}, \quad (12.1)$$

where T_a is the adiabatic temperature, k^c is the thermal conductivity, and $r = r_c$ is the CMB radius. The total heat flow on each side of the CMB is

$$Q^j = -4\pi r_c^2 \left(k^j \left. \frac{\partial T^j}{\partial r} \right|_{r=r_c} - \sum_i R_i^j \mathbf{n} \cdot \mathbf{i}_i^j \right),$$

where $j \in \{c,m\}$ and superscript m denotes the mantle side of the CMB. Continuity at r_c of heat flux and the mass flux per unit area $\mathbf{i}_i$ of element i requires that the conductive fluxes, $Q_c^j = -4\pi r_c^2 k^j (\partial T^j / \partial r)|_{r=r_c}$, are related by

$$Q_c^c = -4\pi r_c^2 k^m \left. \frac{\partial T^m}{\partial r} \right|_{r=r_c} - 4\pi r_c^2 \sum_i \left[R_i^c - R_i^m \right] \mathbf{n} \cdot \mathbf{i}_i$$

$$Q_c^c = Q_c^m + Q_h \quad (12.2)$$

(Davies et al., 2020), where $\mathbf{n}$ is the outward unit normal to the CMB and $\left[R_i^c - R_i^m \right] < 0$ (>0) is the amount of heat released (absorbed) as one formula unit of i is transferred from the core to the mantle or vice versa (Pozzo et al., 2019). Here, $R_i = \mu_i - T(\partial \mu_i / \partial T)_{P,T}$ is the heat of reaction coefficient with μ_i the chemical potential of element i and P the pressure. If the heat of reaction $Q_h < 0$, corresponding for example to an exothermic reaction with accompanying mass transfer into the core ($\mathbf{n} \cdot \mathbf{i} < 0$), then the heat flow available to core convection is reduced below the heat Q^m conducted through the lower mantle boundary layer, while $Q_h > 0$ acts as a heat source, increasing Q^c for a given Q^m. If $Q^c > Q_a^c$ then thermal convection probably occurs throughout the core. Conversely, if $Q^c < Q_a^c$ then a thermally stratified layer exists below the CMB in which heat is transported by conduction and vertical motion is strongly impeded. Depending on the radial variation of $k^c(r)$ and the distribution of buoyancy sources within the core, which are both uncertain at present, it is possible to produce stratification at intermediate depths (Gomi et al., 2013). In this review, we will mainly consider the case where stratification arises directly below the CMB.

The total chemical flux I_i of species i at the CMB is given by

$$I_i = -4\pi r_c^2 \rho D_i \left. \frac{\partial w_i^c}{\partial r} \right|_{r=r_c} + 4\pi r_c^2 \alpha_i^c \alpha_i^D g, \quad (12.3)$$

where D_i, w_i^c, α_i^c, and α_i^D are respectively the self-diffusion coefficient, mass fraction, compositional expansion coefficient, and barodiffusion coefficient of species i and g is gravity. I_i is continuous at the CMB if the small effect of core contraction is neglected (Gubbins et al., 2003; Davies et al., 2020). In equation (12.3), the second term on the right-hand side is the barodiffusion, representing transport of light element down the hydrostatic pressure gradient $dP/dr = -\rho g$, while element transport along the temperature gradient (thermodiffusion) is small and has been omitted (Gubbins et al., 2004). I is very hard to estimate because global mass balance constrains the bulk chemical composition of the core and mantle but not the compositional gradient at the CMB. Therefore, much recent work has focused on establishing the equilibrium chemical conditions at the CMB, which relate compositions on either side of the interface (e.g., Fischer et al., 2015; Badro et al., 2018; Pozzo et al., 2019). If $I_i < 0$, then light elements leave the mantle, which almost certainly results in chemical stratification below the CMB since the minute chemical anomalies associated with core convection are probably insufficient to mix the anomalously light fluid downward (Buffett & Seagle, 2010; Davies et al., 2018, 2020). Conversely, $I_i > 0$ implies that light elements precipitate out of solution (as oxides) and underplate onto the base of the mantle; the residual fluid, slightly iron-rich compared to the fluid below, will sink via Rayleigh-Taylor instability thus helping to drive core flow (O'Rourke & Stevenson, 2016).

The lower mantle is thermally and chemically heterogeneous and so heat and mass exchange should vary with location on the CMB. Lateral variations in CMB heat flow are expected from seismic tomography, geodynamic simulations (see, e.g., Gubbins, 2003; Olson et al., 2015, for reviews), and inferred spatial variations in thermal conductivity (Deschamps & Hsieh, 2019), which drive baroclinic flows at the top of the core (e.g., Zhang, 1992) that might affect the observed magnetic field (Gubbins et al., 2007; Aubert et al., 2007). CMB heat flow heterogeneity can also lead to penetration of a preexisting stable layer (e.g., Olson et al., 2017; Christensen, 2018; Cox et al., 2019) or even induce regional stratification if the anomalies are strong enough to make the heat flow locally subadiabatic (Olson et al., 2018; Mound et al., 2019). Lateral variations in chemical flux also seem likely if LLVPs are compositionally distinct (Garnero et al., 2016), though this effect does not appear to have been studied to date.

The existence of stratification and/or precipitation has important implications for the dynamics and evolution of the core. Stratified layers suppress radial motion and may support strong toroidal fields (Hardy et al., 2020) and distinct classes of wave motions (Braginsky, 1999) that are consistent with observed periodic variations of the geomagnetic field (Buffett, 2014; Buffett et al., 2016). Such a layer also acts to preferentially filter small-scale and rapidly time-varying fields that are generated in the bulk core (Christensen, 2006; Gastine et al., 2020), effectively filtering our view of the dynamo process, which is primarily based on observations that only probe CMB field. Precipitation has recently been advocated as the primary long-term power source for Earth's magnetic field (O'Rourke & Stevenson, 2016; Hirose et al., 2017; Wilson et al., 2022), while precipitation products may have been incorporated into the mantle via Rayleigh-Taylor instability (Helffrich et al., 2018). However, at present, a definitive observation of either stratification or precipitation is lacking. Therefore, in this review, we focus on predictions from modeling studies, such as the thickness and strength of stratification and the thermal and magnetic history of the core, that add further constraints to complement the observational evidence.

Broadly speaking, there are presently two scenarios for thermo-chemical core-mantle interactions that depend to a large extent on the core thermal conductivity k^c (see Table 12.1). In the "low conductivity" scenario, the core cooled slowly over geological time, powering the geomagnetic field by thermal convection until the onset of inner core freezing around 1 billion years ago, which provided additional power for field generation through release of latent heat and light elements (e.g., Buffett et al., 1996; Labrosse et al., 2001; Gubbins et al., 2003, 2004; Nimmo et al., 2004). Due to the low conductivity the present adiabatic heat flow is predicted to be around 4–6 TW, and hence thermal convection probably operated throughout the core until the present-day. In this scenario, some thermal history models (e.g., Nimmo et al., 2004) indicate

Table 12.1 Two scenarios for core-mantle evolution described in the text. The CMB heat flow is estimated as $Q^c = 7$–17 TW (Nimmo, 2015a), which is constrained independently of the core conductivity.

	Low conductivity	High conductivity
k^c	$\lesssim 50$ W m^{-1} K^{-1}	$\gtrsim 90$ W m^{-1} K^{-1}
Q_a^c	4–6 TW	14–16 TW
Core cooling rate	Slow	Fast
Inner core age	~1 Gyr	~0.5 Gyr
Thermal stratification	Never	Likely at present
Basal magma ocean (BMO)	Maybe, possibly long-lived	Likely, potentially short-lived
Pre-inner core dynamo power	Secular cooling	Secular cooling, but precipitation maybe also required
Chemical exchange	Efficient with BMO	Efficient only in early times

that the ancient core temperature could have exceeded the mantle solidus, and a long-lived Basal Magma Ocean (BMO) could have formed via mantle crystallization that proceeded from the middle outward (Labrosse et al., 2007; Stixrude et al., 2009). With low k^c, models predict that the BMO can survive to the present day as a thin (<1 km) mushy layer at the base of the mantle (Labrosse et al., 2007) while still drawing enough heat from the core to sustain the magnetic field (Blanc et al., 2020). This situation would facilitate efficient long-term chemical exchange between the core and mantle owing to the much higher self-diffusion coefficients of chemical species in the silicate liquid (e.g., Posner et al., 2018; Caracas et al., 2019).

The second scenario for thermo-chemical core-mantle evolution corresponds to a high thermal conductivity exceeding around 90 W m^{-1} K^{-1}. In order to maintain the geomagnetic field for the last 3.5 Gyrs, the core must cool faster to offset the enhanced power losses from thermal conduction, leading to an estimated inner core age of ~0.5–0.7 Gyrs (Driscoll & Bercovici, 2014; Davies, 2015; Labrosse et al., 2015; Nimmo, 2015a). The high conductivity values predict $Q_a^c = 14$–16 TW, comparable to the upper estimates of Q^c at the present day and suggesting thermal stratification of the upper core. Rapid cooling further implies early core temperatures that far exceeded current estimates of the lower mantle solidus and hence the presence of a BMO. However, since release of latent and radiogenic heat in the BMO stifle heat loss from the core (Labrosse et al., 2007), maintaining the early magnetic field with high k^c may require that the BMO was short-lived (Davies et al., 2020). Alternatively, high electrical conductivity of silicate liquids could have enabled the BMO itself to host the early dynamo (Ziegler & Stegman, 2013; Stixrude et al., 2020; Blanc et al., 2020).

The major problem posed by the high conductivity scenario is illustrated by parameterized models of coupled core-mantle evolution (Driscoll & Bercovici, 2014; O'Rourke et al., 2017; Driscoll & Davies, 2023) and could have been appreciated from the early study by Nimmo et al. (2004). With high k^c, classical parameterized mantle evolution models based on boundary layer theory predict an approximately exponential decline in CMB heat flow over time, which can lead to a loss of power to the dynamo before inner core nucleation around 1–2 Ga, in contradiction with paleomagnetic data (Biggin et al., 2015; Bono et al., 2019). However, the obvious remedy, increasing CMB heat flow and hence core cooling rate, leads to an old inner core that grows larger than its present size as determined by seismology. The apparent contradiction between observations and the fundamental model of core evolution has been termed the "new core paradox" (Olson, 2013). The term "paradox" is used because higher k^c generally implies higher electrical conductivity in metals (Chester & Thellung, 1961) and hence weaker magnetic diffusion, which should be beneficial to dynamo action. Driscoll and Du (2019) show that the ratio of magnetic induction to diffusion declines in both high and low electrical conductivity limits and suggest that Earth's core came close to this "no dynamo" state prior to inner core nucleation (ICN). Thermal history models have attempted to overcome the new core paradox by invoking additional effects such as a significant amount of radiogenic heating (e.g., from ^{40}K; Driscoll & Bercovici, 2014) or gravitational power provided by the precipitation of MgO (O'Rourke et al., 2017) or SiO$_2$ (Hirose et al., 2017; Wilson et al., 2022), though the viability of all of these processes has been questioned (Xiong et al., 2018; Du et al., 2019; Arveson et al., 2019).

In this review, we first discuss the material properties of the core that are required to model the processes of stratification and precipitation, focusing on the composition on either side of the CMB and the core thermal conductivity (section 12.2). This motivates us to consider the high conductivity scenario in the remainder of the review. In section 12.3, we describe recent studies of core-mantle chemical equilibrium and discuss constraints on the onset and rate of chemical precipitation and stratification below the CMB. Section 12.4 reviews thermal and chemical stratification at the top of the core, while section 12.5 discusses recent studies of chemical precipitation. In section 12.6, we discuss potential resolutions to the "new core paradox." We summarize our main conclusions in section 12.7.

12.2. MATERIAL PROPERTIES OF THE CORE

The standard tools for investigating coupled core-mantle evolution on Gyr timescales are thermal history models, which are 2-D (radius and time) parameterizations of the complex 4-D processes that arise in direct numerical simulation (DNS) of core and mantle dynamics. The primary core-side constraints on these models, and their predictions of stratification and precipitation processes, are (1) the continuous generation of a magnetic field for at least the last 3.5 Gyrs (Tarduno et al., 2010); and (2) the present-day radius r_i of the inner core, $r_i = 1{,}221$ km. Therefore, constraining the evolution of the CMB region requires knowledge of the material properties of the whole core. The main quantities used in this review are defined in Table 12.2.

The challenge of estimating core material properties arises from the extreme conditions that must be replicated. The pressure ranges from $P = 135$ GPa to $P = 330$ GPa across the liquid core (Dziewonski & Anderson, 1981), T is several thousands of Kelvin, while

Table 12.2 Core material properties for pure iron and three Fe-O-Si mixtures denoted by their molar concentrations in the header line

Symbol	100%Fe	82%Fe-8%O-10%Si	79%Fe-13%O-8%Si	81%Fe-17%O-2%Si
$\Delta\rho$ (kg m^{-3})	240	600	800	1,000
w_O^s	–	0.0002	0.0004	0.0006
w_{Si}^s	–	0.0554	0.0430	0.0096
w_O^l	–	0.0256	0.0428	0.0559
w_{Si}^l	–	0.0560	0.0461	0.0115
C_p (J/kg/K)	715—800	–	–	–
$L_h(r_i)$ (MJ/kg)	0.75	–	–	–
$T_m(r_i)$ (K)	6,350	5,900	5,580	5,320
$\left.\frac{dT_m}{dP}\right\|_{r_i}$ (K/GPa)	9.01	9.01	9.01	9.01
$\alpha_T(r_i)$ ($\times 10^{-5}$/K)	1.0	–	–	–
$T_a(r_c)$ (K)	4,735	4,290	4,105	3,910
$\left.\frac{\partial T_a}{\partial P}\right\|_{r_i}$ (K/GPa)	6.96	6.25	6.01	5.81
$\left.\frac{\partial T_a}{\partial r}\right\|_{r_c}$ (K/km)	−1.15	−1.03	−1.00	−0.96
k^c (W/m/K)		See	Text	
D_O ($\times 10^{-8}$ m^2/s)	–	1.31	1.30	–
D_{Si} ($\times 10^{-8}$ m^2/s)	–	0.52	0.46	–
ν ($\times 10^{-7}$ m^2/s)	6.9	6.8	6.7	–
α_O^D ($\times 10^{-12}$ kg/m^3/s)	–	0.72	0.97	1.11
α_{Si}^D ($\times 10^{-12}$ kg/m^3/s)	–	1.19	1.10	0.40
		O	Si	
α_i^c	–	1.1	0.87	
$R_i^c - R_i^m$ (eV/f.u.)	–	−2.5		

Superscripts c have been suppressed for clarity. Gravity g, pressure P, and gravitational potential ψ are derived from the PREM density ρ. Quantities in the first section define the core chemistry model, where w_i denotes mass fraction of species i. Numbers in the second section determine the core temperature properties in the third section, which are given for the present day. The core temperature is assumed to follow an adiabat, denoted T_a, and the melting temperature of the core alloy is denoted T_m. L_h denotes the latent heat of fusion, C_p is the specific heat capacity, and α_T is the thermal expansion coefficient. CMB values for viscosity ν, thermal conductivity k^c, self-diffusion coefficients D_i, and barodiffusion coefficients α^D (defined in Gubbins and Davies, 2013) calculated along the corresponding adiabats are given in the fourth section. The CMB radius is denoted $r_c = 3{,}480$ km, the present-day ICB radius is $r_i = 1{,}221$ km. α_i^c are the compositional expansion coefficients, and $R^c - R^m$ is the heat of reaction coefficient [equation (12.2)] from Pozzo et al. (2019). Source: Adapted from Davies et al. (2015).

the mass fractions w_i^c of light element i are themselves determined by partitioning behavior at high P and T. The main experimental tool used to access these conditions is the laser-heated diamond anvil cell. Here, the challenges include minimizing temperature gradients across small samples (Sinmyo et al., 2019), identifying melting (Anzellini et al., 2013), and the potential for oxidation of the sample at high P-T (Frost et al., 2010).

Ab initio calculations can sample core P-T conditions, but also contain uncertainties such as the form of the exchange-correlation functional and must ultimately be ground-truthed by experiments. Here, we do not systematically treat the uncertainties on material properties and instead focus on models of the core that are consistent with seismic observations (Badro et al., 2014; Davies et al., 2015).

Present-day constraints on P, T, and w_i^c come from the liquid core density ρ, which is about 10 wt% lighter than pure iron, and also from the density jump $\Delta\rho$ at the inner core boundary (ICB). Fluctuations in ρ due to convection are small (Stevenson, 1987), while time variations in core composition are tiny (Davies, 2015), and so the pressure gradient is determined from hydrostatic balance with ρ and gravity g derived from 1-D seismic models of the core (Dziewonski & Anderson, 1981; Irving et al., 2018). Part of the observed density jump, $\Delta\rho_m = 240$ kg m^{-3} (Alfè et al., 2002c), arises from the phase change at the ICB; the rest determines the excess concentration of light elements in the liquid core compared to the solid core. Matching candidate compositions derived from partitioning behavior at ICB conditions to observational constraints on $\Delta\rho$ allows to estimate the present core composition and hence the melting temperature T_m of the iron alloy (e.g., Alfè et al., 2002a). The core temperature T is usually assumed to vary adiabatically outside thin boundary layers and stable regions. The anchor point for T is the value of T_m at the ICB. The chemical properties α_i^c, α_i^D, and R_i are calculated from chemical potentials at fixed P, T, and composition. Finally, transport properties such as the core viscosity ν, self-diffusion coefficients D_i, and thermal conductivity can be calculated for specified composition at points along core P-T curves (e.g., Pozzo et al., 2013).

The ICB density jump $\Delta\rho$ is rather uncertain (see Wong et al., 2021, for a recent review). In this work, we take the range obtained from normal modes of $\Delta\rho = 800 \pm 200$ kg m^{-3} (Masters & Gubbins, 2003) and consider the three values $\Delta\rho = 600$ kg m^{-3}, 800 kg m^{-3}, and 1,000 kg m^{-3}. The parameter values for each $\Delta\rho$ are listed in Table 12.2. These are generally taken from Davies et al. (2015) where more details can be found. Next we review constraints on the core and magma ocean compositions that are relevant for understanding mass exchange at the CMB. We then consider the core temperature structure and sketch a derivation of the core energy balance before discussing recent estimates of core thermal conductivity.

12.2.1. Bulk Composition of the Core and Basal Magma Ocean

The chemical constitution of the core and the nature and abundance of mineral phases at the base of the mantle are still uncertain (Hirose et al., 2013; Garnero et al., 2016). Core formation models suggest that O, Si, and S are likely to partition into metal (Rubie et al., 2015a; Badro et al., 2015), though at very high temperatures other elements such as Mg can also become siderophile (O'Rourke & Stevenson, 2016). Carbon has also been considered (Rubie et al., 2015a), but recent work suggests C partitions weakly into metal at high P and T (Fischer et al., 2020). Calculations attempting to match the present-day core mass and $\Delta\rho$ show that O and C partition almost exclusively into liquid at ICB conditions (Alfè et al., 2002a; Li et al., 2019) and so matching the overall mass of the core requires another element that partitions evenly such as S or Si (Alfè et al., 2002a). Umemoto and Hirose (2020) calculated density and compressional wave speed for a number of liquid binary alloys and mixtures and found that the composition providing the best fit to PREM involves oxygen as the primary element at high temperature, with hydrogen becoming the primary element as T decreases. The main stable phase in the present lower mantle is (Mg,Fe)SiO$_3$ silicate perovskite, with an additional ~15% ferropericlase and some calcium silicate perovskite (Garnero et al., 2016). Bridgmanite composition is dominated by the oxides SiO$_2$ and MgO (Garnero et al., 2016). In addition, LLVPs may be enriched in iron (FeO) and bridgmanite or calcium silicate perovskite compared to ambient mantle (Vilella et al., 2021).

Much recent work has focused on the partitioning of Mg, Si, and O between the core and mantle. Mg and Si are of interest because they may become saturated in the core as the planet cools, precipitating as oxides MgO and SiO$_2$, respectively, which releases gravitational energy that is available to power the geodynamo (O'Rourke & Stevenson, 2016, Badro et al., 2016, Hirose et al., 2017, Mittal et al., 2020; Wilson et al., 2022). The study of FeO has attracted attention because it provides a mechanism for oxygen to enter the core, either from FeO in ferropericlase in the present Earth (Frost et al., 2010) or from an FeO-enriched BMO in the past (Davies et al., 2020), which leads to a stable stratification below the CMB (Buffett & Seagle, 2010; Davies et al., 2020). We will therefore focus on the interactions between Fe, Mg, Si, and O in the remainder of this review. Note that the material properties listed in Table 12.2 were obtained without Mg, though the error is probably not significant since the fraction of Mg dissolved in the core is probably much less than Si or O.

The initial bulk compositions of the core and mantle were set during planetary differentiation. Recent multistage core formation models find broadly consistent initial oxygen concentrations in the range 2–5 wt% (Badro et al., 2015; Rubie et al., 2015b), but diverge on the estimated silicon content with Badro et al. (2015) finding 2–3.6 wt% Si while Rubie et al. (2015b) obtained 8–9 wt% Si. The difference is partly due to the inferred oxidation state (oxidizing or reducing) of accretion materials, though other uncertainties in the core formation process mean that initial Si and O core concentrations in the range

1–10 wt% cannot be ruled out (Fischer et al., 2017). Partitioning of Mg has generally been omitted in core formation studies. Badro et al. (2016) ran multistage core formation models and found that 0.8 wt% MgO could be delivered to the core without a late giant impact, while 1.6–3.6 wt% MgO could be delivered depending on the mass of a late impactor. O'Rourke and Stevenson (2016) estimated 0.5 wt% Mg in the core for a single-stage model with equilibration at 3500 K (similar to the 0.3 wt% Mg obtained by Helffrich et al., 2020) while a two-stage model with a second equilibration at higher T permitted up to 2 wt% Mg in the core. The initial BMO composition is also hard to constrain. Andrault et al. (2017) concluded that deep mantle melts near the eutectic temperature may have had compositions similar to pyrolite, that is, 40 mol% SiO_2, 50 mol% MgO, and 10 mol% FeO (Eggins et al., 1998). Caracas et al. (2019) calculated that increasing the fraction of bridgmanite crystallized in the magma ocean from 0 to 30% reduced the SiO_2 and MgO content of the melt by 10 mol% and 5 mol%, respectively, while the FeO content of the melt increased by 37 mol%.

12.2.2. Core Temperature and Energy Balance

The ICB temperature T^i is obtained from the melting point of pure iron, T_m^{Fe}, depressed by an amount ΔT to account for the presence of impurities. We take $T_m^{Fe} = 6350$ K from the *ab initio* study of Alfè et al. (2002a), which is consistent with the experimental results of Anzellini et al. (2013), though higher than recent estimates of 5500 K from Sinmyo et al. (2019). The gradient of the melting curve, dT_m^{Fe}/dP, is more important for thermal history calculations, which is more consistent between the Sinmyo et al. (2019) and Anzellini et al. (2013) studies when accounting for uncertainties in extrapolating the experimental results to ICB pressure. For ΔT, we employ the linear melting point depression derived by Alfè et al. (2002a) using a truncated expansion of the chemical potentials at ICB conditions. The total ΔT is assumed to be a linear combination of the values for O and Si (ignoring any effect from Mg). The adiabatic temperature gradient is given by

$$\frac{\partial T_a}{\partial r} = -\frac{\alpha_T g T_a}{C_p}, \quad (12.4)$$

where α_T is the thermal expansion coefficient and C_p is the specific heat capacity.

In thermal history models, the bulk of the core is assumed to be hydrostatic, adiabatic, and compositionally well mixed, while within a stable layer diffusion is assumed to control the radial temperature and compositional profiles. With these assumptions, it can be shown (Nimmo, 2015a) that the power released by stored heat Q_s, latent heat at the ICB Q_L, and gravitational energy Q_g are respectively

$$Q_s = \int \rho C_p \frac{DT}{Dt} dV, \quad Q_L = 4\pi r_i^2 \rho(r_i) L_h \frac{dr_i}{dt},$$
$$Q_g = \int \rho \psi \alpha_i^c \frac{Dw_i^l}{Dt} dV_s. \quad (12.5)$$

Here, ψ is the gravitational potential, L_h is the latent heat coefficient, V_s is the volume of the convecting core, the rate of growth of the inner core is

$$\frac{dr_i}{dt} = \frac{1}{(dT_m/dP)_{r=r_i} - (\partial T_a/\partial P)_{r=r_i}} \frac{1}{\rho(r_i)g(r_i)}$$
$$\times \frac{T^i}{T^{cen}} \frac{dT^{cen}}{dt}, \quad (12.6)$$

where T^{cen} is the temperature at the center of the core (Gubbins et al., 2003; Greenwood et al., 2021), and the rate of change of light element fraction in the core is

$$\frac{Dw_i^l}{Dt} = \frac{4\pi r_i^2 \rho(r_i)}{M_{oc}} \left(w_i^l - w_i^s\right) \frac{dr_i}{dt}, \quad (12.7)$$

(Gubbins et al., 2004), where M_{oc} is the mass of the outer core and the superscripts l and s here define quantities in the liquid and solid cores, respectively.

Together with the power Q_p produced by precipitation (defined precisely below), Q_s, Q_L, and Q_g are the dominant terms in the core energy balance (Gubbins et al., 2004; Nimmo, 2015a), which can be written

$$Q^c = Q_s + Q_L + Q_g + Q_p. \quad (12.8)$$

In the absence of stable layers, all four terms on the RHS of equation (12.8) may be related to the cooling rate at the CMB, that is, $Q^c = A dT^c/dt$, where A represents integrals over core properties that can be calculated from Table 12.2. Note that we have also neglected radioactivity since ^{40}K is not thought to partition significantly into the core (Xiong et al., 2018). The magnetic field appears in the entropy budget, which can be written symbolically (again neglecting small terms) as

$$E_J + E_\alpha + E_k = E_s + E_L + E_g + E_p. \quad (12.9)$$

Here, E_α is the entropy due to molecular diffusion of light elements, E_k is the entropy due to thermal conduction (which depends on the thermal conductivity), and E_J is the entropy production by Ohmic dissipation.

12.2.3. Core Thermal Conductivity

A detailed comparison of different methodologies for determining k^c is both beyond the scope of this chapter and the expertise of the authors, and so we refer the reader to Williams (2018), Zhang et al. (2020), Pourovskii et al. (2020), and Pozzo et al. (2022) for recent discussions. We consider experimental studies comprising direct determinations of k^c in hcp iron (Konôpková et al., 2016) and solid Fe-Si alloys (Hsieh et al., 2020) and inferences of k^c based on measured electrical conductivity σ of hcp iron (Ohta et al., 2016; Xu et al., 2018; Zhang et al., 2020) and hcp Fe-Si alloys (Inoue et al., 2020) using the Wiedemann-Franz law

$$k = LT\sigma, \qquad (12.10)$$

where L is the Lorenz number. Equation (12.10) assumes that heat is carried by free electrons, which are predominantly scattered elastically by phonons; in the case of perfect scattering, L takes the Sommerfeld value of $L = L_0 = 2.44 \times 10^{-8}$ W Ω K^{-2} (e.g., Secco, 2017). Recent computational studies also include inferences of k^c from the Wiedemann-Franz law (Xu et al., 2018) as well as direct determinations of k^c in liquid iron (Pozzo et al., 2012; de Koker et al., 2012; Pozzo & Alfè, 2016) and iron alloys (de Koker et al., 2012; Pozzo et al., 2013).

Figure 12.1 shows k^c values obtained directly (top) and inferred from the Wiedemann-Franz law (bottom) at the P-T conditions reported in the above studies, that is, without extrapolation to core conditions. Only selected high P-T results are shown and so the P-T trends obtained by individual studies are not represented. When comparing the various data, several factors need to be taken into account. Increases in k^c usually arise from increasing pressure and temperature. Decreasing k^c arises from the solid-liquid transition, presence of impurities, the effect of electron-electron scattering (for calculations), and a nonideal value of L (for electrical conductivity measurements). We consider each of these factors in turn:

Pressure: Pozzo and Alfè (2016) provide the pressure-dependence of electrical conductivity of pure iron at 4350 K. Inoue et al. (2020) show P-dependence of a 4 wt% Si alloy at 300 K and also at the similar temperatures of 1570 K and 1650 K. Converting to k^c values using equation (12.10) with $L = L_0$ yields mean dk/dP values of 0.4 W m^{-1} K^{-1} GPa^{-1} for Pozzo and Alfè (2016) and 0.13 W m^{-1} K^{-1} and 0.5 W m^{-1} K^{-1} GPa^{-1} for Inoue et al. (2020) corresponding to an increase in k^c of 15–20 W m^{-1} K^{-1} from 95 to 135 GPa. We use $dk^c/dP = 0.4$ W m^{-1} K^{-1} GPa^{-1} below.

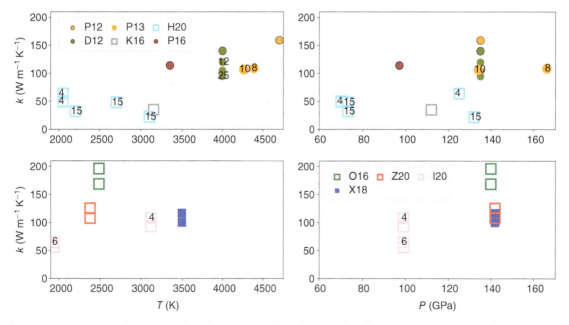

Figure 12.1 Summary of recent studies of core thermal conductivity k^c. The top row shows direct determinations of k^c while the bottom row shows inferences of k^c using electrical conductivity and the Wiedemann-Franz law. Left column shows the dependence on temperature T at the pressure P shown in the right column. Colors distinguish studies; open/closed symbols denote the method employed (experiment, calculation); shape denotes the material (square=solid, circle=liquid). All results are for pure iron except where a number appears inside the symbol, which denotes the molar concentration of Si. In the bottom row, the upper (lower) symbol corresponds to $L = 2.44 \times 10^{-8}$ W Ω K^{-2} ($L = 2.1 \times 10^{-8}$ W Ω K^{-2}). The considered studies are: P12 (Pozzo et al., 2012); D12 (de Koker et al., 2012); P13 (Pozzo et al., 2013); K16 (Konôpková et al., 2016); O16 (Ohta et al., 2016); P16 (Pozzo & Alfè, 2016); X18 (Xu et al., 2018); Z20 (Zhang et al., 2020); I20 (Inoue et al., 2020); H20 (Hsieh et al., 2020).

Temperature: The expected T behavior depends critically on the validity of equation (12.10) and the role of saturation effects (Konôpková et al., 2016; Pozzo & Alfè, 2016). In the absence of saturation, the Bloch-Grüneisen formula predicts that the electrical conductivity due to electron-phonon scattering varies as T^{-1} at high T, and hence $k^c \propto L \sim$ constant according to equation (12.10). Saturation can arise at high T when the electron mean free path becomes comparable to the interatomic distance, at which point σ stops decreasing with temperature and equation (12.10) predicts that k^c increases with T. The relevance of saturation to Earth's core properties was first recognized by Gomi et al. (2013) and has been observed by Pozzo and Alfè (2016) and Inoue et al. (2020), though not by Zhang et al. (2020). As a simple estimate of dk/dT, we use the results from de Koker et al. (2012), who found $dk/dT \approx 0.01$ W m^{-1} K^{-2} for FeO$_3$ at 135 GPa and $dk/dT \approx 0.02$ W m^{-1} K^{-2} for FeO$_7$ in the pressure range 130–160 GPa. In order to produce a conservative increase in k^c, we adopt $dk/dT = 0.01$ W m^{-1} K^{-2} below.

- Phase transition: Zhang et al. (2020) discuss recent literature and invoke a 10% decrease in σ on melting. Pozzo et al. (2013) find a change in σ of 18% − 25%, which is mainly due to the solid structure, but also contains a contribution from the uneven partitioning of elements at the ICB. We take the value of 18% below since this is roughly halfway between the two extremes.
- Impurities: Few studies have systematically compared the effect of different elements on k^c, but those that have find that the identity of the impurity is of secondary importance compared to their abundance as should be expected from relatively insulating impurities acting as disruptions to metallic structure. Inoue et al. (2020) found that up to 6.5 wt% Si could reduce k^c by 10% − 20%, while de Koker et al. (2012), Pozzo et al. (2013), and Zhang et al. (2020) found that various combinations of Si and O could reduce k^c by up to 30%. Hsieh et al. (2020) found that the thermal conductivity of alloys with 4 mol% and 7 mol% Si matched values for pure Fe at similar P-T conditions within uncertainty, but observed a substantial reduction in k^c for alloys with 15 mol% Si; however, such a large amount of Si exceeds the most extreme estimates of core Si composition by Davies et al. (2015) and Badro et al. (2014). Here, we assume a 20% reduction.
- Electron-electron scattering (EES) and nonideal L: EES can reduce both the k^c calculated from classical density functional theory (de Koker et al., 2012; Pozzo et al., 2013) and the L in equation (12.10) below the ideal value L_0. At high P-T for hcp iron, Zhang et al. (2020) find a 20% decrease in σ due to EES and estimate $L \approx 2.0$–2.1×10^{-8} W Ω K^{-2}, while Pourovskii et al. (2020) obtain a 20% decrease in k^c for bcc and hcp iron and estimate $L = 2.28 \times 10^{-8}$ W Ω K^{-2} at ICB conditions. de Koker et al. (2012) also obtain $L \approx 1.8$–2.4×10^{-8} W Ω K^{-2} without EES, indicating nonnegligible inelastic scattering effects. In view of the current uncertainty, we use $L = L_0$ and $L = 2.1 \times 10^{-8}$ W Ω K^{-2} and adopt a 20% drop in k^c due to EES.

Figure 12.2 shows the extrapolated values of k^c for the studies in Figure 12.1. The majority of values fit within the range $70 \leq k \leq 110$ W m^{-1} K^{-1}. This range overlaps

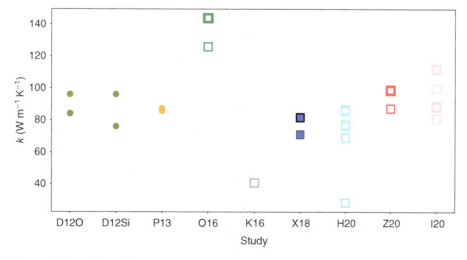

Figure 12.2 Extrapolation of k^c values in Figure 12.1 to CMB pressure of 135 GPa and temperature of 4000 K assuming a liquid iron alloy with a 20% reduction from the pure iron conductivity to account for the presence of impurities. The symbol styles are the same as in Figure 12.1. Thick (thin) edges correspond to electrical conductivity calculations extrapolated using $L = 2.44 \times 10^{-8}$ W Ω K^{-2} ($L = 2.1 \times 10^{-8}$ W Ω K^{-2}).

with that of Pozzo et al. (2022), who demonstrated consistency between the thermal conductivity of Fe-Si alloys in EES-corrected DFT calculations and experiments and inferred a range of $k^c = 75–81$ W/m/K at CMB conditions. Notable low values are the extrapolations from direct conductivity measurements for the pure hcp (Konôpková et al., 2016) and Si-rich Fe-Si solid (Hsieh et al., 2020). Future work is needed to understand the reasons for this, and to better constrain the extrapolation, which is subject to significant uncertainties as discussed above. For the rest of this chapter, we focus on two values of conductivity: $k^c = 70$ W m^{-1} K^{-1} and $k^c = 100$ W m^{-1} K^{-1} as suggested by Figure 12.2. As such we will henceforth focus on the "high conductivity" scenario in Table 12.1.

12.3. MASS TRANSFER AT THE CMB

In general, the chemical compositions of material in contact at the CMB will differ from the bulk compositions of the core and mantle, which gives rise to a chemical flux given by equation (12.3). Mass transfer at the CMB therefore depends on the compositions of the core and mantle, both in the bulk and on either side of the CMB. Since we are primarily interested in the "high conductivity" scenario (see Table 12.1), we will focus on the interaction between the core and silicate melts in a BMO. This scenario is expected to yield greater chemical exchange than the interaction between the core and solid mantle because the significant increase in diffusion coefficient between solid mantle and BMO overwhelms any potential reduction in partition coefficient due to entropic effects in the melt (Pozzo et al., 2019).

Elements are usually assumed to be well mixed by vigorous convection in the proto-core (e.g., Rubie et al., 2015a), though it is possible that a stratified layer developed near the end of core formation (Landeau et al., 2016; Jacobson et al., 2017) as discussed in section 12.4.4. Self-diffusion coefficients of O and Si in the liquid are very small (see Pozzo et al., 2013, and Table 12.2), and so chemical diffusion in a primordial stratified layer was probably too slow to produce significant time variations in the bulk composition. An early BMO was also presumably well mixed (Solomatov, 2015); however, its bulk composition could evolve over time. In the model of Labrosse et al. (2007), the melt becomes depleted in MgO and enriched in FeO as the ocean shrinks (Labrosse et al., 2007). However, different scenarios for BMO evolution, such as compaction of an Fe-depleted mush layer, could produce alternative compositional evolution (Caracas et al., 2019). Therefore, the distinction between precipitation and stratification scenarios depends primarily on the compositional evolution of a BMO and interactions at the CMB.

Chemical stratification of the upper core arises when the equilibrium concentration of an element i at the CMB exceeds its bulk concentration. The flux I_i is negative and light element enters the core. Precipitation arises when the equilibrium concentration of i falls below the bulk concentration; I_i is positive and light element leaves the liquid. If precipitation arises at the CMB then the precipitate, which is lighter than the bulk core liquid, will underplate onto the CMB, leaving behind a residual liquid at the top of the core that is depleted in light element and hence denser than the core fluid below. Owing to the low viscosity of the core, the dense residual liquid will rapidly sink via a Rayleigh-Taylor instability, presumably mixing throughout the core. The gravitational energy released by precipitation of element i is

$$Q_p = \int \rho \psi \alpha_i^c \frac{dw_i^c}{dt} dV_s \approx \int \rho \psi \alpha_i^c \frac{dw_i^c}{dT} \frac{dT}{dt} dV_s. \quad (12.11)$$

The primary quantities of interest are therefore w_i^c, which is critical for determining the onset and evolution of stratification/precipitation, and dw_i^c/dT, which determines the power released by precipitation. w_i^c and dw_i^c/dT are obtained from the equilibrium conditions at the CMB.

In this section, we first present the calculation of equilibrium conditions at the CMB. We demonstrate the case of precipitation ($I_i > 0$) for MgO partitioning and stratification ($I_i < 0$) for FeO partitioning in isolation. Finally, we consider the coupled equilibrium conditions for MgO, FeO, and SiO$_2$. We calculate equilibrium conditions from the end of core formation to the present day and consider a wide range of initial CMB temperatures, T_0, and initial core/BMO compositions since the flux of a given element depends strongly on T and the abundance of other elements. Most studies find that the core started hot, ~6000 K (Driscoll & Bercovici, 2014; Nakagawa & Tackley, 2014; Rubie et al., 2015a; O'Rourke & Stevenson, 2016). Conversely, Andrault et al. (2016) argue that the magma ocean (assumed to freeze from the bottom up) implied by such high T_0 would rapidly cool the core to ~4400 K. In general, we will not assume a value of T_0, but will use 6000 K as the upper limit on T_0.

12.3.1. Chemical Equilibrium at the CMB

Departures from chemical equilibrium for materials in contact at the CMB should be very small since the timescale for diffusion is very short over such small length scales. Chemical equilibrium at the CMB requires equality of chemical potentials μ_i for each species i, while mass conservation (ignoring thermal contraction of the core) implies that the total flux of mass from the mantle I_i

equals the mass added to the core (Braginsky & Roberts, 1995; Davies et al., 2020). These conditions can be written

$$\mu_1^m(P, T, c_1^m, \ldots, c_N^m) = \mu_1^c(P, T, c_1^c, \ldots, c_N^c),$$
$$\mu_2^m(P, T, c_1^m, \ldots, c_N^m) = \mu_2^c(P, T, c_1^c, \ldots, c_N^c),$$
$$\ldots$$
$$\mu_N^m(P, T, c_1^m, \ldots, c_N^m) = \mu_N^c(P, T, c_1^c, \ldots, c_N^c); \quad (12.12)$$

$$I_1^m(P, T, c_1^m, \ldots, c_N^m) = I_1^c(P, T, c_1^c, \ldots, c_N^c),$$
$$I_2^m(P, T, c_1^m, \ldots, c_N^m) = I_2^c(P, T, c_1^c, \ldots, c_N^c),$$
$$\ldots$$
$$I_N^m(P, T, c_1^m, \ldots, c_N^m) = I_N^c(P, T, c_1^c, \ldots, c_N^c). \quad (12.13)$$

where superscripts m and c denote the mantle and core, respectively, $i, = 1 \ldots, N$ represents the number of chemical species, and c_i denotes the mole fraction of species i. Here, the pressure and temperature correspond to conditions at the CMB. Note that equation (12.12) does not imply equality of the chemical compositions.

The key quantity for determining equilibrium conditions at the CMB is the equilibrium constant K, which is defined as

$$K = \frac{\prod_i a_i^{\alpha_i}}{\prod_j a_j^{\alpha_j}} = \frac{\prod_i c_i^{\alpha_i}}{\prod_j c_j^{\alpha_j}} \cdot \frac{\prod_i \gamma_i^{\alpha_i}}{\prod_j \gamma_j^{\alpha_j}} = K_d \cdot \frac{\prod_i \gamma_i^{\alpha_i}}{\prod_j \gamma_j^{\alpha_j}}, \quad (12.14)$$

where K_d is the distribution coefficient, $a_i = c_i \gamma_i$ are the activities, γ_i are the activity coefficients, and α_i are the reaction exponents. Here, i denotes the products that appear on the right side of the reaction and j denotes the reactants. At equilibrium K is related to the Gibbs free energy change across the reaction ΔG_r by

$$K = \exp\left(-\frac{\Delta G_r}{k_B T}\right) = \exp\left(-\frac{\Delta H_r - T \Delta S_r + P \Delta V_r}{k_B T}\right), \quad (12.15)$$

where k_B is the Boltzmann constant and ΔH_r, ΔS_r, and ΔV_r, are respectively the standard state change in enthalpy, entropy, and volume across the reaction. Equation (12.15) is usually written as

$$\log K_d = a + \frac{b}{T} + c\frac{P}{T} - \sum_i (\log \gamma_i^{\alpha_i}) + \sum_j (\log \gamma_j^{\alpha_j}), \quad (12.16)$$

where the coefficients a, b, c, and γ_i are to be determined from recovered phases that are analyzed at known P-T-composition conditions. Note that for consistency with previous work, we have retained the notation for the coefficient c, which should not be confused with mole fraction.

Computer simulations calculate chemical potentials for each species (e.g., Alfè et al., 2002b; Pozzo et al., 2019) and obtain the equilibrium concentrations directly from equations (12.12). Separating out the configurational part of the chemical potential, that is, $\mu_i = k_B T \ln c_i + \tilde{\mu}_i$, the equilibrium becomes

$$k_B T \ln\left[\frac{\prod_j c_j^c}{\prod_i c_i^m}\right] = k_B T \ln K_d = \sum_i \tilde{\mu}_i^m - \sum_j \tilde{\mu}_j^c, \quad (12.17)$$

(Davies et al., 2018; Pozzo et al., 2019). Since the chemical potentials are completely determined, this formulation can be shown to be equivalent to equation (12.15) by separating the chemical potentials as $\mu_i = \mu_i^0 + k_B T \ln a_i$, where μ_i^0 is the value of μ_i in standard state.

The form of K (and K_d) is determined by the nature of the chemical reaction. The reactions that have generally been considered in the literature are dissolution, dissociation, and exchange (e.g., Badro et al., 2018). These are summarized in Table 12.3. In principle numerical simulations could be used to distinguish between the different possibilities; however, the simulation sizes required to obtain meaningful concentrations have traditionally been prohibitively costly in *ab initio* calculations. Another approach is to compare large datasets against the predictions from equation (12.16), which has been done recently for MgO by Badro et al. (2018). We reproduce the workflow of Badro et al. (2018) below to demonstrate the steps involved in obtaining equilibrium concentrations and precipitation rates and to provide a consistent framework with which to compare recent studies. Compositional variations in silicate activity coefficients are neglected and hence the γ_j^m can be absorbed into the parameters a and b; the γ_i below therefore refers to the metal. Silicate activities can be included in the modeling (Frost et al., 2010; Helffrich et al., 2020), but at the expense of introducing more fitting parameters.

12.3.2. Partitioning of MgO at the CMB

The equations determining $\log K$ for MgO dissolution, dissociation, and exchange are respectively

$$\log \frac{c_{MgO}^c \gamma_{Mg}^c \gamma_O^c}{c_{MgO}^m} = \log K_{dl}^{MgO} = a_{dl} + \frac{b_{dl}}{T} + c_{dl}\frac{P}{T}, \quad (12.18)$$

$$\log \frac{a_{Mg}^c a_O^c}{c_{MgO}^m} = \log K_{dc}^{MgO} = a_{ds} + \frac{b_{ds}}{T} + c_{ds}\frac{P}{T}, \quad (12.19)$$

$$\log \frac{a_{Mg}^c c_{FeO}^m}{a_{Fe}^c c_{MgO}^m} = \log K_e^{MgO} = a_e + \frac{b_e}{T} + c_e\frac{P}{T}. \quad (12.20)$$

The activity coefficients in the dissolution reaction arise because it is assumed that dissolved MgO further breaks down into Mg and O; see Badro et al. (2018) for detailed discussion. Equations (12.18)–(12.20) are evaluated using the values of a, b, and c reported in several previous

Table 12.3 Summary of chemical reactions between MgO, SiO$_2$, FeO, and metallic alloys considered in recent literature

Reaction	K_d	References
MgO^m $\Longleftrightarrow$ MgO^c + O^c	$\dfrac{c_{Mg}^c \, c_O^c}{c_{MgO}^m}$	B16 B18 M20 H20
MgO^m $\Longleftrightarrow$ MgO^c	$\dfrac{c_{MgO}^c}{c_{MgO}^m}$	B18
MgO^m + Fe^c $\Longleftrightarrow$ FeO^m + Mg^c	$\dfrac{c_{FeO}^m \, c_{Mg}^c}{c_{Fe}^c \, c_{MgO}^m}$	O16, D17, D19
2MgO^m + Si^c $\Longleftrightarrow$ SiO$_2^m$ + 2Mg^c	$\dfrac{c_{SiO_2}^m \, (c_{Mg}^c)^2}{c_{Si}^c \, (c_{MgO}^m)^2}$	H20
FeO^m $\Longleftrightarrow$ Fe^c + O^c	$\dfrac{c_{Fe}^c \, c_O^c}{c_{FeO}^m}$	F10 O16 M20 F15
SiO$_2^m$ $\Longleftrightarrow$ Si^c + 2O^c	$\dfrac{c_{Si}^c (c_O^c)^2}{c_{SiO_2}^m}$	H17 M20 H20
SiO$_2^m$ $\Longleftrightarrow$ SiO$_2^c$	$\dfrac{c_{SiO_2}^c}{c_{SiO_2}^m}$	
SiO$_2^m$ + 2Fe^c $\Longleftrightarrow$ 2FeO^m + Si^c	$\dfrac{(c_{FeO}^m)^2 \, c_{Si}^c}{(c_{Fe}^c)^2 \, c_{SiO_2}^m}$	O16, F15

The cited studies are Badro et al. (2016, B16), Badro et al. (2018, B18), Du et al. (2017, D17), Du et al. (2019, D19), Fischer et al. (2015, F15), Frost et al. (2010, F10), Helffrich et al. (2020, H20), Hirose et al. (2017, H17), Mittal et al. (2020, M20), and O'Rourke and Stevenson (2016, O16).

studies and reproduced in Table 12.4. When accounting for compositional effects O'Rourke and Stevenson (2016) set all activity coefficients to 1, Du et al. (2019) model the effect of O and Si, while Badro et al. (2018) consider interactions between O, Si, Mg, C, and S. Figure 12.3 shows K^{MgO} for the three reactions using the Badro et al. (2018) dataset, both with and without composition-dependence. It is clear that accounting for composition-dependence produces a significant reduction in data scatter. The importance of oxygen content was noted by Du et al. (2017), while the composition-dependence on joint solubility of Si, Mg, and O is clearly demonstrated in Helffrich et al. (2020).

The γ_i^c are quite sensitive to the values of the parameters ϵ_i^j, which describe the interaction between elements i and j in the liquid (e.g., Badro et al., 2018). Datasets of ϵ_i^j from Badro et al. (2018), Fischer et al. (2015), and Liu et al. (2020) display general consistency, and also differ on some of the important self-interaction parameters, for example, a factor of 7 difference in ϵ_O^O. The challenge is that the number of parameters increases with the number of elements (e.g., an O-Si system requires 3 ϵ_i^j while an O-Si-Mg-C system requires 10 ϵ_i^j), and so very large datasets are required to obtain robust constraints for complex systems.

Table 12.4 Values of the constant parameters used in this study to fit empirically determined distribution coefficients

Study	Reaction	a_{Mg}	b_{Mg}	c_{Mg}
O16	e	0.1	−10851	0
B16	dl	1.23	−18816	0
B18	ds	0.1	−14054	0
B18	e	1.06	−12842	0
D19	e	−3.0	−2314	26
		a_O	b_O	c_O
O16	e	0.6	−3800	22
M20	ds	−0.3	0.0	−36.8
		a_{Si}	b_{Si}	c_{Si}
O16	e	1.3	−13500	0
M20		See text		

The sections show from top to bottom Mg, O, and Si. For O and Si, the values denoted by O16 (O'Rourke and Stevenson, 2016) were obtained from Fischer et al. (2015), while the values for Mg were estimated from experiments in Takafuji et al. (2005). For Mittal et al. (2020, M20), the values for O come from Hirose et al. (2017). Abbreviations "e," "ds," and "dl" denote exchange, dissociation and dissolution, reactions and are used as subscripts in the text.

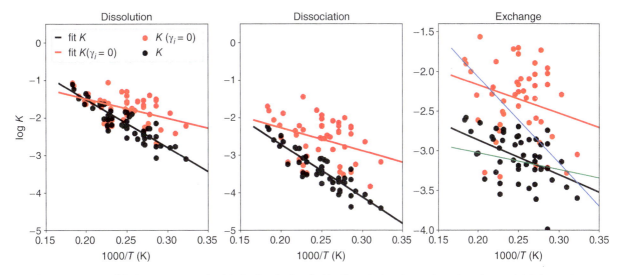

Figure 12.3 Equilibrium constants for MgO dissolution (left), dissociation (center), and exchange (right) reactions using the dataset in Badro et al. (2018, Table S1) and values in Table 12.4. Red points show K with all activity coefficients set to one, while black points show K values calculated as in Badro et al. (2018) with data in their Table S2. Black and red lines are fits to the respective datasets using equations (12.18)–(12.20). Blue and green lines show respectively the models of O'Rourke and Stevenson (2016), and Du et al. (2019 evaluated at $P = 8$ GPa) using values in Table 12.4.

In order to compare results from different assumed reactions and modeling strategies, Figure 12.4 shows the core weight fraction of MgO, w^c_{MgO}, versus temperature. We consider the same compositions as Badro et al. (2018): a constant 3 wt% O and 3 wt% Si in the core and 50 mol% MgO in the mantle. Using the Badro et al. (2018) method and dataset, the dissociation and dissolution reactions produce almost identical results while all three reactions yield similar dw^c_{MgO}/dT as found by Du et al. (2019), though the exchange reaction yields a worse fit to their data (see Badro et al., 2018; Fig. 12.3). O'Rourke and Stevenson (2016) obtain a much larger equilibrium concentration and dw^c_{MgO}/dT than the more recent studies that include composition-dependence on the equilibrium conditions. This result underscores the importance of accounting for the light element content of the core when modeling precipitation rates.

The pressure-dependence of equilibrium is a critical issue because this governs the depth in the core at which precipitation will commence. Badro et al. (2016), Badro et al. (2018), and Du et al. (2017) find that the K^{MgO} are independent of P and hence precipitation must begin at the CMB. Du et al. (2019) obtained a statistically significant pressure variation for K^{MgO}_e, which has a significant impact on the equilibrium behavior obtained from their model. Figure 12.4 shows that at 8 GPa and 10 mol% FeO the equilibrium composition from Du et al. (2019) is almost independent of temperature as advocated in their earlier study (Du et al., 2017). However, when evaluated at CMB pressure this model predicts $dw^c_{MgO}/dT < 0$; therefore, once the core reaches the equilibrium composition, further cooling requires an inward MgO flux and precipitation ceases.

The equilibrium concentrations in Figure 12.4 should be compared to the initial Mg content of the core, estimated to lie in the range 0.3–3.6 wt% (section 12.2). Taking the higher end of these estimates, all studies in Figure 12.4 except O'Rourke and Stevenson (2016) predict that the core was oversaturated in Mg for all temperatures below 6000 K; the bulk core Mg content was then higher than the CMB value corresponding to a positive (outward) flux I_{Mg} and precipitation of MgO from the core. Conversely, using the lowest value, 0.3 wt% Mg, all studies predict that the core was undersaturated in Mg for all temperatures above 4000 K; the bulk core Mg content has then always been lower than the CMB value corresponding to a negative (inward) I_{Mg} and stratification of the uppermost core due to enrichment in Mg. Therefore, for fixed core and mantle compositions, Mg could either dissolve or precipitate at the top of the core within the uncertainties in partitioning behavior and initial core composition.

Focusing on the precipitation case, Figure 12.4 shows that the individual modeling approaches and datasets used by different groups result in a spread of MgO precipitation rates dw^c_{MgO}/dT that span almost 2 orders of magnitude. The high values from O'Rourke and Stevenson (2016) are likely due to their assumption that O and Mg activity coefficients could be set to one, which was reasonable at the time when few experimental data were

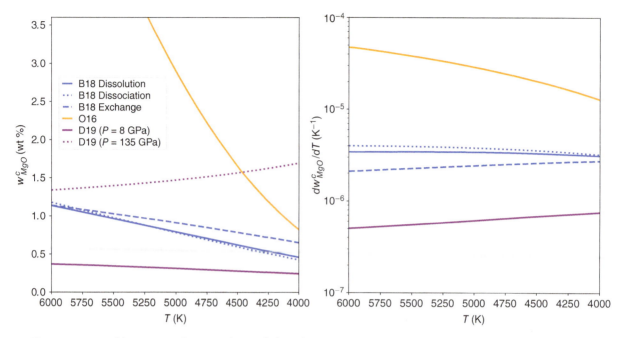

Figure 12.4 Equilibrium mass fraction of MgO (left) and precipitation rate dw^c_{MgO}/dT for a constant core composition of 3 wt% O and 3 wt% Si and a constant BMO composition of 50 mol% MgO and 10 mol% FeO. Considered studies are O'Rourke and Stevenson (2016, O16), Badro et al. (2018, B18), and Du et al. (2019, D19).

available. More recent work suggests lower precipitation rates, which correspondingly reduce the efficiency of precipitation as a mechanism for sustaining the ancient geomagnetic field.

12.3.3. Partitioning of FeO at the CMB

Previous studies have generally modeled FeO transfer using a dissolution reaction with distribution coefficient $K_d^{FeO} = c_{Fe}^c c_O^c / c_{FeO}^m$. As with Mg, the most significant interaction parameters involve Si and O because of their expected high concentrations in the core. However, Fischer et al. (2015) found that their fitted ϵ_{Si}^O and ϵ_O^O values produced an unstable parameterization in which partitioning of O into metal would cause ever more O and Si to enter the core. Considering the interaction between an Fe-O metal and ferropericlase, Davies et al. (2018) found that K_d^{FeO} is a weak function of oxygen concentration in the range $0 \leq c_O^c \leq 30$ mol%, while adding 7.6 mol% Si to the metal produced a strong increase of K_d^{FeO} with c_O^c, consistent with the findings of Tsuno et al. (2013) and Fischer et al. (2015) for the case of silicate melts. Pozzo et al. (2019) performed first-principles molecular dynamics calculations to determine K_d^{FeO} at CMB P-T conditions for a silicate melt comprising 50 mol% SiO_2, 44 mol% MgO, and 6 mol% FeO and a liquid metal comprising 95 mol% Fe and 5 mol% O; however, they were not able to determine the composition-dependence of K_d^{FeO} owing to the large system sizes needed to robustly estimate free energy changes. Here, we ignore the composition-dependence on FeO partitioning and focus on K_d^{FeO}, noting that improved constraints by future studies will be very valuable.

Figure 12.5 shows the temperature and pressure dependence of K_d^{FeO} from a number of recent experimental and computational studies. Davies et al. (2018) have shown that simulations at 134 GPa and 3200 K agree well with experiments at the same conditions with a starting composition consisting of a powdered mixture of pure metal and $Mg_{81}Fe_{19}O$ (Ozawa et al., 2008). Therefore, any discrepancy between the two types of study are likely due to differences in the starting compositions and uncertainties in determining exact P-T conditions. These factors produce a scatter of 0.5–1 log units over much of the moderate T range and are consistent with the differences observed at high T. The results show that K_d^{FeO} increases with both P and T and that O tends to favor the metal as core conditions of $T > 4000$ K are approached.

Figure 12.6 shows the equilibrium concentration of O in the core for different core and BMO Fe concentrations spanning the ranges discussed in section 12.2. Here, K_d^{FeO} has been fit using the black line in Figure 12.5, which yields values on the lower end of the range at high T; higher K_d^{FeO} would therefore increase the equilibrium concentrations in Figure 12.6. The results clearly show that the equilibrium O concentration exceeds all estimates for

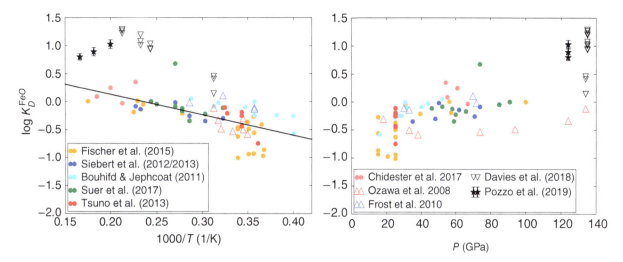

Figure 12.5 Comparison of published FeO distribution coefficients. Panels show values of the distribution coefficient K_d^{FeO} plotted against inverse temperature (left) and pressure (right) for solid-silicate–liquid-metal partitioning (open symbols) and silicate-melt–liquid-metal partitioning (closed symbols). The plotted studies are: Fischer et al. (2015), Siebert et al. (2012), Bouhifd and Jephcoat (2011), Suer et al. (2017), Tsuno et al. (2013), Chidester et al. (2017), Ozawa et al. (2008), Frost et al. (2010), Davies et al. (2018) and Pozzo et al. (2019). Source: Figure adapted from Pozzo et al. (2019).

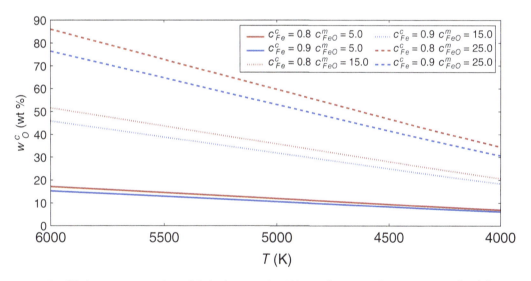

Figure 12.6 Equilibrium concentration of O in the core (wt %) as a function of temperature T for different concentrations of iron in the core, c_{Fe}^c (wt %), and FeO concentrations in the BMO, c_{FeO}^m (wt %).

the bulk core O concentration. Therefore, FeO is expected to partition strongly into liquid iron at high T, creating a stably stratified layer atop the core (Buffett & Seagle, 2010; Davies et al., 2018).

12.3.4. Partitioning of Multiple Species at the CMB

In general, the CMB compositions of the four elements assumed to be in the core (Fe, Si, O, Mg) and the three oxides assumed to comprise the BMO (MgO, FeO, SiO$_2$) can vary over time. The seven equations required to solve the system are obtained from mass balance of the four elements and the equilibrium constants for the three reactions (Rubie et al., 2011). These equations are nonlinear, and hence both the onset and rate of precipitation of a given chemical species will in general depend sensitively on P, T, starting composition and the functional forms of the equilibrium constants.

We calculate equilibrium concentrations following the method of Rubie et al. (2011, see Appendix A). The main limitation of this method is that it is not easily generalized to include composition-dependence of the equilibrium constants. This is clearly an important issue since we have shown above that the equilibrium concentration of Mg is sensitive to the O and Mg concentration in the core. However, given the complexity of multispecies partitioning and significant uncertainties on some of the key parameters, this is a necessary first step. It also simplifies the calculation of precipitation rates, which are needed by core thermal history models. We consider three different cases labeled according to whether the reaction governing transfer of O, Si, and Mg are respectively exchange (E) or dissociation (D):

1. DEE. This case corresponds to that of Rubie et al. (2011), who model oxygen transfer as a dissociation reaction and Si and Ni (here Mg) transfer by exchange reactions. The distributions coefficients are

$$\log \frac{c_{Fe}^c c_O^c}{c_{FeO}^m} = a_O^{ds} + \frac{b_O^{ds}}{T} + c_O^{ds}\frac{P}{T}, \quad (12.21)$$

$$\log \frac{(c_{FeO}^m)^2}{(c_{Fe}^c)^2} \frac{c_{Si}^c}{c_{SiO_2}^m} = a_{Si}^e + \frac{b_{Si}^e}{T} + c_{Si}^e\frac{P}{T}, \quad (12.22)$$

$$\log \frac{c_{FeO}^m}{c_{Fe}^c} \frac{c_{Mg}^c}{c_{MgO}^m} = a_{Mg}^e + \frac{b_{Mg}^e}{T} + c_{Mg}^e\frac{P}{T}. \quad (12.23)$$

2. DED. This case retains the same reactions for Si and O as in case 1, but employs a dissociation reaction for Mg as advocated by Badro et al. (2018).

3. DDD. This case employs dissociation reactions for all three species as done by Mittal et al. (2020).

For cases 2 and 3, the required modifications to the method of Rubie et al. (2011) are explained in Appendix A.

In this section, the values of a, b, and c for MgO are not the same as those in Table 12.4 because we ignore the composition-dependence. We have therefore refit K_d^{MgO} using the Badro et al. (2018) dataset as shown by the red lines in Figure 12.3, obtaining $a = -1.45$ and $b = -3596$ for the exchange reaction and $a = -1.039$ and $b = -6151$ for the dissociation reaction. We have also refit the a and b values from Du et al. (2019) based on a mean 15 mol% O in the core in order to account for the composition-dependence of their parameterization. For Fe and Si, the parameters are given in Table 12.4. For Si, we use the exchange reaction parameterization from Fischer et al. (2015) and the dissociation parameterization of Mittal et al. (2020), who refit the partitioning data of Hirose et al. (2017). Figure 12.7 shows the different temperature-dependences of Kd used in this section. Note that K_d^{Si} from Fischer et al. (2015) is 2–4 log units larger than that of Hirose et al. (2017) across the temperature range 3000–6000 K.

Figure 12.8 shows the DED and DEE cases using the initial compositions of Badro et al. (2018). The general behavior in both cases is very similar to that described in O'Rourke and Stevenson (2016) and Liu et al. (2020) who used slightly different compositions and calculation methods: the core becomes gradually depleted in all light elements and the equilibrium oxide budget is dominated by MgO. The main difference between DED and DEE cases is that the equilibrium Mg core composition and precipitation rate dw_{MgO}^c/dT are increased by a factor of 3 and 2, respectively. Indeed, for the DED case, the results are very similar to those for pure Mg partitioning (Fig. 12.4) because the larger MgO concentration preferred in the multicomponent case is offset by the larger equilibrium core O concentration. The increased w_{MgO}^c in the exchange reaction arises because of the increased

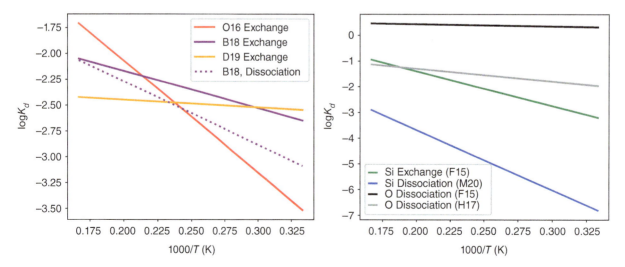

Figure 12.7 Distribution coefficients for Mg (left) and Si/O (right) used in comparison of multi-species precipitation.

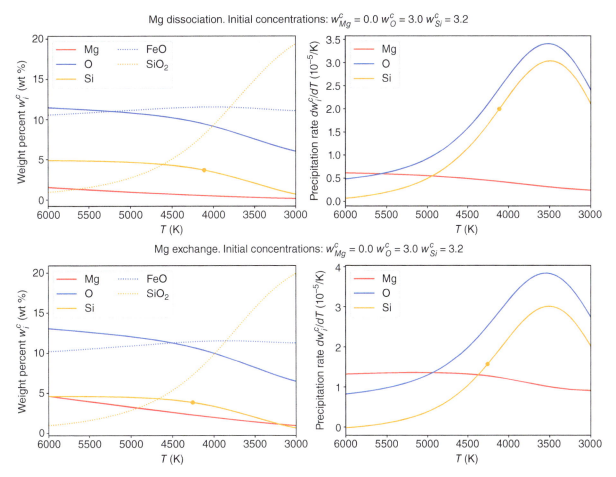

Figure 12.8 Equilibrium concentrations (left) and precipitation rate (right) for core elements and BMO oxides assuming the DED reaction set (top) and DEE reaction set (bottom). Dots mark the temperature at which the equilibrium core composition for element i falls below its concentration in the core. All concentrations are in wt%.

MgO content of the BMO, while the FeO concentration is about the same as assumed in Figure 12.4 when considering only MgO partitioning [see equation (12.23)].

Figure 12.9 compares equilibrium Mg concentrations and precipitation rates from three published thermodynamic models calculated using the same method, reaction set (DEE), and initial compositions. For direct comparison, we have also reproduced a calculation where the O'Rourke and Stevenson (2016) parameters are all reduced by 0.25σ, where σ is the standard deviation quoted in their Extended Table 12.1. The results for the Du et al. (2019) and O'Rourke and Stevenson (2016) 0.25σ parameters are very similar to those reported in Figure 3a of Du et al. (2019). The results using the Badro et al. (2018) parameters differ from those reported by Du et al. (2019), probably because we are considering the exchange reaction, which increases w^c_{Mg} as shown in Figure 12.8.

Figure 12.10 shows the equilibrium concentrations for Mg and Si and the Mg precipitation rate for the three different cases and three initial oxide compositions corresponding to a MgO-rich, FeO-rich, and SiO$_2$-rich BMO. There are three main messages from this figure. First, the combination of reactions is crucial for determining the onset temperature for precipitation, T_o, and dw^c_{Mg}/dT; for certain BMO compositions, dw^c_{Mg}/dT varies by over an order of magnitude, while Mg precipitation can begin anywhere between 6000 and 4000 K. Second, the initial BMO composition produces a factor 2–3 variation in the value of dw^c_{Mg}/dT and a change in T_o of up to 1000 K depending on the chosen reaction set. Third, dw^c_{Mg}/dT is not a monotonic function of T, though it is usually close to its maximum value when $T = T_o$. We also expect solutions to depend strongly on the initial core

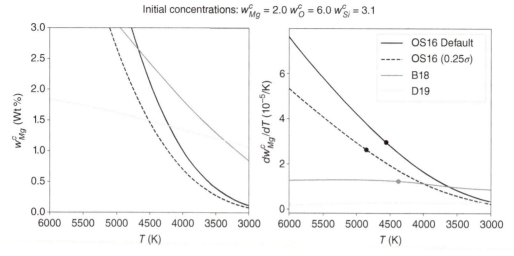

Figure 12.9 Equilibrium Mg concentration w_{Mg}^c (left) and Mg precipitation rate dw_{Mg}^c/dT (right) plotted as functions of temperature T for the DEE reaction set. Parameters are from O'Rourke and Stevenson (2016, O16), Badro et al. (2018, B18), and Du et al. (2019, D19). Also shown is a case where the O'Rourke and Stevenson (2016) parameters are all reduced by 0.25σ, where σ is the standard deviation quoted in their Extended Table 12.1. Dots mark the temperature at which the equilibrium core composition for element i falls below its concentration in the core. All concentrations are in wt%.

composition once the composition-dependence of K_d (which has been ignored here) is taken into account (section 12.3).

Figure 12.11 synthesizes the present results by plotting T_o against precipitation rate at T_o for Mg and Si. In all calculations, we have used an initial 2 wt% Mg in the core and so the values of T_o are probably at the upper end of viable estimates based on core formation studies. As shown by Mittal et al. (2020), the onset and rate of precipitation depend sensitively on several factors including the initial compositions and equilibrium constants. dw_{Mg}^c/dT spans the range $0.3–3 \times 10^{-5}$ K^{-1}, which is broadly consistent with the results for pure Mg partitioning (Fig. 12.4), while dw_{Si}^c/dT spans the range $0.1–8 \times 10^{-5}$ K^{-1}. There is a large spread of T_o values in both cases; however, most models favor onset of Mg precipitation at or below 5000 K while Si precipitation tends to begin at or below 4500 K. Assuming an initial core temperature of 6000 K (section 12.3) suggests that precipitation did not begin

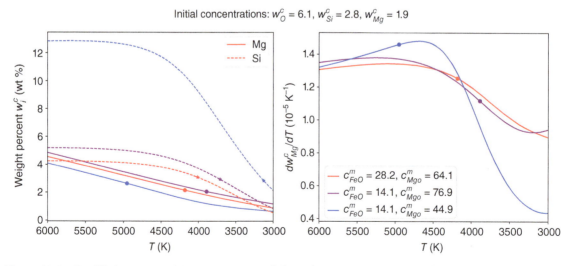

Figure 12.10 Equilibrium Mg and Si concentrations (left) and Mg precipitation rate dw_{Mg}^c/dT (right) plotted as functions of temperature for cases DEE (top), DDD (middle), and DED (bottom) described in the text. Dots mark the temperature at which the equilibrium core composition for element i falls below its concentration in the core. Mass fractions are denoted by w (in wt%) and mole fractions are denoted by c (in mol%).

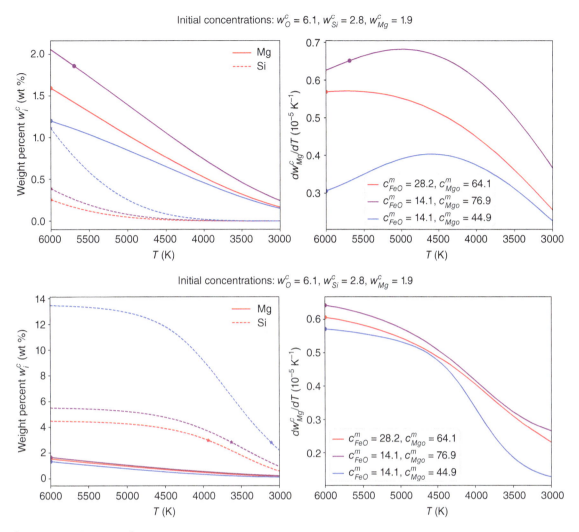

Figure 12.10 (*continued*)

until after core formation (O'Rourke & Stevenson, 2016; Badro et al., 2016). The results in section 12.3.2 suggest that accounting for composition-dependence reduces both T_o and dw_i^c/dT and so we regard the values in Figure 12.11 as upper estimates based on presently available information.

12.4. STRATIFICATION BELOW THE CMB

12.4.1. Modern-Day Observations of Stratification

Observational constraints on stratification at the top of the core have primarily originated from seismic studies. A number of SmKS wave studies (Lay & Young, 1990; Garnero et al., 1993; Helffrich & Kaneshima, 2010; Kaneshima, 2018) find a P-wave velocity reduction and steeper P-wave gradient relative to PREM up to 400 km deep into the core. The strength of stratification is often measured by the Brunt-Väisälä period

$$T_{\rm BV} = \frac{2\pi}{N} = 2\pi \left(-\frac{g}{\rho} \frac{\partial \rho'}{\partial r} \right)^{-1/2}, \quad (12.24)$$

which determines the period of oscillations that arise when a fluid parcel in a stratified region is subjected to vertical displacement. Here, the equation defines the Brunt-Väisälä frequency N and a prime denotes the density perturbation about the adiabatic and well-mixed state. Matching a compositional model to the observed wave speed suggests $T_{\rm BV} = 1.6$–3.4 hours, implying strong stratification (Helffrich & Kaneshima, 2010). Alexandrakis and Eaton (2010) obtained a best fitting velocity model to a dataset of SmKS traveltimes that was very similar to PREM and hence argued that stratification is absent at the top of the core; however, van Tent et al. (2020) showed that the Alexandrakis and Eaton (2010) data do not conflict with a low-velocity region in the uppermost

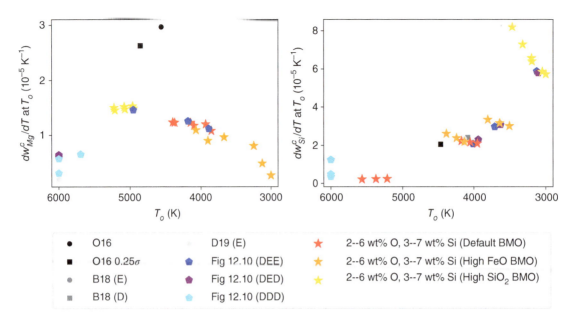

Figure 12.11 Precipitation rate of Mg (left) and Si (right) at the temperature T_o when precipitation began. The corresponding value of T_o is shown on the horizontal axis. All calculations have used an initial Mg core concentration of 2 wt%. Results for O'Rourke and Stevenson (2016, O16), Badro et al. (2018, B18), and Du et al. (2019, D19) are from Figure 12.9 except for the B18 Dissociation (D) case. The results denoted by stars have a default BMO composition of 28 mol% FeO, 64 mol% MgO, and 8 mol% SiO_2. The results denoted by pentagons are from Figure 12.10.

core, suggesting that methodological differences are responsible for the divergent conclusions. Irving et al. (2018) have derived a new 1-D core reference model using normal mode center frequencies, which provide a more direct constraint on density than body waves. The model suggests a lower P-wave velocity and higher density than PREM throughout the core thus reducing, though not eliminating, the stratification signal. van Tent et al. (2020) recently conducted an extensive review and concluded that "both seismological body-wave and normal mode observations require a low-velocity outermost core with respect to PREM, as well as a steeper velocity gradient than PREM." Evidently, there is now a reasonable degree of support for anomalous seismic velocity structure in the uppermost core.

At present, it is not clear whether low seismic velocities in the upper core are a global or local feature. The SmKS data coverage is rather heterogeneous, with large regions of the uppermost core (e.g., under North America and the Indian Ocean) not sampled by available ray path (see Kaneshima, 2018). The distinction is crucial. Low velocities (with respect to PREM) reflect variations in either density or bulk modulus. If a global layer of anomalous fluid exists at the top of the core then this layer must be light, otherwise it would mix back into the bulk core. This implies that the velocities must reflect a greater decrease in bulk modulus than density, for example, due to enrichment in one or more light elements (Helffrich, 2012; Komabayashi, 2014; Brodholt & Badro, 2017). On the other hand, if the velocity anomalies are local then there is no stability requirement since the anomalies could sample part of a large-scale circulation pattern (Mound et al., 2019). However, in both cases, the seismic velocities imply thermo-chemical anomalies greater than those associated with core convection (Helffrich & Kaneshima, 2010) and so some other mechanism is required to explain their existence.

Observations of the geomagnetic secular variation have been used to search for radial motion near the top of the core, which is expected to be absent in a stable layer. In purely toroidal flows, local extrema in the radial magnetic field are time invariant (Whaler, 1980); however, this test for stratification renders inconclusive results owing to large uncertainties on estimates of the CMB field at a point (Whaler, 1986). Gubbins (2007) showed that the present evolution of the South Atlantic anomaly, when attributed to flux expulsion, strongly suggests radial flow in the top 100 km of the core, while Amit (2014) argued that the intensification of high-latitude flux patches is best explained by localized downwelling. Lesur et al. (2015) inverted for the fluid flow at the top of the core and found that purely toroidal flow is not compatible with satellite observations of recent field variations but that a very limited amount of radial motion (comparable to diffusion, which was ignored) allows for acceptable fits. Huguet et al. (2018) argued that the observed variation of the total geomagnetic energy on the CMB is caused by kinetic to magnetic energy transfer, which requires upwelling/downwelling at the top of the core. All of these studies neglected magnetic diffusion (following Roberts & Scott, 1965), which should be accurate on large length scales where advection dominates, though radial diffusion on smaller scales could be significant. Indeed, diffusion can potentially explain much of the observed variation (Metman et al., 2019) with radial motion confined to greater depth (Gubbins, 2007). The flux expulsion mechanism is also complicated because it relies on upwelling to concentrate field below the CMB, where it is then expelled by diffusion (Bloxham, 1986). Finally, steady flow over CMB topography in a stably stratified layer can induce radial motion (Glane & Buffett, 2018), complicating attempts to rule out stratification by searching for radial flow.

Buffett (2014) has shown that simple combinations of axisymmetric Magneto-Archimedian-Coriolis (MAC) waves in a stably stratified layer can explain a 60-year periodic variation of the dipole geomagnetic field and the recent time-dependent evolution of zonal flow at the top of the core. The inferred stratified layer thickness is 130–140 km with a maximum $N/\Omega \sim 1$ (Buffett et al., 2016) or $T_{BV} \sim 24$ hours, implying weaker stratification than inferred from seismology. Subsequent work has shown that these waves can be generated by underlying core convection (Jaupart & Buffett, 2017) and exchange some angular momentum with the mantle though not enough to explain decadal variations in length of day (Buffett et al., 2016). Thus far, models based on MAC waves have assumed a global stable layer at the top of the core.

Another approach to investigating present-day stratification is to calculate the radial variation of buoyancy sources within the core (Davies & Gubbins, 2011; Gomi et al., 2013; Nimmo, 2015a). This method uses energy and mass conservation to balance the CMB heat flow against the sum of power sources inside the core (as outlined in section 12.2.2). The core is assumed to be 1-D and so stratification implicitly arises in the form of a layer. Stratification requires that

$$\alpha_T \left(\frac{dT}{dr} - \frac{dT_a}{dr} \right) + \alpha_i^c \frac{dw_i^c}{dr} > 0 \qquad (12.25)$$

(Landau & Lifshitz, 1987), which serves to define the base of the layer. The main challenge is approximating the gravitational energy since the spatial distribution of ohmic and viscous dissipation is not known (Jackson & Livermore, 2009), so various approaches have been used in the literature. Pozzo et al. (2012) used high k^c and found stable layers up to $O(1,000)$ km thick depending on the imposed CMB heat flow. Gubbins et al. (2015) calculated a maximum present-day stable layer thickness of 740 km assuming high k^c and no dissipation available to generate the magnetic field; however, they dismiss such thick layers as being incompatible with geomagnetic secular variation.

The "buoyancy" approach to assessing present-day stratification is sensitive to a number of uncertain parameters including the CMB heat flow and ICB density jump, but also the depth dependence of thermal conductivity. Labrosse (2015) calculated convective heat flow using the k^c profiles from Gomi et al. (2013) and Pozzo et al. (2012), the latter of which has a slightly shallower gradient. For mildly superadiabatic Q^c, the Gomi et al. (2013) $k^c(r)$ suggests a stratified region within the core, whereas the Pozzo et al. (2012) $k^c(r)$ predicts no stratification anywhere. The present uncertainty on $k^c(r_c)$ (section 12.2.3), let alone the radial profile $k^c(r)$, currently prevents definitive conclusions on the presence of stratified regions within the bulk core.

Overall, there is support from seismology for strongly stratified regions up to 400 km thick at the top of the core. Geomagnetic observations paint a more complex picture and seem to prefer thinner stratified regions or no stratification at all. The observations also do not determine whether the stratification is regional or in the form of a global layer. We therefore turn to computational investigations of core stratification: direct numerical simulations (DNS; section 12.4.2) of short timescale layer dynamics, and parameterized models (Section 12.4.3) of long-term layer evolution. The stratification derives from some combination of thermal and chemical effects and so below we consider these possibilities in turn, focusing on the following questions: How did the stratification form? How has the stratification evolved over time and has it survived to the present day? What is the predicted present-day thickness and stratification strength? Is the stratification global or local?

12.4.2. Direct Numerical Simulations (DNS) and Theory

There is a growing consensus from DNS that strong and thick stable layers are incompatible with the morphology of the present magnetic field. Olson et al. (2017), Olson et al. (2018), Christensen (2018), and Yan and Stanley (2018) performed DNS with thermal and compositional effects combined into a single codensity (see Braginsky & Roberts, 1995) and imposed a variety of CMB codensity gradients, both homogeneous and heterogeneous, promoting varying degrees of stabilizing density gradients. Olson et al. (2017, 2018) examined over 60 dynamo solutions and found that the high-latitude field morphology and the ratio of normal to reversed CMB flux are sensitive to the degree of stratification. They concluded that a weakly stratified 400 km thick layer with $N_0/\Omega \sim 0.5$ ($T_{BV} \sim 48$ hours) is compatible with the simulation results, where

$$\frac{N_0}{\Omega} = \frac{1}{\Omega}\left(\alpha_T g \frac{\partial T'}{\partial r}\right)^{1/2} \qquad (12.26)$$

is the Brunt-Väisälä frequency derived from thermal variations only. Christensen (2018) considered 26 simulations with N_0/Ω in the range 2.4–4. He applied the morphological criteria defined in Christensen et al. (2010) and found that simulations with 400 km thick layers were only marginally compatible with the modern field. Yan and Stanley (2018) showed that the ratio of zonal dipole to octupole Gauss coefficients, g_3^0/g_1^0, is sensitive to the presence of a stable layer. From 33 simulations they found that matching both Earth's g_3^0/g_1^0 over the last 10 kyrs (obtained from the CALS10K.2 model of Constable et al., 2016) and the modern field [according to the Christensen et al. (2010) criteria] entails a trade-off between stratification strength and thickness. Their preferred solutions had layer thicknesses in the range 60–130 km and $N_0/\Omega < 1$. Recently, Gastine et al. (2020) modeled thermal stratification in a suite of 70 simulations with $0 \leq N_0/\Omega \leq 50$ and found that CMB fields become more dipolar and axisymmetric with increasing layer thickness, in line with previous studies (Christensen, 2006; Nakagawa, 2011), and hence generally do not match the modern geomagnetic field [again as assessed by the Christensen et al. (2010) criteria]. They further found that the mismatch between simulated and observed fields increases as the Ekman number is reduced towards Earth-like values, and hence argue against the presence of stratification in Earth's core.

A number of the aforementioned studies combined an imposed stable layer with lateral heat flow variations on the CMB. When the stratification is weak the lateral variations can induce flow below the CMB (Olson et al., 2017), effectively overcoming the mean stabilizing codensity gradient in local regions where the CMB heat flow is anomalously high. However, for thick imposed layers, as the stratification strength increases the influence of the lateral variations is strongly diminished and the stable layer behavior is relatively unaffected by their presence (Christensen, 2018). Using a simple model of nonmagnetic thermal convection, Cox et al. (2019) showed that the transition between these two regimes (boundary-dominated and stratification-dominated) arises when the stratification parameter S, defined as the relative size of boundary temperature gradients to imposed vertical temperature gradients, exceeds unity. However, given uncertainties in estimating S for Earth they were unable to conclude whether the core is currently in the high S or low S regime.

Lateral heat flow variations can induce regional stratification even when the mean CMB heat flow is destabilizing.

Mound et al. (2019) found that thick localized stable regions were ubiquitous in a large suite of nonmagnetic simulations that access the regime of rapid rotation and vigorous convection thought to be most relevant to Earth's core (Long et al., 2020). In these simulations, the lateral extent of the stable regions is set by the imposed boundary anomalies (which were derived from seismic tomography) rather than the small-scale motions associated with vigorous convection in the bulk of the core. Interestingly, 1-D averaging in these models can yield a net stabilizing temperature gradient, giving the impression of global stratification despite the presence of motion in regions of the upper core. Using scaling analysis Mound and Davies (2020) estimated that stable regions in Earth's core could extend up to 350 km depth, similar to the thick layers inferred from seismology. They obtained values of $N_0/\Omega \approx 2$–5, corresponding to $T_{BV} \sim 5$–12 hours, lower than estimates by Helffrich and Kaneshima (2010) but larger than that inferred from MAC waves (Buffett et al., 2016).

A variety of processes besides lateral heat flow variations can act to disrupt or even completely erode a preexisting stable layer. It is well known from oceanography and astrophysics (Turner, 1973; Garaud, 2018) that stable systems where thermal and compositional fields have different diffusivities and adverse gradients are prone to instabilities that can drastically change their behavior. These "double-diffusive" instabilities have recently begun to receive substantial attention in the planetary core context (Monville et al., 2019, Bouffard et al., 2020, Mather & Simitev, 2020; Guervilly, 2022). Heat diffuses faster than light elements in the core (Pozzo et al., 2013) and so the double-diffusive dynamics take the form of "oscillatory convection" if the chemical gradient is stabilizing and the thermal gradient is destabilizing; switching the signs of the gradients gives "finger convection" (Turner, 1973). Given current uncertainties on the superadiabatic heat flow $Q^c - Q_a^c$ and the existence of chemical stratification, both regimes potentially occurred at different stages of core evolution.

At present, it seems premature to apply the results of double-diffusive DNS studies to Earth's core. The simulations are extremely challenging because the Lewis number $Le = \kappa/D_i$, the ratio of thermal and chemical diffusion coefficients, is $O(1,000)$ in Earth's core (Pozzo et al., 2013), which induces a large disparity in scales between thermal and compositional fields. This difficulty has also prompted workers to invoke further simplifications, such as omitting the magnetic field (Monville et al., 2019) or imposing double-diffusive conditions throughout the core (rather than just near the CMB) (Mather & Simitev, 2020). Finally, current double diffusion dynamo simulations are far from the rapidly rotating and low-viscosity conditions of the core, and established scaling relationships of the kind that have recently been devised for pure thermal or compositional driving (Aubert et al., 2017; Wicht & Sanchez, 2019) have not yet been produced (though see the recent non-magnetic study of Guervilly, 2022). This area of research will undoubtedly see significant progress in the coming years.

Simulations suggest that strongly stratified layers can resist penetration and entrainment from the underlying convection. Takehiro and Lister (2001) obtained a scaling law from simulations of rapidly rotating nonmagnetic convection underlying a stable layer (later supported by the dynamo simulations of Gastine et al., 2020) that predicted a penetration depth in Earth's core of O(100) m. Gubbins and Davies (2013) obtained a similar result by balancing buoyancy and Coriolis forces at the top of the core. Bouffard et al. (2020) considered the erosion of a thick (~700 km) preexisting chemically enriched layer by thermal convection in nonmagnetic simulations representative of an early Earth (no inner core). They found greater erosion in the equatorial plane than near the poles and estimated erosion rates (represented as the rate of change of stable layer thickness) of only ~1 km Gyr^{-1} or less, despite considering the endmember case of zero chemical diffusion.

Overall, numerical dynamo simulations incorporating global stratification that have attempted to match geomagnetic observations tend to favor thinner and more weakly stratified layers than those inferred from seismology. Some studies have also argued against the presence of a stable layer. The role of double-diffusive instabilities awaits clarification, but there is general agreement that existing layers are stable to penetration, entrainment, interface instabilities, and lateral variations in CMB heat flow. Regional stratification is another possibility, offering a plausible framework for reconciling both the significant compositional anomalies suggested by seismic studies and the upwelling/downwelling flow near the top of the core that is preferred by a number of geomagnetic studies.

12.4.3. Evolution of Thermal Stratification

Here, we consider solutions to equations (12.8) and (12.9) with $Q_p = E_p = 0$; these terms will be reintroduced when considering precipitation in section 12.5. The main uncertainties in the calculation are the time evolution of the CMB heat flow Q^c, and the ICB density jump $\Delta\rho$ (see section 12.2). The main outputs are the time evolution of the radius of the inner core, stable layer thickness and strength, and E_J, which is required to be positive for dynamo action (Gubbins et al., 2003, 2004; Nimmo, 2015a). The vast majority of previous studies have assumed that the stable layer grows downward from the CMB and so we also make this assumption in the remainder of this section.

Early studies implementing a time-dependent solution for a thermally stable layer in Earth's core (Gubbins et al., 1982; Labrosse et al., 1997; Lister & Buffett, 1998) have been reviewed previously (Nimmo, 2015a; Greenwood et al., 2021); they found that a thermal layer several hundred km thick can be formed over the timescale of inner core growth. Recently, Greenwood et al. (2021) have examined the limits to thermal stratification in the high conductivity scenario (Table 12.1) with $k^c = 100$ W m^{-1} K^{-1} at the CMB and we focus on their results here.

In the absence of precipitation, thermal convection is required to generate the magnetic field prior to inner core nucleation and so high k^c implies that the time during which thermal stratification may grow is limited to the last 0.5–1 Gyrs. Like the earlier studies cited above, Greenwood et al. (2021) did not solve for the mantle evolution; instead they imposed a linear variation in Q^c following inner core formation, approximating the behavior seen in coupled core-mantle evolution models (Driscoll & Bercovici, 2014; Nakagawa & Tackley, 2014; Patočka et al., 2020) in the same high k^c scenario (see Greenwood et al., 2021; Figure 6). Considering a wide range of present-day heat flows and constant dQ^c/dt values, Greenwood et al. (2021) estimated a maximum present-day layer thickness of 700 km. In cases where the recent linear trends in Q^c are consistent with the longer term (3.5 Gyrs) decay in Q^c seen in published coupled core-mantle evolution models, thermal layers are limited to 400 km in size with T_{BV} in the range 8–24 hours.

Several studies have explored the role of entrainment on the long-term evolution of stable layers. Lister and Buffett (1998) assumed that salt finger convection mixes the layer to a uniform composition distinct from the underlying convective region, slowing the advance of the thermal layer. Greenwood et al. (2021) used an equivalent model to Labrosse et al. (1997), where the destabilizing effects of chemistry are assumed negligible. Greenwood et al. (2021) also included a parameterization of layer entrainment, which could be utilized to represent double-diffusive effects. They showed that an entrainment flux corresponding to a >20% departure of the temperature gradient from the adiabatic profile could cause complete erosion of the stable layer. Whether such entrainment fluxes can be reached in the core requires further input from DNS (see section 12.4.2).

We end this section by examining stable layer properties obtained using $k^c = 70$ W m^{-1} K^{-1} at the CMB, the lower values proposed in the "high" conductivity scenario (Table 12.1). We repeat both the methodology and analysis of Greenwood et al. (2021), using the same depth dependence on k^c given in Davies et al. (2015) for ICB density jumps of $\Delta\rho = 600$ and $1,000$ kg m^{-3} and a wide range of dQ^c/dt values. A full list of parameter values is

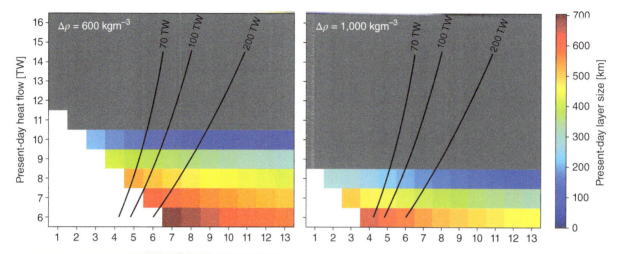

Figure 12.12 Present-day thickness of thermally stratified layers for a parameter search across linear CMB heat flow trends and $\Delta\rho = 600$ and $1{,}000$ kg m^{-3}, assuming $k^c = 70$ W m^{-1} K^{-1} at the CMB. Gray regions are superadiabatic at present and so produce no thermal stratification. Note that the adiabatic gradient is different for each assumed $\Delta\rho$ and so thermal layers form for different values of Q^c (see Davies et al., 2015). White regions indicate solutions where positive dynamo entropy was not maintained across the duration of the run. Contours indicate the CMB heat flow at 3.5 Ga (beyond the simulation time) by extrapolating along an exponential fitted to the present-day Q^c and dQ^c/dt.

given in Table 12.2. Figure 12.12 shows the present-day layer thickness, which can be compared to Figure 8 of Greenwood et al. (2021) who considered $k^c = 100$ W m^{-1} K^{-1} at the CMB. Viable solutions maintaining $E_J > 0$ and matching the present ICB radius (colored region in the figure) are obtained with lower values of the present-day Q^c for $k^c = 70$ W m^{-1} K^{-1} compared to $k^c = 100$ W m^{-1} K^{-1} due to a lower E_k in the entropy balance. Excluding solutions that produce ancient heat flows exceeding 70 TW (see contours in Fig. 12.12) gives a maximum layer thickness of ~500 km with $\Delta\rho = 600$ kg m^{-3} or ~700 km when $\Delta\rho = 1{,}000$ kg m^{-3}, significantly larger than the maximum thickness of ~400 km when $k^c = 100$ W m^{-1} K^{-1}. This increase in the upper bound results from the layer thickness being more sensitive to the ratio of Q^c/Q^c_a, rather than the absolute difference between the two heat flows. A lower Q^c_a, therefore, requires a smaller absolute difference between Q^c and Q^c_a to retain the same layer size, corresponding to lower dQ^c/dt and expected ancient heat flows. If k^c were reduced to below 50 W m^{-1} K^{-1} then $Q^c_a < 6$ TW for $\Delta\rho = 1{,}000$ kg m^{-3} and so no stable layer would grow. For $\Delta\rho = 600$ kg m^{-3} $Q^c_a \leq 7$ TW and so any thermal layers that did form would only be relatively thin for the ranges of Q^c considered here.

The minimum Brunt-Väisälä period (peak N_0), shown in Figure 12.13, is not significantly different to the range in Greenwood et al. (2021) (8–24 hours). Despite the range of core properties and $dQ^c(t)/dt$ values used, the strength of stratification depends predominantly on the ratio Q^c/Q^c_a at present day. Models that are mildly subadiabatic ($Q^c/Q^c_a > 0.8$) give periods similar to those inferred from MAC waves (Buffett et al., 2016) and dynamo models that reproduce morphological features of the geomagnetic field (Olson et al., 2017). Periods inferred from seismology of 1.3–3.5 hours (Helffrich & Kaneshima, 2010) are too low to be matched with thermal stratification in the high k^c regime because further reducing Q^c/Q^c_a produces insufficient power for maintaining the dynamo.

12.4.4. Evolution of Chemical Stratification

Chemical stratification arises when fluid at the top of the core is enriched in one or more light elements, thus reducing the fluid density. The source for this light element enrichment must be either an internal mechanism redistributing light element within the core, or an external mechanism that enables the addition of material from the mantle. Internal mechanisms include the barodiffusion of light elements along the core pressure gradient (Fearn & Loper, 1981; Gubbins & Davies, 2013), immiscibility in the Fe-Si-O system at high pressure and temperature (Arveson et al., 2019), or the accumulation of light fluid parcels emitted from the ICB (Moffatt & Loper, 1994; Bouffard et al., 2019). Komabayashi (2014) found that an increase in O concentration could decrease the seismic velocity in line with observations; however, Brodholt and Badro (2017) found that these simple accumulation mechanisms do not produce layers that are light and slow as required for a global stable layer. Instead Brodholt

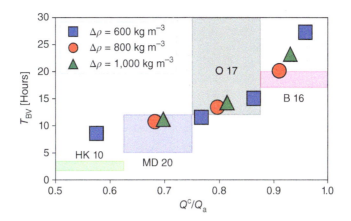

Figure 12.13 Buoyancy period, T_{BV}, for all models producing a stable layer, plotted as a function of the ratio Q^c/Q_a^c. Squares, circles, and triangles denote the ICB density jump used as indicated in the legend. Many models of the same $\Delta\rho$ plot on top of each other since the dominant control on T_{BV} is Q^c/Q_a^c. Also shown are T_{BV} values from other studies (offset such that they do not overlap; they have no relation to the x-axis): HK 10 (Helffrich & Kaneshima, 2010), MD 20 (Mound & Davies, 2020), O 17 (Olson et al., 2017), and B 16 (Buffett et al., 2016). Note that the upper bound provided by Olson et al. (2017) stretches to infinity since they also promote models with no stratification.

and Badro (2017) argued that an exchange of elements is required, for example, by decreasing the Si concentration and increasing the O concentration compared to the bulk core. If one instead considers regional stratification then simple light element accumulation may not be incompatible with observations, though it is not clear how these internal mechanisms could generate enhanced regional chemical concentration on the scales suggested by the seismic observations. Clearly more work is required here; however, in the following, we focus on external mechanisms.

Two external mechanisms for chemical stratification have been proposed. Landeau et al. (2016) used analog experiments to argue that a stable layer of comparable thickness to seismic inferences could have been emplaced toward the end of core formation due to turbulent mixing between a light-element-enriched impactor and the proto-core. Jacobson et al. (2017) showed that changing equilibrium conditions during multistage core formation can lead to the formation of stable chemical layering. Their results indicate that the stable layer could be erased by a late giant impact, such as the hypothesized moon-forming event, though Bouffard et al. (2020) argued based on the results of Landeau et al. (2016) that the mixing efficiencies assumed by Jacobson et al. (2017) are too high and hence the stratification would have survived. A resolution to this issue awaits improved physical descriptions and observational constraints on core formation processes.

The second external mechanism for stratifying the upper core is through chemical interactions with the mantle. As established in section 12.3 and originally shown by Frost et al. (2010) and Buffett and Seagle (2010), the core has likely been undersaturated in oxygen for much of its history and has therefore become progressively enriched in O at the CMB. Other elements such as Si and Mg may also have entered the core following its formation; however, the uncertainties are currently significant (see section 12.3) and so here we focus on FeO partitioning.

Nakagawa (2018) and Buffett and Seagle (2010) modeled the growth of an O-enriched layer, with its growth determined by the balance of downward diffusion and convective entrainment. Entrainment was modeled based on the "oscillatory" double-diffusive instability whereby radial oscillations in a thermally unstable and chemically stable layer develop into distinct convecting staircases (see Turner, 1973; section 12.4.2). Both studies obtained layers of $O(100)$ km thickness and similar stability by the present day, despite Nakagawa (2018) coupling the thermal evolution of the core to that of the layer rather than prescribing it as in Buffett and Seagle (2010). The similarity arises because enhanced oxygen concentrations in the layer correspond to large density anomalies that are relatively insensitive to the heat loss from the core. Nakagawa (2018) demonstrated how the size of the layer is more directly controlled by the diffusivity of O in the core, being $\propto \sqrt{D_O t}$, reaching up to 270 km for their upper bound of $D_O = 4.8 \times 10^{-8}$ m^2 s^{-1}. Layer growth is interrupted when the inner core forms since release of O at the ICB enriches the convecting fluid, however this only reduced the layer size by ~10 km.

Buffett and Seagle (2010) and Nakagawa (2018) fixed the O concentration at the CMB (to the bulk mantle

FeO composition), assuming that mantle convection continually enriches the CMB in oxygen. Alternatively, either advection or diffusion in the mantle may limit the replenishment of O-depleted material at the CMB (Davies et al., 2018). Taking optimistic estimates of $D_{FeO}^m = 10^{-12}$ m² s⁻¹ for the diffusion coefficient of FeO in the solid mantle (Ammann et al., 2010) and a 20 mol% change in FeO composition across the chemical boundary layer in the lower mantle, Davies et al. (2018) obtained a chemical mass flux of $I_{FeO} \sim 1,000$ kg s⁻¹. This value is comparable to the flux due to barodiffusion (Gubbins & Davies, 2013), which produces a ~10% change in concentration at the top of the core over 4.5 Gyrs, a relatively small effect. A similar result is obtained when considering the (Stokes) rise time of a buoyant parcel of mantle material away from the CMB. The actual timescale for the Rayleigh-Taylor instability is more complex and depends on various uncertain quantities such as the length scale of the instability and the viscosity contrast between enriched and depleted layers (Ribe, 1998). Nevertheless, existing studies suggest that it is difficult to produce significant FeO flux through the solid mantle.

The high early core temperatures suggested by thermal history models with $k^c \sim 100$ W m⁻¹ K⁻¹ (Nimmo, 2015a; Davies, 2015; Labrosse, 2015) imply a molten lower mantle, which should significantly enhance chemical exchange with the core (Brodholt & Badro, 2017). Davies et al. (2020) used the data of Pozzo et al. (2019) to model FeO exchange between the upper core and a BMO, extending the models of Buffett and Seagle (2010) and Labrosse et al. (2007). They found that the upper core could become strongly enriched in FeO (sometimes reaching a pure FeO composition) with stable layers of 70–80 km thickness growing in the first 1 Gyr of evolution before reaching up to 150 km thickness at the present day. Furthermore, they found that FeO loss increased the freezing rate of the BMO in order to keep the region on the liquidus. Complete freezing of the BMO occurred in the first 1–3 Gyrs following core formation and hence the BMO did not survive to the present day (at least not in the form of a layer), contrasting with the original results of Labrosse et al. (2007).

Davies et al. (2020) did not calculate the entropy production E_J in the core and hence could not show that their FeO evolution models were consistent with the existence of a dynamo for the past 3.5 Gyrs. Nakagawa (2018) did calculate E_J, but neglected the entropy change due to molecular diffusion in core, E_α. This assumption is reasonable when considering just the well-mixed core (Gubbins et al., 2004), but may fail when strong gradients exist in a chemically enriched layer. E_α is given by

$$E_\alpha = \int \frac{i^2}{\alpha_i^D T} dV \qquad (12.27)$$

(Gubbins et al., 2004). All else being equal, equation (12.9) shows that an increase in E_α reduces E_J, limiting the power available to the geodynamo.

We end this section by providing a plausible scenario for FeO exchange between the core and BMO that will be used in section 12.6. To do this, we have made four changes to the calculations of Davies et al. (2020). First, we calculate the core entropy budget, including E_α, using the standard method in Greenwood et al. (2021). Second, we apply the entrainment boundary condition directly at the base of the time-varying oxygen-enriched layer following Buffett and Seagle (2010) rather than at a fixed distance 400 km below the CMB. This treatment makes little difference to the predicted layer thickness but allows us to self-consistently partition energy and entropy between convecting and stable regions as in Greenwood et al. (2021) (note thermal stratification is not considered). Strictly the double-diffusive instability envisaged by Buffett and Seagle (2010) requires $Q^c > Q_a^c$, though in practice the layer evolution is set by the inward FeO flux (which is stronger than downward entrainment by a factor 1,000 in our calculations) and so conditions at the core-layer interface have little effect. The upper boundary condition on Q^c is given by equation (12.2) with R given in Table 12.2 and the FeO flux calculated by the boundary layer model of Davies et al. (2020).

Third, we solve for the evolution of the solid mantle using the methodology of Driscoll and Bercovici (2014) (with mantle melting ignored), which produces a self-consistent heat flow out of the BMO and allows the calculation to continue to the present day once the BMO fully crystallizes. The heat flow across the solid mantle boundary layer itself depends on the time-varying temperature at the top of the BMO [radius $r_{bmo}(t)$] and so the BMO and solid mantle are thermally coupled as in the classical coupled core-mantle thermal history (Driscoll & Bercovici, 2014). We assume no mass flux between solid and liquid mantle. This modification produces a heat flow at r_{bmo} that is initially larger than that of Labrosse et al. (2007), but decreases more rapidly with time, which is more conducive for dynamo action.

Finally, we raise the initial CMB temperature to 5500 K, the melting temperature of bridgmanite at CMB pressure, which is the liquidus phase in the deep mantle (e.g., Andrault et al., 2017). The presence of impurities would depress the melting point, perhaps by several hundred Kelvin, though this is still potentially within the significant uncertainties on the bridgmanite melting point at these conditions (Stixrude et al., 2009). Higher initial temperatures allow sufficient cooling of the core to enable a dynamo since ~4 Ga while retaining the correct ICB radius. We also increase the initial thickness of the BMO from 400 km (Labrosse et al., 2007) to 600 km, which increases the BMO lifetime, insulating the core from excessive heat loss to the solid mantle, particularly in the first 1 Gyr. The initial thickness of the BMO is poorly constrained; however, values up to $\mathcal{O}(1,000)$ km have been suggested (Stixrude et al., 2009; Blanc et al., 2020). Core properties not already specified are taken from Davies et al. (2015) assuming an inner core density jump of 800 kg m^{-3}.

Figure 12.14 shows two solutions from Davies et al. (2020; Figure 2), which use the default BMO model in Labrosse et al. (2007). One model has no FeO exchange (the original thermal evolution in Labrosse et al., 2007), while the other uses a partition coefficient of $P = K_d^{FeO}/c_{Fe}^c = 10$, mantle FeO molar fraction $c_{FeO}^m = 0.05$, core oxygen molar fraction $c_O^c = 0.05$. For the entropy balance we assume $k^c = 100$ W m^{-1} K^{-1} in both cases. Dynamo action does not occur at any time in both calculations. FeO transfer into the core initially produces $E_\alpha > 1,000$ MW K^{-1}, which quickly falls to 250–500 MW K^{-1}, comparable to the entropy from thermal conduction E_k. Since $E_k \propto k$ the thermal conductivity would need to be more than halved throughout the core in order to promote dynamo action in the case without FeO transfer. We have not taken into account the gravitational energy associated with mixing O entrained from the base of the layer across the well-mixed region given the relatively small fluxes at r_s; doing so would further reduce E_J. In this example, the lifetime of the BMO is reduced from ~4.5 Gyrs to less than 2 Gyrs with FeO loss, which causes the growth of a ~100 km thick chemically stable layer atop of the core.

Figure 12.15 shows results for varying P and c_{FeO}^m in models with the 4 modifications described above. Higher P produces a larger FeO flux into the core, a larger E_α, and hence lower E_J. E_J is initially negative in all models, but becomes positive around 4 Ga before declining toward ICN and subsequently rising during inner core growth. Only models toward the lower range of P or c_{FeO}^m produce a positive E_J just prior to ICN (Fig. 12.15a). At 4 Ga, only solutions with $P = 1$ and $c_{FeO}^m < 0.2$ give $E_J > 0$ (Fig. 12.15b). The decrease of E_J with P is more significant at 4 Ga since oxygen is actively being transferred to the core, producing steeper chemical gradients that have not yet been smoothed out by diffusion.

Figure 12.15c shows that thicker present-day layers are attained for larger P or c_{FeO}^m. However, varying the input parameters causes thickness variations of only ~30 km because the layer growth is limited by diffusion. Finally, Figure 12.15d shows the shortest T_{BV} within the layer at the present day. All models exhibit periods under 1 hour,

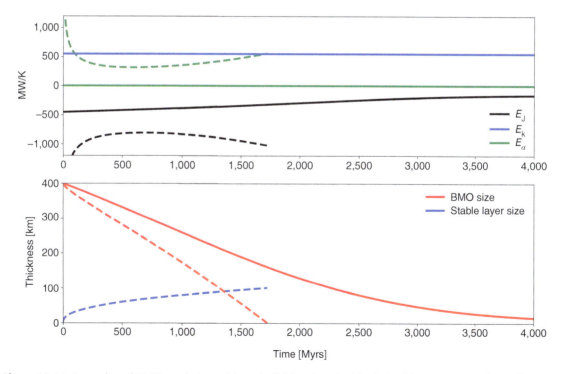

Figure 12.14 Examples of BMO evolution without (solid lines) and with (dashed lines) FeO transfer to the core, equivalent to those shown in Figure 2 of Davies et al. (2020). A partition coefficient of $P = 10$, a mantle FeO molar fraction of 0.05 and oxygen molar fraction of 0.05 in the core are used (see Davies et al., 2020, for a full set of parameters used for the BMO calculation). Top panel shows the entropy sources within the core assuming $k^c = 100$ W m^{-1} K^{-1}, and bottom panel shows the evolution of BMO and core stable layer thickness.

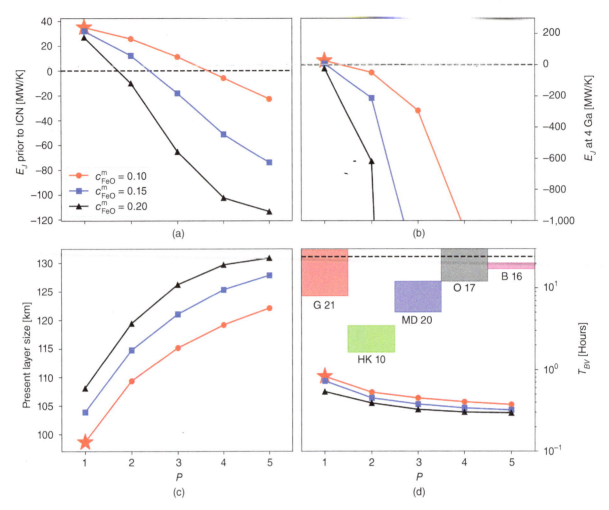

Figure 12.15 Results from the suite of models calculating the coupled evolution of the isentropic core, chemical stable layer, BMO, and solid mantle. All results are plotted with $P = 1$–5 on the horizontal axis, with varying mantle FeO concentrations shown by the colors that are consistent across each panel. Panels show values for E_J immediately prior to ICN (a) and at 4 Ga (b), present-day chemical layer thickness at the top of the core (c), and the minimum Brunt-Väisälä period (peak N), T_{BV}, for the present-day layer (d) [equation (12.24)]. Also in panel (d) are T_{BV} values from the same selection of studies with the same abbreviations as Figure 12.13 with the addition of Greenwood et al. (2021) (G21). The dashed lines in (a) and (b) show $E_J = 0$ and in (d) they show $T_{BV} = 24$ hours. Note the log scale in (d). Stars indicate the model which produces positive E_J for the last 4 Gyrs, which is discussed further in section 12.6.

indicating a very strong density stratification. There is a rapid increase in the periods as P is lowered and so achieving periods within the 1.45–3.5 hours inferred from seismology (Helffrich & Kaneshima, 2010) would require a value of $P < 0.5$. Other studies quoted in Figure 12.15d (Buffett et al., 2016; Olson et al., 2017, Mound and Davies, 2020; Greenwood et al., 2021) all favor much longer periods consistent instead with our previous results on thermal stratification.

In summary, the chemical stratification mechanisms that appear the most likely candidates to explain a thick and strongly stratified layer at the top of Earth's core are incomplete mixing during core formation (Landeau et al., 2016) and FeO exchange with the mantle (Buffett & Seagle, 2010; Brodholt & Badro, 2017). Whether a primordial layer can survive mixing due to late-stage impacts is a key issue that will benefit from improved models of core formation. We find that models of FeO transfer between a BMO and the core require relatively weak partitioning ($P \sim 1$) in order to enable dynamo action in the early core that continues to the present day while also producing present-day stable layers of similar strength to inferences from seismic models. These calculations are limited because they only include FeO partitioning with a constant value of P. Future work will need to couple the reactions of SiO_2 and MgO; however, as with the precipitation case it seems premature to move down this path owing to the significant uncertainties in the equilibrium calculations explained in section 12.3. The multielement calculations in section 12.3.4 suggest that the core is strongly undersaturated in O, while P does not vary significantly when the BMO lifetime is short (and hence there is little variation in T). Therefore, the calculations presented in this section hopefully represent a reasonable starting point for further investigations into coupled chemical core-mantle evolution.

It is notable that thermal stratification produces layers that match the thickness but not the stability strength inferred from seismology, instead predicting T_{BV} values more in line with inferences from geomagnetism. Conversely, FeO transfer produces layers that approximate the stability but not the thickness of the seismic observations, instead predicting layer thicknesses comparable to inferences from DNS and geomagnetism. One potential resolution is that the top of Earth's core comprises a thin and strongly chemically stratified chemically region embedded within a thicker and more weakly thermally stratified layer. This scenario would require high T_{BV} values confined close to the CMB, with geomagnetic observations sampling an average stratification signal in the upper core.

12.5. CHEMICAL PRECIPITATION

In this section, we discuss the effect of precipitation on the thermal and magnetic evolution of the core. The efficiency of precipitation in powering the geodynamo depends crucially on the precipitation rate dw_i^c/dT of oxide i. Simple models assuming high conductivity and constant precipitation rates have shown that precipitation of MgO with $dw_{MgO}^c/dT = 5 \times 10^{-5}$ K^{-1} (O'Rourke & Stevenson, 2016) or precipitation of SiO$_2$ with $dw_{SiO_2}^c/dT = 4 \times 10^{-5}$ K^{-1} (Hirose et al., 2017) can maintain the geomagnetic field over the past 4 Gyrs with similar cooling rates and heat flows to those inferred from conventional low conductivity calculations. On the other hand, Du et al. (2019) found that high heat flows and cooling rates were still required to drive the dynamo using precipitation rates of $dw_{MgO}^c/dT = 6 \times 10^{-6}$ K^{-1} obtained from their experiments. A similar conclusion was reached in a recent study of SiO$_2$ precipitation by Wilson et al. (2022), who found rates of $dw_{SiO_2}^c/dT = 5–10 \times 10^{-4}$ K^{-1} could help to produce successful thermal histories (in the sense of maintaining dynamo action over the last 3.5 Gyrs and matching the present ICB radius and mantle potential temperature and convective heat flow) with present-day CMB heat flows of ~10 TW. Additional power provided by precipitation reduces the core cooling rate required to meet a given entropy production and hence predicts an older inner core age; however, thermal history models with precipitation still predict supersolidus temperatures for the first ~1–3 Gyrs, after core formation (O'Rourke et al., 2017, Mittal et al., 2020; Wilson et al., 2022) and so suggest the existence of a BMO at least in early times.

O'Rourke et al. (2017) conducted a large number of coupled core-mantle evolution models using a standard core setup (Labrosse, 2015) with the addition of precipitation (described in O'Rourke & Stevenson, 2016). Their mantle evolution model is from Korenaga (2006), which produces a much flatter CMB heat flow evolution compared to conventional mantle evolution models based on standard boundary layer theory (e.g., Driscoll & Bercovici, 2014; Jaupart et al., 2015). O'Rourke et al. (2017) focused on the case where $k^c \approx 90$ W m^{-1} K^{-1} at the CMB and varied dw_{MgO}^c/dT between 0 and 8×10^{-5} K^{-1}. For their nominal setup, they found a preferred value of $dw_{MgO}^c/dT \sim 2 \times 10^{-5}$ to ensure E_J is sufficiently large to maintain dynamo action since core formation.

Mittal et al. (2020) modeled the simultaneous precipitation of Mg, Si, and O. They coupled the evolution of the core and solid mantle to an intermediate "interaction layer" comprising precipitated material (MgO, FeO, and SiO$_2$) together with MgSiO$_3$ and FeSiO$_3$. In this model, the interaction layer evolution is governed by a balance between growth due to precipitation and erosion by mantle flow. Mittal et al. (2020) found that a wide range of evolutionary scenarios are possible with different oxides precipitating at different times depending on the properties of the interaction layer (its thickness and erosion rate), the initial compositions, and the parameters defining the equilibrium constants. This behavior is consistent with the simple mass balance calculations presented in section 12.3.

The large number of poorly constrained parameters means that it is difficult to make general statements regarding the thermal and magnetic evolution of the core when precipitation is included. We therefore consider simple scenarios whereby MgO precipitation begins at core formation and proceeds at a constant rate in the range 0.3–1.5 K^{-1} as shown in Figure 12.11. For simplicity, we neglect the effects of SiO$_2$ and FeO and seek the minimum CMB heat flow that will enable dynamo action for the past 3.5 Gyrs. To do this, we follow Nimmo (2015a) and Davies et al. (2015) and prescribe $E_J = 0$ before inner core formation and specify Q^c during inner core growth, which produces conservative estimates of the cooling rate, core temperature, and inner core age and avoids the nonphysical behavior that arises when E_J is fixed for all time (Nimmo, 2015a).

Figure 12.16 shows the predicted inner core age and the CMB temperature and CMB heat flow at 3.5 Ga (the age of paleointensities determined by Tarduno et al., 2010) for different $\Delta\rho$ and $k^c = 70,100$ W m^{-1} K^{-1}. Also shown

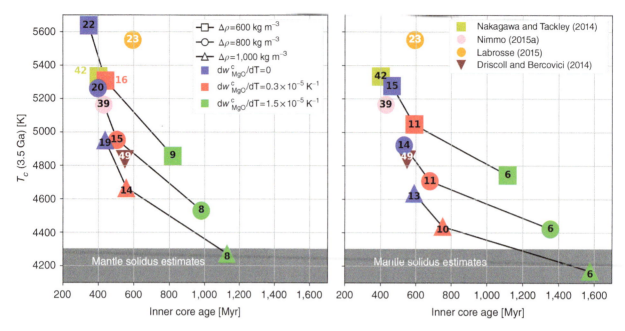

Figure 12.16 Effect of MgO precipitation on inner core age and early CMB temperatures. Left (right) panel shows results using a CMB conductivity of $k^c = 100$ W m^{-1} K^{-1} ($k^c = 70$ W m^{-1} K^{-1}). Symbols denote different core properties based on density jumps at the ICB of 600 (squares), 800 (circles), 1,000 kg m^{-3} (triangles) with parameters taken from Table 12.2. Colors indicate no MgO precipitation (blue), and at a fixed rate of 0.3×10^{-5} K^{-1} (red) and 1.5×10^{-5} K^{-1} (green) as derived from Figure 12.11. Solid lines link models with the same core properties but varying rates of MgO precipitation. Numbers show the CMB heat flow in TW at 3.5 Ga. Results from other studies using a high thermal conductivity are also shown, replicated on both panels for comparison to each of our datasets. Source: Adapted from Davies et al. (2015; Figure 3).

are favored models from Labrosse (2015), Driscoll and Bercovici (2014), Nakagawa and Tackley (2014), and Nimmo (2015a), who considered high k^c but used different model setups and constraints on CMB heat flow. Lower k^c implies an older inner core and lower Q^c and core cooling rates to maintain the dynamo. Increasing $\Delta\rho$ from 600 to 1,000 kg m^{-3} can produce a 600–800 K decrease in the early core temperature and a 200–400 Myrs increase in the inner core age, depending on the details of the model. With $dw^c_{MgO}/dT \leq 0.3 \times 10^{-5}$ K^{-1} we find an inner core age of at most 300–600 Myrs (400–800 Myrs) and minimum CMB heat flows at 3.5 Ga in the range 14–22 TW (10–15 TW) for $k^c = 100$ W m^{-1} K^{-1} ($k^c = 70$ W m^{-1} K^{-1}). With a precipitation rate of 1.5×10^{-5} K^{-1} the maximum inner core age rises to 800–1,100 Myrs (1,100–1,500 Myrs) and required CMB heat flows at 3.5 Ga decrease to 8–9 TW (~6 TW) for $k^c = 100$ W m^{-1} K^{-1} ($k^c = 70$ W m^{-1} K^{-1}). The vast majority of models predict an inner core age of at most 700 million years and early core temperatures exceeding the lower mantle solidus estimate of $4{,}150 \pm 150$ K (Fiquet et al., 2010; Andrault et al., 2011).

Davies et al. (2015) considered how uncertainties in a number of input parameters could affect predictions of inner core age and early core temperature. Within plausible ranges they varied the thermal expansivity, latent heat coefficient, specific heat capacity, and core melting curve and found that the combined variations produced uncertainties on the inner core age of ± 150 Myrs, and the early temperature of ± 400 K. These uncertainties are comparable to the uncertainty in $\Delta\rho$ alone. When combined with the fact that the temperatures and inner core ages in Figure 12.16 are lower bounds, this suggests that while MgO precipitation undoubtedly helps relax the power requirements for the dynamo, some key implications of high core conductivity such as the existence of an early BMO remain even in the presence of precipitation. The inner core is also certainly much younger than the core, though its age is evidently rather uncertain. In particular, these models cannot differentiate between paleomagnetic inferences of ICN at ~0.5 Ga (Bono et al., 2019) and ~1.3 Ga (Biggin et al., 2015).

12.6. TOWARD RESOLVING THE NEW CORE PARADOX

Over the last few years, various proposals have been put forth to resolve the new core paradox. Driscoll and Bercovici (2014) argued for 2 TW of heat produced by ^{40}K, which slows the core cooling rate for a given mantle heat flow and hence helps enable positive E_J before inner core formation. The drawback here is that experiments and simulations suggest that little ^{40}K partitioned into the core during formation (Watanabe et al., 2014; Xiong et al., 2018). Precipitation provides another potential solution, though it introduces a number of uncertain parameters and is difficult to constrain from available observations. Laneuville et al. (2018) suggested a compositionally stratified BMO, which helps retain heat in the core; however, their model still suggests that the dynamo shuts off prior to inner core formation. Recently, Driscoll and Davies (2023) conducted a wide parameter survey using coupled core-mantle evolution models with no radioactivity or precipitation and a thermal conductivity of 70 W m^{-1} K^{-1}. They found very few successful solutions, but were able to maintain dynamo action for the last 3.5 Gyrs and obtain the correct inner core size with a hot initial CMB temperature of ~6,000 K and a central core melting point of ~5,550 K. The problem with this successful thermal history is that it is not robust to changing core material properties such as the melting point within current experimentally-derived uncertainties.

Here, we present another possible resolution to the new core paradox that does not rely on precipitation or radiogenic heating. The approach is to retain the minimum number of physical processes (and hence poorly constrained parameters) while maintaining consistency with the basic predictions of core evolution with high conductivity. The model is described in section 12.4.4.

The early evolution involves coupled thermo-chemical interactions between the core and BMO, as expected from the high temperatures that arise in the high k^c scenario (section 12.5). FeO enters the core (section 12.3.3), causing growth of a chemically stratified layer from the start of the model. We have not included a stratified layer emplaced at core formation (Landeau et al., 2016); however, since the FeO flux rapidly forms a strongly stratified layer, while erosion of chemical layers is expected to be weak (Bouffard et al., 2020), we may anticipate similar long-term behavior in the two cases. The model must match the present ICB radius and predict dynamo generation from 4 Ga.

Figure 12.17 shows the results of one calculation that matches the constraints using $k^c = 70$ W m^{-1} K^{-1} and $\Delta\rho = 800$ kg m^{-3}, corresponding to the model denoted by a star in Figure 12.15. The BMO is initialized at 600 km thick and persists for 2 Gyrs producing a large flux of FeO into the core. The enhanced heat flux out of the BMO arising from our revisions to the original Labrosse et al. (2007) model (section 12.4.4) enables the onset of dynamo action around 4 Ga with high k^c. Once the BMO freezes, the chemical layer continues to thicken by diffusion before the initiation and growth of the inner core around 0.8 Ga begins to erode it back toward the CMB. Prior to inner core formation E_J remains just above zero, and hence the model predicts continuous dynamo action for the last 4 billion years. The present-day heat flow and potential temperature at the top of the convecting

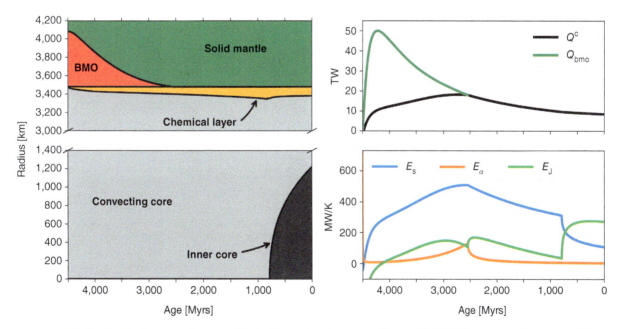

Figure 12.17 Results from our best model, indicated by the star in Figure 12.15. Left column shows a radial cross-section through time of the coupled Earth evolution. Note the break in the y-axis and that both halves of the figure are to scale with each other. The right panels show energy (top) and entropy (bottom) sources from the calculation.

mantle are respectively 35 TW and 1653 K, within current constraints of 35–41 TW and ~1550–1750 K (Jaupart et al., 2015), while the current inner core size is 1,221 km as in Earth.

The results in Figure 12.17 are sensitive to the parameter choices; in particular, increasing k^c above $k^c = 70$ W m^{-1} K^{-1}, which is on the lower end of the estimates presented in section 12.2.3, causes E_J to fall below zero. We have not conducted an exhaustive search of the solution space, but did not obtain viable solutions in the absence of a BMO, using the original BMO setup of Labrosse et al. (2007), or with strong FeO partitioning ($P > 1$). However, while the solution might appear somewhat specialized, there are a large number of parameter combinations that have yet to be tested. Moreover, a large range of successful solutions are clearly available with only a modest additional amount of entropy due to precipitation or radiogenic heating that are within current observational or modeling uncertainties. Assuming a CMB temperature of 5000 K for the onset of Mg and/or Si precipitation (Fig. 12.11), the corresponding onset time for the solution in Figure 12.17 is 2.8 Ga. Prior to this, the dynamo would remain reliant on rapid cooling.

The solution in Figure 12.17 provides a number of predictions that can be tested with past and present observations. First, the Ohmic dissipation displays local minima just prior to inner core formation and completion of BMO freezing and a global minimum around 4 Ga. The minimum at 2.5 Ga results from the cessation of FeO flux into the core, corresponding to a rapid decrease in E_α. These minima might be observable in paleointensity data, though care is needed when translating E_J to an equivalent virtual dipole moment (Driscoll, 2016; Landeau et al., 2017; Driscoll & Wilson, 2018; Davies et al., 2022), and we have also ignored the possibility that dynamo action also occurs in the BMO. The inner core age is 800 Myrs, which sits between the paleointensity changes inferred at ~0.5 Ga by Bono et al. (2019) and ~1.3 Ga by Biggin et al. (2015), while the delayed onset of dynamo action appears (perhaps coincidentally) close to the Hadean paleointensity data of Tarduno et al. (2015). These results will hopefully motivate future attempts to link paleointensity variations to abrupt changes in core evolution. Second, the present-day strength of stratification is strong enough to match the estimates derived from seismic observations (Helffrich & Kaneshima, 2010), but larger than inferences from MAC wave studies and geodynamo simulations. The stable layer thickness is 100 km, which is thinner than some seismic studies (section 12.4.1) but more in line with inferences from geomagnetism and geodynamo simulations (section 12.4.2). Finally, the present-day CMB heat flow is 8.5 TW, which is within the range of 7–17 TW estimated by Nimmo (2015a) and the 5–15 TW suggested by Lay et al. (2009). The core is actually mildly subadiabatic at present ($Q_a^c = 9.4$ TW), though we did not account for the accompanying growth of a thermally stable layer in the model. A potential resolution to the contrasting observational constraints on chemical versus thermal layers may be that a strongly stratified chemical sublayer exists within a broader weakly stratified thermal layer.

It is worth noting that our preferred evolution scenario requires significant core cooling, with the CMB temperature falling by over 1000 K in 4.5 Gyrs. Alternative mechanisms, for example, tidal heating (which we have neglected in this review), have been argued to provide sufficient power to the dynamo while requiring a drop in CMB temperature of only ~300 K (Andrault et al., 2016). With high core conductivity we find rapid cooling is ubiquitous in our models of thermo-chemical dynamo action and have not found a way to match the available constraints on core and mantle evolution with such slow cooling rates.

Resolving the new core paradox requires an interdisciplinary approach combining observations, theory, and modeling. Systematic studies of core thermal conductivity approaching CMB conditions are needed to provide robust methods for extrapolating from lower P-T conditions, while the effects of composition and the discordant results from direct experimental and computational determinations of k^c need to be resolved. The basic operational requirements for the dynamo with low k^c are well understood (see Table 12.1); however, some of the processes we have considered could also occur in the low k^c scenario. Moreover, the improved understanding of processes such as stratification and chemical interactions between the core and mantle, which have arisen in part from the high k^c calculations, are also potentially significant for other terrestrial bodies (e.g., Breuer et al., 2015). We therefore believe that considering both scenarios is crucial to understanding the dynamics and evolution of Earth and other terrestrial planets.

Improved constraints on the temperature- and composition-dependence of partitioning at CMB conditions as well as further systematic comparisons of candidate thermodynamic models (Badro et al., 2018) will help reduce the range of viable precipitation rates and onset times (Fig. 12.11). Future seismic and geomagnetic observations together with high-resolution DNS conducted in dynamical regions approaching Earth's core conditions (Aubert et al., 2017; Wicht & Sanchez, 2019) can help constrain the existence, thickness, and global versus local nature of stable regions below the CMB. Numerical simulations of dynamo action in a BMO coupled to the core are needed to better understand the generation of the early geomagnetic field.

Improved constraints on CMB topography and its effect on the core, for example, by organizing stationary flow structures (Calkins et al., 2012) or coupling with stable stratification to enable angular momentum exchange between the core and mantle, may enable independent observational constraints on dynamics below the CMB. It is also crucial to continue to seek observational evidence for the existence of a basal magma ocean, for example, through its potential links to LLVPs and ultralow-velocity zones (Labrosse et al., 2015), and also for precipitation, perhaps in the form of a thin layer at the CMB or the incorporation of precipitation products into the mantle (Helffrich et al., 2018).

Improved constraints on the ICB density jump $\Delta\rho$ are also clearly needed. Wong et al. (2021) have made a potentially promising step in this direction by combining a theoretical model of a slurry region above the ICB (the so-called F-layer) with seismic observations of 1-D compressional wave speed variations. From a large suite of models that span uncertainties in the main input parameters, Wong et al. (2021) constrain $\Delta\rho \approx 530$ kg m^{-3}, on the lower end of the range of values obtained from normal modes (Masters & Gubbins, 2003). This model also yields an independent constraint on the CMB heat flow that is consistent with our preferred model.

Finally, we note that the structure, dynamics, and evolution of stratified layers within the core depend crucially on the role of myriad instabilities that can lead to partial or complete mixing. Parameterizations of these processes in thermal history models are rather crude (Greenwood et al., 2021) but rely heavily on results from DNS. In particular, future DNS studies will hopefully shed light on the role of double-diffusive instabilities in the formation and survival of layering in the rapidly rotating, turbulent, and magnetic environment that characterizes the core.

12.7. CONCLUSIONS

We have reviewed the high thermal conductivity scenario for core evolution, which predicts a young inner core and early temperatures consistent with the existence of a BMO (Table 12.1). The main conclusions are as follows:

- Consistent extrapolation of thermal and electrical conductivity estimates from a number of recent studies suggests $k^c = 70$–110 W m^{-1} K^{-1} at CMB conditions of 4000 K, 135 GPa and ~ 10 weight percent light element.
- Both the onset time and rate of MgO and SiO$_2$ precipitation are uncertain and depend on a number of factors including temperature, compositions on both sides of the CMB, and the nature of the reactions that govern the equilibrium.
- MgO precipitation may begin anywhere between 3000 and 6000 K with rates between 0.3 and 1.5×10^{-5} K^{-1}. The majority of our calculations suggest a narrower range of onset between 4000 and 5000 K with rates between 1.0 and 1.5×10^{-5} K^{-1}.
- SiO$_2$ precipitation may begin anywhere between 3000 and 6000 K with rates between 0.1 and 8×10^{-5} K^{-1}. The majority of our calculations suggest a narrower range of onset between 3000 and 4500 K with rates between 2 and 8×10^{-5} K^{-1}.
- The core is always undersaturated in O in our calculations, which causes FeO dissolution at all times.
- Our results suggest that light elements dissolved into the core after its formation, forming a stably stratified chemical layer below the CMB. Precipitation was delayed, but once initiated would supply ample power for sustaining the geodynamo.
- Viable core evolution scenarios predict thermally stable layers at most 400–700 km thick. The strength of stratification can match some inferences from geomagnetism but not values derived from seismic observations.
- The minimum requirements for maintaining the dynamo over the last 3.5 Gyrs suggest an inner core age of at most 300–600 Gyrs (400–800 Gyrs) for $k^c = 100$ W m^{-1} K^{-1} ($k^c = 70$ W m^{-1} K^{-1}) and an MgO precipitation rate $\leq 0.3 \times 10^{-5}$ K^{-1}. With a precipitation rate of 1.5×10^{-5} K^{-1} the maximum inner core age is 800–1,100 Gyrs (1,100–1,500 Gyrs) for $k^c = 100$ W m^{-1} K^{-1} ($k^c = 70$ W m^{-1} K^{-1}). The temperature of the early core almost always exceeds present estimates of the mantle solidus, suggesting a BMO event with precipitation.
- We present a solution that overcomes the new core paradox by enabling continuous dynamo generation from 4 Ga to present. This model uses $k^c = 70$ W m^{-1} K^{-1} and matches the present inner core size and heat flow and temperature at the top of the convecting mantle. It predicts a present-day CMB heat flow of 8.5 TW, chemically stable layer of 100 km produced by FeO exchange with the mantle, and a BMO lifetime of 2 Gyrs.

ACKNOWLEDGMENTS

S. Greenwood is supported by NSF-NERC grant NE/T003855/1. CD acknowledges support via NERC grants NE/T000228/1 and NE/V010867/1. We thank five reviewers for supportive and detailed comments that helped to improve the manuscript. We are also indebted to Prof. James Badro for detailed discussions and assistance benchmarking code used to calculate CMB equilibria. Finally, we thank Jon Mound, Alfred Wilson, and Dario Alfè for commenting on various aspects of the chapter.

APPENDIX
MASS BALANCE BETWEEN THE CORE AND MAGMA OCEAN

We implement two differences compared to the algorithm presented in Rubie et al. (2011): (1) Mg replaces Ni in the reaction set; (2) distribution coefficients for Si and Mg are defined by dissociation reactions rather than exchange reactions. We start by considering the reaction

$$[(FeO)_x(MgO)_y(SiO_2)_z] + [(Fe)_a(Mg)_b O_c(Si)_d] \iff \quad (A.1)$$

$$[(FeO)_{x'}(MgO)_{y'}(SiO_2)_{z'}] + [(Fe)_{a'}(Mg)_{b'} O_{c'}(Si)_{d'}] \quad (A.2)$$

which is essentially the reaction considered by Rubie et al. (2011), ignoring elements that do not partition and replacing Ni with Mg. Mass conservation demands

$$a' = x + a - x', \quad (A.3)$$

$$b' = y + b - y', \quad (A.4)$$

$$c' = x + y + 2z + c - x' - y' - 2z', \quad (A.5)$$

$$d' = z + d - z'. \quad (A.6)$$

The distribution coefficients are given in this notation by

$$K_D^O = \frac{c_{Fe} c_O}{c_{FeO}} = \frac{a'c'}{x'} \frac{(x' + y' + z')}{(a' + b' + c' + d')^2} \quad (A.7)$$

$$K_D^{Mg} = \frac{c_{Mg} c_O}{c_{MgO}} = \frac{b'c'}{y'} \frac{(x' + y' + z')}{(a' + b' + c' + d')^2} \quad (A.8)$$

$$K_D^{Si} = \frac{c_{Si} c_O^2}{c_{SiO_2}} = \frac{d'(c')^2}{z'} \frac{(x' + y' + z')}{(a' + b' + c' + d')^3}. \quad (A.9)$$

The procedure of Rubie et al. (2011) starts by guessing a value for x', which gives a' from equation (A.3). Next y' is obtained from the definition of K_D^{Mg}. We note that

$$\frac{K_D^{Mg}}{K_D^O} = \frac{x'b'}{y'a'}, \quad (A.10)$$

which is the same result as equation S12 in Rubie et al. (2011) despite the fact that we are considering different reactions. This arises since the FeO and MgO concentrations in the silicate are determined by the amount of Fe and Mg, respectively. Equation (A.10) allows us to determine y' from an initial guess at x'. Using the definitions of b' and y' gives

$$y' = \frac{x'(y + b)}{(x + a - x')K_D^{Mg}/K_D^O + x'}. \quad (A.11)$$

and hence b' is also determined from equation (A.4).

To obtain z' substitute equations (A.5) and (A.6) into the definition of K_D^O/K_D^{Si}, obtaining

$$\frac{K_D^O}{K_D^{Si}} = \frac{a'c'z'(a' + b' + c' + d')}{x'd'(c')^2}, \quad (A.12)$$

$$= \frac{a'z'(a' + b' + x + y + 3z + c - x' - y' - 3z' + d)}{x'(z + d - z')(x + y + 2z + c - x' - y' - 2z')}. \quad (A.13)$$

Defining

$$\alpha = z + d, \quad (A.14)$$

$$\gamma = a' + b' + x + y + 3z + c - x' - y' + d, \quad (A.15)$$

$$\sigma = x + y + 2z + c - x' - y', \quad (A.16)$$

we can write

$$\frac{K_D^O}{K_D^{Si}} = \frac{a'z'(\gamma - 3z')}{x'(\alpha - z')(\sigma - 2z')}, \quad (A.17)$$

which turns in to a quadratic equation for z'

$$(z')^2 \left[3a' + 2x'\frac{K_D^O}{K_D^{Si}} \right] - z' \left[(2\alpha x') + x'\sigma)\frac{K_D^O}{K_D^{Si}} + a'\gamma \right]$$
$$+ \frac{K_D^O}{K_D^{Si}} x'\alpha\sigma = 0. \quad (A.18)$$

We note here an analytical solution for the special case where exchange of Fe and Si are disallowed. We require that

$$x = x', a = a', z = z', d = d'. \quad (A.19)$$

The mass balance equations reduce to

$$b' = y + b - y' \quad (A.20)$$

$$c' = y - y' + c, \quad (A.21)$$

while the distribution coefficients are

$$K_D^O = \frac{ac'(x + y' + z)}{x(a + b' + c' + d)^2}, \quad (A.22)$$

$$K_D^{Mg} = \frac{b'c'(x + y' + z)}{y'(a + b' + c' + d)^2}, \quad (A.23)$$

$$K_D^{Si} = \frac{d(c')^2(x + y' + z)}{z(a + b' + c' + d)^3}, \quad (A.24)$$

$$\frac{K_D^{Mg}}{K_D^O} = \frac{xb'}{a(y + b - b')}, \quad (A.25)$$

$$\frac{K_D^O}{K_D^{Si}} = \frac{az(a + b' + c' + d)}{xdc'}. \quad (A.26)$$

From the first ratio, we find a solution for b' as

$$b' = \frac{a(y + b)K_{Mg}/K_O}{x + aK_{Mg}/K_O} \quad (A.27)$$

and from the second ratio we get

$$b' = \left[az(a+c-b+d) - \frac{K_O}{K_{Si}}xd(c-b)\right]\left(\frac{K_O}{K_{Si}}xd - 2az\right). \tag{A.28}$$

Equating these two expressions gives a constraint on the input compositions.

REFERENCES

Alexandrakis, C., & Eaton, D. (2010). Precise seismic-wave velocity atop Earth's core: No evidence for outer-core stratification. *Physics of the Earth and Planetary Interiors, 180,* 59–65.

Alfè, D., Gillan, M., & Price, G. (2002a). Composition and temperature of the Earth's core constrained by combining *ab initio* calculations and seismic data. *Earth and Planetary Science Letters, 195,* 91–98.

Alfè, D., Gillan, M., & Price, G. (2002b). Ab initio chemical potentials of solid and liquid solutions and the chemistry of the Earth's core. *Journal of Chemical Physics, 116,* 7127–7136.

Alfè, D., Price, G., & Gillan, M. (2002c). Iron under Earth's core conditions: Liquid-state thermodynamics and high-pressure melting curve from *ab initio* calculations. *Physical Review B, 65,* 165118.

Amit, H. (2014). Can downwelling at the top of the Earth's core be detected in the geomagnetic secular variation? *Physics of the Earth and Planetary Interiors, 229,* 110–121.

Ammann, M., Brodholt, J., Wookey, J., & Dobson, D. (2010). First-principles constraints on diffusion in lower-mantle minerals and a weak D". *Nature, 465,* 462–465.

Andrault, D., Bolfan-Casanova, N., Bouhifd, M. A., Boujibar, A., Garbarino, G., Manthilake, G., et al. (2017). Toward a coherent model for the melting behavior of the deep Earth's mantle. *Physics of the Earth and Planetary Interiors, 265,* 67–81.

Andrault, D., Bolfan-Casanova, N., Lo Nigro, G., Bouhifd, M., Garbarinho, & G., Mezouar, M. (2011). Solidus and liquidus profiles of chondritic mantle: Implication for melting of the Earth across its history. *Earth and Planetary Science Letters, 304,* 251–259.

Andrault, D., Monteux, J., Le Bars, M., & Samuel, H. (2016). The deep Earth may not be cooling down. *Earth and Planetary Science Letters, 443,* 195–203.

Anzellini, S., Dewaele, A., Mezouar, M., Loubeyre, P., & Morard, G. (2013). Melting of iron at Earth's inner core boundary based on fast x-ray diffraction. *Science, 340,* 464–466.

Arveson, S. M., Deng, J., Karki, B. B., & Lee, K. K. (2019). Evidence for Fe-Si-O liquid immiscibility at deep Earth pressures. *Proceedings of the National Academy of Sciences, 116*(21), 10238–10243.

Aubert, J., Amit, H., & Hulot, G. (2007). Detecting thermal boundary control in surface flows from numerical dynamos. *Physics of the Earth and Planetary Interiors, 160,* 143–156.

Aubert, J., Gastine, T., & Fournier, A. (2017). Spherical convective dynamos in the rapidly rotating asymptotic regime. *Journal of Fluid Mechanics, 813,* 558–593.

Badro, J., Aubert, J., Hirose, K., Nomura, R., Blanchard, I., Borenztajn, S., et al. (2018). Magnesium partitioning between Earth's mantle and core and its potential to drive an early exsolution geodynamo. *Geophysical Research Letters, 45,* 13–24.

Badro, J., Brodholt, J., Piet, H., Siebert, J., & Ryerson, F. (2015). Core formation and core composition from coupled geochemical and geophysical constraints. *Proceedings of the National Academy of Sciences, 112,* 12310–12314.

Badro, J., Côté, A., & Brodholt, J. (2014). A seismologically consistent compositional model of Earth's core. *Proceedings of the National Academy of Sciences, 111,* 7542–7545.

Badro, J., Siebert, J., & Nimmo, F. (2016). An early geodynamo driven by exsolution of mantle components from Earth's core. *Nature, 536*(7616), 326.

Biggin, A., Piispa, E., Pesonen, L., Holme, R., Paterson, G., Veikkolainen, T., et al. (2015). Palaeomagnetic field intensity variations suggest Mesoproterozoic inner-core nucleation. *Nature, 526*(7572), 245.

Blanc, N., Stegman, D., & Ziegler, L. (2020). Thermal and magnetic evolution of a crystallizing basal magma ocean in Earth's mantle. *Earth and Planetary Science Letters, 534,* 116085.

Bloxham, J. (1986). The expulsion of magnetic flux from the Earth's core. *Geophysical Journal International, 87*(2), 669–678.

Bloxham, J., & Jackson, A. (1990). Lateral temperature variations at the core-mantle boundary deduced from the magnetic field. *Geophysical Research Letters, 17,* 1997–2000.

Bono, R. K., Tarduno, J. A., Nimmo, F., & Cottrell, R. D. (2019). Young inner core inferred from Ediacaran ultra-low geomagnetic field intensity. *Nature Geoscience, 12*(2), 143–147.

Bouffard, M., Choblet, G., Labrosse, S., & Wicht, J. (2019). Chemical convection and stratification in the Earth's outer core. *Frontiers in Earth Science, 7,* 99.

Bouffard, M., Landeau, M., & Goument, A. (2020). Convective erosion of a primordial stratification atop Earth's core. *Geophysical Research Letters, 47*(14), e2020GL087109.

Bouhifd, M., & Jephcoat, A. (2011). Convergence of Ni and Co metal-silicate partition coefficients in the deep magma-ocean and coupled silicon-oxygen solubility in iron melts at high pressures. *Earth and Planetary Science Letters, 307*(3), 341–348.

Braginsky, S. (1999). Dynamics of the stably stratified ocean at the top of the core. *Physics of the Earth and Planetary Interiors, 111,* 21–34.

Braginsky, S., & Roberts, P. (1995). Equations governing convection in Earth's core and the geodynamo. *Geophysical & Astrophysical Fluid Dynamics, 79,* 1–97.

Breuer, D., Rueckriemen, T., & Spohn, T. (2015). Iron snow, crystal floats, and inner-core growth: modes of core solidification and implications for dynamos in terrestrial planets and moons. *Progress in Earth and Planetary Sciences, 2*(1), 39.

Brodholt, J., & Badro, J. (2017). Composition of the low seismic velocity E' layer at the top of Earth's core. *Geophysical Research Letters, 44,* 2017GL074261.

Buffett, B. (2014). Geomagnetic fluctuations reveal stable stratification at the top of the Earth's core. *Nature, 507*, 484–487.

Buffett, B. (2015). Core-mantle interactions. In G. Schubert (Ed.), *Treatise on Geophysics,* Vol. 8: pp. 213–224. Elsevier, Amsterdam: Core Dynamics.

Buffett, B., Huppert, H., Lister, J., & Woods, A. (1996). On the thermal evolution of the Earth's core. *Journal of Geophysical Research, 101*, 7989–8006.

Buffett, B., Knezek, N., & Holme, R. (2016). Evidence for MAC waves at the top of Earth's core and implications for variations in length of day. *Geophysical Journal International, 204*, 1789–2000.

Buffett, B., & Seagle, C. (2010). Stratification of the top of the core due to chemical interactions with the mantle. *Journal of Geophysical Research, 115*, B04407.

Calkins, M. A., Noir, J., Eldredge, J. D., & Aurnou, J. M. (2012). The effects of boundary topography on convection in Earth's core. *Geophysical Journal International, 189*(2), 799–814.

Caracas, R., Hirose, K., Nomura, R., & Ballmer, M. D. (2019). Melt–crystal density crossover in a deep magma ocean. *Earth and Planetary Science Letters, 516*, 202–211.

Chester, G., & Thellung, A. (1961). The law of Wiedemann and Franz. *Proceedings of the Physical Society London, 77*, 1005–1013.

Chidester, B. A., Rahman, Z., Righter, K., & Campbell, A. J. (2017). Metal-silicate partitioning of U: Implications for the heat budget of the core and evidence for reduced U in the mantle. *Geochimica et Cosmochimica Acta, 199*, 1–12.

Christensen, U. (2018). Geodynamo models with a stable layer and heterogeneous heat flow at the top of the core. *Geophysical Journal International, 215*(2), 1338–1351.

Christensen, U., Aubert, J., & Hulot, G. (2010). Conditions for Earth-like geodynamo models. *Earth and Planetary Science Letters, 296*, 487–496.

Christensen, U. R. (2006). A deep dynamo generating Mercury's magnetic field. *Nature, 444*(7122), 1056–1058.

Constable, C., Korte, M., & Panovska, S. (2016). Persistent high paleosecular variation activity in southern hemisphere for at least 10 000 years. *Earth and Planetary Science Letters, 453*, 78–86.

Cox, G., Davies, C., Livermore, P., & Singleton, J. (2019). Penetration of boundary-driven flows into a rotating spherical thermally-stratified fluid. *Journal of Fluid Mechanics, 864*, 519–553.

Davies, C. (2015). Cooling history of Earth's core with high thermal conductivity. *Physics of the Earth and Planetary Interiors, 247*, 65–79.

Davies, C., & Gubbins, D. (2011). A buoyancy profile for the Earth's core. *Geophysical Journal International, 187*, 549–563.

Davies, C., Bono, R.K., Meduri, D.G., Aubert, J., Greenwood, S. and Biggin, A.J. (2022). Dynamo constraints on the long-term evolution of Earth's magnetic field strength. *Geophysical Journal International, 228*(1), 316–336.

Davies, C., Pozzo, M., Gubbins, D., & Alfè, D. (2015). Constraints from material properties on the dynamics and evolution of Earth's core. *Nature Geoscience, 8*, 678–687.

Davies, C., Pozzo, M., Gubbins, D., & Alfè, D. (2018) Partitioning of oxygen between ferropericlase and Earth's liquid core. *Geophysical Research Letters, 45*, 6042–6050.

Davies, C., Pozzo, M., Gubbins, D., & Alfè, D. (2020). Transfer of oxygen to Earth's core from a long-lived magma ocean. *Earth and Planetary Science Letters, 538*, 116208.

de Koker, N., Steinle-Neumann, G., & Vojtech, V. (2012). Electrical resistivity and thermal conductivity of liquid Fe alloys at high P and T and heat flux in Earth's core. *Proceedings of the National Academy of Sciences, 109*, 4070–4073.

Deschamps, F., & Hsieh, W.-P. (2019). Lowermost mantle thermal conductivity constrained from experimental data and tomographic models. *Geophysical Journal International, 219*(Suppl_1), S115–S136.

Driscoll, P. (2016). Simulating 2 Ga of geodynamo history. *Geophysical Research Letters, 43*(11), 5680–5687.

Driscoll, P., & Bercovici, D. (2014). On the thermal and magnetic histories of Earth and Venus: Influences of melting, radioactivity, and conductivity. *Physics of the Earth and Planetary Interiors, 236*, 36–51.

Driscoll, P. and Davies, C. (2023). The "New Core Paradox:" Challenges and potential solutions. *Journal of Geophysical Research: Solid Earth*, p.e2022JB025355.

Driscoll, P. E., & Du, Z. (2019). Geodynamo conductivity limits. *Geophysical Research Letters, 46*(14), 7982–7989.

Driscoll, P. E., & Wilson, C. (2018). Paleomagnetic biases inferred from numerical dynamos and the search for geodynamo evolution. *Frontiers in Earth Science, 6*, 113.

Du, Z., Boujibar, A., Driscoll, P., & Fei, Y. (2019). Experimental constraints on an MgO exsolution-driven geodynamo. *Geophysical Research Letters, 46*(13), 7379–7385.

Du, Z., Jackson, C., Bennett, N., Driscoll, P., Deng, J., Lee, K., et al. (2017). Insufficient energy from MgO exsolution to power early geodynamo. *Geophysical Research Letters, 4*, 2017GL075283.

Dumberry, M. (2018). Earth Rotation, Excitation, Core. In E. Grafarend (Ed.), *Encyclopedia of Geodesy.* pp. 1–5. Cham: Springer International Publishing.

Dziewonski, A., & Anderson, D. (1981). Preliminary reference Earth model. *Physics of the Earth and Planetary Interiors, 25*, 297–356.

Eggins, S., Rudnick, R., & McDonough, W. (1998). The composition of peridotites and their minerals: A laser-ablation ICP-MS study. *Earth and Planetary Science Letters, 154*(1-4), 53–71.

Fearn, D., & Loper, D. (1981). Compositional convection and stratification of Earth's core. *Nature, 289*, 393–394.

Fiquet, G., Auzende, A., Siebert, J., Corgne, A., Bureau, H., Ozawa, H., et al. (2010). Melting of peridotite to 140 gigapascals. *Science, 329*, 1516–1518.

Fischer, R. A., Campbell, A. J., & Ciesla, F. J. (2017). Sensitivities of Earth's core and mantle compositions to accretion and differentiation processes. *Earth and Planetary Science Letters, 458*, 252–262.

Fischer, R. A., Cottrell, E., Hauri, E., Lee, K. K., & Le Voyer, M. (2020). The carbon content of Earth and its core. *Proceedings of the National Academy of Sciences, 117*(16), 8743–8749.

Fischer, R. A., Nakajima, Y., Campbell, A. J., Frost, D. J., Harries, D., Langenhorst, F., et al. (2015). High pressure metal-silicate partitioning of Ni, Co, V, Cr, Si, and O. *Geochimica et Cosmochimica Acta*, *167*(Suppl C), 177–194.

Frost, D., Asahara, Y., Rubie, D., Miyajima, N., Dubrovinsky, L. S., Holzapfel, C., et al. (2010). Partitioning of oxygen between the Earth's mantle and core. *Journal of Geophysical Research*, *115*, B02202.

Garaud, P. (2018). Double-diffusive convection at low Prandtl number. *Annual Review of Fluid Mechanics*, *50*, 275–298.

Garnero, E., McNamara, A., & Shim, S.-H. (2016). Continent-sized anomalous zones with low seismic velocity at the base of Earth's mantle. *Nature Geoscience*, *9*, 481–489.

Garnero, E. J., Helmberger, D. V., & Grand, S. P. (1993). Constraining outermost core velocity with SmKS waves. *Geophysical Research Letters*, *20*(22), 2463–2466.

Gastine, T., Aubert, J., & Fournier, A. (2020). Dynamo-based limit to the extent of a stable layer atop Earth's core. *Geophysical Journal International*, *222*(2), 1433–1448.

Glane, S., & Buffett, B. (2018). Enhanced core-mantle coupling due to stratification at the top of the core. *Frontiers in Earth Science*, *6*, 171.

Gomi, H., Ohta, K., Hirose, K., Labrosse, S., Caracas, R., Verstraete, V., et al. (2013). The high conductivity of iron and thermal evolution of the Earth's core. *Physics of the Earth and Planetary Interiors*, *224*, 88–103.

Greenwood, S., Davies, C., & Mound, J. (2021). On the evolution of thermally stratified layers at the top of Earth's core. *Physics of the Earth and Planetary Interiors*, 318.

Gubbins, D. (2003). Thermal core-mantle interactions: Theory & observations. In V. Dehant, K. Creager, S. Karato, & S. Zatman (Eds.), Earth's core: Dynamics, Structure, Rotation. Geodynamics Series 31. *American Geophysical Union*, 162–179.

Gubbins, D. (2007). Geomagnetic constraints on stratification at the top of Earth's core. *Earth Planets Space*, *59*, 661–664.

Gubbins, D., Alfè, D., Davies, C., & Pozzo, M. (2015). On core convection and the geodynamo: Effects of high electrical and thermal conductivity. *Physics of the Earth and Planetary Interiors*, *247*, 56–64.

Gubbins, D., Alfe, D., Masters, G., Price, G., & Gillan, M. (2003). Can the Earth's dynamo run on heat alone? *Geophysical Journal International*, *155*, 609–622.

Gubbins, D., Alfè, D., Masters, G., Price, G., & Gillan, M. (2004). Gross thermodynamics of two-component core convection. *Geophysical Journal International*, *157*, 1407–1414.

Gubbins, D., & Davies, C. (2013). The stratified layer at the core-mantle boundary caused by barodiffusion of Oxygen, Sulphur and Silicon. *Physics of the Earth and Planetary Interiors*, *215*, 21–28.

Gubbins, D., Thomson, C., & Whaler, K. (1982). Stable regions in the Earth's liquid core. *Geophysical Journal of the Royal Astronomical Society*, *68*, 241–251.

Gubbins, D., Willis, A., & Sreenivasan, B. (2007). Correlation of Earth's magnetic field with lower mantle thermal and seismic structure. *Physics of the Earth and Planetary Interiors*, *162*, 256–260.

Guervilly, C. (2022). Fingering convection in the stably stratified layers of planetary cores. *Journal of Geophysical Research: Planets*, *127*(11), e2022JE007350.

Hardy, C. M., Livermore, P. W., & Niesen, J. (2020). Enhanced magnetic fields within a stratified layer. *Geophysical Journal International*, *222*(3), 1686–1703.

Helffrich, G. (2012). How light element addition can lower core liquid wave speeds. *Geophysical Journal International*, 1065–1070.

Helffrich, G., Ballmer, M. D., & Hirose, K. (2018). Core-exsolved SiO_2 dispersal in the Earth's mantle. *Journal of Geophysical Research*, *123*(1), 176–188.

Helffrich, G., Hirose, K., & Nomura, R. (2020). Thermodynamical modeling of liquid Fe-Si-Mg-O: molten magnesium silicate release from the core. *Geophysical Research Letters*, *47*(21), e2020GL089218.

Helffrich, G., & Kaneshima, S. (2010). Outer-core compositional stratification from observed core wave speed profiles. *Nature*, *468*, 807–809.

Hernlund, J., & McNamara, A. (2015). The Core-Mantle Boundary Region. In G. Schubert. (Ed.), *Treatise on Geophysics*, Vol. 7. pp. 461–519. Elsevier.

Hirose, K., Labrosse, S., & Hernlund, J. (2013). Compositional state of Earth's core. *Annual Review of Earth and Planetary Sciences*, *41*, 657–691.

Hirose, K., Morard, G., Sinmyo, R., Umemoto, K., Hernlund, J., Helffrich, G., et al. (2017). Crystallization of silicon dioxide and compositional evolution of the Earth's core. *Nature*, *543*(7643), 99–102.

Hsieh, W.-P., Goncharov, A. F., Labrosse, S., Holtgrewe, N., Lobanov, S. S., Chuvashova, I., et al. (2020). Low thermal conductivity of iron-silicon alloys at Earth's core conditions with implications for the geodynamo. *Nature Communications*, *11*(1), 1–7.

Huguet, L., Amit, H., & Alboussière, T. (2018). Geomagnetic dipole changes and upwelling/downwelling at the top of the Earth's core. *Frontiers in Earth Science*, *6*, 170.

Inoue, H., Suehiro, S., Ohta, K., Hirose, K., & Ohishi, Y. (2020). Resistivity saturation of hcp Fe-Si alloys in an internally heated diamond anvil cell: A key to assessing the Earth's core conductivity. *Earth and Planetary Science Letters*, *543*, 116357.

Irving, J. C., Cottaar, S., & Lekić, V. (2018). Seismically determined elastic parameters for Earth's outer core. *Science Advances*, *4*(6), eaar2538.

Jackson, A., & Livermore, P. (2009). On Ohmic heating in the Earth's core I: Nutation constraints. Geophys. *Journal of International*, *177*, 367–382.

Jacobson, S., Rubie, D., Hernlund., J., Morbidelli, A., & Nakajima, M. (2017). Formation, stratification, and mixing of the cores of Earth and Venus. *Earth and Planetary Science Letters*, *474*, 375–386.

Jaupart, C., Labrosse, S., & Mareschal, J.-C. (2015). Temperatures, heat and energy in the mantle of the Earth. In G. Schubert. (Ed.), *Treatise on Geophysics*, Vol. 7. pp. 223–270. Amsterdam: Elsevier.

Jaupart, E., & Buffett, B. (2017). Generation of MAC waves by convection in Earth's core. *Geophysical Journal International*, *209*(2), 1326–1336.

Kaneshima, S. (2018). Array analysis of SmKS waves and stratification of Earth's outermost core. *Physics of the Earth and Planetary Interiors*, *276*, 234–246.

Koelemeijer, P., Deuss, A., & Ritsema, J. (2017). Density structure of Earth's lowermost mantle from Stoneley mode splitting observations. *Nature Communications*, *8*(1), 1–10.

Komabayashi, T. (2014). Thermodynamics of melting relations in the system Fe-FeO at high pressure: Implications for oxygen in the Earth's core. *Journal of Geophysical Research*, *119*(5), 4164–4177.

Konôpková, Z., McWilliams, R., Gómez-Pérez, N., & Goncharov, A. (2016). Direct measurement of thermal conductivity in solid iron at planetary core conditions. *Nature*, *534*, 99–101.

Korenaga, J. (2006). Archean geodynamics and the thermal evolution of Earth. *Geophysical Monograph-American Geophysical Union*, *164*, 7.

Labrosse, S. (2015). Thermal evolution of the core with a high thermal conductivity. *Physics of the Earth and Planetary Interiors*, *247*, 36–55.

Labrosse, S., Hernlund, J., & Coltice, N. (2007). A crystallizing dense magma ocean at the base of the Earth's mantle. *Nature*, *450*, 866–869.

Labrosse, S., Hernlund, J. W., & Hirose, K. (2015). Fractional melting and freezing in the deep mantle and implications for the formation of a basal magma ocean. In J. Badro, M. Walter. (Eds.), *The early Earth: Accretion and differentiation*, Ch. 7, pp. 123–142. AGU.

Labrosse, S., Poirier, J.-P., & Le Moeul, J.-L. (1997). On cooling of the Earth's core. *Physics of the Earth and Planetary Interiors*, *99*, 1–17.

Labrosse, S., Poirier, J.-P., & Le Moeul, J.-L. (2001). The age of the inner core. *Earth and Planetary Science Letters*, *190*, 111–123.

Landau, L., & Lifshitz, E. (1987). *Fluid mechanics (course of theoretical physics, volume 6)*, 2nd Edn. Permagon Press.

Landeau, M., Aubert, J., & Olson, P. (2017). The signature of inner-core nucleation on the geodynamo. *Earth and Planetary Science Letters*, *465*, 193–204.

Landeau, M., Olson, P., Deguen, R., & Hirsh, B. H. (2016). Core merging and stratification following giant impact. *Nature Geoscience*, *1*(Sept), 1–5.

Laneuville, M., Hernlund, J., Labrosse, S., & Guttenberg, N. (2018). Crystallization of a compositionally stratified basal magma ocean. *Physics of the Earth and Planetary Interiors*, *276*, 86–92.

Lau, H. C., Mitrovica, J. X., Davis, J. L., Tromp, J., Yang, H.-Y., & Al-Attar, D. (2017). Tidal tomography constrains Earth's deep-mantle buoyancy. *Nature*, *551*(7680), 321–326.

Lay, T., Hernlund, J., & Buffett, B. (2009). Core-mantle boundary heat flow. *Nature Geoscience*, *1*, 25–32.

Lay, T., & Young, C. (1990). The stably-stratified outermost core revisited. *Geophysical Research Letters*, *71*, 2001–2004.

Lesur, V., Whaler, K., & Wardinski, I. (2015). Are geomagnetic data consistent with stably stratified flow at the core-mantle boundary? *Geophysical Journal International*, 929–946.

Li, Y., Vočadlo, L., Alfè, D., & Brodholt, J. (2019). Carbon partitioning between the Earth's inner and outer core. *Journal of Geophysical Research*, *124*(12), 12812–12824.

Lister, J., & Buffett, B. (1998). Stratification of the outer core at the core-mantle boundary. *Physics of the Earth and Planetary Interiors*, *105*, 5–19.

Liu, W., Zhang, Y., Yin, Q.-Z., Zhao, Y., & Zhang, Z. (2020). Magnesium partitioning between silicate melt and liquid iron using first-principles molecular dynamics: Implications for the early thermal history of the Earth's core. *Earth and Planetary Science Letters*, *531*, 115934.

Long, R., Mound, J., Davies, C., & Tobias, S. (2020). Scaling behaviour in spherical shell rotating convection with fixed-flux thermal boundary conditions. *Journal of Fluid Mechanics*, *889*, P.A7.

Masters, G., & Gubbins, D. (2003). On the resolution of density within the Earth. *Physics of the Earth and Planetary Interiors*, *140*, 159–167.

Mather, J. F., & Simitev, R. D. (2020). Regimes of thermo-compositional convection and related dynamos in rotating spherical shells. *Geophysical & Astrophysical Fluid Dynamics*, 1–24.

Metman, M. C., Livermore, P. W., Mound, J. E., & Beggan, C. D. (2019). Modelling decadal secular variation with only magnetic diffusion. *Geophysical Journal International*, *219*(Suppl_1), S58–S82.

Mittal, T., Knezek, N., Arveson, S. M., McGuire, C. P., Williams, C. D., Jones, T. D., et al. (2020). Precipitation of multiple light elements to power Earth's early dynamo. *Earth and Planetary Science Letters*, *532*, 116030.

Moffatt, H., & Loper, D. (1994). The magnetostrophic rise of a buoyant parcel in the Earth's core. *Geophysical Journal International*, (117), 394–402.

Monville, R., Vidal, J., Cébron, D., & Schaeffer, N. (2019). Rotating double-diffusive convection in stably stratified planetary cores. *Geophysical Journal International*, *219*(Suppl_1), S195–S218.

Mound, J., Davies, C., Rost, S., & Aurnou, J. (2019). Regional stratification at the top of Earth's core due to core-mantle boundary heat flux variations. *Nature Geoscience*, *12*(7), 575–580.

Mound, J. E., & Davies, C. (2020). Scaling laws for regional stratification at the top of earth's core. *Geophysical Research Letters*, *47*(16), e2020GL087715.

Nakagawa, T. (2011). Effect of a stably stratified layer near the outer boundary in numerical simulations of a magnetohydrodynamic dynamo in a rotating spherical shell and its implications for Earth's core. *Physics of the Earth and Planetary Interiors*, *187*, 342–352.

Nakagawa, T. (2018). On the thermo-chemical origin of the stratified region at the top of the Earth's core. *Physics of the Earth and Planetary Interiors*, *276*, 172–181.

Nakagawa, T. (2020). A coupled core-mantle evolution: Review and future prospects. *Progress in Earth and Planetary Science*, *7*(1), 1–17.

Nakagawa, T., & Tackley, P. (2014). Influence of combined primordial layering and recycled MORB on the coupled thermal evolution of Earth's mantle and core. *Geochemistry, Geophysics, Geosystems*, *15*, 619–633.

Nimmo, F. (2015a). Energetics of the core. In G. Schubert. (Ed.), *Treatise on Geophysics* 2nd Edn, Vol. 8, pp. 27–55. Amsterdam: Elsevier.

Nimmo, F. (2015b). Thermal and compositional evolution of the core. In G. Schubert. (Ed.), *Treatise on Geophysics* 2nd Edn, Vol. 9, pp. 209–219. Amsterdam: Elsevier.

Nimmo, F., Price, G., Brodholt, J., & Gubbins, D. (2004). The influence of potassium on core and geodynamo evolution. *Geophysical Journal International*, 156, 363–376.

Ohta, K., Kuwayama, Y., Hirose, K., Shimizu, K., & Ohishi, Y. (2016). Experimental determination of the electrical resistivity of iron at Earth's core conditions. *Nature*, 534(7605), 95.

Olson, P. (2013). The new core paradox. *Science*, 342, 431–432.

Olson, P., Deguen, R., Rudolph, M., & Zhong, S. (2015). Core evolution driven by mantle global circulation. *Physics of the Earth and Planetary Interiors*, 243, 44–55.

Olson, P., Landeau, M., & Reynolds, E. (2017). Dynamo tests for stratification below the core-mantle boundary. *Physics of the Earth and Planetary Interiors*, 271, 1–18.

Olson, P., Landeau, M., & Reynolds, E. (2018). Outer core stratification from the high latitude structure of the geomagnetic field. *Frontiers in Earth Science*, 6, 140.

O'Rourke, J., Korenaga, J., & Stevenson, D. (2017). Thermal evolution of Earth with magnesium precipitation in the core. *Earth and Planetary Science Letters*, 458, 263–272.

O'Rourke, J. G., & Stevenson, D. J. (2016). Powering Earth's dynamo with magnesium precipitation from the core. *Nature*, 529(7586), 387–389.

Ozawa, H., Hirose, K., Mitome, M., Bando, Y., Sata, N., & Ohishi, Y. (2008). Chemical equilibrium between ferropericlase and molten iron to 134 GPa and implications for iron content at the bottom of the mantle. *Geophysical Research Letters*, 35, L05308.

Patočka, V., Šrámek, O., & Tosi, N. (2020). Minimum heat flow from the core and thermal evolution of the Earth. *Physics of the Earth and Planetary Interiors*, 305, 106457.

Posner, E. S., Schmickler, B., & Rubie, D. C. (2018). Self-diffusion and chemical diffusion in peridotite melt at high pressure and implications for magma ocean viscosities. *Chemical Geology*, 502, 66–75.

Pourovskii, L., Mravlje, J., Pozzo, M., & Alfè, D. (2020). Electronic correlations and transport in iron at Earth's core conditions. *Nature Communications*, 11(1), 1–8.

Pozzo, M., & Alfè, D. (2016). Saturation of electrical resistivity of solid iron at Earth's core conditions. *SpringerPlus*, 5(1), 1–6.

Pozzo, M., Davies, C. and Alfè, D. (2022). Towards reconciling experimental and computational determinations of Earth's core thermal conductivity. *Earth and Planetary Science Letters*, 584, 117466.

Pozzo, M., Davies, C., Gubbins, D., & Alfè, D. (2012). Thermal and electrical conductivity of iron at Earth's core conditions. *Nature*, 485, 355–358.

Pozzo, M., Davies, C., Gubbins, D., & Alfè, D. (2013). Transport properties for liquid silicon-oxygen-iron mixtures at Earth's core conditions. *Physical Review, B 87*, 014110.

Pozzo, M., Davies, C., Gubbins, D., & Alfè, D. (2019). The FeO Content of Earth's Core. *Physical Review, X 9*, 041018.

Ribe, N. (1998). Spouting and planform selection in the Rayleigh–Taylor instability of miscible viscous fluids. *Journal of Fluid Mechanics*, 377, 27–45.

Roberts, P., & Scott, S. (1965). On analysis of the secular variation. I. a hydromagnetic constraint: Theory. *Journal of geomagnetism and geoelectricity*, 17, 137–151.

Rubie, D., Nimmo, F., & Melosh, H. (2015a). Formation of Earth's Core. In G. Schubert. (Ed.), *Treatise on Geophysics* 2nd Edn, Vol. 9., pp. 43–79. Amsterdam: Elsevier.

Rubie, D. C., Frost, D. J., Mann, U., Asahara, Y., Nimmo, F., Tsuno, K., et al. (2011). Heterogeneous accretion, composition and core-mantle differentiation of the Earth. *Earth and Planetary Science Letters*, 301, 31–42.

Rubie, D. C., Jacobson, S. A., Morbidelli, A., O'Brien, D. P., Young, E. D., de Vries, J., et al. (2015b). Accretion and differentiation of the terrestrial planets with implications for the compositions of early-formed Solar System bodies and accretion of water. *Icarus*, 248, 89–108.

Secco, R. A. (2017). Thermal conductivity and Seebeck coefficient of Fe and Fe-Si alloys: Implications for variable Lorenz number. *Physics of the Earth and Planetary Interiors*, 265, 23–34.

Siebert, J., Badro, J., Antonangeli, D., & Ryerson, F. J. (2012). Metal-silicate partitioning of Ni and Co in a deep magma ocean. *Earth and Planetary Science Letters*, 321(Suppl C), 189–197.

Sinmyo, R., Hirose, K., & Ohishi, Y. (2019). Melting curve of iron to 290 GPa determined in a resistance-heated diamond-anvil cell. *Earth and Planetary Science Letters*, 510, 45–52.

Solomatov, V. (2015). Magma oceans and primordial mantle differentiation. In G. Schubert. (Ed.), *Treatise on Geophysics* 2nd Edn, Vol. 10., pp. 81–104. Amsterdam: Elsevier.

Stevenson, D. (1987). Limits on lateral density and velocity variations in the Earth's outer core. *Geophysical Journal International*, 88, 311–319.

Stixrude, L., de Koker, N., Sun, N., Mookherjee, M., & Karki, B. B. (2009). Thermodynamics of silicate liquids in the deep Earth. *Earth and Planetary Science Letters*, 278(3-4), 226–232.

Stixrude, L., Scipioni, R., & Desjarlais, M. P. (2020). A silicate dynamo in the early Earth. *Nature Communications*, 11(1), 1–5.

Suer, T.-A., Siebert, J., Remusat, L., Menguy, N., & Fiquet, G. (2017). A sulfur-poor terrestrial core inferred from metal-silicate partitioning experiments. *Earth and Planetary Science Letters*, 469(Suppl C), 84–97.

Takafuji, N., Hirose, K., Mitome, M., & Bando, Y. (2005). Solubilities of O and Si in liquid iron in equilibrium with (Mg,Fe)SiO$_3$ perovskite and the light elements in the core. *Geophysical Research Letters*, 32(6), L06313.

Takehiro, S.-i., & Lister, J. R. (2001). Penetration of columnar convection into an outer stably stratified layer in rapidly rotating spherical fluid shells. *Earth and Planetary Science Letters*, 187(3-4), 357–366.

Tang, F., Taylor, R. J., Einsle, J. F., Borlina, C. S., Fu, R. R., Weiss, B. P., et al. (2019). Secondary magnetite in ancient zircon precludes analysis of a Hadean geodynamo. *Proceedings of the National Academy of Sciences*, 116(2), 407–412.

Tarduno, J., Cottrell, R., Bono, R., Oda, H., Davis, W., Fayek, M., et al. (2020). Paleomagnetism indicates that primary magnetite in zircon records a strong Hadean geodynamo. *Proceedings of the National Academy of Sciences*, 117(5), 2309–2318.

Tarduno, J., Cottrell, R., Watkeys, M., Hofmann, A., Doubrovine, P., Mamajek, E., et al. (2010). Geodynamo, solar wind, and magnetopause 3.4 to 3.45 billion years ago. *Science*, *327*, 1238–1240.

Tarduno, J. A., Cottrell, R. D., Davis, W. J., Nimmo, F., & Bono, R. K. (2015). A Hadean to Paleoarchean geodynamo recorded by single zircon crystals. *Science*, *349*, 521–524.

Tauxe, L., & Yamazaki, T. (2015). 5.13-paleointensities. In *Treatise on Geophysics*. Elsevier, pp. 461–509. https://www.elsevier.com/books/treatise-on-geophysics/schubert/978-0-444-53802-4.

Tilgner, A. (2015). Rotational Dynamics of the Core. In G. Schubert. (Ed.), *Treatise on Geophysics*, 8.07., pp. 183–212. Amsterdam: Elsevier.

Tsuno, K., Frost, D. J., & Rubie, D. C. (2013). Simultaneous partitioning of silicon and oxygen into the Earth's core during early Earth differentiation. *Geophysical Research Letters*, *40*, 66–71.

Turner, J. (1973). *Buoyancy Effects in Fluids*. Cambridge: University Press.

Umemoto, K., & Hirose, K. (2020). Chemical compositions of the outer core examined by first principles calculations. *Earth and Planetary Science Letters*, *531*, 116009.

van Tent, R., Deuss, A., Kaneshima, S., & Thomas, C. (2020). The signal of outermost-core stratification in body-wave and normal-mode data. *Geophysical Journal International*, *223*(2), 1338–1354.

Vilella, K., Bodin, T., Boukaré, C.-E., Deschamps, F., Badro, J., Ballmer, M. D., et al. 2021. Constraints on the composition and temperature of LLSVPs from seismic properties of lower mantle minerals. *Earth and Planetary Science Letters*, *554*, 116685.

Watanabe, K., Ohtani, E., Kamada, S., Sakamaki, T., Miyahara, M., & Ito, Y. (2014). The abundance of potassium in the Earth's core. *Physics of the Earth and Planetary Interiors*, *237*, 65–72.

Whaler, K. A. (1980). Does the whole of the Earth's core convect? *Nature*, *287*, 528–530.

Whaler, K. A. (1986). Geomagnetic evidence for fluid upwelling at the core-mantle boundary. *Geophysical Journal of the Royal Astronomical Society*, *86*, 563–588.

Wicht, J., & Sanchez, S. (2019). Advances in geodynamo modelling. *Geophysical & Astrophysical Fluid Dynamics*, *113*(1-2), 2–50.

Williams, Q. (2018). The thermal conductivity of Earth's core: A key geophysical parameter's constraints and uncertainties. *Annual Review of Earth and Planetary Sciences*, *46*, 47–66.

Wilson, A.J., Pozzo, M., Alfè, D., Walker, A.M., Greenwood, S., Pommier, A. and Davies, C. (2022). Powering Earth's ancient dynamo with silicon precipitation. *Geophysical Research Letters*, *49*(22), e2022GL100692.

Wong, J., Davies, C., & Jones, C. A. (2021). A regime diagram for the slurry F-layer at the base of Earth's outer core. *Earth and Planetary Science Letters*, *560*, 116791.

Xiong, Z., Tsuchiya, T., & Taniuchi, T. (2018). Ab initio prediction of potassium partitioning into Earth's core. *Journal of Geophysical Research*, *123*(8), 6451–6458.

Xu, J., Zhang, P., Haule, K., Minar, J., Wimmer, S., Ebert, H., et al. (2018). Thermal conductivity and electrical resistivity of solid iron at Earth's core conditions from first principles. *Physical Review Letters*, *121*(9), 096601.

Yan, C., & Stanley, S. (2018). Sensitivity of the geomagnetic octupole to a stably stratified layer in the Earth's core. *Geophysical Research Letters*, *45*(20), 11–005.

Zhang, K. (1992). Convection in a rapidly rotating spherical shell at infinite Prandtl number: transition to vacillating flows. *Physics of the Earth and Planetary Interiors*, *72*, 236–248.

Zhang, Y., Hou, M., Liu, G., Zhang, C., Prakapenka, V. B., Greenberg, et al. (2020). Reconciliation of experiments and theory on transport properties of iron and the geodynamo. *Physical Review Letters*, *125*(7), 078501.

Ziegler, L. B., & Stegman, D. R. (2013). Implications of a long-lived basal magma ocean in generating Earth's ancient magnetic field. *Geochemistry, Geophysics, Geosystems*, *14*, 4735–4742.

INDEX

^{182}Hf-^{182}W system   193
^{3}He/^{4}He ratio   193
^{57}Fe   193

Ab initio calculation   223
activity coefficient   229
adiabatic mantle temperature   137
Alpha (α) decay   4
amorphous boron   197
anisotropy
   cylindrical   166, 179, 183–185
   inner core   37, 38, 49
   lower mantle (shear wave splitting)   38, 41, 165
anorthosite   106–110
asteroid belts   126
atomistic diffusion   128
attenuation (Q)
   inner core   36, 38, 39, 41, 42, 49
   outer core   37, 39
   mantle   38, 47, 49
average geotherm   137

barodiffusion   123, 147, 149, 150, 220
Basal Magma Ocean (BMO)   127, 219, 222, 224, 228, 232, 233, 235, 244, 246–251
basalt   106, 107, 109
Beta decay   4
Bloch-Grüneisen formula   227
boron-doped diamond heater   198
bridgmanite   134
Brunt-Väisälä
   frequency   146, 237, 240
   period   242
Bullen's parameter   146
bulk modulus   146, 148
buoyancy flux   147, 148, 151, 152, 158
Birch's law   124

CAI (Calcium-aluminum rich inclusion)   118
CALS10.2K   240
Ca(OH)$_2$ 64
(CaSiO$_3$) perovskite (CaPv)   140
carbonaceous chondrite   118, 121, 126
CaSiO$_3$ perovskite (Ca-Pv)   62
CCD/CMOS camera   63

Central America   137
Central Pacific   137
chemical expansion coefficient   148
chemical flux   148, 154, 157, 160, 161
chemical potential   148, 150, 220, 228
chemical stratification   146
Chondrite   5
Clapeyron slope   137
CMB heat flux   137
coherent scattering   204
compositional flux   150
COMSOL Multiphysics   137
convective entropy flux   151
core   191
core formation process   192
core-mantle boundary (CMB)   45, 59, 103, 104, 109, 111, 133, 191
core-mantle chemical coupling   145–149, 153, 158, 160, 161
core-mantle interaction   193
core-mantle separation   117
core melting curve   248
Coble creep   70
creep flow law   70
creep strength   68
cubic boron nitride   197

D″   32, 34, 36–38, 40
   discontinuity   137
   layer   128, 134
D111 deformation multi-anvil press   59
d-DIA geometry   60
Debye-Scherrer ring   65
Debye sound velocity   199
deep earth structure   31, 32
deformation experiment   61
deformation-T-cup   61
density   37
Depleted MORB Mantle (DMM)   76, 79
DFT (Density Functional Theory)   228
diamond   197
diamond anvil cell (DAC)   134, 194
diffusion coefficient   123, 124
diffusion creep   70
dipole moment   151, 156–158
Dirac delta function   150, 161

direct numerical simulation (DNS) 222, 239–241, 251
discontinuity, seismic 36–38, 42, 45, 46, 137
dissociation 230, 231
dissolution 125, 146, 230, 231
distribution coefficient 229
double-crossing 137
double-diffusive instability 240, 243
double-stage semispherical diamond anvil 194

elastic anisotropy 66
electrical conductivity 125, 126, 227
electron antineutrino 5–7
electron microscopic analysis 193
electron-phonon scattering 227
electron probe micro-analyzer (EPMA) 201
energy-dispersive X-ray spectroscopy (EDS) 201
Enriched Mantle (EM1 and EM2) 76, 79
entropy
    flux 151
    transport 151
eutectic composition 201
external resistive heating 197

Fe_2S 198
$Fe_{47}Ni_{28}S_{25}$ 209
$Fe_{52}Ni_{10}Si_{38}$ 209
$Fe_{60}Ni_{10}Si_{29}$ 209
Fe_7C_3 198
$Fe_{70}Ni_4Si_{26}$ 210
$Fe_{75}P_{25}$ 209
$Fe_{77}Ni_4S_{19}$ 210
$Fe_{80}S_{20}$ 209
$Fe_{84}C_{16}$ 209
$Fe_{84}Si_{16}$ 209
$Fe_{88}C_{12}$ 210
Fe-wetted grain-boundaries 124–126
Fick's law 122, 148
field-emission type electron-gun 201
finger convection 240
finite element method calculation 137
first-principle
    calculation 127
    molecular dynamics calculation 232
F-layer 140, 191
focused ion beam (FIB) system 197
Focus zone (FOZO) 77, 79

Gamma distribution 21, 23, 27
geodynamo 147, 161
geomagnetic field 146, 220, 221, 247
geoneutrino 5, 11
geoneutrino detectors 6
    Andes 7, 8
    Baksan 7, 8
    Borexino 7, 8, 11, 12
    Jinping 7, 8, 12
    JUNO 7, 8, 12
    KamLAND 7, 8, 11, 12
    Liquid scintillation spectrometers 6, 8
    Ocean Bottom Detector (ODB) 6, 13
    SNO+ 7, 8, 12
giant impact 119
global energy and mass balance 149
gravitational constant 148
gravitational energy release 150
Grüneisen parameter 148

harzburgite 106, 107
hcp (hexagonal closed packing) 228
hcp-Fe 208
heat capacity 149
heat conduction equation 137
heat flow 133
heat flux 133
Heat Producing Elements (HPE) 5, 10
Helium isotopes 85, 87
heterogeneity
    chemical 104, 105, 107–110
    thermal 107
    deep mantle 161
hot geotherm 137
hotspot 48, 104
Highly siderophile elements (HSEs) 88, 118, 119, 121, 122, 126–128
    Chondritic overabundance 92, 94
    Fractionation 89
    Peridotites 90, 91
    Pyrolite 92
High-μ mantle (HIMU) 76, 77, 79
hydrostatic pressure gradient 220

impactor 119
inner core 33, 34, 40, 42, 45, 49, 140, 191
    anisotropy 33, 39, 41, 42
    anisotropy strength 166, 167, 185
    anisotropy coefficients 178–181, 184
    anisotropy direction 185, 186
    boundary (ICB) 34, 36–38, 48, 140, 191
    growth 147, 148
    hemispheric structure 37, 39, 41, 166
    nucleation 222
    parameters of anisotropy, *see* inner core, anisotropy coefficients
in situ synchrotron X-ray measurement 193, 198
internally resistive heated DAC, (IH-DAC) 197
iron isotopic fractionation 193
iron spin crossover 136
isentropic heat flow 150, 160
isobaric heat capacity 137
isotopic fractionation coefficient 199

Jupiter 118

Kawai-type multi-anvil press   61, 197
Kimberlites   85
kinetic energy production   147, 160
Kolmogorov-Smirnov test   25
KREEP   104, 106, 108, 110, 111

Large Low Shear Velocity Province (LLSVP)   32, 41, 48, 137
Large Low Velocity Provinces (LLVPs)   104, 219–221, 224, 251
large volume press   197
laser-heated diamond anvil cell (LH-DAC)   194, 223
latent heat   150, 248
lateral heat flux variation   140
lateral heterogeneity   161
lateral variations   221
later or late stage bombardment   126
late veneer   118, 121, 126
lattice thermal conductivity   134
light element in the core   192
Log-normal distribution   20, 23, 26, 27
longitudinal acoustic (LA) mode   206
Lorenz number   226
lower mantle   34, 40, 41, 44–49, 103, 104
lower mantle (velocity structure)   170, 173, 175, 177, 183
low-velocity region   123, 124

magma ocean
    earth   111
    moon   104, 106
magnetic permeability   151
magnetic Rossby wave   146
magneto-archimedes-coriolis (MAC) wave   146, 239, 240, 242, 250
mantle plume   32, 47, 111
mantle radiogenic heat flux   11, 12
Markov Chain Monte Carlo   179–181
maximum penetration assessment   151, 153, 155
mechanical mixing   107
melt   104–106, 108–111
melting temperature of iron and iron alloys   199
melt-solid separation   126
MERMAID   50
metallization   141
metal-silicate partitioning   146, 149, 150
(Mg,Fe)O ferropericlase   136
MgO periclase   136
mid-ocean ridge basalt (MORB)   59, 76, 105, 106, 108, 110, 111, 140
minimum penetration assessment   151, 153, 155
morphological instability   124, 125, 127, 128
multi-anvil (MA) apparatus   61, 194, 197

Nabarro-Herring creep   70
nano-polycrystalline diamonds (NPD)   194
Neutrino Geoscience   13
new core paradox   219, 222, 250
noise magnitude   180, 181

normal
    distribution   23
    mean   18, 19, 23, 27
    median   18, 19, 27
    modes   224
nuclear resonant inelastic X-ray scattering (NIS) measurement   194

ocean bottom seismographs (OBS)   50, 51
Oceanic crust   76
Ocean Island basalt (OIB)   76, 79, 193
    Sr-Nd-Pb isotopic composition   77
    Tungsten isotopes   85
Ohmic dissipation   250
olivine   64
optical absorption measurement   136
Osmium isotopes   84, 93
outer core   32, 34, 36, 38, 39, 48, 191
    topmost   191
Oxygen fugacity (f_{O_2})   89

paleomagnetic measurements   158
Paris-Edinburgh press   197
Partition coefficients   82
pebbles   118
percolation   126
perovskitic model   137
phase, mineral   105–107
phonon dispersion relation   206
phonon DOS spectra   199
photon factory at KEK   61
pile (chemical)   109, 110
piston-cylinder apparatus   195
Platinum group elements (PGEs)   88, 89
polycrystalline   125
post-perovskite (PPv) phase transition   135
preferred orientation   165, 186
PREM (Preliminary Reference Earth Model)   191, 219, 223, 224, 237, 238
Prevalent mantle (PREMA)   77, 79
Primitive mantle (PM)   5, 76
Primordial energy   5
Pt marker foil   64
pulsed laser   196
pulsed-light heating thermoreflectance technique   134
P-wave velocity (V_P)   207
pyrolite   105, 106, 109–111
pyrolitic lower mantle   136

q* value   140

radial distribution function   205
radiative thermal conductivity   136
radioactive isotope   193
radiogenic energy   5
radiogenic heating   222

ramp compression 198
Rayleigh-Taylor instability 119, 221
reaction coefficient 220
receiver function 104
refractory elements 4, 5
rheological behavior of Ca-Pv 63

Saturn 118
scanning electron microscopy (SEM) 201
scatterer 32, 36, 37, 40, 42, 44, 46, 48
secular cooling 150
secular mantle cooling 139
secular variations 146
seismic data availability 50
seismic heterogeneity 31
self-diffusion coefficient 220, 222–224
self-exciting dynamo 152
SEM (scanning electron microscopy), micrograph 125
Serpentinites 81
shear modulus 67
shock (dynamic) compression experiment 198
single-crystal X-ray diffraction 194
SiO_2 glass 64
slab 32, 41, 47, 49, 103, 104
slowness 48
SmKS 145, 237, 238
solid-sample background method 204
solid-silicate-metal partitioning 233
solidus 104, 106
sound velocity of liquid iron 192, 207
specific heat capacity 248
spectrometer for temperature determination 196
stably stratified region 145, 146
StagYY 106
standard Gibbs free energy 202
standard enthalpy 202
static compression 193
Stokes' flow model 139
strain 65
  marker 62
  rate 65
stratification 32
stress-strain data processing 65
structure factor 204
subadiabatic 221, 250
subduction zone 103, 137
sub-isentropic shell 146
sulfide melt 90
sulfur fugacity (f_{S2}) 89
Sun 118
superadiabatic heat flow 220
superionic conduction 141
surface heat flux 6

temperature 103
temperature-diffuse scattering 204

thermal boundary layer (TBL) 59, 104, 134
thermal conductivity 134, 146, 149, 152, 160, 221, 222, 225, 249, 251
thermal diffusivity 119
thermal equation of state 207
thermal expansion 149
thermal insulation 197, 203
thermally activated creep flow law 70
thermally stable layer 146
thermo-chemical core-mantle coupling 220
tomography 37–39, 41, 47, 48, 103, 104, 175, 176
toroidal field 221
trace element 122
transmission electron microscopy (TEM) 201
transition zone 47, 103
transverse acoustic (TA) mode 206
transverse isotropy 166
travel times
  absolute 166, 168, 170, 172, 173, 181
  differential 166, 167, 169, 172, 173
TTG 105, 106, 108–111
Tungsten-carbide (WC) anvil 197
Tungsten isotopes 84, 94
^{182}W anomaly 85–87
Type-D thermocouple 64

U and Th abundance in continents 10
U and Th concentration 19, 23
  Amphibolite 19
  Gabbro 19
  Granite 19
  Shale 19
  Tonalite 19
ULVZ (ultra low velocity zone) 32, 37, 38, 44, 46–48, 76, 104, 173
  chemical composition 80
  dehydration conditions 81
  recycling 80
ultrasonic sound velocity measurement 198
U-Pb 118

viscosity 127
volatility 121

wavelength dispersive X-ray spectroscopy (WDS) 202
W-Hf 118
wollastonite 64

X-ray absorption spectroscopy (XAS) 199
X-ray diffraction measurement 198
X-ray diffuse scattering 198
X-ray free-electron laser 198
X-ray radiographic image 62

zonal flow 239